Photovoltaics for Commercial and Utilities Power Generation

Photovoltaics for Commercial and Utilities Power Generation

By

Anco S. Blazev

LONDON AND NEW YORK

Published 2020 by River Publishers
River Publishers
Alsbjergvej 10, 9260 Gistrup, Denmark
www.riverpublishers.com

Distributed exclusively by Routledge
4 Park Square, Milton Park, Abingdon, Oxon OX14 4RN
605 Third Avenue, New York, NY 10017, USA

First issued in paperback 2023

Library of Congress Cataloging-in-Publication Data

Blazev, Anco S., 1946-
Photovoltaics for commercial and utilities power generation / by Anco S. Blazev.
p. cm.
Includes bibliographical references and index.
ISBN-10: 0-88173-652-X (alk. paper)
ISBN-13: 978-8-7702-2294-5 (electronic)
ISBN-13: 978-1-4398-5631-4 (Taylor & Francis : alk. paper)
1. Photovoltaic power systems. 2. Solar power plants. I. Title.

TK1085.B58 2011
621.31'244--dc23

2011026609

Photovoltaics for commercial and utilities power generation by Anco S. Blazev
First published by Fairmont Press in 2012.

Routledge is an imprint of the Taylor & Francis Group, an informa business

Publisher's Note
The publisher has gone to great lengths to ensure the quality of this reprint but points out that some imperfections in the original copies may be apparent.

ISBN 13: 978-87-7022-909-8 (pbk)
ISBN-10: 0-88173-652-X (The Fairmont Press, Inc.)
ISBN-13: 978-1-4398-5631-4 (hbk)
ISBN-13: 978-8-7702-2294-5 (online)
ISBN-13: 978-1-0031-5163-0 (ebook master)

Table of Contents

Preface

Presently most of the electric energy used worldwide is generated by burning fossil fuels. In 2009 over 35% of the produced energy came from crude oil, 29% from coal, 24% from natural gas, and the other 12% come from nuclear, hydro and biomass power. Wind and solar energy generation account for less than 0.02% of the total electric energy generated worldwide. Their contribution, however, is quickly increasing. For example, the US and EU are on track to provide up to 15-20% of their respective electrical demand via wind and solar electricity by the year 2020. If these goals are met, crude oil imports will be significantly decreased and CO_2 emissions reduced by 1.0 billion tons/annum, while at the same time 6.3 million jobs will be created to support the development and operation of new energy sources. These are guestimates, of course, which could be argued, and anything can happen to change the direction, or size of their growth with time. These goals are, however, achievable, and vitally important.

This book is a comprehensive summary of the solar industry's products, processes, and devices, and the recent achievements in the solar field, in the context of global energy developments. We will focus on the developments in the commercial and utilities solar power generation sector in the US and world's energy markets, with emphasis on large scale power generation (large solar power plants).

The US solar industry has gone through a series of "boom and bust" cycles, starting at the onset of the Oil Embargo of 1973. At the whims of oil price setting OPEC sheikhs and those of Oil companies and other big business enterprises in the US and abroad, the solar industry was essentially "tricked" into existence in times of desperation and then "kicked" back into a corner and forgotten until the next energy crisis. The solar industry has endured several ups and downs, instigated by promises of governments and big businesses, only to be broken time after time. As a result, during the present solar "boom" we are firmly grounded in the knowledge of the good and bad that political and socio-economic changes can bring, and so we are moving forward cautiously, but optimistically and steadily.

The present day unsustainable demand for energy, combined with concerns for global warming and environmental pollution—supported by headlines of mining and oil rig accidents--give our present situation more depth and hope than similar "awakening" cycles of the 1970s and 1980s. So we are more hopeful than ever that solar is here to stay.

This book is NOT a "how to" instruction manual or a technical reference. Instead, it is a comprehensive summary of the solar industry's products, processes and equipment, as well as a brief review of recent worldwide developments in the solar energy industry and markets. We focus on commercial and utility type solar energy products and markets in the US, with emphasis on large-scale power generation (utility type power plants to be installed and operated in the deserts.) We analyze the performance and other issues of various PV technologies, considering both manufacturing and field use, to provide the reader (regardless of his or her technical level or professional interests) with the most comprehensive up-to-date information on everything solar.

Acknowledgments

Writing a book of such broad and complex nature requires a complete understanding of all subjects at hand. Since no one can claim absolute expertise in all subject areas herein, we do appreciate the full and friendly cooperation of our capable co-authors, helpers, assistants and editors. We also would like to thank all the collaborators and well-meaning solar professionals and enthusiasts, who gave us permission to use their materials in this book. Without them, the book would not have been possible in its present, and presumably complete, form.

Due to the quickly developing solar energy industry and the related fast-growing body of R&D and field data, as well as the ever-changing political, regulatory and socioeconomic environments in the US and abroad, we were not able to use the materials as originally intended, and numerous last-minute changes were necessary.

Our thanks go to our many supporters and collaborators. Just to mention a few: Dr. Anand Gupta, Dr. Michael Sauer, William Summers, Mark Herbst, Joseph Berwind, John Music, Mitko Nenkov, Linda Blazev. Many others contributed hands-on assistance, support and encouragement during the planning, information-gathering, sorting, writing and editing processes.

We do appreciate the immense contribution of Christiana Honsberg and Stuart Bowden for allowing us to use valuable information from their PVCDROM program at www.pveducation.org which we recommend to all readers interested in obtaining additional technical details on solar cells and modules. PVCDROM is the most complete, comprehensive technical PV database to date and worth a close look.

Very special thanks are due to Dr. Govindasamy Tamizhmani (Dr. Mani) from Arizona State University and TUV Rheinland PTL, LLC (PV test and certification lab at ASU) in Tempe, Arizona, and his team for allowing us to use the large body of pertinent and very important data from lab and field tests with different PV modules, done through the years.

We do also sincerely thank the scientists and staff at Sandia National Laboratory (Sandia) and the National Renewable Energy Laboratory (NREL) for their continuous work in the field and for their kind permissions to reprint some of that work and the results thereof.

NOTE: The Sandia, NREL-developed text and figures, as well as any other materials in this book are not to be used without written permission, or to promote any commercial product or service, or to imply an endorsement by Sandia, NREL, the U.S. Department of Energy, the author, or the publisher.

We do hope that our collaborators and assistants will continue working with us on completing the tasks outlined in the book. The key issues discussed here must be put on the table for open, sincere and professional discussion and debate—with you, the reader, as judge and jury. This is absolutely vital to further the goals of the fledgling PV industry and ensure its progress now and through the 21st century.

A. Blazev
2011

Objective

The object of this book is to describe the state-of-the-art of photovoltaic (PV) materials, equipment, processes, products and technologies; focusing on the related issues at hand, and outlining the possible paths and areas of development for the near future. The main emphasis of this writing is on the design, manufacturing, application and function of PV devices and systems, and their use in commercial and utilities type solar power generating systems and power plants.

PV technologies are already a significant part of our lives, and their presence will continue to increase. It is our responsibility to future generations to apply this technology safely and efficiently. This will happen only through a thorough understanding of the structure and function of the different components and systems.

A single solar technology by itself cannot solve the present energy and environmental crisis. Several different new technologies will be needed to fill the gaps in the energy markets. Only by knowing exactly what the different technologies offer can we make good decisions for filling the energy needs of the future.

This book is a guide through the intricate mechanisms and processes of PV devices and systems, from their proper design and manufacturing to efficient installation and profitable operation. Our goal is to expose the actual function and the related advantages and disadvantages of the different solar technologies, in order to provide a better understanding of the issues at hand and how to deal with them. Only by thorough understanding of the intricacies of the different technologies will we be able to identify existing and potential problems with their safety, efficiency, and proper and profitable operation, thus being able to eliminate the problems before they grow too big to handle.

Although we cannot offer an adequate solution to all issues and challenges in the energy sector, and the PV technologies in particular, we expect that this book will shed enough light on the issues at hand and bring them out in the open, fueling discussions on several key subjects. This, we honestly believe, is a good step towards facilitating the progress of the PV industry, thus allowing the world's energy future to shine sooner and brighter.

Chapter 1

Introduction To Solar Energy

"I'd put my money on the sun and solar energy. What a source of power! I hope we don't have to wait until oil and coal run out before we tackle it."

Thomas Edison

BACKGROUND

Looking at the early morning sky we are greeted by Earth's oldest friend, the sun, which we have known, enjoyed and mostly taken for granted since our earliest childhood. We are well aware that our wellbeing, and in fact our very lives, depends upon its daily presence in the sky, and yet we seldom give it more than a glance and a passing thought. However, it never fails us; it is out there shining brightly every day whether or not we see it or care. It is one of the few things in this life on which we can count. But if the sun's rays were blocked from reaching the Earth for longer than usual, life as we know it would quickly change and even cease to exist. So we can only hope that this will not happen anytime soon.

Mr. Edison recognized and appreciated the power of the sun almost 100 years ago. Even then, he was worried that oil and coal would prevent solar energy from being widely used. And he was right. Coal and oil are still winning the race, and the awesome power of the sun is kept out of our reach by oil, coal and other big business enterprises and political interests. In our estimate, this situation won't change until the last drop of oil is pumped out of the Earth and the last shovel of coal is thrown into the furnace—regardless of the price of energy and deteriorating environmental conditions.

Our goal here is to consider and reconsider the sun's power as an alternative for achieving a renewable energy future, and to help us clean up the environment. So let us take this opportunity to examine more closely the facts concerning this source of life and energy, keeping in mind that virtually all energy on Earth has been (and still is) one way or another created by "solar" power—the power of light coming from the sun.

The fundamental energy that sustains human life—food energy—comes from plants using photosynthesis to convert sunlight and CO_2 into living tissue. Even the hamburger you had for dinner last night can be traced back to the grasses and grain eaten by the cows—and that again is energy converted by sunlight thru photosynthesis. And what about the gasoline you put in your car? It comes from petroleum, formed when plants and animals (full of sunlight (again, energy as a result of photosynthesis) were submerged underground and transformed over thousands of years into hydrocarbons which we extract and burn. Just think of the immense supply of energy stored underground in the form of crude oil, coal and natural gas. This is energy that the sun provided millions of years ago to life on Earth, energy which has been carefully stored until now, and which we are determined to deplete within the next 30-40 years.

What about wind energy? Well, without the heating rays of the sun there would be no wind or changing weather on Earth. The sun heats the air, and this results in wind currents which are transformed into wind energy, which can then be captured by wind mills and converted into electricity. What about hydroelectric energy? Without the sun heating the oceans and drawing up water into the clouds, which then falls as rain in the mountains, there would be no rivers and therefore nothing for us to dam in order to capture the energy from falling water. So you see, with very few exceptions, all forms and shapes of energy on Earth are related to solar energy.

Here is an interesting fact to consider: the sunlight that reaches the Earth's surface in one 24-hour period contains enough energy that, if it could be converted into usable electricity, it would equal the needs of the entire world for one full year. Imagine that! So, then, why are we digging into the depths of the Earth to excavate and pump to the ground dinosaur remains (petroleum, coal and such) to use for fuel and electric energy generation? Why are we exposing human health and life to the dangers of the deep and toxic by-products? Why do we continue making oil rich nations richer by buying their remaining oil and gas reserves? Why not do what we should have done decades ago: harness the abundant, free and renewable energy of the wind and sun, thus generating free, clean and renewable energy?

The sooner these renewable energy technologies are perfected, the better life will be for us and future generations. Also, the sooner this is accomplished the sooner we will be able to begin to reverse the damages caused by our greed and negligence which result in global warming and other disasters.

It has been the dream of mankind through the centuries to be able to capture the bright warm rays of the sun and harness their power, but our ancestors were limited in their know-how and technical abilities. We are now, however, the generation that has the knowledge, experience, materials and equipment to harness a large portion of the sunlight coming down to Earth and to use it as we wish. Nevertheless, we continue to use fossil fuels to transport, cool and warm us. How long will these fossil fuels last? What excuse do we have to postpone the inevitable? What excuse are we going to give the next generations if we fail to address the misuse of fossil fuels?

Most of the energy we use comes from burning dirty coal, and many die in the process of mining it and breathing its by-products. We still use crude oil as if there were an unlimited supply, pumping millions of gallons daily and saturating the atmosphere with harmful by-products. We still plan the construction of dangerous nuclear power plants to supplement our energy gluttony, and then we hastily hide the toxic waste materials in weird places, leaving future generations to figure out what to do with them. And in the process of doing all this, we waste 99% of the free and clean solar power coming to us.

Just think, 100 square miles of Nevada desert could power the entire USA, and 3% of the world's unused desert lands could power the entire world. A small part of the world's deserts covered with PV modules (or other solar power generating equipment), could power the entire world. Imagine that! We have a chance to use that free sunlight in unprecedented efficient and cost-effective ways. Why not take full advantage of it now, while we can and while we still have time?

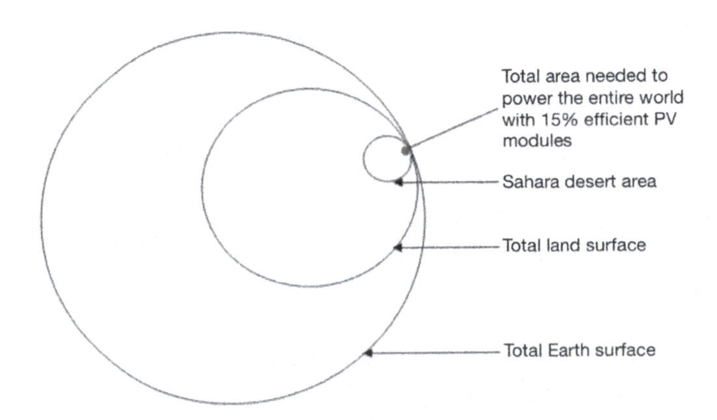

Figure 1-1. How little is needed to provide power to the world. (Not to scale)

COMMERCIAL AND UTILITIES POWER GENERATION

The variations of solar power generation that we will be discussing here are the technologies capable of producing large amounts of electric power to be used for commercial applications, and / or utilities-scale power generation. These are basically solar power plants, dedicated to generating electricity which is to be used locally, or sent into the power grid. These are herein referred to as commercial and utilities power generation technologies in order to distinguish them from residential and small agricultural types of solar power generating systems. These are usually referred as "large-scale" PV installations, which are basically utility-type grid connected installations.

Commercial power generating systems consist of arrays of water heaters, or PV modules, installed on roofs of small manufacturing and service facilities (or ground-mounted nearby) and used to generate power—electricity, hot water, or both. Most often these are PV based systems, which produce electricity for local use, or which are plugged into the grid and running the electric meter backwards when not in use. These are usually partial or supplemental sources of energy of up to 10 MWp in size. Utility-scale power generating systems are ground-mounted PV modules, or other PV technologies, the primary objective of which is to provide electric power to the electric utility they serve. These are large facilities of varying sizes.

We will take a look at these applications and the technologies used in them, focusing on the larger scale installations, and especially those for utilities use. These require special review and consideration because, in our opinion, they will be the primary solar energy generators used in the future, capable of bringing us energy independence and a cleaner environment.

The technologies that are suitable for large-scale PV power generation have special characteristics and requirements which need to be thoroughly understood and considered during the planning, design, installation, operation and maintenance stages of each PV system and power plant. The main goal of this book is to provide a comprehensive level of understanding of the PV technologies and methodologies in current use by the solar industry. The book also attempt to bring the key technical and administrative issues into the open, analyzing some of them as needed for the reader to consider and draw the appropriate conclusions.

But before we plunge into the particularities of the different solar technologies, their function, use and related issues, let us review some of the basics of the source of it all; sunlight, its properties and applications.

SUNLIGHT BASICS AND KEY CONCEPTS

Life on Earth depends on the sun's energy providing warmth and light. The sun is a large star at the center of our solar system and is approximately 93 million miles away, but its life-giving energy arrives on Earth every day.

This distance between the sun and the Earth is crucial to maintaining life. The Earth's orbit and its very appropriate distance from the sun are responsible for providing climate conducive for life to exist here. The other planets are either too close (thus too hot), or too far away (and too cold) for organic life as we know it to exist. The Earth also has an appropriate atmosphere, which contains enough oxygen, carbon dioxide, and water to maintain life. All life-maintaining elements are in the right proportion, form and shape needed for life to flourish on Earth. These are also in a very delicate balance, so any changes or modifications to this precise balance would be detrimental to life on Earth. Sunlight is one of the key, very precise and delicate variables that is responsible for keeping humans, animals, and vegetation alive and thriving on the Earth' surface.

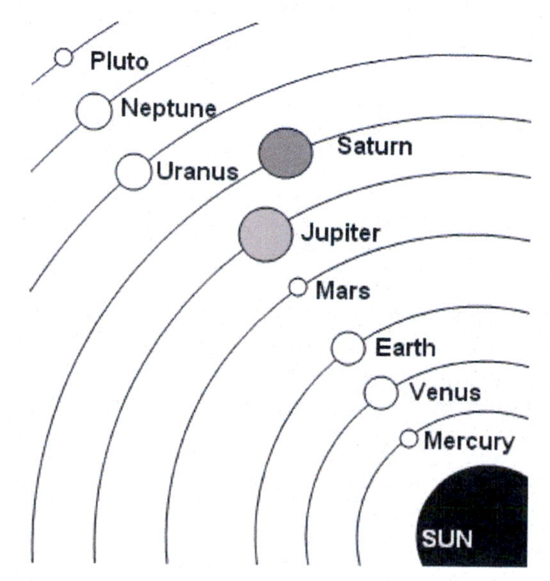

Figure 1-2. The sun and the planets in the solar system.

Sunlight essentially consists of a range of energy bands, which we generally refer to as the solar spectrum. We can see some of these bands, but most are invisible to the human eye. We can also feel some of these when they impact our skin and are perceived as heat. So, sunlight is radiation of the electromagnetic type, with what we see and feel being only a small part of its entire spectrum of energy particles on their respective levels and energy potentials.

From our perspective, we think of visible light as the most important component of the solar spectrum, because we can see it. The infrared part of the solar spectrum is also noticeable since we feel it as heat. However, these are just small parts of the entire light (sunlight) spectrum, and they are clearly distinguished by their different wavelengths and respective properties. As the study of light has advanced, physicists have found that while it is usually best to consider light as energy traveling through space as a wave, in some situations light behaves as if it were made up of tiny "packets" of energy or "particles" that have no mass and always travel at the same speed, which we call the speed of light. When light is discussed in this manner the packets of energy are called photons, which is the term we will use in this text.

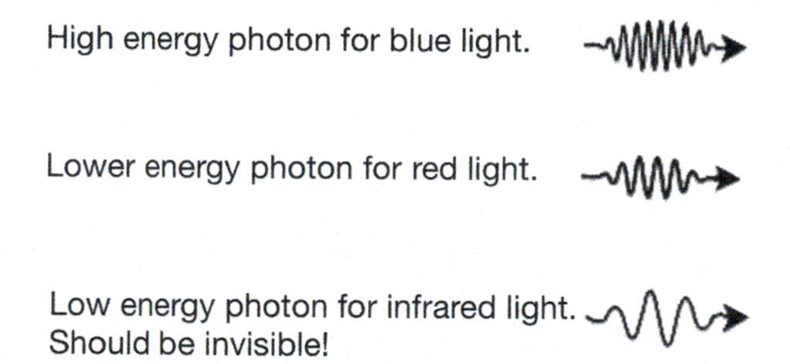

Figure 1-3. Paths of sunlight energy particles (photons)(1)

The Solar Spectrum

The solar spectrum shows the range of energy coming from the sun. It is subdivided into sections and classifications based on the wavelengths of their particular energies.

The key components of the solar spectrum are as follow:

1. Ultraviolet (UV) radiation:

This part of the spectrum accounts for less than 10% of the energy from the sun that reaches the Earth's surface. Most ultraviolet energy is filtered out by our atmosphere. Ultraviolet energy is divided into three parts:

a. Ultraviolet C or (UVC) spans the range of 100 to 280 nm. The term ultraviolet refers to the fact that

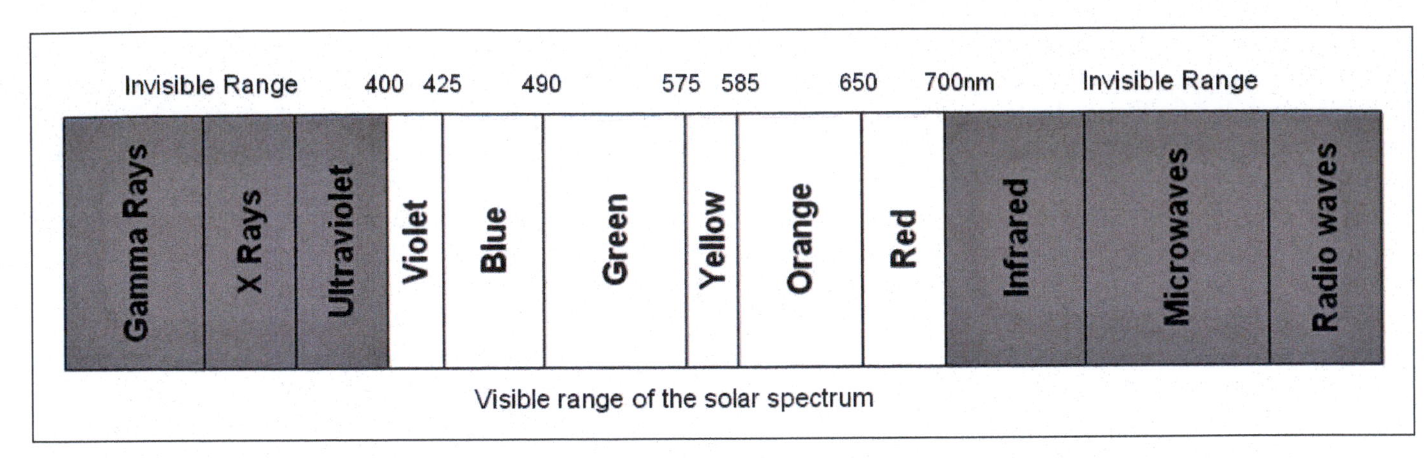

Figure 1-4. The solar spectrum

the radiation is at a higher frequency than violet light (and hence invisible to the human eye). Due to absorption by the atmosphere, very little UVC reaches the Earth's surface. This spectrum of radiation has germicidal properties and its property of killing bacteria and viruses is used in germicidal lamps.

b. Ultraviolet B or (UVB) range spans 280 to 315 nm. This part of the spectrum is of interest to us because these are the rays that cause our skin to tan and burn, and our bodies to produce vitamin D. It is also absorbed by the atmosphere; along with UVC it is responsible for the photochemical reactions leading to the production of the ozone layer.

c Ultraviolet A or (UVA) range spans 315 to 400 nm. This part of the spectrum also has rays that cause our skin to tan, although it has been traditionally held as less damaging to the DNA (does not cause skin burns) and hence is used in tanning beds and UVA therapy for psoriasis.

2. Visible radiation

Most of the energy reaching the Earth' s surface from the sun is in the visible (and some in the infrared) part of the spectrum. The visible radiation spans 400 to 700 nm. As the name suggests, it is this range that is visible to the naked eye. This is also the part of the spectrum that is most useful to power generation, since it is readily captured by PV equipment and turned into electricity.

3. Infrared (IR) radiation

The infrared range spans 700 nm to 10^6 nm. It is responsible for an important part of the electromagnetic radiation that reaches the Earth. This is the part of the spectrum that heats water in solar thermal technologies which is then used for heating or to generate electricity. IR radiation is considered "parasitic" energy in most PV technologies, since it heats and overheats the PV cells and modules, thus rendering them inefficient and even causing them to fail.

Figure 1-5. The sun energy

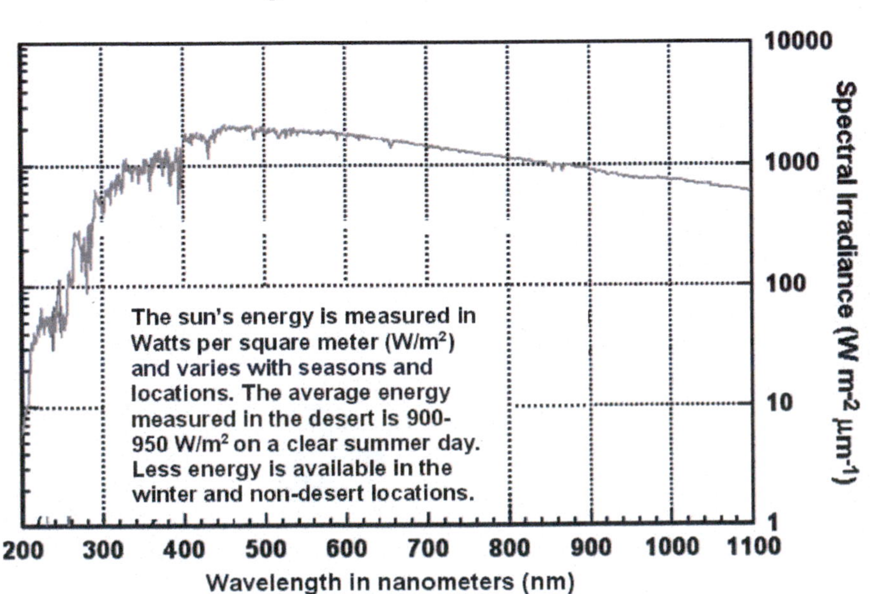

The different wavelengths of the solar spectrum have different energies, with the most energetic being in the 400 to 700 nm range. Some wavelengths, like the IR, are even harmful to PV devices, because they generate heat within the PV cells and modules.

In case the reader is interested in learning more about the nature of light we recommend an excellent book on the subject, *QED: The Strange Theory of Light and Matter*, by Richard Feynman, where QED stands for quantum electro-dynamics. In it Feynman lays down the basics of light and its interactions with matter in a very clear and entertaining way. Feynman cautions the audience that they may not understand what he will be saying, not because of technical difficulty, but because they may be unable to believe it or unable to accept it. "The theory of quantum electrodynamics describes Nature as absurd from the point of view of common sense. And it fully agrees with experiment. So I hope you can accept Nature as she is—absurd."

For all practical purposes, most PV cells capture only certain limited wavelength ranges of the solar spectrum, but some PV technologies, using special non-silicon substrates and multi-junction solar cells are capable of capturing most of the wavelengths, including IR radiation, falling upon them and efficiently converting them into electric energy. We consider these devices the precursor of the PV technologies of the future, since they would be able to use fully the incoming sunlight, while at the same time being able to withstand the excess desert insolation and heat.

Factors Influencing Solar Power Reaching the Earth

Now that we have examined the solar spectrum and have discussed its properties, let's move on to see how much of that energy can be captured and converted into electric energy by our PV devices. These are the properties and characteristics of the sunlight that are very important for calculating and designing PV projects,

Sunlight travels from the sun to Earth in approximately 8 minutes, and while it loses some of its energy during this journey, most of it arrives here safe and sound and ready to serve us. Measured at the top of the atmosphere, we find the highest power density reaching up to 1,367 W/m2. That is to say that if we could mount a PV module that was a square meter in an area 200 miles above the Earth's surface, and if that PV module were 100% efficient (which is not possible as yet), we would be able to generate 1,367 W DC electric power from it. The problem of how to transport the produced electricity back to Earth is a separate issue for which we have no answer at this time.

Several factors used to characterize sunlight and measure its properties are: (1)

1. Spectral radiance.

The spectral irradiance as a function of photon wavelength (or energy), denoted by *F*, is the most common way of characterizing a light source (sunlight in our case). It gives the power density at a particular wavelength. The units of spectral irradiance are in $Wm^{-2}\mu m^{-1}$. The Wm^{-2} term is the power density at the wavelength $\lambda(\mu m)$. Therefore, the m^{-2} refers to the surface area of the light emitter and the μm^{-1} refers to the wavelength of interest.

In the analysis of solar cells, the photon flux is often needed as well as the spectral irradiance. The spectral irradiance can be determined from the photon flux by converting the photon flux at a given wavelength to W/m2. The result is then divided by the given wavelength, as shown in the equation below.

$$F = \left(\frac{W}{m^2 \mu m}\right) = q\Phi \frac{1.24}{\lambda^2(\mu m)} = q\Phi \frac{E^2(eV)}{1.24}$$

where:

F is the spectral irradiance in $Wm^{-2}\mu m^{-1}$;

Φ is the photon flux in # photons $m^{-2}sec^{-1}$;

E and λ are the energy and wavelength of the photon in eV and μm respectively; and

q, h and c are constants

2. Photon energy

Sunlight consists of many photons traveling together as a photon flux. Each photon in the flux is characterized by either a wavelength, denoted by λ or equivalently energy denoted by *E*. There is an inverse relationship between the energy of a photon (*E*) and the wavelength of the light (λ) given by the equation:

$$E = \frac{hc}{\lambda}$$

where: *h* is Planck's constant and
c is the speed of light.

The above inverse relationship means that light consisting of high energy photons (such as "blue" light) has a short wavelength. Light consisting of low energy photons (such as "red" light) has a long wavelength. When dealing with "particles" such as photons or electrons, a commonly used unit of energy is the electron-volt (eV) rather than the Joule (J). An electron volt is the energy required to raise an electron through 1 volt, thus 1 eV = 1.602×10^{-19} J.

By expressing the equation for photon energy in terms

of eV and μm we arrive at a commonly used expression which relates the energy and wavelength of a photon, as shown in the following equation:

$$E = \frac{1.24}{\lambda(\mu m)}$$

The exact value of $1 \times 10^6(hc/q)$ is 1.2398 but the approximation 1.24 is sufficient for most purposes.

3. Photon flux

The photon flux (and its quality and quantity) determine the intensity of the sunlight reaching our PV devices. The photon flux is defined as the number of photons per second per unit area:

$$\Phi = \frac{\#\ of\ photons}{\text{sec m}^2}$$

The photon flux is important in determining the number of electrons which are generated, and hence the current produced from a solar cell. As the photon flux does not give information about the energy (or wavelength) of the photons, the energy or wavelength of the photons in the light source must also be specified. At a given wavelength, the combination of the photon wavelength or energy and the photon flux at that wavelength can be used to calculate the power density for photons at the particular wavelength.

4. Power density

The power density is calculated by multiplying the photon flux by the energy of a single photon. Since the photon flux gives the number of photons striking a surface in a given time, multiplying by the energy of the photons comprising the photon flux gives the energy striking a surface per unit time, which is equivalent to a power density. To determine the power density in units of W/m^2, the energy of the photons must be in Joules. The equation is:

$$H\left(\frac{W}{m^2}\right) = \Phi \times \frac{hc}{\lambda}\,(\text{J}) = q\Phi\,\frac{1.24}{\lambda(\mu m)}$$

where:

Φ is the photon flux.

One implication of the above equations is that the photon flux of high energy (or short wavelength) photons needed to give a certain radiant power density will be lower than the photon flux of low energy (or long wavelength) photons required to give the same radiant power density. In the animation, the radiant power density incident on the surface is the same for both the blue and red light, but fewer blue photons are needed since each one has more energy.

The total power density emitted from a light source (sunlight in this case) can be calculated by integrating the spectral irradiance over all wavelengths or energies of interest. However, a closed form equation for the spectral irradiance for a light source often does not exist. Instead the measured spectral irradiance must be multiplied by a wavelength range over which it was measured, and then calculated over all wavelengths. The following equation can be used to calculate the total power density emitted from a light source.

$$H = \int_0^{\infty} F(\lambda)\, d\lambda = \sum_{i=0}^{\infty} F(\lambda)\, \Delta\lambda$$

where:

H is the total power density emitted from the light source in W m^{-2};

$F(\lambda)$ is the spectral irradiance in units of $Wm^{-2}\mu m^{-1}$; and

$d\lambda$ or $\Delta\lambda$ is the wavelength.

As the sunlight travels down and gets close to Earth, it has to travel through the atmosphere, where it collides with dust and water vapor particles, thus losing some of its power. So, on a perfectly clear day on the equator we could measure up to 1,110 W/m^2 (down from 1,367 W/m^2 measured above the atmosphere). In the Arizona desert we measure 900-1100 W/m^2 on a clear summer day. Not bad for generating useful power from a piece of otherwise "useless" desert land during the daylight periods, when we need it most, especially during the summer months.

Clouds, fog and dust will rob some of the power that sunlight is trying to deliver to Earth, but even then most PV modules will be able to convert some of that energy reaching their surface to electric power (10-90%, depending on the cloud cover density). Although that is a greatly reduced amount, it is still enough to produce a lot of usable energy. Ask the German people, where under mostly cloudy conditions the number of PV installations have risen at an unbelievably high pace lately, and continues to rise as we speak.

The curvature of the Earth also influences the amount of energy that strikes its surface at any given location. The intensity of the sunlight is greater in the area of the equator between the tropics, and it loses intensity towards the South and North poles. This is because the sunlight, due to the Earth's curvature, has to travel much greater distance at a sharper angle through the atmosphere before reaching the Earth's surface.

In addition, since the Earth is tilted on its axis, the

time of year also influences the amount of energy which strikes its surface at different locations and times.

And so, the important characteristics of the incident solar energy are:

1. The spectral content of the incident light (visible, UV or IR light content),
2. The average radiant power density of sunlight (W/m^2) at the location,
3. The angle at which the incident solar radiation strikes a PV module,
4. The seasonal sunlight energy (summer vs. winter), and the local variations, and
5. The atmospheric and weather conditions (clouds, fog, smog etc.).

All these properties and characteristics of sunlight are very important for calculating and designing PV projects, and are used extensively by design engineers and installers alike, thus making them integral parts of the proper execution of any solar project. We will take a closer and much more detailed look at all these parameters in the text below.

Solar Energy Characteristics and Use

In order to produce electricity, sunlight has to be captured and converted into thermal or electric energy suitable for human consumption. As we discussed in the previous section, sun energy races toward the Earth at very high speed, and if its path was not obstructed by space junk or clouds and dust in the atmosphere, it would arrive to us at full power. We call this "beam" or "direct" radiation. Arriving "directly" from the sun it is most powerful and measures around 1367 W/m^2 just above our atmosphere. Its power drops after crossing the atmosphere to approximately 900-1,100 W/m^2 as measured at noon in the deserts during the summer months, and much less than that in other parts of the globe and during different seasons of the year.

When the sunlight hits clouds, dust, or man-made gasses in the air, it gets scattered and we call that "diffused" radiation. Diffused radiation has properties very different from direct radiation. It contains less energy, and thus PV modules will produce less power under diffused radiation, in some cases much less. Particularly, concentrating thermal or PV equipment is affected by this diffusion, and this will cause it to lose its focus and operate well below its maximum efficiency rating, if at all, under extreme conditions.

Sunlight hitting the Earth is reflected from its surface, and we call this effect "albedo." Different materials have different reflecting properties, but most do reflect and some reflect a lot. Take fresh snow, for example. It will reflect almost 80% of the light falling on its surface. Water, on the other hand, absorbs most of the sunlight, instead, and gets heated in the process. Thus, reflected sunlight can be captured by our PV modules installed nearby as well. The albedo always has an effect on PV module performance, so it should be taken into consideration, especially in areas with snow cover or other highly reflective ground surface cover.

As we discussed in the previous section, another factor that affects the amount of energy available for conversion into electricity is the distance that the sunlight travels once it enters the Earth's atmosphere. At certain times of the day and year, the sun seems to be overhead at a 90-degree angle to the Earth's surface, and this is when sunlight travels the shortest path and is the strongest. We call this Air Mass = 1 (or AM 1). AM 0 is measured above the atmosphere and is much stronger than AM 1. In the early morning and later afternoon, the sun is at a sharper angle and sunlight travels a longer path through the atmosphere, so the angle decreases (approaching 45 degrees), the sunlight has a longer path to travel and the AM number increases: AM 1.15, 1.5 etc. depending on the angle. The sharper the angle of the sun rays, the larger the AM number and the objects' shadows. AM 1.5 is measured at an angle close to 45 degrees.

The air mass number can be determined by the formula:

$$AM = \sqrt{1 + \left(\frac{s}{h}\right)^2}$$

where:

h is the object's height, and
s is its shadow length

The revolution of the Earth around the sun and its rotation on its axis produces seasonal and daily effects, which vary by location on the globe. This location is measured on the world map in terms of longitude (east-west direction), and latitude (south-north direction). The intersection of these provides us with a precise point on the map.

All these components taken together represent what we call "global radiation," which is a very important factor in the proper design, installation and operation of solar energy generating systems. Solar professionals need to be very familiar with it, if a properly designed PV system is their goal.

The solar spectrum components and the quantity of sunlight traveling to Earth at any moment are constant, with only slight variations. The sunlight reaching the

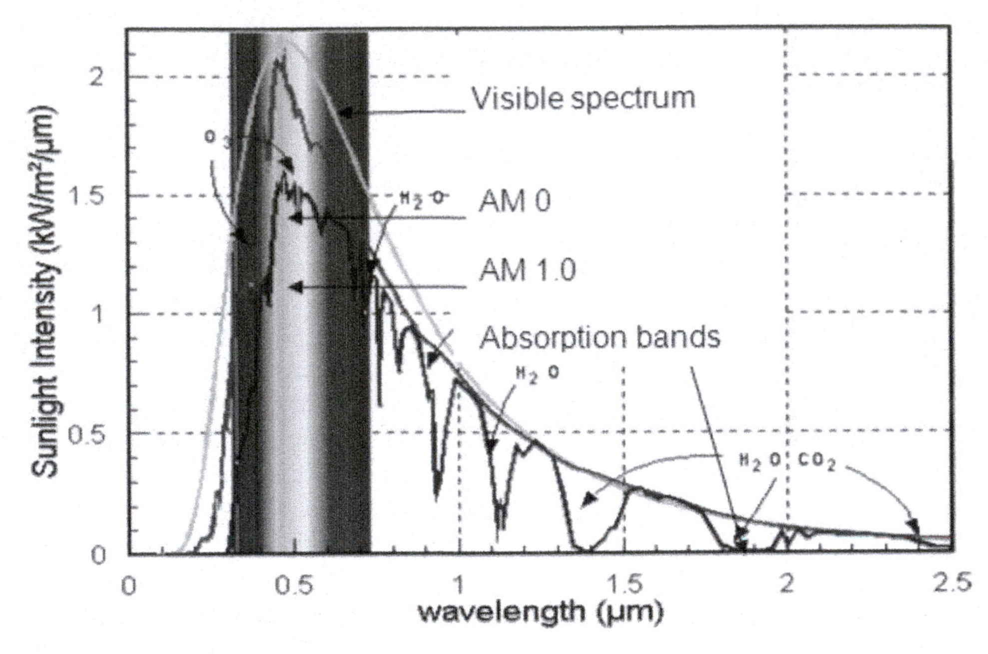

Figure 1-6. Sunlight wavelength vs. power measurements. (1)

Earth's atmosphere (but before entering) is represented by AM 0 in Figure 1-6 and is a constant with some very small and predictable variations. When the sunlight hits the atmosphere it becomes a variable, depending mostly on the contents of the atmosphere, the seasons, weather, and the time of day. Thus the amount of sunlight is a variable, which is mostly out of our control, but which we must take into consideration when working with PV equipment.

Water in the atmosphere (clouds), dust, CO_2 and other gasses absorb radiation, thus diminishing the amount and level of energy that makes it to Earth. So the amount of water vapor and CO_2 (of which there is a lot in the atmosphere) in the sunlight's path will determine how much energy will reach the Earth's surface. The influence of all these variables is significant and must be well understood and taken into consideration in our solar calculations and designs, if we are to optimize the performance of our PV devices. There is not much we can do about these variables, for they obey higher orders, and so we can call them "fixed" variables. We can, however, work around them and build and use renewable energy technologies that are best suited for local conditions.

Major variables over which we have no control are the local weather conditions—clouds, rain, snow, fog, smog, dust, humidity, etc. These must also be well understood and looked at very carefully, when designing, installing and operating solar energy generating systems, be it thermal, or photovoltaic. We cannot control the weather, dust, smog and other natural phenomena, but we can study the historical weather data and anticipate their behavior, in order to compensate for their effects, thus obtaining the best possible results at a given location with the highest possible power output.

Ideally, large-scale solar power plants should be installed in locations with the least cloud, fog and humidity levels. However since there is no place with perfectly clear skies all through the year, we must provide for the clouds and other effects of weather, estimate their influence, and design the systems accordingly. There are variables which we can and should control such as the position of trees, buildings, smoggy factories, dusty fields, and other obstructions which will diminish system performance. Tall trees, buildings, etc. near a power field will reduce the output, so they must be avoided, removed, or trimmed. Air pollution generated by industrial activities can hinder the performance of PV devices, so it has to be considered. The effects of smog and fog from nearby populated centers or large bodies of water must be also considered and calculated as accurately as possible.

Other variables worthy of mention, due to their effects on power output of PV devices, are the air and ground temperature. Our Earth provides a marvelous thermodynamic balance, where the temperature increase due to incoming sunlight during the day is balanced by the outgoing heat during the cool of night, so that balance is maintained. During the summer months, however, the balance in some areas, like the deserts, is temporarily altered. The air and ground can get extremely hot during the day, heating up the solar energy generating equipment to levels they cannot easily handle. The glass and metal structures of the solar power fields could reach temperatures well over the operating limits of the PV devices in them (temperatures near 200°F have been measured), thus causing drastic power output decrease and outright electro-mechanical failures.

Prolonged exposure to extreme heat can cause deterioration, damage and even destruction to most present day photovoltaic devices, which is why serious consideration must be given to the structure and proper use of these devices in extreme climates. Their function and behavior must be well understood and proven, well before the design stages and definitely before installing and exposing them to harsh deserts and other extreme climate conditions.

Finally, we must also understand and take into consideration the environmental consequences of our efforts. Installing acres upon acres of solar collectors (thermal or PV) in the deserts, for example, could have serious effects on the delicate ecosystems of these lands. Unfortunately, there is not enough data right now to predict these effects. There may be some positive effects too–the structures will provide shade which could cool the ground in the summer months. However the Earth has a delicate and even fragile environmental balance, and we have learned during the past several decades that every action people take for their own comfort and convenience has an effect on the environment—and usually not for the better. So, covering a large land mass with any solar technology will have an effect, which we must evaluate and consider before acting.

In all cases, all variables must be taken into consideration and their impact must be incorporated into the system design.

BRIEF HISTORY OF SOLAR AND PV POWER GENERATION

Let's take some time now for a brief "walk" through history to see how our ancestors harnessed sunlight for their energy needs and see what led to the more recent developments in the field. We have to go very far back indeed to come to the first human use of sunlight as a "tool."

Solar energy is the foundation of life on Earth because every living being depends on the sun to provide the warmth and energy required to sustain life. Since the beginning of time our ancestors have used sunlight to keep track of time, stay warm, dry clothes, grow and dehydrate vegetables and animal products. They knew well the value of sunlight and appreciated its usefulness, since in many cases it meant the difference between a full and empty stomach, and even between life and death.

Early on, people began to realize that sunlight was beneficial to their health. In fact, vitamin D, which is produced in the body when exposed to direct sunlight (prevalent in the UV-B band of the sun spectrum), is a necessary component in maintaining good health. People who are deprived of exposure to sunlight are deficient in vitamin D and exhibit health problems, one of which is Ricketts (softening of the bones). Without enough sunlight some people experience a type of depression known as Seasonal Affective Disorder (SAD). On the other hand, the ancients also knew well that excessive exposure to sunlight leads to unwanted effects, such as burnt skin, skin discoloration, skin cancer, etc.

The industrial revolution brought more opportunities for the use of the sun's energy because of an increase in knowledge and tools such as mirrors, lenses and optics. The use of these tools was limited to producing enough heat (thermal energy) for some practical purposes.

Early Solar Energy Developments

The scientific drive to capture and harvest the sun's energy dates back to the 18th century, when Sir Edmund Becquerel discovered, during his studies of the solar spectrum, that sunlight can be captured by different materials where it can be converted into useful energy. However, this ingenious observation faded into obscurity in the annals of history not very long after Becquerel's death.

Later in the same century, Mr. Auguste Mouchout, a French mathematics teacher, invented a steam engine which ran on solar energy. He got the attention of the French government and was awarded funds to continue his research. He built the first solar powered steam generator. The scheme worked quite well, and he was able to run a small 0.5 horse power steam engine using solely sunlight as a power source. As an indicator of things to come, however, his invention was abandoned when his natural, renewable power source was replaced because coal became less expensive and was easily mined in England's deadly coal mines. The fact that coal is the dirtiest energy source did not stop people then, and it is not stopping them now, even though they are well aware of the consequences of using mass quantities of coal daily.

The first recorded PV conversion of sunlight into electric power is attributed to Sir Willoughby Smith, whom some consider to be the father of photovoltaic solar energy. In 1873 he was experimenting with many materials, while looking for suitable candidates for making cables. He noticed that Selenium metal is very sensitive to sunlight, and was able to capture the electric power produced by the first known solar cell-like device.

By the end of the 18th century Sir William Grylls Adams published a book called *Substitutes for Fuel in Tropical Regions*. Mr. Adams experimented with reflecting sunlight from mirrors to power a small steam engine. The 2.5 horse power engine (the most powerful renewable energy powered mechanical device to date) worked quite well and amazingly enough is in use today under the name of Power Tower. He also did work in which he discovered that illuminating a junction between selenium and platinum produces a photovoltaic effect.

Mr. Charles Tellier was a brilliant Frenchman who is famous for being the inventor of cold storage refrigeration and for outfitting the first steam ship with refrigeration so that meat could be transported by ship. In the 1880s he designed and built the first roof-mounted solar water heater. He simply installed pipes and plates on the roof of his house and was able to heat water for everyday use. However, he abandoned his invention in pursuit of more lucrative refrigeration ventures.

Enter the Americans. In the late 1800s one of the most brilliant mechanical engineers ever, Mr. John Erickson (Swedish immigrant to the States), was the first American renewable energy advocate. In 1877, he described using concave mirrors to gather sun radiation strong enough to run an engine. He also invented "sun engines," which collected solar heat for a hot-air engine. Apparently he was also something of a prophet because he is credited with the words, we (solar enthusiasts and specialists) still often repeat. He essentially said, "In a couple of thousand years the coal fields of Europe will be exhausted, unless we use heat from the sun." Wow, what a prophetic thought! However, he did not know that there is a lot of coal in America, and did not anticipate the significant role of the other fossil fuels. Most importantly, he did not have a way to foresee how fast we will go through the coal and oil fields in one single century, so his 2,000-years prophesy must be revised to several decades before the coal and oil fields are exhausted.

The American Solar Revolution continued across the boundaries of the 18th and 19th centuries thanks to Mr. Aubrey Eneas, who established the commercial aspect of the solar industry by creating the first solar energy company in the world. The Solar Motor Co. sold their system for approximately $2,000 to Dr. A.J. Chandler of Mesa, Arizona. It was destroyed by a windstorm shortly after it was installed. When their second system was also destroyed by bad weather (a hailstorm) the company went out of business.

Large-scale commercial solar plants (which this book pays special attention to) were initiated by another American visionary, Mr. Henry Willsie, who built large solar thermal plants with nighttime storage in California in the early 1900s, a remarkable vision for those days. The plants used flat plate collectors to heat water. The water was stored in insulated collectors; then tubes filled with sulfur dioxide ran through the tanks, and were transformed into high pressure vapor. The vapor was then used to run an engine. This was the first large attempt to use stored heat at night. Unfortunately, the government and state incentives and subsidy programs of those days were not sufficient to support such a great endeavor, and the company went bankrupt.

Another large and most effective solar plant was built by Mr. Frank Shuman. His company, Sun Power Co., built a solar energy system—the largest generator to date. Shuman built the world's first solar thermal power station in Meadi, Egypt (1912-1913). The plant used parabolic troughs to power a 60-70 hp engine that pumped 6,000 gallons of water per minute from the Nile River to adjacent cotton fields. This system included a number of technological improvements, including absorption plates with dual panes separated by a one-inch air space. Although the outbreak of World War I and the discovery of cheap oil in the 1930s discouraged the advancement of solar energy, Shuman's vision and basic design were resurrected in the 1970s with a new wave of interest in solar thermal energy, when US Department of Energy (DOE) poured billions into R&D and testing of new solar thermal and PV technologies.

In 1954 Bell Laboratories engineers Calvin Fuller, Gerald Pearson, and Daryl Chaplin were working on something that had nothing to do with solar energy, when they accidentally discovered that silicon is a different material, the special properties of which can be used to make semi-conductor devices. One thing led to another and soon enough they made the first recorded solar (PV) cell and then a solar panel. Its efficiency was only 6%, but that didn't discourage them and they continued working on the materials and processes, until other US and Japanese companies more focused on profit beat them to the Solar Gold Rush of the 1970s.

In the late 1950s the first solar cells were deployed in space. We don't know the efficiency vs. cost of these first solar cells, but we dare guess that the ratio was astronomical. The Vanguard I satellite is credited with the deployment of first solar powered space communications, using a 0.1 Wp, 100 cm^2 solar panel. This venture ended with communication failure later (we dare guess due to deterioration of the silicon solar cells—and which problem still haunts the technology today). Vanguard I satellite, and its malfunctioning solar cells, are still part of the space debris that circles the Earth, and it serves as a daily reminder that our man-made solar technology still has a long way to go on the path to perfection. A number of US, Russian and other solar powered space crafts followed, through the years, most of which functioned properly and completed their missions.

The infamous 1970 OPEC oil embargo shook the American public to the core. Those of us who were adults at the time remember that we suddenly became aware that our way of life was dependent on decisions made by people sitting thousands of miles away who did not have our interests in mind. We suddenly realized that something drastic needed to be done, if we were going to be energy independent. This picture is etched in the American psyche and has been a most powerful dose of reality and a constantly nagging reminder that in order to be free, we need energy independence.

So OPEC's embargo has been the engine behind the development of renewable energies in the US and the world. It sparked a great push in Congress and in the hi-tech community to take a close look at alternative fuels, such as wind and solar. A country-wide effort to develop and deploy such technologies was planned and implemented. However, when gas prices went down to pre oil-crisis levels the nation heaved a collective sigh of relief, and once again solar and wind technologies were put on the back burner, and many alternative energy companies went out of business.

On the positive side of things, during this short lived 1970-1980 Alternative Energy Renaissance in the US, a large number of solar companies came into existence. Several large companies such as IBM, Motorola and Shell Oil, among others, were the leaders in the solar energy R&D for cell and panel manufacturing. A large number of equipment manufacturers took off under the solar banner as well. Silicon solar cells and panels were the predominant product in those days and the efficiency of the cells was increasing along with the optimism of the companies involved. The US government financed support through DOE and, with the technical assistance of its national labs, was the driver of the effort to bring solar energy into the US energy market.

Nowadays the solar revolution is on a scale much larger than ever, so solar is finally here to stay and prosper. Or is it...? We will review the different possibilities in this text too.

The PV Revolution

PV devices were manufactured since the 1950s, as part of the semiconductor processes development, but the real PV invasion started with the manufacturing of the first PV module. See the short list below of the development of the PV industry.

1963 Sharp Corporation manufactures the first silicon PV modules.

Japan installs a 242 W PV array on a lighthouse, the world's largest at that time.

1964 NASA launches the first Nimbus spacecraft powered by a 470-watt PV array.

1965 Peter Glaser conceives the idea of the satellite solar power station.

1966 NASA launches the first Orbiting Astronomical Observatory, powered by a 1-kilowatt PV array, in order to provide astronomical data in the ultraviolet and X-ray wavelengths filtered out by the Earth's atmosphere.

1969 The Odeillo solar furnace, located in Odeillo, France, was constructed. This featured an 8-story parabolic mirror.

1970s Dr. Elliot Berman, with help from Exxon Corporation, designs a cost-effective solar cell at $20 a watt. Solar cells are used in navigation warning lights and horns on many offshore gas and oil rigs, lighthouses, railroad crossings and domestic applications.

1972 The French install a cadmium sulfide (CdS) PV system to operate an educational television at a village school in Niger.

1972 The Institute of Energy Conversion is established at the University of Delaware to perform research and development on thin-film PV and solar thermal systems, becoming the world's first laboratory dedicated to PV research and development.

1973 The University of Delaware builds "Solar One," one of the world's first PV powered residences.

1976 The NASA Lewis Research Center starts installing 83 PV power systems on every continent except Australia.

1976 David Carlson and Christopher Wronski, RCA Laboratories, fabricate the first amorphous silicon PV cells.

1977 The U.S. Department of Energy launches the Solar Energy Research Institute, a federal facility dedicated to harnessing power from the sun.

1977 Total PV manufacturing production exceeds 500 kilowatts.

1978 1978 NASA's Lewis Research Center dedicates a 3.5-kilowatt PV system it installed on the Papago Indian Reservation in southern Arizona—the world's first village PV system.

1980 ARCO Solar becomes the first company to produce more than 1 megawatt of PV modules in one year.

1980 At the University of Delaware, the first thin-film solar cell exceeds 10% efficiency using copper sulfide/cadmium sulfide.

1981 Paul MacCready builds the first solar-powered aircraft (the Solar Challenger) and flies it from France to England across the English Channel.

1982 The first PV megawatt-scale power station goes on-line in Hisperia California. It has a 1-megawatt capacity system, developed by ARCO Solar

1982 Australian Hans Tholstrup drives the first solar-powered car (the Quiet Achiever) almost 2,800 miles between Sydney and Perth in 20 days—10 days faster than the first gasoline-powered car to do so.

1982 Volkswagen of Germany begins testing PV arrays mounted on the roofs of Dasher station wagons, generating 160 watts for the ignition system.

The Florida Solar Energy Center's "Southeast Residential Experiment Station" begins supporting the DOE's PVs program in the application of systems engineering.

Worldwide PV production exceeds 9.3 megawatts.

1983 ARCO Solar dedicates a 6-megawatt PV substation in central California. The 120-acre, unmanned facility supplies the Pacific Gas & Electric Company's utility grid with enough power for 2,000-2,500 homes.

Solar Design Associates completes a stand-alone, 4-kilowatt powered home in the Hudson River Valley.

Worldwide PV production exceeds 21.3 megawatts, with sales of more than $250 million.

1984 The Sacramento Municipal Utility District commissions its first 1-megawatt PV electricity generating facility.

1985 The University of South Wales breaks the 20% efficiency barrier for silicon solar cells under 1-sun conditions.

1986 ARCO Solar releases the G-4000—the world's first commercial thin-film power module.

1988 Dr. Alvin Marks receives patents for two solar power technologies he developed: Lepcon and Lumeloid. Lepcon consists of glass panels covered with a vast array of millions of aluminum or copper strips, each less than a micron or thousandth of a millimeter wide.

1989 The first high concentration PV (HCPV) tracker was designed and manufactured by Alpha Solarco, Inc. with the financial help of DOE and with the technical assistance of engineers and scientists of its National laboratories. The tracking system was installed in the Nevada desert, where it was operated successfully over 10 years.

1991 President George Bush redesignates the DOE's Solar Energy Research Institute as the National Renewable Energy Laboratory (NREL).

1992 University of South Florida develops a 15.9% efficient thin-film PV cell made of cadmium telluride, breaking the 15% barrier for the first time for this technology.

A 7.5-kilowatt prototype dish system using an advanced stretched-membrane concentrator becomes operational.

1993 Pacific Gas & Electric completes installation of the first grid-supported PV system in Kerman, California. The 500-kilowatt system was the first "distributed power" effort.

Alpha Solarco, with the assistance of DOE and NREL, redesigns the world's first full-scale HCPV tracking system with the latest state-of-the-art CPV cells and Fresnel lenses, which allowed it to achieve world record of 18% efficiency at the time.

1994 NREL (formerly the Solar Energy Research Institute) completes construction of its "Solar Energy Research Facility," which was recognized as the most energy-efficient of all U.S. government buildings worldwide.

The first solar dish generator using a free-piston Stirling engine is tied to a utility grid.

The National Renewable Energy Laboratory develops a solar cell made from gallium indium phosphide and gallium arsenide that becomes the first one to exceed 30% conversion efficiency.

1996 The world's most advanced solar-powered airplane, the Icare, flew over Germany. The wings and tail surfaces of the Icare are covered by 3,000 super-efficient solar cells, with area of 21 m^2.

1998 The remote-controlled, solar-powered aircraft, "Pathfinder" sets an altitude record, 80,000 feet, on its 39th consecutive flight on August 6, in Monrovia, CA.

Subhendu Guha, a noted scientist with his pioneering work in amorphous silicon, led the invention of flexible solar shingles, a roofing material and state-of-the-art technology for converting sunlight to electricity.

1999 Construction was completed on 4 Times Square, the tallest skyscraper built in the 1990s in New York City. It incorporates more energy-efficient building techniques than any other commercial skyscraper and also includes building-integrated PV (BIPV) panels on the 37th through 43rd floors on the south- and west-facing facades that produce portion of the building's power.

1999 Spectrolab, Inc. and NREL develop a PV solar cell that converts 32.3 percent of the sunlight that hits it into electricity. The high conversion efficiency was achieved by combining three layers of PV materials into a single solar cell. The cell performed most efficiently when it received sunlight concentrated to 50 times normal.

NREL achieves a new efficiency record for thin-film PV solar cells. The measurement of 18.8 percent efficiency for the prototype solar cell topped the previous record by more than 1 percent.

Cumulative worldwide installed PV capacity reaches 1000 megawatts.

2000 First Solar begins production in Perrysburg, Ohio, at the world's largest PV manufacturing plant with estimated annual capacity of 100MW.

At the International Space Station astronauts begin installing solar panels on what will be the largest solar power array deployed in space. Each "wing" of the array consists of 32,800 solar cells.

Sandia National Laboratories develops a new inverter for solar electric systems that will increase the safety of the systems during a power outage.

Two new thin-film solar modules, developed by BP Solarex, break previous performance records. The company's 0.5-square-meter module achieves 10.8 % conversion efficiency, the highest in the world for thin-film modules of its kind. Its 0.9-square-meter module achieved 10.6% conversion efficiency and a power output of 91.5 watts, the highest power output for any thin-film module in the world.

A family in Morrison, Colorado, installs a 12-kilowatt solar electric system on its home—the largest residential installation in the United States to be registered with the U.S. Department of Energy's "Million Solar Roofs" program. The system provides most of the electricity for the 6,000-square-foot home and family of eight.

2001 Home Depot begins selling residential solar power systems in three of its stores in San Diego, California. A year later it expands sales to include 61 stores nationwide.

2001 NASA's solar-powered aircraft Helios sets a new world record for non-rocket powered aircraft: 96,863 feet, more than 18 miles high.

The National Space Development Agency of Japan, or NASDA, announces plans to develop a satellite-based solar power system that would beam energy back to Earth. A satellite carrying large solar panels would use a laser to transmit the power to an airship at an altitude of about 12 miles, which would then transmit the power to Earth.

TerraSun LLC develops a unique method of using holographic films to concentrate sunlight onto a solar cell, instead of using Fresnel lenses or mirrors.

This capability allows the modules to be integrated into buildings as skylights.

PowerLight Corporation places online in Hawaii the world's largest hybrid system that combines the power from both wind and solar energy. The grid connected system is unusual in that its solar energy capacity—175 kilowatts—is actually larger than its wind energy capacity of 50 kilowatts.

British Petroleum (BP) and BP Solar announce the opening of a service station in Indianapolis that features a solar-electric canopy. The Indianapolis station is the first U.S. "BP Connect" store, a model that BP intends to use for all new or significantly revamped BP service stations. The canopy is built using translucent PV modules made of thin films of silicon deposited onto glass.

2002 NASA successfully conducts two tests of a solar-powered, remote-controlled aircraft called Pathfinder Plus. In the first test in July, researchers demonstrated the aircraft's use as a high-altitude platform for telecommunications technologies. Then, in September, a test demonstrated its use as an aerial imaging system for coffee growers.

Union Pacific Railroad installs 350 blue-signal rail yard lanterns, which incorporate energy saving light-emitting diode (LED) technology with solar cells, at its North Platt, Nebraska, rail yard—the largest rail yard in the United States.

ATS Automation Tooling Systems Inc. in Canada starts to commercialize an innovative method of producing solar cells, called Spheral Solar technology, based on tiny silicon beads bonded between two sheets of aluminum foil—promises lower costs due to its greatly reduced use of silicon relative to conventional multicrystalline silicon solar cells.

The largest solar power facility in the Northwest—the 38.7-kilowatt White Bluffs Solar Station—goes online in Richland, Washington.

And then the real PV revolution started in the mid-2000, with Asian companies taking the lead in manufacturing enough PV modules to cover every house roof in the world.

CONCLUSIONS

In conclusion, solar energy is still abundant and readily available, waiting to be harnessed for the good of mankind. We have seen the solar industry's ups and downs of the past decades and have learned much from the lessons of the past. Starting now we should make solar energy a priority for our energy future, until we achieve the goals of energy independence and a clean environment.

Technological and market conditions were not ripe for the full implementation of solar energy in energy markets until recently, and there were other barriers, too. Nowadays, however, the environmental and energy situations in the US and the world have changed so that there is no going back—solar is here to stay! Existing and future solar energy companies will be on the forefront of alternative energy development, thus bringing bright, efficient and safe solar energy and clean environment to future generations.

A major focus presently is (and will be for the foreseeable future) on access to the coveted desert lands with their abundant sunlight. This opens a huge market, where millions and billions of PV modules could be installed, so the battle is just now starting. The goal of this book is to provide an unbiased review of the technological aspects of the competing solar technologies and bring their advantages and disadvantages out into the open.

These are exciting times, and we all should participate in the energy revolution using our God-given gifts and abilities to move it ahead quickly, thus ensuring its progress and ultimate success. Future generations will definitely appreciate our efforts and the results thereof.

Notes and References

1. PVCDROM by C.B. Honsberg and S. Bowden, www.pveducation.org, 2010.
2. History of Solar, EERE, http://www1.eere.energy.gov/solar/pdfs/solar_timeline.pdf.

Chapter 2

Solar Thermal Technologies

"The most incomprehensible thing about the world is that it is comprehensible."

Albert Einstein

BACKGROUND OF SOLAR THERMAL AND THERMO-ELECTRIC POWER GENERATION

Converting sunlight into heat or electricity is nothing new, but only recently have we seen some serious developments in this area of the world's energy markets. Solar thermal power generation is finally receiving the attention it deserves as a significant source of power for both heat and electric generation. From the developments of late we can confidently say that it is the energy of the future too. It is the new and fastest way to our energy independence and to a cleaner environment.

Solar thermal (heat generation) and thermo-electric (electric generation) technologies are the most reliable and mature solar energy conversion technologies nowadays, so no doubt they will help us achieve these goals quicker.

Definition of Solar Thermal and Solar Thermo-electric Technologies

Solar thermal equipment uses sunlight to convert its energy directly into heat which is then used for heating, or for electric power generation. The heat can be used for heating homes, or as a heat source in commercial processes. Most often, however, the heat is converted into electric power which is then sent into the electric grid.

The conversion of sunlight to heat is a straightforward process, while the conversion of heat to electricity is somewhat more complex and requires expensive equipment and large installations to make it cost effective. This conversion is usually done at the so-called "utility scale power plants," using concentrated solar power (CSP) equipment which is the technology of choice nowadays.

There are currently four different types of solar thermal systems. The first type, *flat plate water heater* solar thermal energy generator is in its own category, because it generates heat only and is the simplest and cheapest of the bunch. It can be mounted on the roofs of houses and businesses and is used only to heat water (or other liquids) to a moderate temperature. This technology has been successfully used by commercial operations, such as restaurants, laundromats, canning facilities, etc. for several decades.

Smaller size, roof-mounted, parabolic troughs were also popular in the Southwest USA in the 1980s, and are making a slow comeback nowadays, while generally speaking, the major CSP technologies are large, ground-mounted, grid-connected systems.

The other types of thermal solar systems, *Stirling engine, parabolic trough* and *power tower* are in the category of concentrated solar power (CSP) technologies, because they do capture the sunlight and concentrate (focus) it onto a receiver. The heat is then normally used to make electricity by several different methods which we will review below as well.

The CSP technologies require direct (clear sky) sunlight for their proper and efficient operation, as needed for the optics to reflect and concentrate the reflected light onto a receiver which converts it into heat. The heat can be used for heating or for electricity generation. Relatively flat land is best for CSP systems, with slopes not exceeding 3 percent being recommended in most cases. The area of land required depends on the type of solar plant, but on average it is about 5-6 acres per produced megawatt (MW) of electric energy. So, cost-effective utility scale CSP power plants are 100 MW in size or larger, requiring minimum of 500-600 acres land for each installed 100 MWp. This large land base requirement involves significant surface disturbance (digging, land leveling and other modifications) with associated potential impact on a variety of resources on public and private lands.

These types of facilities also require roads, water source, wind protection, security fencing and such for their safe and efficient operation. The generated electricity is sold to the local utilities under a power purchase agreement (PPA), or other long-term power sale agreements.

Introduction

The sun's energy falling on the Earth's surface for just 60 minutes is equivalent to the entire annual global energy consumption. From that fact alone we can easily conclude that the potential for getting free energy from the sun is virtually unlimited. The deployment of concentrating solar thermal power (CSP) technologies is a good example of our attempts to capture that potential.

CSP's capacity is expected to increase exponentially over the next several years. The worldwide installed CSP capacity estimates vary widely from 20-35 GW by 2025 and reaching 1000-3000 GW by 2050. These estimates might be too optimistic, especially in light of the developments of late, reflected in decrease of government subsidies, water shortages and the growing trend of conversion of CSP to PV power plants. Some restrictions in EU, such as the law passed recently in Spain which reduced the government subsidies for solar energy, will have a profound effect on the CSP industry as a whole. We do believe that it will grow as the needs and the energy markets of developing countries continue to expand, but the pace of this growth is uncertain.

CSP technology is facing increasing challenges from PV competitors who have leveraged PV's declining costs and adaptability, to create a large global market. CSP will have trouble competing directly with PV on a cost per kWh basis in the near future but might be able to occupy niche markets with its ability to provide more stable power by providing after-hours energy via on-site thermal storage. The need for cooling water is a great problem facing the CSP industry. PV doesn't have this problem and is taking full advantage of this fact presently.

We estimate that the total solar power (CSP and PV) produced around the world will continue to grow exponentially, but anything can happen at any time to alter the growth pattern one way or another. Technology types, proper design, manufacturing, installation and operation have a lot to do with it, but other—even greater and unrelated—forces will be shaping the overall future of solar energy generation in the future. These forces are future demand and supply balance, material prices, energy costs, financing options, land availability and permitting, transmission and interconnection, socio-economic, political, and a number of other factors that contribute to the complexity and the degree of difficulty in deploying solar power generating equipment.

There are a number of thermal solar energy converting technologies, and we will review the major ones, focusing on those which we consider most likely to take major part in, and have largest impact on, the commercial and utilities type power generation development and use in the 21st century.

So, the major solar thermal power generating technologies suitable for commercial and utilities applications today, and presumably for the rest of the 21st century are:

1. Solar thermal technologies:
   a. The flat plate water heater
   b. The Stirling engine dish
   c. Parabolic troughs,
   d. Power towers, and
   e. New and exotic solar thermal technologies

2. Solar photovoltaics technologies
   a. Crystalline silicon PV modules,
   b. Thin film PV modules
   c. HCPV trackers, and
   d. New and exotic PV technologies

In this chapter we will present a quick review of the solar thermal technologies, and will then take a much closer look at the competing PV technologies in the following chapters.

The key solar thermal technologies used today are discussed below.

KEY SOLAR THERMAL AND THERMO-ELECTRIC TECHNOLOGIES

We will review the solar thermal and thermo-electric technologies, focusing on the CSP technologies, since they are most suitable for commercial and utilities power generation.

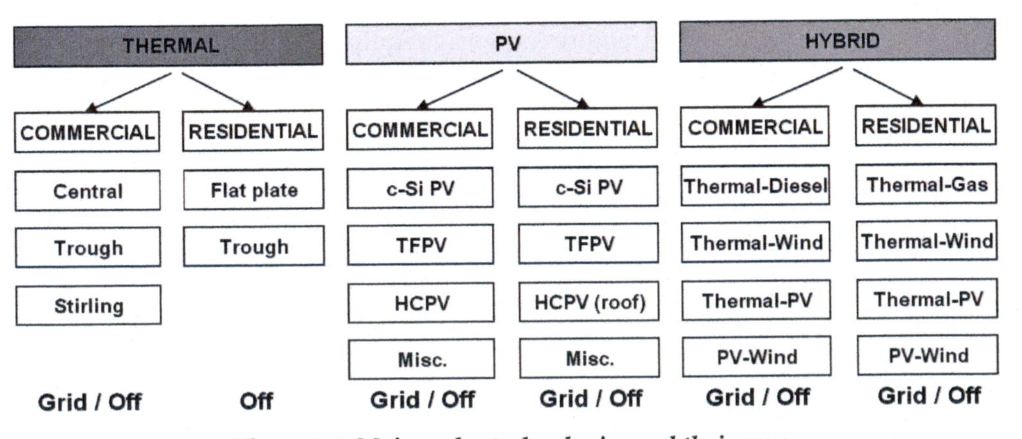

Figure 2-1. Major solar technologies and their uses

Flat Plate Solar Water Heater

Flat plate water heating systems have been, and still are, used in residential, commercial and industrial applications, primarily for heating water in laundromats, restaurants, public parks, car washes, and canning and bottling facilities. These heating systems could be used practically anywhere where low temperature hot water is needed during the day. Adding a storage tank could provide water for use during the night and/or cloudy days. In all cases, they are truly "thermal" systems designed to provide hot water.

They are the simplest and cheapest energy conversion devices today, consisting of a frame onto which a heat exchanger plate (or some modification of) is mounted. Water runs through the heat exchanger plate and absorbs the sunlight energy, thus heating the plate and the water (or other heat absorbing liquid) running through it.

The materials, as well as the manufacturing, installation and operation procedures are straightforward and relatively inexpensive. The return on investment (ROI) is one of the highest in the industry, if the systems are properly designed, installed and operated. There are also a number of incentives today which make it even more feasible and desirable to own and operate such a renewable energy system.

Background of CSP Technologies

With the flat plate water heater covered as much as needed for our purposes and filed in the category of thermal (heat) generation for residential and small commercial applications, we will now concentrate on the three major types of thermo-electric systems presently used

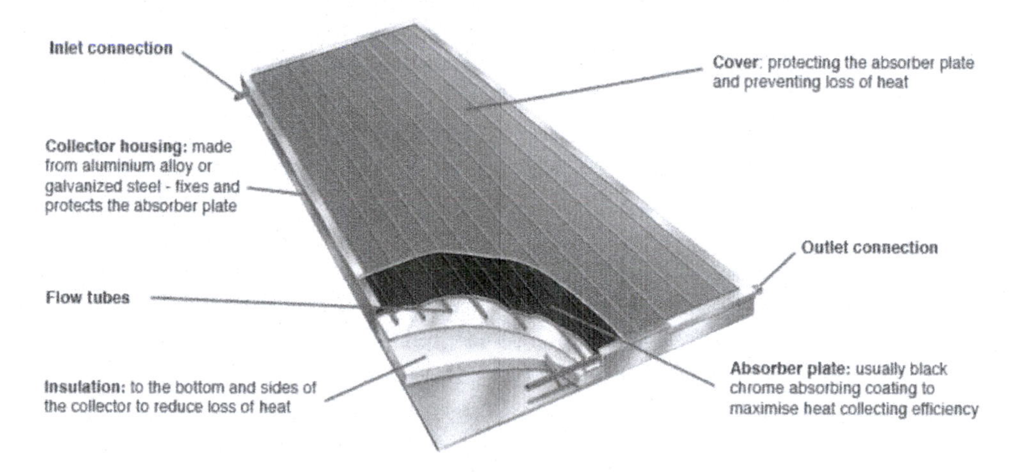

Figure 2-2. Flat plate solar water heater

Table 2-1. Advantages and disadvantages of flat plate solar water heaters

Flat Plate Water Heaters
Advantages
+ Flat plate water heaters are simple and cheap to manufacture and operate solar energy conversion devices
+ They are made out of everyday materials (metals and glass), which contributes to their simplicity and cost effectiveness.
+ They are easy to install and operate, and require no special procedures for proper operation..
Disadvantages
- In most cases, the complete systems require a special heat exchanger (or special hot water tank) in order to provide constant and controllable temperature of hot water usage.
- Some of the system components could rust with time, and the hoses, connections and other components could start leaking, thus requiring replacement which adds to the Cost of Operation (COO).
- The top glass plate requires periodic washing and cleaning in order to provide maximum efficiency, which requires treated water and labor.

for large-scale solar applications. These are: the Stirling engine dish, the parabolic trough, and the power tower.

These three technologies have one thing in common: they all use trackers and optics of some type or another to optimize their efficiency. They also require relatively flat land with slopes not exceeding three percent to accommodate the solar collectors. The area of land required depends on the type of plant, but it is about five acres per installed megawatt (MW). It is anticipated that a commercial scale CSP facility of any type, would be in the range of 100 MW or larger and will require in excess of 500 acres plus whatever else is needed for their installation and proper infrastructure.

Unlike solar photovoltaic technologies which use semiconductors to convert sunlight directly into electricity, CSP plants generate electricity by converting sunlight into heat first. Much like a reflective mirror their reflectors focus sunlight onto a receiver. The heat absorbed by the receiver is used to move an engine piston (Sterling engine), or generate steam that drives a turbine to produce electricity (parabolic troughs and power tower). Power generation after sunset is possible also by storing excess heat in large, insulated tanks filled with molten salt during the day and using it at night. Since CSP plants require high levels of direct solar radiation to operate efficiently, deserts make ideal locations. As a matter of fact, these types of systems cannot operate efficiently in any other environment.

A study by Ausra, a solar energy company based in California, indicates that more than 90% of fossil fuel-generated electricity in the U.S. and the majority of U.S. oil usage for transportation could be eliminated by using solar thermal power plants and will cost less than it would cost to continue importing oil. The land requirement for the CSP plants would be roughly 15,000 square miles in the SW USA deserts, or the equivalent of 15% of the land area of Nevada. While this may sound like a large tract of land, in the long run CSP plants use less land per equivalent electrical output than large hydroelectric dams when flooded and wasted land is included and less than coal plants when factoring in the land used for mining and waste disposal. Another study, published in *Scientific American*, proposes using CSP and PV plants to produce 69% of U.S. electricity and 35% of total U.S. energy including transportation by 2050.

The major CSP technologies are:

1. The Stirling engine-dish tracker
2. The parabolic trough tracker, and
3. The power tower (central receiver)

We will take a close look and discuss each of these below, focusing on their technological advancements and use in large-scale solar installations.

The Stirling Engine

One of the most elegant and flexible solar thermal power conversion technologies today is the Stirling engine dish system. It consists of mirrors mounted on a frame which is continuously tracking the sun all through the day. The mirrors focus the reflected sunlight onto the receiver of the Stirling engine mechanism which is activated by the heat and turns on a shaft, connected to the rotor of an electric generator similar to that of the alternator of your car. The generator rotor turns with the engine shaft and generates electric power while the sun is shining and the receiver is hot enough to activate the engine and rotate the shaft.

A Stirling engine system is actually a solar electricity generator because the heat produced by the mirrors attached to it is converted into electric energy on the spot so small installations of a few units are possible; something that is just not practical with the other CSP technologies. The Stirling engine needs cooling just like a car engine for more efficient operation and in order to cool the engine walls, bearings and other moving parts.

The mirror, or mirrors, are mounted on a metal frame which is driven by two motor-gear assemblies (x-y drives), programmed to move the frame in such a fashion that it follows the sun's movement precisely all day long, thus providing accurate focusing of the sunlight onto the heating plate of the Stirling engine. When the plate gets hot enough, the air in one of the cylinders in it is compressed and forces the piston in it to move up. This action forces the piston in the other cylinder (which is simultaneously cooled) to move down. Eventually, the compression in the second cylinder increases to the point that its piston is forced to go back, thus forcing the piston in the first cylinder to assume its initial position. The cycle repeats over and over while there is enough heat to maintain the process.

The Stirling engine function, under ideal operating conditions, can be represented by four cycles, or *thermodynamic process segments*, of interaction between the working gases, the heat exchanger, pistons and the cylinder walls.

The Stirling engine is a very ingenious and efficient piston engine, without the noise and exhaust of internal combustion engines. As a matter of fact, it can be classified as an "external combustion" engine. The gasses inside the cylinders are not exhausted, so there is no pollution and there are few moving parts with very little noise, so it can be used virtually anywhere.

Since its invention in 1816 by the Scottish inventor Dr. Robert Stirling, the Stirling engine has been considered

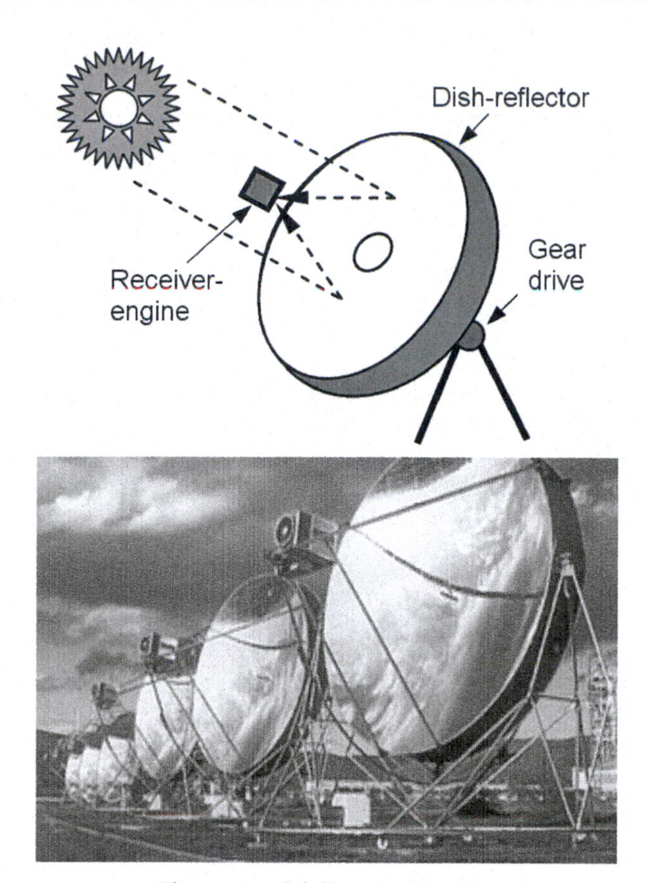

Figure 2-3. Stirling engine dish

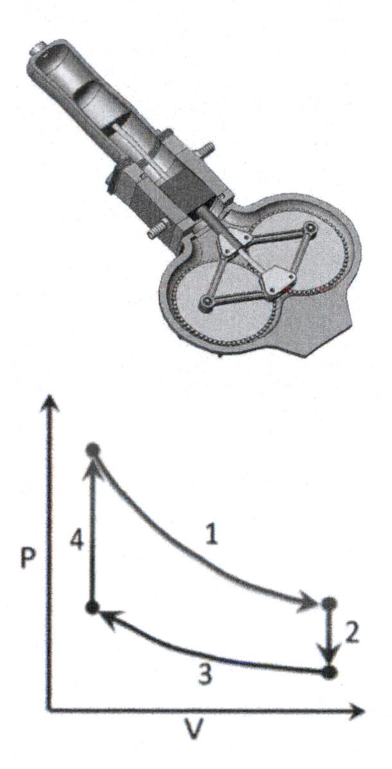

Figure 2-4. Stirling engine and its operating cycles (at ideal conditions)

and proposed for use in many different applications. Presently it is used in some specialized applications, where quiet and clean (no exhaust) operation is required. Some fancy and special purpose submarines and vessels use Stirling engines part of the time under special conditions.

Unfortunately, mass-market application for the Stirling engine is not found as yet, although many scientists and inventors are working on it. Its use in solar power generating equipment might be a good start in that direction. There are currently a number of installations using this technology, but most of them are smaller, demo type systems. No large CSP power plant using Stirling engine technology is in operation today, although there are plans for construction of large-scale Stirling dish power plants in the California and Arizona deserts in the near fugure.

The Parabolic Trough

Parabolic trough solar systems consist of frame in a parabolic trough shape in which glass, metal or plastic reflectors are mounted to focus the sun's energy onto a receiver pipe running above and in parallel to the trough's length. The receiver pipe, or heat collection element (HCE), is centered at the focal point of the reflectors and is heated by the reflected sunlight to very high temperatures. Liquid of some sort is pumped through the receiver pipe and is heated in the process.

The HCE of the parabolic trough units is usually composed of a metal pipe with a glass tube surrounding it, and with the space between these evacuated to provide low thermal losses from the pipe. The pipe is coated with a material that improves the absorption of solar energy. Several improvements have been made or are underway to improve performance, the most significant of which is the seal between the glass and the pipe, which seal has not been as reliable as desired and development of better seal materials/seal configuration is still underway.

Parabolic troughs can focus the sunlight many times its normal intensity on the receiver pipe, where heat transfer fluid (HTF—usually mineral or synthetic oil) flowing through the pipe is heated. This heated fluid is then used to generate steam which powers a turbine that drives an electric generator. The collectors are aligned on an east-west axis and the trough is rotated north-south, following the sun as needed to maximize the sun's energy input to the receiver tube.

Parabolic trough power plants, also called solar electric generating systems (SEGS), represent the most mature CSP technology, with the most installed capacity of all CSP technologies. The first SEGS solar trough plant

Table 2-2. Advantages and disadvantages of Stirling Dish systems

Stirling Engine CSP systems:
Advantages + Stirling engine systems have relatively high efficiency. Theoretically, up to 40% of the total energy input could be converted into electricity, which is usually reduced due to thermodynamic and electro-mechanical losses. + They use few bearings and seals, which require less lubricant and last longer than steam turbines, + Waste heat can be captured, and reused as heat source in some applications + They use an external heat source, which could be anything that generates heat: gasoline, diesel, coal, biofuel, or preferably solar energy. Because of that they are well suited for hybrid power generation by using the external heat generated by the different primary heat sources. + Stirling engines are well suited for installation on not perfectly leveled terrains, where most other solar technologies require more expensive land leveling and other such modifications.
Disadvantages - The major disadvantage is the need for cooling. Efficient cooling is needed, especially in hot desert regions, where fortunately or unfortunately the Stirling engine technology operates best - They need a lengthy warm up period when starting, so operation in a partially cloudy day, when the engine must stop and start frequently might be challenging - The working gas will eventually leak out and will have to be replaced, which means lost time, resources and additional expense - They have a number of moving parts, which require lubrication and maintenance. Periodic maintenance operations increase the cost of operation (COO) by virtue of additional salaries and lost power generation time. - Hot and wet parts rust with time, which requires expensive maintenance and replacement procedures. - Mirrors used in these systems have to be of very high quality, because the optical properties of cheap mirrors reduce the efficiency and deteriorate quickly with time. - Mirrors could get damaged and/or become non-reflective enough with time, thus requiring expensive replacement procedures. - Basically, like with any engine that operates on and off all day long, the parts will wear down and have to be replaced, or the engine has to be rebuilt, which is an expensive operation.

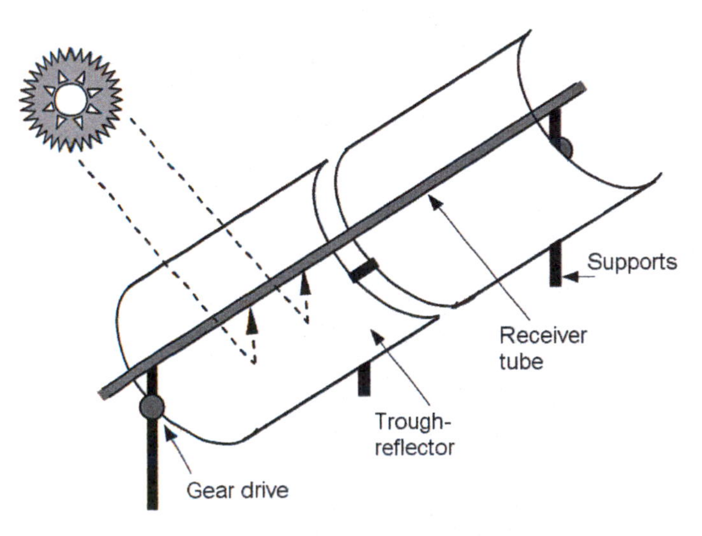

a) Parabolic trough, side view

b) Parabolic trough

Figure 2-5. Parabolic trough technology

were constructed in California by Luz International and started operating in 1984, with the last one coming on line in 1991. Altogether, nine such plants were built, SEGS I–VII at Kramer Junction and VIII and IX at Harper Lake and Barstow respectively. In February 2005, all but two (I and II) of the Kramer Junction SEGS plants were acquired by FPL Energy and Carlyle/Riverstone and are still operating.

A natural gas system added to the plant "hybridizes" it and contributes up to 25% of the output. This feature also allows operation later at night or on cloudy days to meet the requirements of the grid. FPL now runs these systems, making it the largest solar power generator in the United States. All of the power generated from the SEGS projects is sold to Southern California Edison under long-term contracts negotiated by Luz back in the 1980s.

There are a number of such plants operating around the world and many others are planned.

One advantage of CSP systems is their ability to generate power after the sun has gone down. In these cases, the HTF fluid going through the receiver pipe is routed through a thermal storage system which permits the plant to keep operating for several hours after sunset while the electrical demand is still relatively high. The thermal storage system consists of a "hot" storage tank equipped with heat exchanger where HTF circulates and gives up a portion of its heat to heat the storage solution in the tank during the day. At night, the hot storage solution flows through the same heat exchanger heating up HTF which is sent to the steam turbines for generating power. The cooled-down storage solution flows from the heat exchanger to a "cold" storage tank where it stays until daytime when it is reheated and returned to the "hot" storage tank. And the cycle is repeated every night.

Linear Fresnel Reflector

Linear Fresnel reflectors (LFR) systems are similar to the parabolic trough, but use an array of nearly flat Fresnel reflectors instead. These reflectors concentrate solar radiation onto elevated inverted linear receivers. Water, or other liquid, flows through the receivers and is converted into steam. This system is also line-concentrating with the advantages of low costs for structural support

Table 2-3. Advantages and disadvantages of parabolic trough systems

Parabolic Trough Systems
Advantages + Parabolic troughs' components (trough frame, mirrors, receiver, turbine etc.) are well established commercial products, thus risk and uncertainty factors are limited and well controlled + This technology is also very well supported by well financed companies and governments (including the US government and its agencies, such as DOE, NREL, SNL etc.) + They have relatively high efficiency. Theoretically, up to 40% of the total energy input could be converted into electricity, but thermodynamic and mechanical losses usually decrease the total efficiency + Waste heat can be captured and can be used as secondary heat source, and for other thermal and hybrid applications as well.
Disadvantages - Heat transfer fluids heat is limited to 400°C, which offers only moderate steam qualities - Steam turbines need cooling, usually lots of water, which is simply unavailable in the desert. The cooling source could be replaced by an AC (cryogenic) system but they reduce the energy output of the systems - They have a lengthy warm up period at start, so operation in partially cloudy days, when the system must stop and start frequently, might be quite challenging and at reduced efficiency - The troughs and the steam turbines have large number of moving parts, which require lubrication and periodic maintenance, which require additional parts, labor and lost production time **- Hot and wet parts rust with time, which requires expensive maintenance and replacement** procedures, and causes increase in lost power generation. - Miles of hydraulic and liquid transferring pipes, hoses, connectors, expanders etc. contribute to leaks and malfunctions, which increases the O&M cost as well. - Mirrors used in these systems have to be of very high quality, and need to be kept clean, because the optical properties of cheap and dirty mirrors affect the efficiency and longevity - Mirrors could get damaged and/or become non-reflective enough with time, thus requiring expensive replacement procedures

and reflectors, fixed fluid joints, a receiver separated from the reflector system and long focal lengths that allow the use of flat mirrors.

The LFR technology is seen as a potentially lower-cost alternative to trough technology for the production of solar process heat. Planned commercial applications are estimated at a size from 50 to 200 MW. Linear Fresnel applications are mostly at the experimental stage for now. Companies working in the field claim higher efficiency and lower costs per kWh than its direct competitor, parabolic trough, due to high density of mirrors. Fresnel mirror is available at little more than EUR 7.00 per m^2. According to Ausra, this technology can generate electricity for EUR 0.10 per kWh now and under EUR 0.08 per kWh within next 3 years.

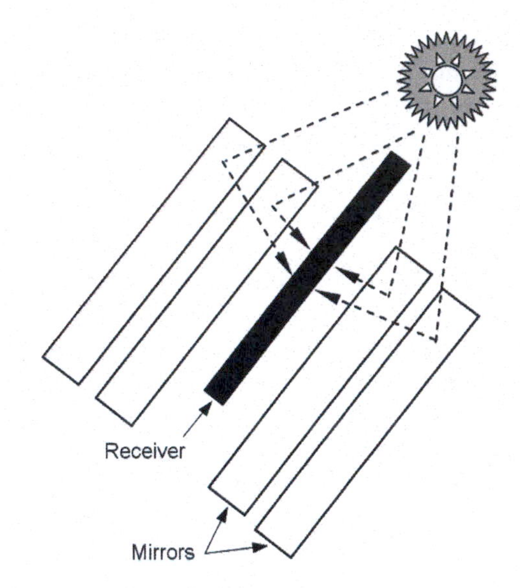

Figure 2-6. Linear Fresnel plant

The Fraunhofer Institute has contributed greatly in making the key components such as the absorber pipe, the secondary reflectors, primary reflector array and their control ready for operation. Based on theoretical investigations and the specific conditions found in sunny climates, Fraunhofer researchers have calculated that the electricity production costs will not rise above EUR 0.12 per kWh.

The linear Fresnel CSP technology derives its name from a type of optical system that uses a multiplicity of small flat optical faces, invented by the French physicist Augustin-Jean Fresnel who, while Commissioner for Lighthouses, invented the segmented lighthouse lens. Flat moving reflectors follow the path of the sun and reflect its radiation to the fixed pipe receivers above. Molten salt or other operating liquid powers a steam turbine, or is stored for night use. The technology itself is simple; the biggest challenge is setting mirrors to track the sun and reflect rays effectively. Flat mirrors are much cheaper to produce than parabolic ones, so this is a bonus.

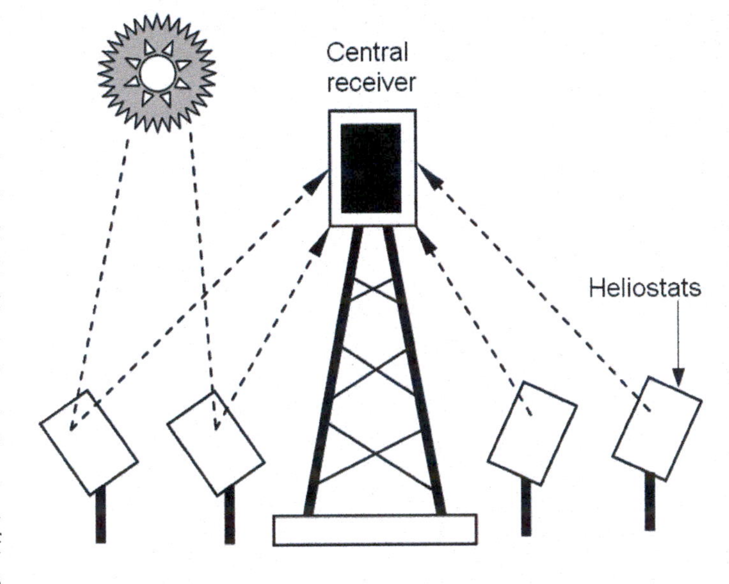

Another advantage of the compact linear Fresnel reflector CLFR is that it allows for a greater density of reflectors in the array. In addition, Fresnel technology is less sensitive to wind loads and allows parallel land use to a large extent.

The LFR technology is more competitive economically due to:

- More effective land use than rival technologies;
- Low visual impact on landscape;
- Lower infrastructure costs due to its design;
- Lighter base, less steel used, flat instead of curved mirrors.

The Power Tower

The power tower (or central receiver) power generation uses methods of collection and concentration of solar power based on a large number of sun-tracking mirrors (heliostats) reflecting the incident sunshine to a receiver

Figure 2-7. Power tower

(boiler) mounted on the top of a high tower, usually in the middle of the collection field. 80 to 95 percent of the reflected energy is absorbed into the working fluid which is pumped up the tower and into the receiver. The heated fluid (or steam) returns down the tower and is fed into a thermal electrical power plant, steam turbine, or an industrial process that uses the heat.

The difference between the central receiver concept of collecting solar energy and the trough or dish collectors discussed previously, is that in this case all of the solar energy to be collected in the entire field is transmitted optically to a relatively small central collection region rather than being piped around a field as hot fluid. Because of this central receiver systems are characterized by large power levels (100 to 500 MW) and higher temperatures (540 to 840°C) of the working fluids which allows the creation of high quality superheated steam which is more efficient for electricity generation.

Power tower technology for generating electricity has been demonstrated in the Solar One pilot power plant at Barstow, California, since 1982. This system consists of 1818 heliostats, each with a reflective area of 39.9 m^2 (430 ft2) covering 291,000 m^2 (72 acres) of land. The receiver is located at the top of a 90.8 m (298 ft) high tower and produces steam at 516°C (960°F) at a maximum rate of 42 MW (142 MBtu/h).

The reflecting element of a heliostat is typically a thin, back (second) surface, low-iron glass mirror. This heliostat is composed of several mirror module panels rather than a single large mirror. The thin glass mirrors are supported by a substrate backing to form a slightly concave mirror surface. Individual panels on the heliostat are also canted toward a point on the receiver. The heliostat focal length is approximately equal to the distance from the receiver to the farthest heliostat. Subsequent "tuning" and optimization of the closer mirrors is done upon installation.

Another heliostat design concept, not so widely developed, uses a thin reflective plastic membrane stretched over a hoop. This design must be protected from the weather but requires considerably less expenditure in supports and the mechanical drive mechanism because of its light weight. Membrane renewal and cleaning appear to be important considerations with this design. In all cases, the reflective surface is mounted on a pedestal that permits movement about the azimuth and elevation axis. Movement about each axis is provided by a fractional-horsepower motor through a gearbox drive. These motors receive signals from a central control computer that accurately points the reflective surface normal halfway between the sun and the receiver.

System design and evaluation for a central receiver application is performed in a manner similar to that when other types of collectors are used. Basically, the thermal output of the solar field is found by calculating collection efficiency and multiplying this by the solar irradiance falling on the collector (heliostat) field, minus some optical, transmission and other losses.

The major components of power tower systems are:

1. Tracking and Positioning

The heliostats must follow the sun all day long in order to focus the sunlight on the tower receiver. This is achieved by means of two electric motor-drive assemblies on each unit. In order to keep parasitic energy low, fractional horsepower motors with high gear rations are used to move the heliostat about its azimuth and elevation axes. This produces a powerful, slow, steady and accurate tracking motion. Under emergency conditions, however, rapid movement of the heliostats to a safe or stow position is an important design criterion. A typical minimum speed requirement would be that the entire field defocus to less than 3 percent of the receiver flux in 2 minutes. Higher speed is desired in case of impending disasters, such as high wind, hail and such, in order to protect the mirrors from mechanical damage.

Since it is currently considered best to stow the heliostats face-down during high wind, hail storms, and at night, an acceptable time to travel to this position from any other position would be a maximum of 15 minutes. The requirement for inverted stow is being questioned since it requires that the bottom half of the mirror surface be designed with an open slot so that it can pass through the pedestal. This space reduces not only the reflective surface area for a given overall heliostat dimension, but also the structural rigidity of the mirror rack. However, face-down stow does keep the mirror surface cleaner and safer.

2. Receivers

The receiver, placed at the top of a tower, is located at a point where reflected energy from the heliostats can be intercepted most efficiently. The receiver absorbs the energy being reflected from the heliostat field and transfers it into a heat transfer fluid. Taking a closer look at the receivers, we see that there are two basic types of receivers, external and cavity receivers.

a. External receivers normally consist of panels of many small (20-56 mm) vertical tubes welded side-by-side to approximate a cylinder. The bottoms and tops of the vertical tubes are connected to headers that supply heat transfer fluid to the bottom of each tube and collect the heated fluid from the top of the tubes. The tubes are made of Incoloy 800 and are coated on the exterior with high-absorption black paint.

External receivers typically have a height to diameter ratio of 1:1 to 2:1. The area of the receiver is kept to a minimum to reduce heat loss. The lower limit is determined by the maximum operating temperature of the tubes and hence the heat removal capability of the heat transfer fluid.

b. Cavity receivers are an attempt to reduce heat loss from the receiver by placing the flux absorbing surface inside an insulated cavity, thereby reducing the convective heat losses from the absorber. The flux from the heliostat field is reflected through an aperture onto absorbing surfaces forming the walls of the cavity. Typical designs have an aperture area of about one-third to one-half of the internal absorbing surface area.

Cavity receivers are limited to an acceptance angle of 60 to 120 degrees Therefore, either multiple cavities are placed adjacent to each other, or the heliostat field is limited to the view of the cavity aperture. The aperture size is minimized to reduce convection and radiation losses without blocking out too much of the solar flux arriving at the receiver. The aperture is typically sized to about the same dimensions as the sun's reflected image from the farthest heliostat, giving a spillage of 1-4%.

3. Heat Transfer Fluids

The choice of the heat transfer fluid to be pumped through the receiver is determined by the application. The primary choice criterion is the maximum operating temperature of the system followed closely by the cost-effectiveness of the system and safety considerations. The heat transfer fluids with the lowest operating temperature capabilities are heat transfer oils. Both hydrocarbon and synthetic-based oils may be used, but their maximum temperature is around 425°C (797°F). However, their vapor pressure is low at these temperatures, thus allowing their use for thermal energy storage. Below temperatures of about -10°C (14°F), heat must be supplied to make most of these oils flow. Oils have the major drawback of being flammable and thus require special safety systems when used at high temperatures. Heat transfer oils cost about $0.77/kg ($0.35/lb).

Water has been studied for many central receiver applications and is the heat transfer fluid used in many power tower plants. Maximum temperature applications are around 540°C (1000°F) where the pressure must be about 10 MPa (1450 psi) to produce a high boiling temperature. Freeze protection must be provided for ambient temperatures less than 0°C (32°F). The water used in the receiver must be highly deionized in order to prevent scale buildup on the inner walls of the receiver heat transfer surfaces. However, its cost is lower than that of other heat transfer fluids. Use of water as a high-temperature storage medium is difficult because of the high pressures involved.

Nitrate salt mixtures can be used as both a heat transfer fluid and a storage medium at temperatures of up to 565°C (1050°F). However, most mixtures currently being considered freeze at temperatures around 140 to 220°C (285 to 430°F) and thus must be heated when the system is shutdown. These mixtures have good storage potential because of their high volumetric heat capacity. The cost of nitrate salt mixtures is around $0.33/kg ($0.15/lb), making them an attractive heat transfer fluid candidate.

Liquid sodium can also be used as both a heat transfer fluid and storage medium, with a maximum operating temperature of 600°C (1112°F). Because sodium is liquid at this temperature, its vapor pressure is low. However, it solidifies at 980°C (208°F), thereby requiring heating on shutdown. The cost of sodium-based systems is higher than the nitrate salt systems since sodium costs about $0.88/kg ($0.40/lb).

For high-temperature applications such as Brayton cycles, the use of air or helium as the heat transfer fluid and operating temperatures of around 850°C (1560°F) at 12 atm. pressure are being proposed. Although the cost of these gases would be low, they cannot be used for storage and require very large diameter piping and expensive compressors to transport them through the system.

CSP Technologies' Future

The dark areas in Figure 2-8 show where the EU community is planning to install a large number of wind, CSP, and PV power plants. CSP technologies will take a major part in this effort which will bring the CSP industry to a new and much higher level by 2020. This, in addition to plans for many additional GWs of CSP installations in the US and Asia, is a very exciting development which paints bright picture for the future of the CSP technologies as a world-class electric power generator.

We have witnessed the quick and successful development of the CSP industry of late and have seen the estimates that place it at the top of the list of large-scale power generation. The fast pace of the technological developments nowadays, and the quickly changing socio-

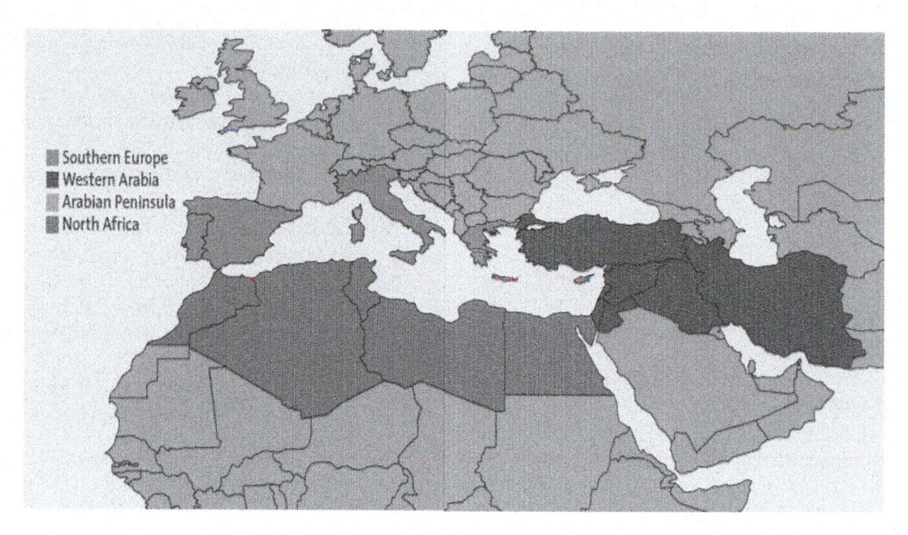

Figure 2-8. CSP technologies: future applications

Table 2-4. Advantages and disadvantages of power tower systems

Power Tower System
Advantages + Power tower components (frame, mirrors, drives, receiver, turbine etc.) are well established commercial products, thus risk and uncertainty factors are limited and well controlled. + The technology is supported by large companies and governments (including the US government and its agencies, such as DOE, NREL, SNL etc.) + They have relatively high efficiency. Theoretically, up to 40% of the total energy input could be converted into electricity, some of which is reduced by thermodynamic and mechanical losses, + Waste heat can be captured also and can be used as secondary heat source and / or for thermal power generation as well. + Mirrors do not require much maintenance once installed, however they must be cleaned periodically at a small cost (water and workforce). Production of mirrors when compared to PV cells is less energy-intensive and more environmentally friendly. + The noisiest part of the system is the steam turbines, but the plants as a whole are quiet.
Disadvantages - Power towers need cooling, but water in the desert is scarce and this is a big problem. AC (cryogenic) systems could be used, but they use a lot of power which decreases energy output. - They use many tracking mirrors, so the terrain before installation needs to be leveled. - Significant part of light beams from the mirrors is reflected by the collector, which might be dangerous to the safety of traffic and other activities in the vicinity. - They need a lengthy warm up period when starting, so operation in a partially cloudy day, when the system must stop and start frequently might be challenging. - They have a large number of moving parts, which require tune-up, lubrication and maintenance. These procedures increase the cost of operation (COO) by adding labor costs and lost power. - Hot and wet parts rust with time, which requires expensive maintenance procedures. - Hydraulic, steam and liquid handling apparatus contribute to leaks and malfunctions which also add to the O&M expenses - Mirrors used in these systems have to be of very high quality, because the optical properties of cheap mirrors cause decease of efficiency and their quality deteriorates quickly with time. - Mirrors could get damaged and/or become non-reflective enough with time, thus requiring expensive replacement procedures - The towers and mirrors have a high profile and can be perceived as a visual and esthetic pollution.

economic and political climates around the world, make it hard to predict what will happen exactly, but we know for sure that CSP is here to stay and that it will grow steadily in the future.

Notes and References

1. Mora Associates, October 2009. http://www.moraassociates.com/publications/0903%20Concentrated%20Solar%20Power.pdf
2. Renewable Energy UK: http://www.reuk.co.uk/First-European-Solar-Power-Tower.htm
3. Environment News Service: http://www.ens-newswire.com/ens/mar2007/2007-03-30-02.asp
4. APS http://www.aps.com/main/green/Solana/About.html
5. Stirling Energy Systems company website: http://www.stirlingenergy.com/
6. National Renewable Energy Laboratory (NREL): dish/stirling report
7. JC Winnie environmental blog: http://www.innovationsreport.com/html/reports/energy_engineering/report-82659.html
8. IEEE Spectrum Online: http://www.spectrum.ieee.org/oct08/6851
9. Global Greenhouse Warming.com: http://www.global-greenhouse-warming.com/solar-parabolictrough.html
10. National Renewable Energy Laboratory (NREL): http://www.nrel.gov/csp/troughnet/
11. Nevada Solar One official website: http://www.nevadasolarone.net/the-plant
12. CSP Today: http://social.csptoday.com/index.php
13. SolarPaces. http://www.solarpaces.org/News/Projects/projects.htm

Chapter 3

Crystalline Silicon Photovoltaic Technologies

"Everything should be made as simple as possible, but not simpler."
Albert Einstein

ABSTRACT

Crystalline silicon (c-Si) based photovoltaic (PV) technologies are a major part of the energy markets today and promise to be even more so in the future. They are the most mature of all PV technologies and compare successfully against the competitors.

c-Si PV technologies also compete successfully with concentrating solar thermal (CSP) and wind energy generators, simply because they are efficient enough and because they can be used in more versatile ways. Because of that, the c-Si PV technologies future looks very promising now.

This success is reflected in Figure 3-1, with c-Si solar cells taking 75% of the world market, and c-Si modules 71%, while the other PV technologies are in the single digits. The total quantity of c-Si cells and modules sold in 2009 was remarkable, keeping in mind that there were only a few MW of PV modules made and sold just 4-5 years ago. The ratios are changing somewhat today with thin film modules taking a more prominent role in the world's PV installations, but there is no question that silicon will lead the pack for the foreseeable future.

The bright future of c-Si in the world's PV energy markets comes at a price and has a number of clouds hanging over it in the shape of unresolved technical issues some of which are quite significant and need to be addressed and resolved.

Our goal, and sincere intent, is to provide a clear down-to-earth description of the c-Si technologies, their materials, processes, and function, and related issues such as their manufacturing and suitability for and application in commercial and utilities power generation projects. We will focus on their use in the world's deserts and other extreme climate areas, which is where most of the large-scale power plants will be located. Since many such installations are presently planned for implementation in the US and abroad, we see this as a perfect and timely opportunity to bring the issues out of the closet for thorough discussion and analysis.

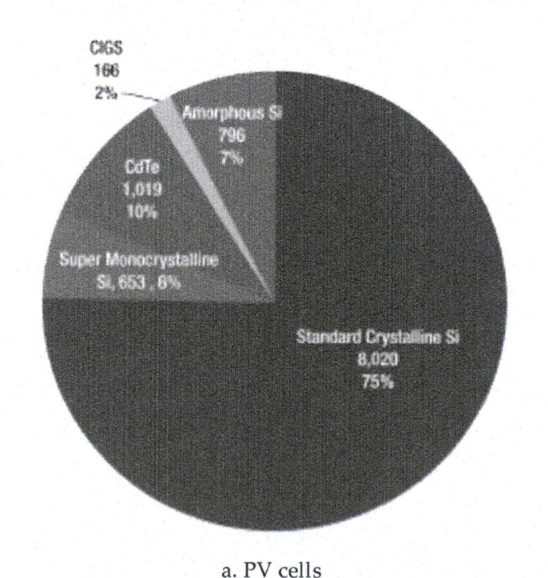

a. PV cells

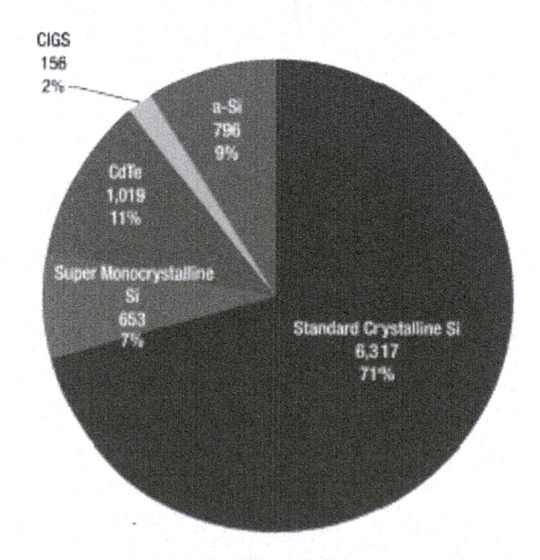

b. PV modules

Figure 3-1. PV Cells and modules production (MWp) in 2009 (1)

GENERAL

We are now getting into the meat of the matter—photovoltaics, or PV, technologies. We all know what PV equipment (solar cells and modules) look like and what they do, but many of us would have a hard time explaining in some detail their structure and function. Not to worry! By the time you finish this book you'll know all there is to know about PV products, processes, function and all. Including their suitability (or not) for utilization in different PV power generating plants. We have a lot of subjects, with lots of ground to cover, so let's go.

Definitions

To make things clearer and easier, we will start by clarifying some of the basic energy terms and key terminologies related to them, as well as those related to photovoltaic power generation and usage.

1. *PV technology components*

The following are the basic definitions of key PV terms used in the solar industry and the energy markets today:

Solar energy is the energy we can get from the sun and convert into thermal or PV power for everyday use.

Photovoltaics, or PV, is the branch of the solar energy generating industry that deals with direct conversion of solar energy into electricity. This is different from thermal electric power generation, where the solar energy is first converted into heat and then into electricity.

Metallurgical grade (MG) silicon is the material from which solar grade silicon is obtained via special processing and refining.

Solar grade (SG) silicon is the base material from which solar wafers and cells are made.

Crystalline silicon (c-Si) is the general category to which all (mono, poly, ribbon, etc.) crystalline silicon materials, wafers, cells and modules belong.

Single crystal, or mono-crystalline (sc-Si) silicon has the original symmetrical silicon material structure which is obtained by growing (pulling) a seed of the pure elemental silicon into large rods which are then sliced into thin round wafers.

Multi-crystalline (m-Si) silicon is lower quality silicon (from a solar point of view) with asymmetrical structure consisting of different type and shape strands of c-Si mixed into its bulk. It is made by melting SG Si chunks into large square blocks, letting the blocks cool, sawing them into smaller pieces and then slicing square wafers from those.

Poly-crystalline (p-Si) silicon, or poly, is a thin film material that is deposited on semiconductor devices. It is used in the solar industry for depositing thin films on special solar cells, as we will see below.

NOTE: There is some confusion about the use of "multi" vs. "poly" silicon terminology, terms that are used interchangeably, so we need to clarify the difference at the onset. "Poly-silicon," or "poly" is actually a thin film used in the semiconductor industry, but "poly" has been widely accepted to refer to "multi-crystalline" silicon widely used in the solar industry. We will also use the terms "multi" and "poly" interchangeably in this text to mean one and the same—multicrystalline silicon.

c-Si solar *wafers* are thin slices of crystalline silicon upon which solar cells are built. They are similar to semiconductor wafers in shape, but are of lower quality in terms of impurities, thus of much lower cost too.

c-Si solar *cells* are devices made out of silicon wafers, which when exposed to sunlight produce a certain amount of electric energy.

c-Si solar *modules* are flat plates (trays) onto which c-Si solar cells are arranged and sealed in, so that when exposed to sunlight they produce electric energy.

c-Si solar *arrays* or systems are groups of c-Si PV modules positioned and interconnected in such a way as to produce a maximum amount of electricity.

c-Si solar *power plants* are groups (strings) of c-Si PV arrays arranged so as to produce a maximum amount of electricity.

2. *PV power generation*

Another set of definitions are needed to clarify the power generation aspects of solar energy devices and systems, as follow.

Commercial and utilities power generation refers to any large (non-residential) PV installations designed to provide power to the grid, or to power commercial enterprises.

Watt (W)—the basic unit of electrical *power*, defining the level and quantity of electricity generation and consumption. It is basically the power developed when electric current of one Amp flows through a circuit with potential difference of one Volt.

1 Watt = 3.4Btu = 1 Joule/sec = 1 N/m = 1 kg/m/sec = 0.00135hp.

Watt peak (Wp)—the basic unit for measuring electric power output from a PV cell, or module, at its peak power output, as measured under an industry standardized light source (1000 W/m^2, at 25°C ambient temperature and AM 1.5). A designation of 100 Wp on a PV module means that this particular PV module produces 100 Watts of DC power under the above-mentioned condition. This is also the module's power rating.

Kilowatt (kW)—the basic and most used unit of measuring the power output produced by a normal size power generating system. A kilowatt is basically the total of 1,000 Watts of electric power.

One Kilowatt is equal to: 1.34 horsepower, 56.87 Btu/minute, 44253.72 foot/pound/minute, or 737.56 foot/pound/second.

Mega Watt (MW)—the basic and most used unit of measuring the power produced by a large size power generating plant. 1.0 MW is basically 1,000 kW, or 1,000,000 W of electric power.

A designation of 100.0 MWp DC power means that the PV plant will produce 100 MWp DC electric power under the design conditions at peak hours.

A designation of 100.0MW AC power plant means that the plant is expected to produce 100MW AC power under the design conditions.

Kilowatt per hour (kWh)—this is equivalent to 1,000 Watts, or 1.0 kW, energy generated during one hour of operation. Or, the amount of electric energy equal to 1 (one) 100 Watt PV module operating under full sunlight for 10 hours (1 module x 100 W x 10 hours = 1 kWh).

Kilowatt hours per year (kWh/y)—this is basically the number of kW produced during one year of operation; i.e., one 1.0 kWp module, installed in the Arizona desert, operating 6 hrs./day at full power will see approximately 300 full sunshine days in one calendar year. So, then 1.0 kWp x 6 hrs./day x 300 days/year = 1,800 kWh/y.

Nameplate capacity—this is the amount of electricity that the solar plant is designed to produce. 1.0 MWp nameplate means that this plant will produce 1.0 MWp under the design conditions at peak hours.

3. Practical examples

a. **Power generation**

1 MWp fixed mount PV modules in Arizona's desert would generate 100% of its nameplate power (1.0 MWp), or close to it, around the summer noon hour. During the rest of the day, the sunlight intensity is lower, thus the produced power is much less, and decreasing quickly, the further from the noon hour we go. So, although there is sunlight 12-14 hours a day, only part of the solar energy is gathered during daylight hours, which averages to the equivalent of ~6-7 hrs. of full power generation per day. Also, on average, only ~300 days per year offer full sunlight in the desert, while the other 65 days are cloudy or partially cloudy. This, of course, varies from year to year.

So, 1.0 MWp x 6 hrs/day x 300 days/year = 1,800 MW/h/year. Or 1.8 GW/h/year electric power can be expected to be produced by our 1.0 MWp PV system in the deserts. This amount will go up and down during the months and years, according to local weather variations and the PV modules' performance. Much less power is expected from a similar system installed in the coastal areas of Washington state, or Northern Germany, where clouds and rain are the norm most of the time. The calculations in these cases are more complex, due to the weather.

b. **Energy conversion**

1 gal. #2 fuel oil = 138,874 Btu = 40 kWh

This means that in order to get the equivalent of the energy contained in 1 gallon of #2 fuel oil, we need to run a 40 kW PV system for one hour on a sunny day, or run a 1 kW PV system for 40 hours at the same location.

A barrel of crude oil (42 gal) contains the energy equivalent of 1,700 kWh, or the power of 1 kWp PV system running non-stop for 1,700 hrs. on a sunny day, or a 1,700 kW system running for one hour.

1 m^2 of parabolic trough with 30% efficiency, irradiated by 1,000 W sunlight will produce 1,020 Btu heat (300 W x 3.4 Btu). This heat can be then converted into electricity at a significant loss of efficiency during transport and conversion in the steam turbine.

PV technologies

The most common, commercially available PV products today are crystalline silicon (c-Si) solar cells and modules, and thin film (TF) cells and modules, as follow.

As seen in Figure 3-2, there are a number of different PV technologies which we will take a look in the text below, but in this chapter we will focus on crystalline silicon (c-Si) based materials, wafers, solar cells and modules. We will do this in some detail, because c-Si technology is the most mature and most widely used PV technology today, which dominates the PV markets by far now and will do so in the foreseeable future.

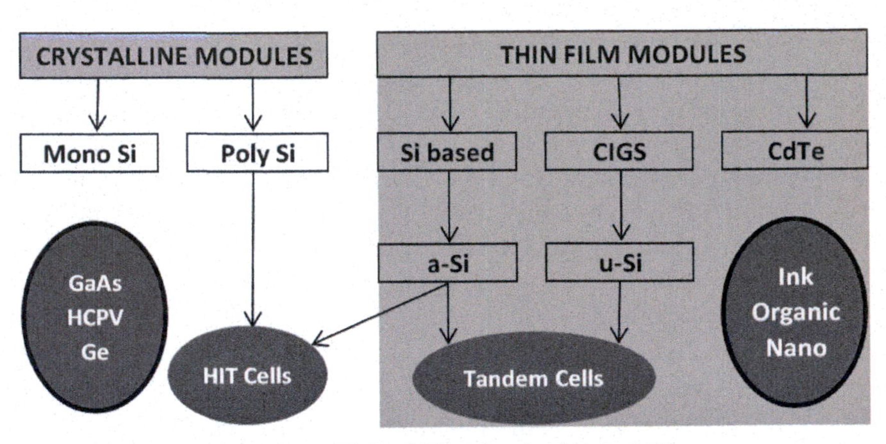

Figure 3-2. Major PV technologies in 2010

SILICON MATERIAL

c-Si solar cells and modules are made from silicon semiconductor material, which is made from SiO_2, or sand. It is actually a special kind of sand with a minimum amount of impurities, but still one of the most abundant raw materials in the world. Its mining, transport, and purification do not require hi-tech equipment, but still use extremely energy- and labor-intensive and otherwise expensive processes, since very large quantities are needed and extracted daily.

Below we take a closer look at the silicon material properties, manufacturing and issues.

Silicon Material Basics

The chemical element silicon in its purest form consists of a large number of atoms, each atom with 4 electrons orbiting its periphery.

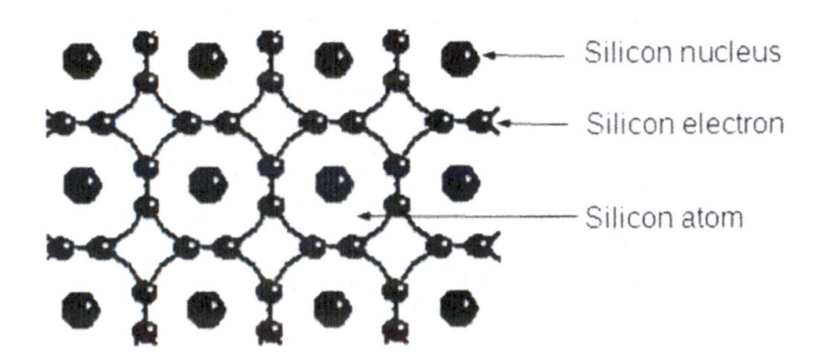

Figure 3-3. The silicon atom

Since adjacent atoms share electrons, it appears that each atom is surrounded by 8 electrons. Each individual atom has a nucleus, which consists of positively charged protons and neutral particles called neutrons, with the 8 electrons hovering around the nucleus. See Figure 3-3. There are equal numbers of electrons and protons in the atom, so that the overall charge of each atom is zero (neutral). The electrons are arranged so that they occupy different energy bands (levels) around the nucleus, which are determined by a number of forces and circumstances.

The four electrons in a silicon atom are held together by a covalent bond, which simply means that two adjacent atoms have the ability to share an electron and the force that holds them together. The electrons in these covalent bonds are held in place by the energy of the bond and are kept close to each other and the nucleus by a significant force. Since the electrons are kept close together by the covalent bond forces, they cannot move up and down the energy levels, and are therefore not free and cannot move with the electric flow (if any) around them under normal conditions. So silicon in its natural form cannot conduct electricity and acts as an insulator. If, however, the temperature is increased sufficiently, an electron can gain enough energy to break the bond and escape, at which point it could potentially conduct electricity. This ability to shed electrons under certain conditions is what makes silicon a semiconductor material and why it is perfect for both semiconductor devices and solar cells manufacturing.

The energy needed to break from the bond and keep an electron free from the covalent forces, is called *band-gap* energy, and sunlight provides this additional energy. Silicon is an excellent material for manufacturing solar cells since its indirect energy band-gap is approximately 1.1 eV at room temperature, which is close to the energy of some of the photons in the solar spectrum. This is the energy that allows the creation of free electrons and holes, thus initiating and propagating the photoelectric process in solar cells. So, the more sunlight that falls on the cell, the more free electrons and current will be created, up to the physical and electronic limits of the solar cell to free electrons from their bonds and conduct electric current through its lattice. See Figure 3-4.

When an electron leaves its place in the covalent bond, it leaves a *hole* (or a positive charge) behind, giving the appearance of positive charges moving thought the lattice.

When sunlight hits the surface of the silicon solar cell it impinges enough energy onto the bonded electrons, freeing some of them from their captivity. The higher the number of free electrons (carriers), the more electric current they conduct. The number, or concentration, of free

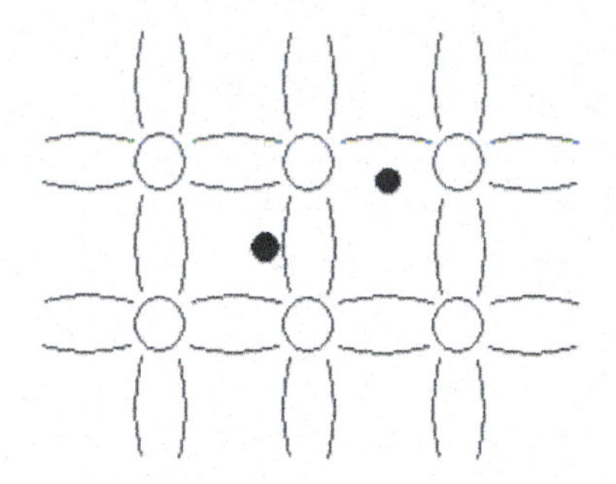

a) Free electrons in Si lattice

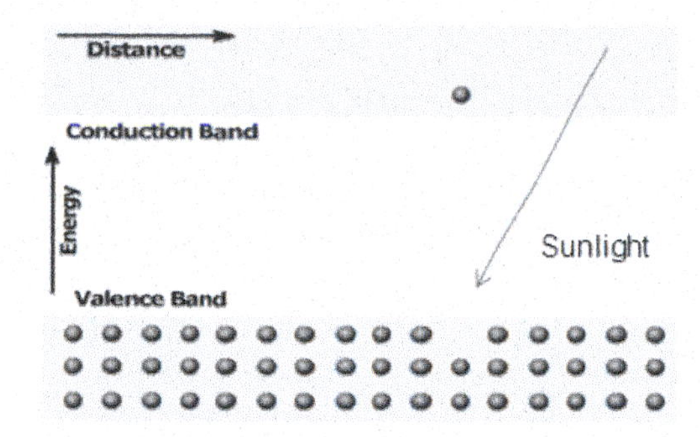

b) A photon frees an electron

Figure 3-4. The PV power generating effect

electrons is called intrinsic carrier concentration. The bonded electrons reside in what is called the valence band (or low energy band), while the free electrons jump into the conduction band (high energy band) where they move freely and could be eventually extracted as electric current. The holes remain mainly in the valence band of the cell. To optimize this process, we create the so called p-n junction in the silicon material by saturating a narrow surface area between the valence and conduction bands. With this addition, the silicon material (silicon wafers in most cases) is then on its way to becoming a solar cell as we will see below.

At the atomic level, silicon has a number of special qualities that make it a very good solar cell material. One particularly useful effect is the way it limits the energy levels which the electrons can occupy, and the way they move about the silicon crystal, all of which facilitate the initiation and propagation of the photoelectric effect, which is the foundation of the solar cell function.

Solar Silicon Material Manufacturing

Silicon is found in large quantities in nature in the form of sand, or silicon oxide (SiO_2), which actually forms 1/3 of the Earth's crust. Most sand, however, contains so much dirt (like the sand in the deserts) that it is useless for all practical purposes. Silicon is environmentally friendly material, and its waste does not pose any special problems. It can be easily melted, shaped and formed into mono-crystalline or poly-crystalline ingots and wafers. Devices made from silicon can operate at up to 125°C air temperatures (with some loss of efficiency), which allows the use of silicon based semiconductor devices and solar cells even in harsh environments, with some exceptions and restrictions.

The purity of silicon used in semiconductor devices is at min. 99.99999999%, and somewhat less when used for solar cells. 99.999% to 99.9999% pure silicon is used in today's manufacturing operations. The actual number of 9s is used to refer to silicon's quality, i.e., 4 nines, six nines etc., and determines the final quality, performance and longevity of the solar cells. Since it is impossible to change the initial silicon material quality at a later stage, it is critical that a high quality production starts with high quality (with minimum possible impurities) silicon material.

The basic process for making silicon to be used in the manufacturing of c-Si PV cells is as follows:

High purity silica sand found in large quantities in several areas around the world is dug out, sifted and loaded in huge furnaces, where it is melted for removal of the oxygen in the silica molecule through a reaction with carbon (coal or charcoal) added to the mix while heating it to 1500-2000°C in huge electrode-arc type furnaces.

$$SiO_2 + C \rightarrow Si + CO_2$$
$$SiO_2 + 2C \rightarrow Si + 2CO$$

The molten MG Si material is usually treated further for impurities removal by the addition of different additives:

$$SiO_2 + 2SiC(\text{and other additives}) \rightarrow 3Si + 2CO$$

The resulting silicon material is metallurgical grade (MG) silicon, still containing a lot of impurities—well over 2-3%—and as such it cannot be used for solar cell manufacturing. At this state it is good for use as an additive in the metallurgical industry only. Additional purification must take place if it is to be used for solar cells, and even more purification is needed if it is to be used in the production of semiconductor devices. After solidification, it is crushed and shipped to a solar grade silicon processing plant for refining as needed to bring its quality to that needed for manufacturing solar cells.

Here the MG silicon chunks are reacted with hydrochloric acid (HCl) to make silicon containing gases:

$$Si + 4HCl \rightarrow SiCl_4 + 2H_2$$

And/or

$$Si + 3HCl \rightarrow SiHCl_3 + H_2$$

The process chemicals go through the distillation and purification processes in order to obtain and isolate the main silicon containing gas $SiHCl_3$ trichlorosilane (TCS) which is then purified and sent for preparation of solar grade (SG) silicon material.

The TCS gas is processed in plasma reactors and solidified in the shape of large ingots of SG multi-silicon (poly) as follow:

$SiCl_4 + 2Zn \rightarrow Si + 2ZnCl_2$ (Du Pont process)

or

$2SiHCl3 \rightarrow Si + 2HCl + SiCl_3$ (Siemens process)

or

$4HSiCl_3 \rightarrow 3SiCl_4 + SiH_4$, where $SiH_4 \rightarrow Si + 2H_2$ (REC process)

Thus obtained ingots of SG silicon are still not good enough for making solar cells, so they are crushed, packed and shipped to be melted into high purity doped ingots from which solar wafers will be sliced and processed into solar cells.

Silicon Material Quality

The quality of the SG grade silicon material quality is extremely important, because it will determine the final quality of the solar cells made from it.

Table 3-1 offers a close look at the types and amounts of impurities contained in semiconductor grade silicon (left column), vs. those in MG silicon (right column), and reveals huge differences, which translate into drastic differences in quality as well. Semiconductor grade Si is several times purer than MG Si, which means that a lot of energy, time and effort must be spent on complex processes, in order to purify it to this degree. Solar grade (SG) c-Si could also benefit from such high purity as that of the semiconductor grade silicon, but the benefits do not justify the steep price increase. Because of that, SG Si is somewhat less pure than its semiconductor grade cousin and is somewhere between the MG and semiconductor grade silicon quality in Table 3-1.

Some of the impurities play a much larger role in the final quality of solar cells and modules, so when we say that 6 nines SG silicon is suitable for solar cell manufacturing we must also qualify the actual types and amounts of the different impurities; i.e., if the amount of some harmful metal impurities is too high, the final product might be of questionable quality as well.

Table 3-1. Different silicon qualities

Chemical Impurity	Semiconductor Grade Si (ppm)	MG-Si (ppm)
B	4.157	14.548
C	14.264	107.565
O	17.554	66.706
Mg	<0.001	8.204
Al	<0.005	520.458
Si	Matrix	Matrix
P	6.801	21.762
S	<0.044	0.096
K	<0.007	<0.036
Ca	<0.007	44.849
Ti	<0.001	47.526
V	<0.001	143.345
Cr	<0.001	19.985
Mn	<0.001	19.938
Fe	<0.005	553.211
Co	<0.002	0.763
Ni	<0.002	22.012
Cu	<0.001	1.724
Zn	<0.002	0.077
As	<0.002	0.007
Sr	<0.0003	0.353
Zr	<0.0003	2.063
Mo	<0.001	0.790
I	<0.0002	<0.001
Ba	<0.0002	0.266
W	<0.0003	0.024

So basically, SG silicon has to be pure enough, and the type and quantity of each impurity must be known, because excess impurities of some types will invariably result in lower efficiency and potentially shorter life span of the cells made from it. This in turn will shorten the life of the PV modules in the field, so a deep understanding of the differences and strict adhesion to quality control procedures by manufacturers is paramount for ensuring the quality of the final product.

SILICON WAFERS

Here we will be looking at the properties and manufacturing process of solar grade wafers, which are used for making solar cells.

Mono-Crystalline Vs. Multi-Crystalline Silicon Solar Cells

The major types of silicon solar cells are mono-crystalline and multi-crystalline (poly).

Figure 3-5 is a graphic representation of silicon material processed by "pulling" as a single crystal (mono-crystalline) material via the so-called Czochralski process. On the right is silicon material produced by melting silicon chunks in a crucible.

Single-crystal Silicon (sc-Si)

Single-crystal silicon (sc-Si) wafers are produced by melting purified SG silicon in special furnaces, called *ingot pullers*, where a long, cylindrical silicon ingot is "pulled"

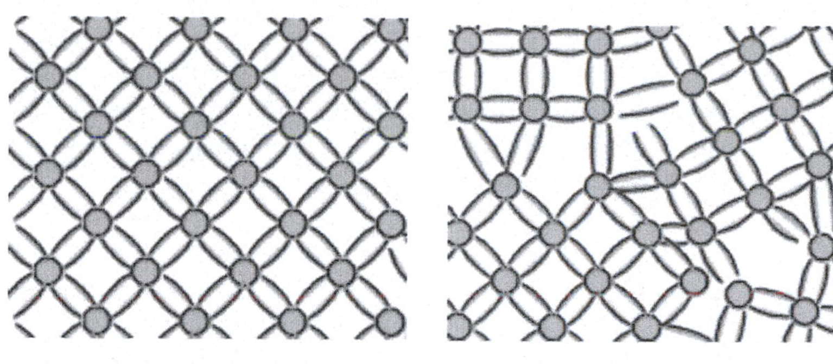

a) Mono-crystalline silicon wafer b) Multi-crystalline silicon wafer

Figure 3-5. Silicon wafer materials

up from the melt as a single crystal silicon. The ingot, 4, 5 or 6" in diameter, is sliced into thin wafers which are then processed into solar cells. Due to the more sophisticated production process, these wafers and the cells made from them tend to be more expensive and more efficient than poly cells. The efficiency increase is due to the uniform structure of the single crystal materials which facilitates the photoelectric effect and provides a good level of reliability and longevity.

sc-Si wafers are cut from cylindrical ingots and then squared, a process which creates substantial waste of silicon material. Their uneven shape also leaves unused gaps in the PV modules' surface which lowers the efficiency. Like most other types of solar cells these suffer a reduction in output at temperatures over 110°F, where a 10-15% drop of output can be expected, but which is actually lower than that of polysilicon cells and some other types of solar cells. Monocrystalline solar panels are first-generation solar technology and have been around a long time, providing evidence of their durability and longevity. The technology, installation, performance and all other issues are well understood. Several of the early sc-Si PV modules installed in the 1970s are still producing electricity today, albeit at reduced output levels, and some have even withstood the rigors of space travel.

Multi-crystalline silicon (poly)

Multi-crystalline silicon (poly) material is produced by melting SG silicon in a large cube-like crucible. The solidified cube of poly material is split into 4-, 5-, or 6-inch-wide and 18- to 36-inch-long bars which are then sliced into thin wafers. Since it is not "pulled" as a single-crystal but rather as randomly mixed columns, the resulting material and the wafers sliced from it, look non-uniform with visible boundaries in the bulk which play a significant role in the solar cells' performance. Large randomly oriented grains (columns) crisscross the bulk giving the poly wafers their distinguished and esthetically pleasing look, but significantly reducing the efficiency of the solar cells made from it.

These grains can be thought of as separate pieces of single crystal silicon, scattered in the bulk and separated by boundaries. They are actual physical and electrical barriers which- create obstacles in the operation of these types of solar cells. The boundaries prevent electrons and holes from freely moving across, and contribute to the recombination process, which minimizes the photoelectric effect, thus reducing the efficiency of the solar cells.

The grains are usually so large that they extend all through the wafers when cut from the solidified silicon block. The incorporation of hydrogen during device processing plays an important role in passivating the inter-grain boundaries which contributes to improving the efficiency and operating stability of the solar cells.

The grain boundaries create a number of effects manifested as sub-grain boundaries, slip deformations, and twinning. These are basically additional lower level boundaries in the form of single dislocations or web of dislocations with different orientation, Burger's vectors and similar abnormalities, all of which complicate the photoelectric process, reduce the efficiency of the solar cells, and contribute to latent (delayed) failure modes.

Advantages of using multi-crystalline growth over the Czochralski method include:

a. Lower capital costs,
b. Higher throughput,
c. Less sensitivity to the quality of the silicon feedstock used, and
d. Higher packing density of cells to make a module because of the initial square or rectangular shape of the material prior to final wafering and processing.

The best modules made from multi-crystalline silicon generally have efficiencies 2-5% less than those made of mono-crystalline silicon and cost approximately 15-25% less to produce.

Crystalline silicon devices and modules dominate the markets lately, but sc-Si cells and modules sales have been significantly reduced due to increased competition from poly cells and modules. The market share of the poly Si products continues to increase as we speak. The aesthetic appearance of the multi-crystalline silicon modules with their dark-blue to black brilliance is considered to be more appealing than their counterparts which is a major consideration in some applications.

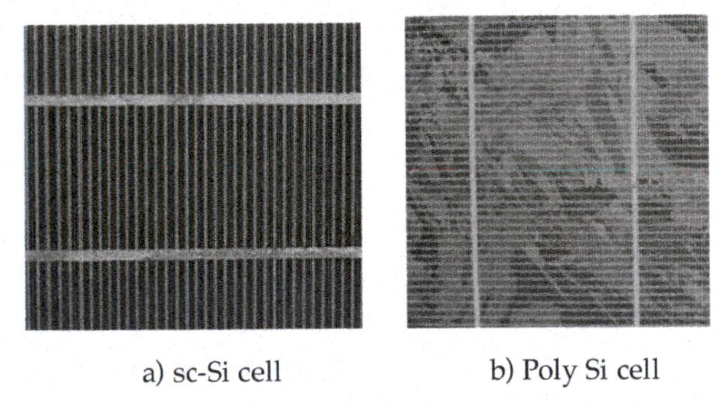

a) sc-Si cell b) Poly Si cell

Figure 3-6. Mono and poly crystalline silicon solar cells

Silicon Wafers Manufacturing Process

Silicon solar wafers are made out of solar grade (SG) silicon, via melting the chunks and "pulling" or "casting" the molten material, thus creating mono, or poly Si materials respectively. The resulting material is cooled down and sliced into very thin, 200-300 microns (0.007"-.0.012") slices (wafers). These wafers are then processed into solar cells.

The solar wafers manufacturing process sequence below was used by Alpha Solarco, Inc., during the late 1990s at their solar cells manufacturing facility in Phoenix, Arizona. Variations of this process are used presently.

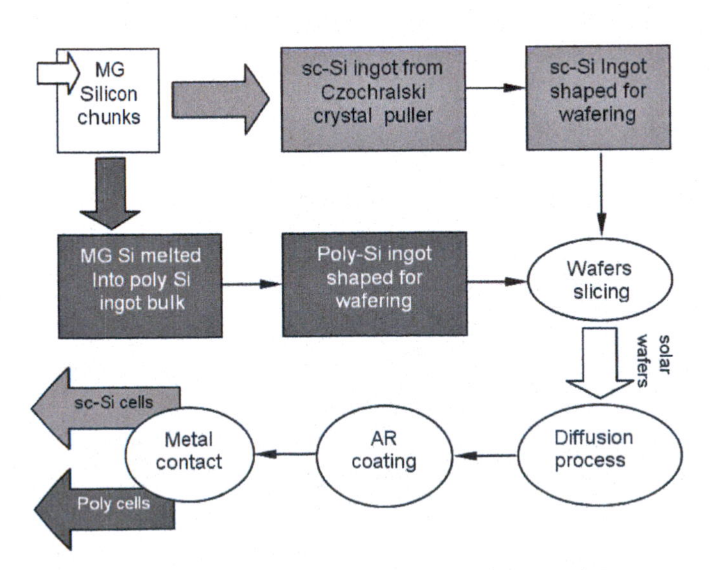

Figure 3-7. sc-Si and poly Si solar wafers manufacturing process

The polysilicon wafers manufacturing sequence is as follows:

1. Sorting, inspection and cleaning the SG silicon

   Silicon chunks, purchased directly from the manufacturers or indirectly from dealers, are received, documented, inspected, sorted by size and tested for resistivity, metal and other contaminants.

2. Bulk silicon cleaning and surface prep

   The sorted Si chunks which pass the initial QC procedures are sent into the wet chemistry room for cleaning with detergents and etching with acids. This is needed in order to remove all dirt, oils and impurities from the material's surface. The chunks are then rinsed, dried and taken to the wafer process room for melting in the poly Si ingot HEM (heat exchange method) furnace.

3. Prepare Ingot Crucible

   The HEM furnace uses a special crucible to melt the silicon in. Before loading the crucible, its inner surface is spray coated with a proprietary solution. This eliminates possible chemical reactions between the silicon material and the silica crucible in the HEM furnace at high temperatures. After application of the coating, the crucible is baked in a kiln for 24 hours to dry the sprayed solution. It is then ready for loading in the HEM furnace.

4. Load HEM Furnace or Puller

   Assemble graphite support plates around the sides of the crucible before loading it into the furnace. Load crucible by adding 80 KG of silicon meltstock and calculated amount of dopant (Boron salt). After loading, clean exterior of furnace, control module, and area adjacent to furnace from any residue or overspray.

5. Grow Ingot

   Apply vacuum to furnace and backfill with Argon gas. Re-apply vacuum to ensure that all moisture has been eliminated inside the furnace chamber. The control instrumentation on the HEM furnace is designed to automatically control the ingot growth process. No manual intervention is allowed during the entire process, except in emergency.

6. Cool and Remove Ingot

   Ingot is annealed prior to cool down. Annealing is achieved by maintaining a furnace temperature slightly below the melting temperature of silicon. Stress relief of the ingot is achieved over a 12-hour period. Ingot is then removed from the crucible and ready for cutting. Cutting is done in two steps, block sawing and wafer slicing.

7. Block Sawing

After growing an ingot, its surfaces must be shaved square using a special ingot saw. The smooth squared block of silicon is then sectioned into bars that measure 10 cm x 10 cm or as desired and variable length. Some additional proprietary processing of the ingot sides follow. It is intended to reduce some of the surface damage of the ingot and the resulting wafers.

8. Wafer Slicing

Wire saws are used to slice the wafers from the bars. Each bar is attached to a glass sheet with a proprietary epoxy adhesive which is dried a minimum of 24 hours. This glass/silicon bar is then attached to a metal work piece (holder) and placed under the cutting wires of the wire saw. The cutting oil and abrasive mix must be prepared and properly mixed in the mixer of the slurry tank at the bottom of the wire saw. Wafers are sliced to the desired thickness and removed from the wire saw with the holder.

9. Pre-cleaning

After slicing, the wafers are separated from the holder and loaded into cassettes. They are then taken to a special degreasing area for gross removal of surface contamination, prior to cleaning. Cutting oils and abrasive residues left on the wafers' surfaces from the slicing process are properly and thoroughly removed via proprietary cleaning formulations. The wafers are stored in a water tank in order to keep them wet.

10. Wafers Uniformity Test

Some wafers are taken out of the batch, cleaned thoroughly and tested at five different locations on the surface of each wafer to ensure that the desired degree of thickness uniformity and resistivity have been achieved by the slicing process. Additional inspection and testing procedures are carried out as needed to ensure the quality of the wafers prior to processing into solar cells.

11. Wafer Storage

The wafers are loaded in cassettes for ease of handling during the rest of the processing sequence. The cassettes loaded with wafers are stored in water tanks where they are kept until taken for final cleaning.

12. Process Controls and Quality Control

Special instrumentation and procedures are used at each process step to ensure proper processing and to check tight adherence to the specs.

13. DI Water System

A special deionized (DI) water generating station is used to make DI water for rinsing the wafers. Proper rinsing of the silicon material (chunks) and the resulting wafers is accomplished, leaving the wafers with clean surfaces.

14. Clean Room

The critical steps of the solar cells wafer and cells manufacturing process are executed in a specially designed area, called a "clean room." It is built and operated to keep any environmental disturbances such as wind, humidity, heat and cold, particles, etc. isolated from the work environment. Special HEPA filters deliver purified and deionized air, and this way the solar cells are processed under near-ideal conditions and without any outside contamination.

The monocrystalline silicon manufacturing sequence is similar to the above, except that the ingot furnace is quite different, as follows:

1. Clean the ingot puller furnace and process crucible.

2. Fill crucible with silicon material and add exact amount of doping chemical.
   Note: Crucible surface tends to dissolve so it has to be made of pure silica and kept at maximum cleanliness possible.

3. Melt the silicon chunks in the crucible and keep the temperature close to the melting point.
   Note: Convection in the silicon melt is suppressed by correct application of magnetic fields around the crucible. The strength and duration of these fields is a proprietary know-how.

4. Insert the silicon seed crystal into the melt and stabilize the temperature

5. Start withdrawing the seed crystal slowly by rotating the seed crystal and the crucible.
   Note: This process step is proprietary for each operation, but basically, fast withdrawal during the initial stages is needed to reduce the diameter of the growing crystal a few mm. This is needed to ensure that most seed-crystal induced dislocations will be removed.

6. The withdrawal rate must be decreased now in order to increase the ingot diameter slightly above

the desired size.

Note: The impurity concentration (including dopants and oxygen) in the melt will change due to segregation, and the resulting crystal properties will change too, usually by increasing impurities concentration from top to bottom of the ingot. The temperature profile will vary at this stage and has to be adjusted. The homogeneity of the crystal depends on the correct temperature and speed of rotation and withdrawal regimes. These parameters are also proprietary and not much actual information can be found for free.

7. When almost all silicon has left the crucible the ingot is complete.

   Note: Some molten silicon must be left in the crucible, because it is where the impurities are concentrated due to their low segregation coefficients.

8. Pull the ingot out, but beware because the thermal shock introduces temperature gradients which cause stress gradient, and dislocations are nucleated into the crystal.

   Note: The new dislocations will interact with previous dislocations, causing serious damage. The withdrawal rate at this point is critical and is usually gradually increased. This leads to a reduced diameter in cone-like shape of the tail end of the ingot.

9. Remove the ingot slowly and place it on a suitable clean surface for cooling.

10. Shape the ingot on a special lathe to give it more uniform surface and bring it to the exact outside diameter—4, 5 or 6" diameter.

11. Inspect and test the ingot for mechanical defects and chemical impurities.

12. Clean and slice it into wafers of the desired thickness.

13. Inspect, test, sort and pack the wafers.

The resulting monosilicon wafers are used for in-house solar cells manufacturing, or are shipped to solar cells and modules manufacturing facilities around the world.

Silicon Material Types and Processing

Despite never ending attempts to make better solar cells by using new and exotic materials, the reality is that the photovoltaics market is still dominated by silicon-based solar cells. This is the oldest (over 50 years in the making) PV technology which is well understood and no surprises are expected under normal manufacturing or operating conditions in most cases. Nevertheless, a large body of research is underway all over the world to improve the quality, lower the cost, and increase the conversion efficie-ncies and longevity of c-Si products without an exorbitant increase in production cost.

Some of the objects of the research effort on silicon wafer-based processes are discussed below.

Silicon (SG) Materials

One way of reducing the cost of PV cells and modules is to develop cheaper methods of producing cheaper silicon material that is sufficiently pure. Silicon is a very common element, but the best quality is normally found in silica, or silica sand. Processing silica (SiO_2) to produce silicon is an energy-consuming and expensive process. It takes up to two years for a conventional solar cell under full power to generate as much energy as was used to produce the silicon it is made of.

More energy efficient methods of synthesis are not only beneficial to the solar industry, but also to industries surrounding the silicon technology, and the world, as a whole, so efforts to achieve this balance continue. The current industrial production of silicon is via the reaction between carbon (charcoal) and silica at a temperature around 1700°C. In this process, known as carbo-thermic reduction, each ton of silicon (metallurgical grade, about 98% pure) is produced with the emission of about 1.5 tons of CO_2. Solid silica can be directly converted (reduced) to pure silicon by electrolysis in a molten salt bath at a fairly mild temperature (800 to 900°C). While this new process is in principle the same as the FFC Cambridge Process which was first discovered in late 1990s, the interesting laboratory finding is that such electrolytic silicon is in the form of porous silicon which turns readily into a fine powder, with a particle size of a few micrometers, and may therefore offer new opportunities (raw materials) for development of silicon based solar cell technologies.

Upgraded Metallurgical Silicon

Recently, several R&D groups have been testing a new method called upgraded metallurgical grade (UMG) silicon. This is a somewhat lower quality but much lower cost material than the usual solar grade (SG) silicon. UMG is nevertheless able to produce solar cells of 16% efficiency. The method consists of simply mixing some percent of UMG with higher quality SG silicon as needed to balance the efficiency and price ratios. Since almost half of the cost of the finished solar cell is in the cost of the raw silicon, this method will become even more important in the future as the price per Watt wars continue.

Thinner Silicon Wafers

A major part of the final cost of a traditional bulk c-Si PV module is related to the high cost of the silicon wafers (about $ 0.4/Wp) so there exists substantial drive to make Si wafers and solar cells thinner for material savings' sake. This can be done by slicing the silicon ingots in a more efficient way, in order to produce thinner wafers, which pushes the limits of the technology. There are presently techniques that can produce much thinner wafers but these thinner wafers are much more fragile than standard wafers, therefore they are much more susceptible to breakage during handling, transport and processing. Improvements in fully automated wafers manufacturing and solar cells processing lines might help to optimize these processes.

Wafer Slivers

Another approach is to reduce the amount of silicon used by micromachining wafers into very thin, virtually transparent layers that could be used, among other things, as transparent architectural coverings. The technique involves taking a silicon wafer, typically 1 to 2 mm thick and making a multitude of parallel transverse slices across the wafer, creating a large number of slivers that have a thickness of 50 microns and a width equal to the thickness of the original wafer. These slices are rotated 90 degrees, so that the surfaces corresponding to the faces of the original wafer become the edges of the slivers.

As a result, the electrical doping and contacts that were on the face of the wafer are located at the edges of the sliver, rather than at the front and rear as in the case of conventional wafer cells. This has the interesting effect of making the cell sensitive from both the front and rear of the cell (a property known as bi-faciality). Using this technique, one silicon wafer is theoretically enough to build a 140 Watt panel compared to about 60 wafers needed for conventional modules of the same power output. Micromachining, however, will increase the price of the device, and will reduce the production yield significantly as well.

Silicon Ribbon

Several different techniques are used to produce a thinner silicon substrate, some of which are:

a. EFG or "edge defined film fed growth" method—a graphite dye is immersed into molten silicon, making it rise into the dye by capillary action. It is then pulled as a self-supporting sheet of silicon which hardens in the air above the dye.

b. The "dendritic web growth process" consists of two dendrites, which are placed into molten silicon and then withdrawn quickly between them, causing the silicon to exit and solidify as a thin sheet. A modification of this method is used today, called "string ribbon method," where two graphite strings are used (instead of the dendrites). This makes process control much easier.

In all cases the resulting silicon is multi-crystalline with a quality approaching that of the directionally solidified material. Solar modules made using silicon sheets produced with these methods generally have efficiencies in the range 10-12%. The major problem here is the difficulties in handling and processing the silicon sheets and wafers, due to their non-uniformity and fragility.

Super Monocrystalline Silicon

This is a new, purer type of silicon with more perfect crystalline structure, which exhibits reduced phonon-phonon and phonon-electron interactions. These phenomena increase certain transport properties, resulting in 40-60% better room temperature thermal conductivity than natural SG silicon. Super silicon is finding increased use in the semiconductor industry presently, but is very seldom used for solar cells because of its complex processing and higher cost. Future equipment and process developments might allow its use in the solar industry, which will lead to solar cells and modules of superior qualities.

Wafers Reclamation Process

A number of wafer recycling facilities in the US and abroad buy rejected wafers from semiconductor and solar devices operations in order to recycle the wafers and sell them for reprocessing into solar cells. The front and back surfaces (front and back metal contacts, diffusion layer, BSF etc. foreign layers on the wafers) are mechanically polished or chemically etched, and this recycled wafer batch is tested for doping and metal levels, cleaned, packed and shipped to a solar manufacturer.

The problem with recycled wafers is that it is hard to know exactly what is on and in the wafers. Residual doping or metallization (which is dangerous) varies from batch to batch and even from wafer to wafer in the same batch. These variations might result in unpredictable quality of the finished product, which will influence its performance and longevity in a similarly unpredictable way.

c-Si Solar Cells Structure and Function

Crystalline silicon (c-Si) solar cells (Figure 3-8) are relatively simple at first glance, but their physical and chemical composition, structure, and electrical properties are quite complex.

We will take a closer look at the c-Si solar cell's structure in order to get a good understanding of the different aspects of its function. We will also review the major issues and failure mechanisms related to its manufacturing, installation and operation.

The p-n Junction

The p-n junction is an integral part of the solar cell operation, since it is where the photoelectric process starts and where the electric current is generated, so let's take a close look at it:

Figure 3-9 shows a cross section of a solar cell with its top surface (N-type region) on the right-hand side and the bottom (P-type region) on the left. The p-n junction is in the middle and represents an electro-chemical boundary between the P and N regions. The N region and the resulting p-n junction are created during the diffusion process by doping the N side (usually the top of the wafer) with phosphorous or other such element, thus saturating a very shallow area (less than a micron) below the top surface of the cell with phosphorous atoms. These atoms have excess electrons, which are loosely attached and are readily knocked out to facilitate the electric power generation, if and when energized by sunlight. The p-n junction is located at the border line between the phosphorus saturated area and the original silicon bulk (P type region) which was very lightly doped with boron (or other similar chemical atoms) during the final silicon ingot melting process. The P region will provide the holes (+ charge) when the electron-hole pairs are broken by the incoming sunlight and the electrons extracted during the photoelectric effect.

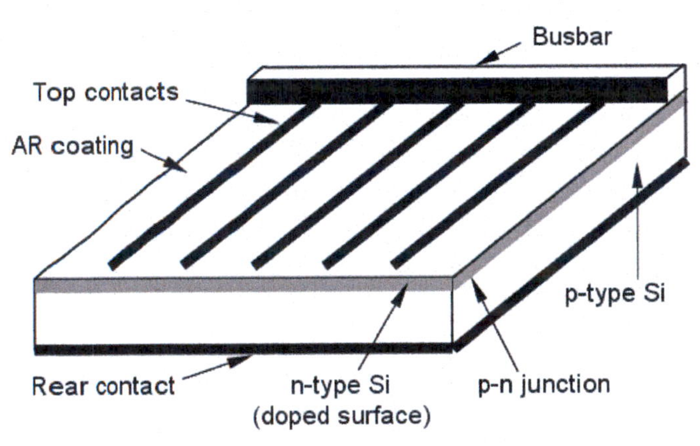

Figure 3-8. Silicon solar cell

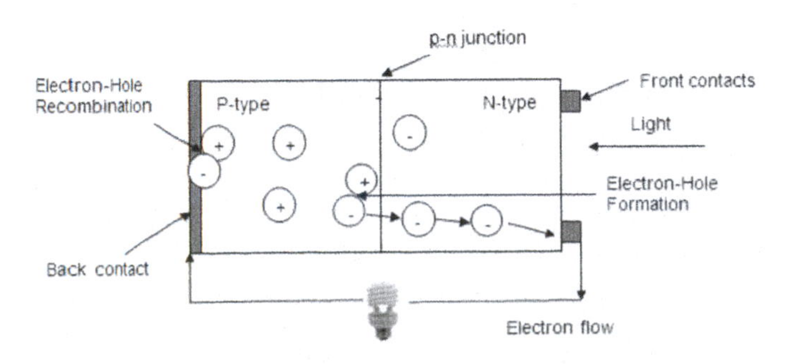

Figure 3-9. The p-n junction function

When at rest, the p-n junction and the areas around it are static with no meaningful activities in or across it. When photons from sunlight with proper energy levels impinge onto the solar cell surface they penetrate into the cell material and impact onto the electrons (in the electron-hole pairs) thus transferring energy to them and breaking the electron-hole pairs. This creates free electrons (-) which start moving around, while the holes (+) from the corresponding electron-hole pairs are mostly left in their original place. Thin areas on both sides of the p-n junction, called the "depletion" region effectively separate the holes from the electrons which are forced towards the N-type region and are finally extracted from the cell as electric current.

The constant movement of electrons back and forth across the p-n junction and through the layers is quite complex, but the final and practical result of it is the creation of DC electric current which flows in an outside circuit attached to the cell. A large part of the electrons do leave the cell through the metal fingers and bars on the N-type side (usually on top of the solar cell), and if we close the circuit, electric current will flow through it. The electrons in the outside circuit will re-enter the cell via the back side metallization on the P-type side of the cell. Here they will recombine with excess holes in the region and the process repeats indefinitely, or at least while there is enough sunlight to keep it going.

The electron flow provides the current (I) in Amperes and the cell's electric field creates a potential, or voltage (V) in Volts. With both current and voltage above zero we have power (P) in Watts, which is a product of the two and which is what we use to define the PV cells and modules power output.

$$P_{Watts} = I_{Amperes} * V_{Volts}$$

When an external load (such as an electric bulb or a battery) is connected between the front and back contacts of the cell, DC electricity flows through the cell and the external circuit, and powers the load connected to the closed external circuit.

NOTE: Remember that the actual electric current of the external circuit flows in the opposite direction of the electron flow.

Band Gap Energy

The key elements in the conversion of sunlight into electric energy are, a) the availability and intensity of incoming sunlight, and b) the ability and efficiency of the solar cell to capture and process the available sunlight into electric energy via the "photovoltaic" effect which takes place close to the top surface of the solar cell in and around the p-n junction as discussed above.

Basically, when sunlight hits the solar cell its photons with energies greater than the energy band-gap of the semiconductor material (1.1eV for Si) are absorbed, thus promoting electrons from the valence band to the conduction band (P-type to N-type respectively). This action leaves a corresponding number of holes in the valence band which then recombine with the electrons returning at the back of the cell.

The energy "band gap" is the energy difference between the top of the *valence band* (electrons at rest) and the bottom of the *conduction band* (electrons with extra energy). When sunlight hits the solar cell, the energy of the electrons in the valence band (where they reside when not activated) is increased, due to internal (parasitic heat) or external (sunlight) energy transfer due to bombardment from photons (light) and phonons (heat) contained in the sunlight. When the energy of the incoming photons in the sunlight reaches the bang gap energy of the material, the electrons in it are activated and are able to jump from the valence (rest) to conduction (energized) energy band. This is only IF they get enough energy for the transition effort. Remember that the band gap energy has different values for different materials and that for silicon it is 1.1eV. This means that the incoming photons must have at least as much energy as the band gap energy of the material (preferably a bit more) to activate the electrons in it. This energy and the resulting interactions are very important since they are the engine of the photoelectric effect.

The silicon energy band gap at 300°K (80.33°F) is calculated as:

$$E_g(300\ K) = E_g(0\ K) - \frac{\alpha T^2}{T+\beta} = 1.166 - \frac{0.473 \times 10^{-3} \times (300)^2}{300 + 636} = 1.12\ \text{eV}$$

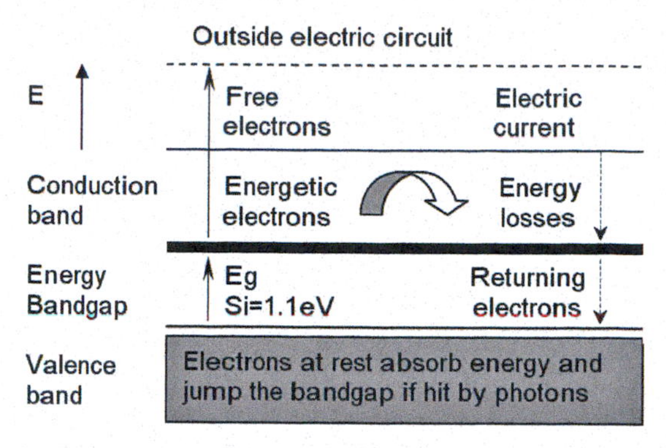

Figure 3-10. Energy band gap

where:

T is the temperature, and

$E_g(0)$, α and β are fitting parameters for different materials I.e., $E_g(0)$ for Si=1.116eV, Ge=0.744e; α for Si=4.73×10^{-4} eV/K, Ge=4.77×10^{-4} eV/K and β for Si=636K, Ge=235.

With the energy exchange initiated by the sunlight falling on the solar cell, its temperature increases and the amplitude of atomic vibrations increases which in turn leads to larger spacing between the atoms in the silicon lattice. The interaction between the lattice phonons and the free electrons and holes affects the band gap too, and its magnitude changes proportionally with temperature increase, so this relationship is reflected by Varshni's empirical expression:

$$E_g(T) = E_g(0) - \frac{\alpha T^2}{T+\beta}$$

where:

T is the temperature

$E_g(0)$, α and β are fitting parameters for different materials.

If the electron-hole pairs are generated within the depletion region of the p-n junction, the electric field in the depletion region separates the pairs and drives the electrons through an external load as DC electricity. Maximum power is generated when the solar cell receives the maximum amount of sunlight, and thus generated power is delivered to the load when its impedance matches that of the illuminated device. The total output, however, will start decreasing as the temperature of the cell increases. We'll take a closer look at the temperature phenomenon below.

Power Generation

Solar cells are characterized by the power they generate, which is at maximum when the current and voltage are at their maximum levels. This can happen only when the cell is fully operational (no defects), and receives maximum solar insolation.

$$P_{max} = I_{max} * V_{max}$$

With these terms at zero, the conditions

$$V = V_{oc}/I = 0 \text{ and } V = 0/I = I_{sc}$$

also represent zero power.

A combination of maximum current and maximum voltage maximizes the generated power and is called the "maximum power point" (MPP).

$$MPP = I_{max} * V_{max}$$

So the MPP of a solar cell with 3.0 A current and 0.5 V voltage would be 1.5 W. A 100 pc. PV module made out of these solar cells connected in series would generate 150 Wp under full solar insolation of 1000 W/m^2

The solar conversion efficiency η of a PV cell or module is used most commonly to express and compare performance. The efficiency is given by:

$$\eta = V_{oc} I_{sc} FF / P$$

Where:

Voc is the open circuit voltage (the voltage generated when the load resistance is infinite, or there is no resistance),

Isc is the short circuit current (the current generated when the load resistance is zero,

FF is the fill-factor, calculated as follows:

$$FF\% = \frac{I_{max} * V_{max} \quad \text{(actual measurements)}}{MPP \quad \text{(maximum obtainable power)}}$$

This is simply the ratio of the actual measurements (Voc and Isc generated by the cell under the specific testing conditions) divided by the maximum power the cell can generate. In other words, the efficiency is basically the ratio of the amount of power a solar cell or module could produce vs. the total amount of power contained in the incoming sunlight and how efficiently the cell converts it into electric power.

So, if a PV module with 1.0 m^2 active surface area is rated at 15% efficiency which is average for c-Si PV modules, we can quickly deduct that it theoretically could produce 150 Watts DC power under 1000 W/m^2 solar insolation (its maximum power). So if it generates 50 V and 3 A then its FF will be 1. If it produces 50 V and only 2 A (or 100 W), then its FF will be (50 * 2)/150 W, or FF = 0.67.

It's a mouthful of concepts, but a good simplified description of the major practical effects of solar power generation which are the foundation of solar cells' and modules' ability to convert sunlight into useful DC power.

SILICON PV CELLS DESIGN CONSIDERATIONS

Even the best quality solar cells operating under the best conditions experience anomalies and go through degradation and deterioration processes which must be well understood and controlled. Usually, field operating conditions are far from best which accelerates the negative processes in any type of PV cell and module. On top of that, and even more importantly, not all PV cells modules are of best quality—people make mistakes, material quality and processes get out of spec and quality becomes iffy. These anomalies will result in poor quality of the final product which aggravates further the natural degradation and other destructive processes during long-term field operation, and must be taken into consideration when designing or evaluating solar cells and modules.

Having efficiently and reliably operating solar cells is a must for the quality and profitable operation of any PV power plant and we cannot emphasize enough the importance of thorough understanding of their structure, characteristics and related issues. There are a number of issues which we intend to bring out. We don't have an answer for all of them, but we'll try to provide educated guesses to at least start the discussion.

Solar Cell Types

We looked at silicon material and wafer types above, so we just have to remind the reader that the category "crystalline silicon," c-silicon, or c-Si is the general designation of all types of crystalline silicon-based products, including c-Si wafers, solar cells and PV modules.

The major types of silicon solar cells therefore are as follow:

Single Crystal Solar Cells

Single crystal silicon, also called mono-crystalline, mono-silicon, or mono-Si, or sc-Si, is a type of silicon that was grown by the very special and expensive Czochralski (or CZ) method, or via the float zone (FZ) method. Both methods use similar equipment and production methods and end up with the best, most efficient, and much su-

CRYSTALLINE Si	THIN FILMS	OTHER
Mono-Crystalline	Amorphous Si	GaAs
Multi-Crystalline	Epitaxial Si	InP
Micro-Crystalline	CdTe	Other III-V
Super c-Si	CIGS	Germanium
Si Ribbon	Organic/Polymer	CPV Cells

perior silicon material for semiconductor and solar cells device manufacturing. Solar cells and panels made out of CZ or FZ silicon material have the highest efficiency and longevity of all silicon based PV devices, primarily due to the uniform, stable and predictable nature of the bulk material.

Multi-Crystalline Silicon

Multicrystalline, mc-silicon, or mc-Si is the most widely used silicon material. It is most often called "poly," "polysilicon," or "polycrystalline" silicon (which is what we will call it in this text too) because it consists of many (poly) strings instead of one single crystal. Poly is made by melting and casting silicon chunks into large blocks, splitting the blocks into smaller rectangular blocks and slicing these into thin, square-shaped wafers.

Poly-Crystalline Silicon Solar Cells

Poly crystalline silicon, also called poly silicon, or poly, or pc-Si is a thin film of silicon, deposited via CVD, or LPCVD processes on semiconductor type wafers, to be used as a gate material in MOSFET transistors and CMOS microchips. The solar industry uses similar equipment and processes to deposit very thin layers of silicon (pc-Si and a-Si) onto polysilicon or other substrates. The resulting devices are of lower efficiency, as compared to sc-Si or mc-Si.

NOTE: There is a confusion created by the term "poly" as it is used widely to identify PV cells modules made out of multi-crystalline silicon, instead of its actual use in the semiconductor thin film. Since we cannot change the decades-long use of the term "poly" in the solar industry to identify multi-crystalline silicon products, we will continue using it too with a certain degree of caution and with due clarification when needed.

Amorphous Silicon Solar Cells

Amorphous silicon is also thin film silicon, alpha silicon, or a-Si is used in p-i-n type solar cells. Typical a-Si modules include front side glass, TCO film, thin film silicon, back contact, polyvinyl butyral (PVB) encapsulant and back side glass. a-Si has been used to power calculators for some time now, mostly because it is easily and cheaply deposited on any substrate.

Micro-Crystalline Silicon Solar Cells

Micro crystalline silicon, also called nano-crystalline silicon, uc-Si, or nc-Si, is a form of silicon in its allotropic form, very similar to a-Si. nc-Si has small grains of crystalline silicon within the amorphous phase, and if grown properly can have higher electron mobility due to the presence of the silicon crystallites. It also shows increased absorption in the red and infrared wavelengths.

Super Monocrystalline Silicon

This is a new purer type of silicon with more perfect crystalline structure, which exhibits reduced phonon-phonon and phonon-electron interactions. This phenomenon increases certain transport properties, resulting in 60% better room temperature thermal conductivity than natural silicon.

In this chapter we will focus on mono and polycrystalline silicon solar cells and modules, their structure, function and the related properties and issues.

Theoretical Design Effects

Design and manufacturing engineers, installers, investors and owners alike must be aware of the important parameters c-Si solar cells and modules. Due to the complexity of these, and the intricacy of their operation when exposed to the elements, a complete understanding of their structure and function is a must for any PV project team.

Below is a list of some of the key effects encountered in solar cells and modules, as referenced in PVCDROM (2). These effects and their influence on the operation of solar cells and modules must be thoroughly understood and applied in the successful design, installation and operation of PV components and power generating facilities.

Light Absorption

The intensity of sunlight incident on a solar cell varies with time, which has profound effect on its performance. The light intensity changes the short-circuit current, the open-circuit voltage, the FF, and the efficiency and impact of series and shunt resistances. The level of light intensity on a solar cell is called "number of suns," where 1 sun corresponds to standard illumination of 1.0 kW/m2 at AM1.5, while 10.0 kW/m2 incident on the solar cell means that it is operating at 10 suns, or at 10X. Concentrating PV systems operate at up to 1000 suns, which require special materials (non-silicon) and more sophisticated equipment and processes.

Solar cells experience daily variations in light intensity, with the incident power from the sun varying between 0 in the early morning and 1.0 kW/m2 and higher at noon in some cases. At low light levels, the effect of the shunt resistance becomes increasingly important. As the light intensity decreases, the bias point and current through the solar cell also decrease and the equivalent resistance of the solar cell may begin to approach the shunt resistance.

When these two resistances are similar, the fraction of the total current flowing through the shunt resistance increases, thereby increasing the fractional power loss due to shunt resistance. Consequently, under cloudy conditions, a solar cell with a high shunt resistance retains a greater fraction of its original power than a solar cell with a low shunt resistance. In either case, the efficiency and final output of the solar cell are directly proportional to the amount and quality of sunlight incident on its surface, minus the above mentioned parasitic effects. Thin film cells and modules seem to perform better under cloudy conditions for these and other reasons.

One characteristic of PV cells and modules is that under any sunlight level some photons falling on the solar cell surface will be reflected, some will be absorbed in the material and some will go right through it. But only photons which are absorbed in the substrate will generate power. If the photon is absorbed it will give its energy to the electron from the valence band to transfer it to the conduction band and generate power into the external circuit. More photons are absorbed, more electrons will be transferred and more power will be generated by the cell.

A key factor in determining if a photon is absorbed or transmitted is the energy of the photon. Photons falling onto a semiconductor material can be divided into three groups based on their energy compared to that of the semiconductor band gap:

a. Photons with energy less than the band gap energy, will interact only weakly with the semiconductor material, passing through it as if it were transparent.
b. Photons with just enough energy to create an electron hole pair and which are efficiently absorbed to generate free electrons and produce power as a result.
c. Photons with energy much greater than the band gap that are strongly absorbed will not contribute to freeing electrons and might even be harmful

The absorbed photons create both majority and minority carriers. In many photovoltaic applications the number of light-generated carriers is on an order of magnitude lower than the number of majority carriers already present in the solar cell due to doping. Consequently, the number of *majority* carriers in an illuminated semiconductor does not alter significantly. However, the opposite is true for the number of minority carriers. The number of photo-generated *minority* carriers outweighs the number of minority carriers existing in the solar cell in the dark, and therefore the number of minority carriers in an illuminated solar cell can be approximated by the number of light generated carriers.

Understanding these effects is critical for the proper design and use of PV cells and modules, for they determine the efficiency and overall behavior of the devices.

Absorption Coefficient

The absorption coefficient is the variable that determines how far into the silicon material the sunlight of a particular wavelength can penetrate before it is absorbed. In a material with a low absorption coefficient, light is only poorly absorbed, and if the material is thin enough, it will appear transparent to that wavelength. The absorption coefficient depends on the material and also on the wavelength of light which is being absorbed. Semiconductor materials have a sharp edge in their absorption coefficient, since light which has energy below the band gap does not have sufficient energy to raise an electron across the band gap. Consequently this light is not absorbed.

For those photons with energy level above the band gap, the absorption coefficient is not constant but depends strongly on the prevailing wavelength. The probability of absorbing a photon depends on the likelihood of having a photon and an electron interact in such a way as to move from one energy band to another. For photons which have energy very close to that of the band gap, the absorption is relatively low since only those electrons directly at the valence band edge can interact with the photon to cause absorption. As the photon energy increases, there are a larger number of electrons which can interact with it and result in its full energy absorption. In c-Si solar cells, the photon energy greater than the band gap is wasted as electrons quickly thermalize back down to the band edges.

The absorption coefficient, α, is related to the extinction coefficient, k, by the following formula:

$$\alpha = \frac{4\pi k}{\lambda}$$

Where:

λ is the light wavelength and

k is the extinction coefficient (~0.06 for silicon and photon energies less than 3.0 eV)

The dependence of absorption coefficient on wavelength causes different wavelengths to penetrate different distances into a semiconductor before most of the light is absorbed. The absorption depth is given by the inverse of the absorption coefficient, or α^{-1}. The absorption depth is a useful parameter which gives the distance into the material at which the light drops to about 36% of its original intensity, or alternately has dropped by a factor of 1/e. Since high energy light has a large absorption coefficient, it is absorbed in a short distance (for silicon solar cells within a few microns) of the surface, while red is absorbed less strongly. Even after traveling several hundred microns, not all red light (IR) is absorbed in silicon.

Light Trapping

The optimum device thickness is not controlled solely by the need to absorb all the light. For example, if the light is not absorbed within a diffusion length of the junction, then the light-generated carriers are lost to recombination. In addition, a thinner solar cell which retains the absorption of the thicker device may have a higher voltage. Consequently, an optimum solar cell structure will typically have "light trapping" in which the optical path length is several times the actual device thickness, where the optical path length of a device refers to the distance that an unabsorbed photon may travel within the device before it escapes out of the device. This is usually defined in terms of device thickness. For example, a solar cell with no light trapping features may have an optical path length of one device thickness, while a solar cell with good light trapping may have an optical path length of 50, indicating that light bounces back and forth within the cell many times.

Light trapping is usually achieved by changing the angle at which light travels in the solar cell by having it be incident on an angled surface. A textured surface will not only reduce reflection as previously described but will also couple light obliquely into the silicon, thus giving a longer optical path length than the physical device thickness.

The angle at which light is refracted into the semiconductor material is, according to Snell's Law, as follows:

$$n_1 \sin \theta_1 = n_2 \sin \theta_2$$

where:

θ_1 and θ_2 are the angles for the light incident on the interface relative to the normal plane of the interface within the mediums with refractive indices n_1 and n_2 respectively.

If light passes from a high refractive index medium to a low refractive index medium, there is the possibility of total internal reflection (TIR). The angle at which this occurs is the critical angle and is found by setting θ_2 in the above equation to 0.

$$\theta_1 = \sin^{-1}\left(\frac{n_2}{n_1}\right)$$

Using total internal reflection, light can be trapped inside the cell and make multiple passes through the cell, thus allowing even a thin solar cell to maintain a high optical path length.

Generation Rate

The generation rate gives the number of electrons generated at each point in the device due to the absorption of photons. Neglecting reflection, the amount of light which is absorbed by a material depends on the absorption coefficient (α in cm^{-1}) and the thickness of the absorbing material. The intensity of light at any point in the device can be calculated according to the equation:

$$I = I_0 e^{-\alpha x}$$

where:

α is the absorption coefficient typically in cm^{-1};

x is the distance into the material at which the light intensity is being calculated; and

I_0 is the light intensity at the top surface.

The above equation can be used to calculate the number of electron-hole pairs being generated in a solar cell. Assuming that the loss in light intensity (i.e., the absorption of photons) directly causes the generation of an electron-hole pair, then the generation G in a thin slice of material is determined by finding the change in light intensity across this slice. Consequently, differentiating the above equation will give the generation at any point in the device.

Hence:

$$G = \alpha N_0 e^{-\alpha x}$$

where:

N_0 = photon flux at the surface (photons/unit-area/sec.);
α = absorption coefficient; and
x = distance into the material.

The above equations show that the light intensity exponentially decreases throughout the material and further that the generation at the surface is the highest at the surface of the material. For photovoltaic applications, the incident light consists of a combination of many different wavelengths, and therefore the generation rate at each wavelength is different.

Carrier Lifetime

The minority carrier lifetime is a measure of how long a carrier is likely to stay around before recombining. It is often just referred to as the "lifetime" and has nothing to do with the stability of the material. Stating that "a silicon wafer has a long lifetime" usually means minority carriers generated in the bulk of the wafer by light or other means will persist for a long lifetime before recombining.

Depending on the structure, solar cells made from wafers with long minority carrier lifetimes will usually be more efficient than cells made from wafers with short minority carrier lifetimes. The terms "long lifetime" and "high lifetime" are used interchangeably.

With low level injected material (where the number of minority carriers is less than the doping) the lifetime is related to the recombination rate by:

$$\tau = \frac{\Delta n}{R}$$

where:

τ is the minority carrier lifetime,
Δn is the excess minority carriers concentration and
R is the recombination rate.

Recombination Losses

Recombination losses effect both the current collection (and therefore the short-circuit current) as well as the forward bias injection current (and therefore the open-circuit voltage). Recombination is frequently classified according to the region of the cell in which it occurs. Typically, the main areas of recombination are at the surface (surface recombination) or in the bulk of the solar cell (bulk recombination). The depletion region is another area in which recombination can occur (depletion region recombination).

Current Losses

In order for the *p-n* junction to be able to collect all of the light-generated carriers, both surface and bulk recombination must be minimized. The two conditions commonly required for current collection are:

1. The carrier must be generated within a diffusion length of the junction, so that it will be able to diffuse to the junction before recombining; and

2. In the case of a localized high recombination site (such as at an unpassivated surface or at a grain boundary in multicrystalline devices), the carrier must be generated closer to the junction than to the recombination site. For less severe localized recombination sites (such as a passivated surface), carriers can be generated closer to the recombination site while still being able to diffuse to the junction and be collected without recombining.

The presence of localized recombination sites at both the front and the rear surfaces of a silicon solar cell means that photons of different energy will have different collection probabilities. Since blue light has a high absorption coefficient and is absorbed very close to the front surface, it is not likely to generate minority carriers that can be collected by the junction if the front surface is a site of high recombination. Similarly, a high rear surface recombination will primarily affect carriers generated by infrared light, which can generate carriers deep in the device. The quantum efficiency of a solar cell quantifies the effect of recombination on the light generation current.

Voltage Losses

The open-circuit voltage is the voltage at which the forward bias diffusion current is exactly equal to the short circuit current. The forward bias diffusion current is dependent on the amount recombination in a *p-n* junction, and increasing the recombination increases the forward bias current. Consequently, high recombination increases the forward bias diffusion current, which in turn reduces the open-circuit voltage. The material parameter which gives the recombination in forward bias is the diode saturation current. The recombination is controlled by the number of minority carriers at the junction edge, how fast they move away from the junction, and how quickly they recombine. Consequently, the dark forward bias current, and hence the open-circuit voltage is affected by the following parameters:

1. The number of minority carriers at the junction edge. The number of minority carriers injected from the other side is simply the number of minority carriers in equilibrium multiplied by an exponential factor which depends on the voltage and the temperature. Therefore, minimizing the equilibrium minority carrier concentration reduces recombination. Minimizing the equilibrium carrier concentration is achieved by increasing the doping.

2. The diffusion length in the material. A low diffusion length means that minority carriers disappear from the junction edge quickly due to recombination, thus allowing more carriers to cross and increasing the forward bias current. Consequently, to minimize recombination and achieve a high voltage, a high diffusion length is required. The diffusion length depends on the types of material, the processing history of the wafer, and the doping in the wafer. High doping reduces the diffusion length, thus introducing a trade-off between maintaining a high diffusion length (which affects both the current and voltage) and achieving a high voltage.

3. The presence of localized recombination sources within a diffusion length of the junction. A high recombination source close the junction (usually a surface or a grain boundary) will allow carriers to move to this recombination source very quickly and recombine, thus dramatically increasing the recombination current. The impact of surface recombination is reduced by passivating the surfaces.

Surface Recombination

Any defects or impurities within or at the surface of the semiconductor promote recombination. Since the surface of the solar cell represents a severe disruption of the crystal lattice, the surfaces of the solar cell are a site of particularly high recombination. The high recombination rate in the vicinity of a surface depletes this region of minority carriers. A localized region of low carrier concentration causes carriers to flow into this region from the surrounding, higher concentration regions.

Therefore, the surface recombination rate is limited by the rate at which minority carriers move towards the surface. A parameter called the "surface recombination velocity," in units of cm/sec, is used to specify the recombination at a surface. In a surface with no recombination, the movement of carriers towards the surface is zero, and hence the surface recombination velocity is zero. In a surface with infinitely fast recombination, the movement of carriers towards this surface is limited by the maximum velocity they can attain, and for most semiconductors is on the order of 1×10^7 cm/sec.

The solar cells' surface area, regardless of the surface finish (mechanical polishing or chemical etch) is a most critical area, for it is on or close to it where the photoelectric effect takes place. The surface area is often ignored, or improperly cleaned and processed, and the resulting solar cells will exhibit low performance. In worst cases, the defective surface area will cause the cells to deteriorate quickly or fail after some time in the field (latent effect).

Note: Latent effects in this context are hidden defects, triggered by a set of conditions, usually a long time after the manufacturing process has been completed. These are especially pertinent and dangerous in the case of PV cells and modules, because they are exposed to harsh environmental conditions (excessive heat, freeze, chemical attacks, high wind loads, etc.) for many years of non-stop field operation. Any of these conditions could trigger a latent effect with time.

Practical Design Considerations

The actual solar cell design and evaluation processes encompass a number of different theoretical considerations, some of which were discussed above. Other more general considerations, related to the cells' functionality are reviewed below.

We agreed above that when sunlight hits a solar cell it breaks apart electron-hole pairs on impact. Each photon with enough energy (band-gap energy of 1.1 eV) frees exactly one electron, which frees a hole in the process. Millions of photons of this energy level will generate millions of electron-hole pairs and millions of electrons released from the pairs' bond will generate current which will flow through the solar cell and into an external load. Photons with much lower or much higher energy levels do not contribute much to the overall energy generation and might even impact it negatively by interfering with the generation process and/or increasing the cell temperature.

How much sunlight energy does our PV cell absorb that is converted in useful electric power? Unfortunately, it is not much. Regular silicon cells (arranged in PV modules) convert only 15-18% of the incoming sunlight today. Thin film cells and modules convert 8-9%, while some exotic cells and modules convert 4-6%. On the other end of the efficiency spectrum, some specialized multi-junction cells convert over 40% of the sunlight

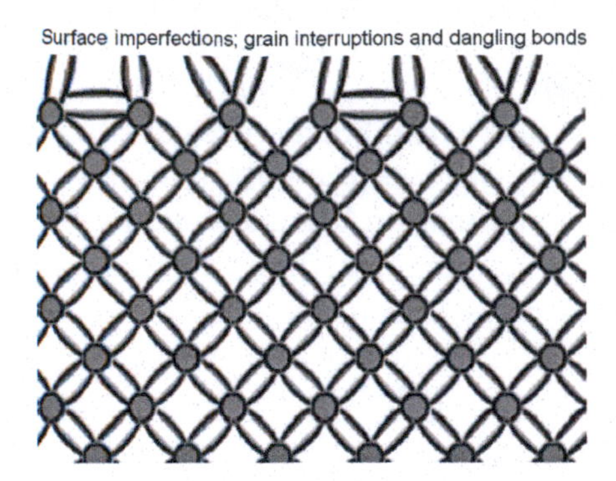

Figure 3-11. Silicon solar cell surface area

they see. Why so much difference? We will review these characteristics below.

Energy Interactions and Losses

Sunlight can be separated into different wavelengths, which we can see in the form of a rainbow. Visible light is what we see in the rainbow, but it is only part of the sunlight spectrum, while the other sunlight components are made up of a range of different wavelengths with different energy levels and overall behavior. The problems are that:

a. Not all photons have enough energy to generate electron-hole pairs, so they'll simply pass through the cell as if they were transparent.

b. Other photons have too much energy which could create negative effects. If a photon has more energy than required, some of the extra energy is lost. A photon with twice the required energy can create more than one electron-hole pair, but this effect is negligible in most cases and could even be harmful in larger quantities.

The above two negative effects account for 70-80% of the loss of incoming energy on the solar cell.

Material with lower band gap can use more of the photons (which have lower energy) but the band-gap also determines the potential (the voltage) of the newly generated electric field, and if it's too low it will make up current by absorbing more photons. In this case, the voltage might be too low and not practical for everyday use. The optimal band gap, balancing these two effects, is around 1.3-1.5 eV for single junction, single material cells of any type.

Temperature Losses

Temperature losses are expressed by the so-called temperature coefficient which is basically the rate of change of the generated power (and other measurable parameters) with respect to increase of temperature, usually measured in degrees C above STC (25°C). Changes can be measured and calculated for: short-circuit current (Isc), maximum power current (Imp), open-circuit voltage (Voc), maximum power voltage (Vmp), and maximum power (Pmp), as well as the fill factor (FF) and efficiency (h).

ASTM standard methods for performance testing of cells and modules addresses only two temperature coefficients, one for current and one for voltage. Actual field characterization of PV modules and arrays performance indicates that four temperature coefficients for Isc, Imp, Voc, and Vmp, are necessary to sufficiently and accurately model electrical performance for a wide range of operating conditions.

ASTM also specifies that temperature coefficients are determined using a standard solar spectral distribution at 1,000 W/m^2 irradiance, but from a practical standpoint they need to be applied at other irradiance levels as well, in order to get a complete picture of the cell/module behavior; i.e., some PV modules operate under the mostly cloudy skies of central Europe or the northeastern US, while others are exposed daily to the ferociously bright and hot deserts of the Southwest.

It is obvious that there is a huge difference in the above operating conditions, thus different test and measurement parameters must be used to determine which type PV modules are best for which climate and operating condition. This is extremely important, because the solar cells and modules must be designed and manufactured with the goal in mind to operate properly under this full sunlight exposure, which could bring cell temperature up to 195°F. This is almost 3 times the STC test standard and will have a profound effect on the cells' and modules' performance and longevity.

The actual cell temperature inside the module determines the drop of output and is therefore a most critical part of the temperature degradation. The cell temperature can be calculated using a measured back-surface temperature and a predetermined temperature difference between the back surface and the cell.

$$T_c = T_m + E/E_o \cdot \Delta T$$

where:

T_c = Cell temperature inside module, (°C)
T_m = Measured back-surface module temperature, (°C)
E = Measured solar irradiance on module, (W/m^2)
E_o = Reference solar irradiance on module, (1000 W/m^2)

ΔT = Temperature difference between the cell and the module back surface at an irradiance level of 1000 W/m2. This temperature difference is typically 2 to 3°C for flat-plate modules in an open-rack mount. For flat-plate modules with a thermally insulated back surface, this temperature difference can be assumed to be zero. For concentrator modules, this temperature difference is typically determined between the cell and the heat sink on the back of the module.

For c-Si solar cells and modules, a 0.5% drop of power output per each degree C increase of temperature above STC (25°C) has been accepted as an average; i.e., a c-Si solar cell in a module operating in the Arizona desert could reach 85°C internal temperature. This is 60°C above STC, or (60 x 0.5) = 30% drop of output. This is a serious reduction of power which we must be fully aware of and anticipate during the design and evaluation stages. Different types of solar cells behave differently under these conditions but they all show reduced power output under extreme heat conditions, so we just have to know how to find out and estimate the power reduction with the seasons and for the duration.

The negative effect of elevated operating temperature is a well known and understood phenomenon. Increasing the temperature reduces the band gap of most semiconductors, thereby affecting most of the key operating parameters and especially the open-circuit voltage. So when cell temperature goes up (due to internal problems or external excess operating temperatures), the voltage goes down. With that, cell efficiency and output decrease proportionally as well.

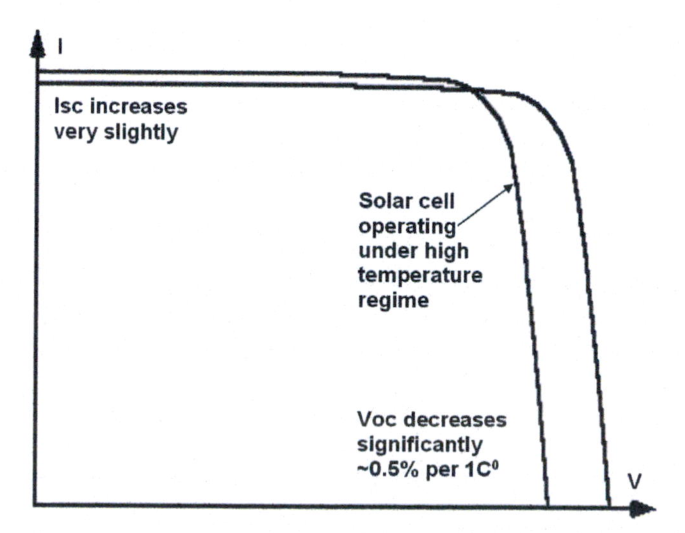

Figure 3-12. Elevated temperature effect

The elevated temperatures decrease the voltage of the cell, which translate into loss of power output.

Resistive Effects

Some of the major negative effects caused by material properties and defects, and/or improper processing, among others, are the so-called resistive (parasitic) effects which reduce the efficiency of the solar cell by dissipating power while increasing the internal resistances. The most common and most important of these are the series and shunt resistance. The influence of temperature and aging on these, as well as the overall solar cells' performance, needs very close investigation as well.

A closer look at the resistive parasitic effects in c-Si solar cells is as follows:

a. Series resistance, R_s, in a solar cell is expressed first by the movement of current through the emitter and base of the solar cell; second, the contact resistance between the metal contacts and the bulk silicon; and finally the resistance of the top and rear metal contacts themselves.

   The main impact of elevated series resistance is reducing the fill factor, although excessively high values may also reduce the short-circuit current and ultimately result in overheating and destruction of the cell and the module housing it.

b. Significant power losses caused by lower than normal shunt resistance, R_{SH}, are typically due to manufacturing defects, rather than poor solar cell design. Low shunt resistance causes power losses in solar cells by providing an alternate current path for the light-generated current. Such a diversion reduces the amount of current flowing through the solar cell junction and reduces the voltage from the solar cell.

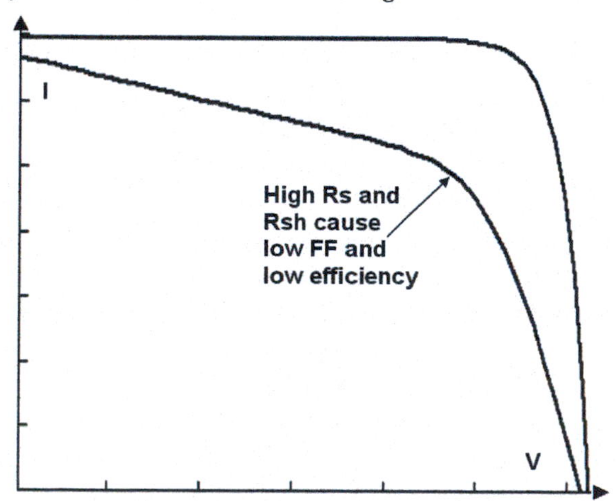

Figure 3-13. High Rs and Rsh effect

The high Rs and Rsh reduce the FF and the efficiency of the cells, which results in power output decrease. The effect of shunt resistance is particularly severe at low light levels, since there will be less light-generated current. The loss of this current to the shunt therefore has a larger impact. In addition, at lower voltages, where the effective resistance of the solar cell is high, the impact of a resistance in parallel is large.

Other Losses

The electrons are forced to flow from one side of the cell to the other through an external circuit going to and from the load (battery, motor, lamp etc.). This means that we need good contacts on the top and bottom of the cell in order to collect and recapture the electrons flowing to and from the external circuit. We can cover the bottom of the cell with a conducting metal, but we cannot cover the top completely because that will stop the electrons from reaching the cell surface. So we deposit appropriately spaced metallic grid on top of the cells to shorten the distance electrons travel from the bulk material to the metal grid, where they are collected and sent into the electrical circuit to provide electrical energy to a load. At the same time, we try to cover only a small part of the top cell surface with this metal grid, because we need a maximum amount of exposed top surface, where the photons can fall onto and generate free electrons (electricity).

The metal grid is just wide and long enough to capture a maximum number of electrons, while not covering (shading) too much of the top active surface. Even so, some photons are still blocked by the grid, while others can't reach it before they recombine with available holes and go back to a state of rest, where they are useless until energized by a photon impact again.

Thickness

The indirect energy band-gap results in a low optical absorption coefficient, and this means that the silicon solar cells need to be several hundred microns thick if they are to absorb most of the incident sunlight (and its photons) in order to prevent greater losses. But the cells cannot be too thick either, because that will increase the distance that charged particles need to travel, and which, combined with other problems will decrease the efficiency. The amount of light absorbed depends on the optical path length and the absorption coefficient. Thicker cells also cost more, and are heavier, so the PV modules get more expensive and heavier, both of which are highly undesirable factors.

Top Contacts

Metallic top contacts are necessary to collect the current generated by a solar cell. "Busbars" are connected directly to the external leads, while "fingers" are finer areas of metalization which collect current for delivery to the busbars. The key design trade-off in top contact design is the balance between the increased resistive losses associated with a widely spaced grid and the increased reflection caused by a high fraction of metal coverage of the top surface.

Contact resistance losses occur at the interface between the silicon solar cell and the metal contact. To keep top contact losses low, the top N^+ layer must be as heavily doped as possible. However, a high doping level creates other problems. If a high level of phosphorus is diffused into silicon, the excess phosphorus lies at the surface of the cell, creating a "dead layer," where light-generated carriers have little chance of being collected. Many commercial cells have a poor "blue" response due to this "dead layer." Therefore, the region under the contacts should be heavily doped, while the doping of the emitter is controlled by the trade-offs between achieving a low saturation current in the emitter and maintaining a high emitter diffusion length.

Combining the effects for resistive losses allows us to determine the total power loss in the top contact grid. For a typical screen printed cell type, the metal resistivity will be fixed and the finger width is controlled by the screen size. Typical values for the specific resistivity of silver are 3×10^{-8} Ω m. For non-rectangular fingers the width is set to the actual width and an equivalent height is used to get the correct cross sectional area.

The rear contact is much less important than the front contact since it is much further away from the junction and does not need to be transparent. The design of the rear contact, however, is becoming increasingly important as overall efficiency increases and the cells become thinner. The types of metals, their adhesion and thickness are important and are under intense investigation.

Annual Degradation

Finally, the power output from most solar cells and modules decreases with time as well. Studies by independent researchers have revealed an average of 0.50 to 1.5% of power output is lost every year by cells and PV modules operating under full sunlight. This degradation is due to a number of phenomena, but part of it is caused by gradual and permanent increase of the internal Rs and Rsh resistances over time for a number of reasons. And this is why most PV module manufacturers issue a long-term guaranty of only 80% of the original power to be generated by year 20 of their modules' on-sun operation.

Surface Area Design Consideration

The solar cells' top surface is one of the most critical areas of the overall cell design that needs special consideration, for it is where most mistakes are made and many problems are created or go undetected. Some of the most important aspects of the top surface design are as follow:

Surface Cleanliness and Defects

Both surfaces of as-cut wafers are dirty, rough and badly damaged by the mechanical action of the wafer saws. The damage is so bad in most cases, that if solar cells are made from these wafers, the resulting efficiency and longevity of the cell will be simply unacceptable. In order to maximize these parameters, the surface has to be "leveled" and cleaned. This is done by chemically etching the surface by dipping the wafers in cleaning solutions where the surface is cleaned from any residue from the previous steps, providing a uniformly clean surface for the next steps. The wafers are then chemically etched to remove some of the damaged top layers. The chemical action dissolves microns from all surfaces, thus removing most of the severe surface damage. In many cases micron-size pyramids are etched in the surface in order to increase the surface area. This method also provides the cleanest and most damage-free surface possible. The wafers are then rinsed and dried, after which they are ready for processing into solar cells.

These are extremely important processes, because they establish the base upon which the cell's active components will be laid. The surface must be perfectly prepared and cleaned for the resulting cells to achieve maximum efficiency and longevity.

AR Coating

There are many antireflective (AR) coatings available today. The type, color, and thickness of the AR coating need to be designed and executed properly so that it will not cause problems later on. The AR coating type will determine how efficient the cell operation is, and how long it will last. The AR color is the visible part of the cell/module surface. It is also one of the major reasons for the success of polysilicon modules, because people like their attractive dark-blue color which is determined by the AR coating type and thickness.

The deposition process parameters determine how well the AR film will perform, and how well adhered to the substrate it will be. Poor adhesion will cause it to peel off, thus causing visual and power degradation and premature failure. Deep penetration into the cell surface, on the other hand, will provide good adhesion but might puncture the diffusion layer and shunt the p-n junction.

Top Metal Contacts

To start with, we will assume that the depth of the metal contacts into the wafer surface is optimized, so that it is not too deep (to cause electrical shorts) and not too shallow (to cause adhesion problems). These are process design and manufacturing issues that have to be resolved before any other tests and optimizations can be considered.

The design of the top contacts is one of the most important solar cell design parameters. It seems like a simple thing to place contacts on the wafer surface, but doing it right is a science. Minimization of the fingers' and busbars' resistance and the overall reduction of losses associated with the top contacts is a major goal here. The possible losses include resistive losses in the emitter, resistive losses in the metal top contact, and shading losses.

The critical features of the top contact design which determine the magnitude of these losses are the metal height-to-width aspect ratio, finger and busbar spacing, the minimum metal line width, and the resistivity of the metal.

a. Aspect ratio is the relation between the height and width of the contacts. The proper selection of this parameter determines the performance characteristics of the solar cell. The aspect ratio is related to the other surface contacts design parameters discussed below.

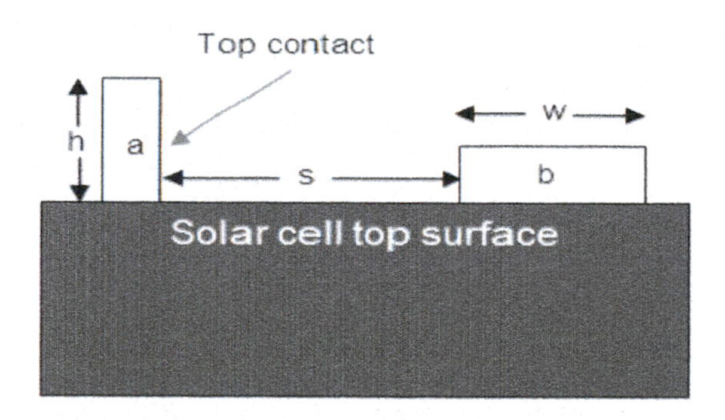

Figure 3-14. Surface contacts parameters

a - High aspect ratio top contact
b - Low aspect ratio top contact
s - Contacts spacing
h - Height of contact
w - Width of contact

$$\text{Aspect ratio} = \frac{h}{w}$$

b. Finger spacing on top surface
A key factor in top contact design is that of controlling the resistive losses in the emitter. We know from experience that the power loss from the emitter depends on the cube of the line spacing, and therefore a short distance between the fingers is desirable for a low emitter resistance. Too short a distance, however, will result in too much shading (coverage) of the front surface, which will prevent light from reaching the active layer of the cell. Optimization, via theoretical calculations, actual tests and trials, is needed, in order to determine the proper spacing of contacts and bus bars for each different type and size of solar cell.

c. Metal grid resistance
The grid resistance is determined by the:

— Resistivity of the metal that the contact is made of,

— The pattern of the metallization, and

— The aspect ratio of the metallization scheme.

Basically, low resistivity and a high metal height-to-width aspect ratio are desirable in solar cells, but in practice are limited by the capability of the particular fabrication technology used to make the solar cell and the cost of such effort. A tapered contact has lower resistive losses than one of constant width.

d. Shading Losses
Shading losses are caused by the presence of metal fingers and busbars on the top surface of the solar cell, which prevents light from entering the active layer. Shading losses are also determined by the transparency of the top surface which, for a planar top surface, is defined as the fraction of the top surface covered by metal. The optimum width of the busbar is achieved when its resistive and shadowing losses are equal.

e. The "transparency"
The transparency of the front surface is determined by the width of the metal lines on the surface and their spacing. An important practical limitation is the minimum line-width associated with a particular metallization technology. For identical transparencies, a narrow line-width technology can have closer finger spacing, thus reducing emitter resistance losses.

General Specs

a. Cell Thickness and Doping Level
Silicon solar cells are usually 250 to 450μm thick, as needed for safe handling and manufacturing. The doping level is kept at average 1.0 ohm/square.

b. Doping
N-type silicon is usually the diffusion layer so it is usually on the top of the cell where most of the light is absorbed. The top layer needs to be thin since a large fraction of the carriers generated by the incoming light are created within a diffusion length of the *p-n* junction. 100 ohm/square is the average emitter doping diffusion level.

c. Reflection Control
AR coating is used in most cases to reduce the reflection. The top surface of monocrystalline silicon cells is usually textured for the same reason.

d. Grid Pattern
Low resistivity metal contacts are deposited on the top surface of the cells to conduct away the current, but the contacts shade the cells from incoming light, so a compromise between light collection and resistance of the metal grid has to be reached. Usually fingers 20 to 200 μm width are deposited 1 to 5 mm apart.

e. Rear Contact
The rear contact is much less important than the front contact since it is much farther away from the junction and does not need to be transparent. The design of the rear contact, however, is becoming increasingly important as overall efficiency increases and the cells become thinner. Full aluminum back with silver fingers (grid) pattern is standard.

SILICON SOLAR CELLS MANUFACTURING

Solar cells have been mass manufactured since the 1950s, starting with 1.0" diameter silicon wafers and 2-3% efficiencies. The energy crisis of the 1970s shifted the attention of the US government and public to PV technologies which allowed the quick development of more efficient solar cells and modules, and the related production equipment and process. Still, the progress was seriously hindered by technological issues such as the small size of the silicon wafers, lack of adequate processing equipment and unresolved issues at the different process steps.

Today there are almost unlimited combinations and permutations of different materials, equipment options, processing sequences and techniques which allow cost-effective and efficient manufacturing of solar cells and modules. The manufacturers keep most of their process specifics secret, but the overall equipment configuration, process scheme, and different procedures are unchanged since the 1970s.

Manufacturing Process Considerations

Keeping all functional and operational parameters of the solar cells in mind, design engineers must consider a large number of process and application factors in order to come up with a practical, efficient, reliable and profitable final product with long lifetime.

Some of the main considerations are as follow:

1. The solar cell's material is of primary consideration. Its electro-mechanical and chemical properties, impurities, availability, price, ease and cost of processing, etc. factors are thoroughly evaluated and compared with competing materials and technologies. Starting with high quality silicon, and considering all process materials and related issues is the first and most important job of the design team.

2. Actual solar cell structure and process sequence design considerations include:
   a. Evaluating the quality of cleaning, texturizing, doping, edge etch chemicals, gasses, and related equipment and process steps specs
   b. Optical losses calculations and AR coating type and process specs
   c. Selection of materials and processes for metallization of the front and back contacts
   d. Quality control specs and inspection procedures for all process steps
   e. In-process and final tests and sorting procedures selection

3. The actual solar cells and modules manufacturing process is done best by using the most appropriate combination of materials, process equipment, chemicals and other consumables. Adequate knowledge and understanding of the cradle-to-grave process steps is paramount.

   We need to know all there is to know about:
   a. Silicon making, and ingot pulling or casting
   b. Slicing ingot into wafers, testing and sorting
   c. Wafer surface cleaning and texturing
   d. Diffusion chemicals and processes
   e. Edge isolation (wet chem. or plasma etch)
   f. Anti-reflection coatings and processes
   g. Metallization of front and rear contacts
   h. Testing and sorting of finished solar cells
   i. Stringing and encapsulation of solar cells into PV modules
   j. Final test and packing for shipping

There are a number of different manufacturing, test and quality control procedures, specifications, and instructions that accompany these steps, each of which has to be perfectly designed and executed.

Raw Materials and Consumables

The quality of the starting materials, silicon and the different process consumables, and chemicals and gasses used in the manufacturing sequence plays a significant role in determining the quality of the finished product. Most of the materials, manufacturing techniques, and quality control procedures used in the semiconductor devices manufacturing process apply to, and should also be used when processing solar cells. Unfortunately, this is not the case and the solar industry is suffering from lack of uniform standards, specs and procedures. Because of this, we need to be fully aware of and understand the different process specs and product characteristics and the issues related to these, discussed in this text.

Raw Silicon

First on the list of concerns to watch for when evaluating silicon solar cells and modules is the quality of the initial raw materials from which these were made. Silicon material used for manufacturing solar cells and modules today must be of highest possible quality, which means no defects (mechanical, electrical, or chemical) and no variations from batch to batch. The silicon quality determines the quality of the finished product in unpredictable fashion. Unwanted impurities in the silicon material could cause problems which could be detected during manufacturing, or shortly thereafter. In worst cases, they can lay dormant for years until finally triggered by mechanical, electrical or chemical changes during their many years of field operation.

The semiconductor industry uses super-pure silicon, but in order to achieve price and quality equilibrium the PV industry allows certain compromises in the quality of the raw silicon used to make solar cells. Nevertheless, if the minimum materials purity specs are met and the process sequence is properly executed, the resulting solar cells will be of good quality and will enjoy long trouble-free life. One of the problems is that some PV cells manufacturers do purchase raw silicon to make wafers in which case they have certain, if not total control over the quality of the wafers. Others, however, purchase the ready-made wafers and have no control over their quality whatsoever. Both types of manufacturers are faced with a number of quality issues they must identify and control as much as possible.

Process Materials and Consumables

A great number of different types and brands of materials and consumables are used in the solar cells and modules manufacturing process. Some of these are:

a. Cleaning and etching chemicals (degreasers, H2O2, HNO3, HF, H2SO4, HCL, CH3COOH etc.) are used extensively during the manufacturing process in different concentrations and combinations. Any impurities in these could imbed and diffuse into the surface and modify its chemical and electric characteristics. These impurities and other defects caused by process materials could affect the light absorbance, and create adhesion and other problems with time.

b. Doping chemicals (POCl3, BBr3, etc.) and gases (Ar, O2, N2 etc.) are used to provide the proper depth and concentration of the p-n junction. Any impurities in the process chemicals and gases will result in extraneous effects that might be detrimental to the efficient and reliable operation of the finished cells. Extreme care must be taken so that only high quality materials are used during this step. The process equipment must also be very clean and well maintained, and the process steps must be properly executed.

c. Metallization pastes (Ag, Al, etc.) and gases (N2, Ar, O2, etc.) are used to deposit a layer of metal on the top and bottom of the cells. They serve a dual purpose—collecting the electrons during the photoelectric conversion of sunlight to electricity, and conducting it to the external load, via bands or wires soldered to the metalized areas. The metal conductor bands and wires are made out of aluminum, silver or different alloys containing these metals. Their elemental composition determines their proper function and failure modes. Cheap metal compositions are prone to corrosion which ultimately reduces the efficiency of the solar cells and shortens their lifetime, especially when exposed to the elements. A broken module glass will allow moisture inside it, where the metallization is the first to get attacked and damaged. Low quality materials will allow quick and complete destruction of the contacts, in such cases, leading to deterioration and eventual failure of the module.

d. Soldering materials are also important as their quality determines the quality of the solder junction, and with that the efficiency and reliability of the solar cells and modules.

e. Antireflective (AR) coating chemicals (and gasses) under a large number of proprietary formulations are used to deposit a thin layer of inorganic chemicals on the solar cells, to prevent reflection of the incident sunlight, thus increasing its absorption by the top semiconductor material. The quality of the AR film is important because it determines the performance of the cells. AR films deposited from cheap chemicals tend to change with time and heat, and provide poor adhesion in some cases.

f. Top glass cover plays a number of important roles in the PV module operation, mainly providing good mechanical and chemical protection for the cells and good optical transmission for the incoming sunlight. It is, therefore, very important for the glass material to be of the highest quality. Lower quality glass might result in a shattered or discolored front cover, both of which would be detrimental to the module's performance and reliability.

g. Aluminum back cover and side frame are key elements of the module's performance and lifetime reliability. Low-grade aluminum alloys will oxidize and deteriorate excessively with time, which will cause decrease of efficiency and even failure of the affected modules.

Materials, processing and quality control procedures in the solar industry are not standardized, while those that are standardized are not uniformly enforced. This creates a peculiar situation where manufacturers are allowed to improvise and compromise with materials, process steps and quality control processes to meet production schedules and/or price restrictions. The results of such actions are uncertain, and could lead to unforeseen behavior of their PV modules during 30 years exposure to the elements. Remember, PV modules are not disposable items. They are energy generating devices, which are supposed to operate a specified 30+ years under harsh climate conditions. Quality of materials and consumables, therefore, is concern #1 for their efficient and reliable field operation.

c-Si Solar Cell Manufacturing Sequence

The basic c-Si solar processing sequence, as used in the late 1990s by an associate company is outlined below. This process, or a variation of it, is used by most world class c-Si PV cells and modules manufacturers today.

The major process steps are as follow:

Wafers Inspection and Sorting

Wafers are placed on inspection tables and are inspected visually and with optical equipment. Any wafers with visible mechanical defects are rejected. The wafers are then tested with a 4-point probe, and are sorted according to their resistivity. Wafers, or wafer samples are sent to an outside lab for metal and organic contamination analysis. Results from these tests determine the level of quality of the finished cells.

Wafers Cleaning and Etch

The wafers that pass all initial inspections and tests go to the wet cleaning line and are chemically processed in special chemicals where they are cleaned and etched to remove damage and oxide formed on the surface. The wafers are then rinsed with de-ionized (DI) water and dried via spin dryer.

Surface Etch (Chemical Etch)

This process is used only on single-crystal silicon with 1-0-0 orientation. (Polycrystalline wafers cannot be textured, because the different strings of silicon have different orientation and the resulting surface is only partially and unevenly textured, if at all.) A controlled chemical solution (composition, concentration, temperature and time) etches the pyramid-like structures in the wafer surface and the surface takes on a dark-gray appearance. The pyramids blend into each other and block excess light reflection. Each pyramid is approximately 4-10 microns high. This step has critical process parameters. The wafers are then rinsed, spin-dried, and stored in special containers for processing.

Note: In a variation of this process, wafers are loaded in a fixture two-by-two with the backs of each pair touching, so that the pyramid structure is formed only on their front surfaces.

Diffusion for P-N Junction Formation

The clean wafers are oven dried, placed in the diffusion furnace at 900-950°C in reactive carrier gasses to impregnate the wafers. POCl3 gas is used for n-type diffusion, which diffuses phosphorous atoms in the wafer's surfaces. This creates a p-n junction in the lightly boron-doped wafers.

Note: This method is easier to control and has more uniform distribution of dopant than using the spray-on diffusion liquid and belt furnace diffusion process used by many companies today. This step has critical process parameters, so any compromise will be reflected in the cells' overall performance and longevity. In a variation of this process, wafers are loaded in a fixture two-by-two with the backs of each pair touching, so that the diffusion layer is formed only, or mostly, on their front surfaces. This facilitates the processing of the back surface later on.

Mass production solar cell operations use a different method in which the wafers are sprayed with a dopant chemical and run through a conveyor belt type furnace, where the dopant is diffused in the wafers' surface. Both methods have advantages and disadvantages.

Plasma Etch for Removal of Edge Layer (diffusion etc.)

The diffusion process implants P dopant in the wafers' side edges, causing an electrical short circuit between the top and bottom (negative and positive) surfaces of the cell, so it is necessary to remove the dopant with a wet chemistry or plasma etch. Wafers are coin-stacked and etched for a brief period in an RF plasma etch reactor; only the edges of the wafer are exposed to the plasma which removes the diffusion coating.

Wafers are then etched gently in a bath of dilute hydrofluoric acid to remove any oxides formed during the plasma etch step, rinsed with DI water, and finally spin dried.

Anti-reflective Coating

AR coating is deposited on the front surface of the wafers. The purpose of the AR coating is to reducing the amount of sunlight reflected from the finished cell surface. The AR coating is deposited via chemical vapor deposition (CVD) or by spraying the chemicals on the wafers and then baking. Both methods achieve similar outcome of enhancing the solar cells' output and giving them the distinctive dark blue color (for poly solar cells).

NOTE: Different manufacturers deposit and fire the AR coating using different process parameters and sequence order. This is an important process, nevertheless, so its proper design and execution will determine the final, most important aesthetic and performance aspects of the cells.

Printing (Metallization)

Several screen printing steps are used to apply the metallization on the front and back of the wafers. First, silver paste is printed on the top surface which then becomes the front metal pattern (top contacts, or fingers). The paste is dried and the wafers are flipped for printing the back surface with aluminum paste. After drying, silver paste is printed in special slots in the dried aluminum and the wafers are then transported into the firing furnace.

Metal Firing

Thus metalized on both sides, wafers are run slowly through an IR-heated furnace where the metal pastes on the top and bottom sides of the wafers diffuse into the substrate, to make an electric contact with the p-n junction and the back surface. This step has critical process parameters.

Note: The firing of the front contacts is a very delicate process, where time and temperature are controlled to achieve the desired depth of penetration of the metal into the silicon surface. The depth of penetration determines the electro-mechanical properties of the finished cell. Specially designed automated printing-firing equipment is available for more precise and consistent process control.

Inspection and Quality Control

Solar wafers and cells are inspected and tested at several stages of the process sequence. This is done by eye inspection, using magnification and other instrumentation. Electrical tests are also performed at some steps of the process. The final inspection is the most important step and must be performed by well trained and experienced operators.

Cell Flash Testing and Sorting

A certain percent of the completed wafers are placed on a test stand in the solar simulator and are illuminated for a period of time. I/V curve is generated for each cell and the output data are used to sort the cells into groups according to the I/V curve characteristics, prior to soldering and lamination into modules.

Cell Storage

The cells are finally loaded into cassettes, or coin-stacked (with protective material in between) and packed for ease of handling, transportation, or storage prior to laminating into solar modules or shipping to another location.

Discussion of the Solar Cells Manufacturing Process

Now we know the basics of what it takes to make a solar cell, so let's take a closer look at the manufacturing process from a quality point of view.

The first step in the solar cell manufacturing process is to clean, rinse and dry the incoming silicon wafers to obtain nearly perfect surfaces. The final etch-clean-dry steps are extremely important, because any contaminants or water spots left on the surface could create a number of problems later on.

Note:

1. The incoming wafers usually are very dirty and badly damaged from the wafer sawing (with saw marks and imbedded slurry particles), so the cleaning and etch chemical baths wear out quickly. Effective stain and saw damage detection using IR CCD and special lighting systems is needed to prevent damaged wafers from entering the process sequence, but seldom done at this stage.
2. 1-2% of the incoming wafers have micro-cracks which are hard to detect, but are detrimental to the quality and longevity of the cells. Micro-crack detection must be performed at this stage, but is ignored in most solar cell manufacturing operations. Thus, wafers doomed to fail are introduced in the process and only good luck will prevent them from breaking during processing—or worse, failing in the field.

Even if the above inspection steps were implemented, only a small part of each batch will go through them. Since there is no way to check the surface cleanliness of every wafer, process design, execution, related tests, and quality control procedures are extremely important.

The wafers are then heated to over 1000°C in the presence of gases such as phosphorous in inert atmosphere (Ar, N_2 etc.) to form a p-n junction. Phosphorous gas atoms penetrate uniformly into the boron doped silicon wafer, but are concentrated just below the surface area. If the wafers are not protected on one side, then both sides and the edges will be doped. Now the concentration of phosphorous atoms is much greater than the boron atoms in the bulk, which in effect creates two adjacent N and P doped layers.

The area where the N and P materials meet is known as the *p-n junction*, where "p" stands for p-type semiconductor (boron doped silicon) and "n" stands for n-type doped (phosphorous doped silicon). The p-n junction is responsible for creating the conditions for the photoelectric effect to be initiated and maintained. The quality of the p-n junction (depth, saturation level, quality and purity of doping materials, and bulk material) determine the efficiency and often the longevity of the solar cells. Improperly diffused cells will degrade and fail much faster; especially under harsh climatic conditions.

The diffused area (across the p-n junction) shifts the balance of electrons in the lattice, to where there are now many free electrons. These are called *majority* carriers, while the holes in this case are the minority carrier. These can be reversed if different materials, doping agents and processes are used. The photovoltaic effect is now much easier to initiate and maintain, with lots of electrons and holes floating freely in the lattice. This is so, provided there is enough sunlight with an appropriate energy level of photons to impact the electrons enough to generate an adequate number of electron-hole pairs. The diffused wafers are usually treated with acids at this stage to remove the heavy layer of surface oxidation.

Then we remove the diffusion layer from the edges of the wafers by coin-stacking them into an RF plasma reactor which strips several microns of the exposed edges. This process eliminates the p-n junction in the critical edge areas and mechanically separates the P and N areas of the p-n junction. We call this process edge isolation.

After this, the AR (antireflective) layer is deposited on the front surface of the wafers by using different deposition methods: Si_xN_y:H deposition with SiH_4 and NH_3 precursor gases in a CVD reactor, or spraying TiO_2 mixture on the wafers' surface and baking them in horizontal furnaces. Thus formed, AR coating plays several important roles, but its primary function is to provide a surface for better absorption of sunlight. The CVD AR process also passivates the surface, which adds to the cells' performance and longevity. Because of the importance of the AR layer, its type, color, thickness, and adhesion are tightly controlled during the deposition process. The color and thickness will determine the performance characteristics of the resulting solar cells, while poor adhesion could cause delamination and significant performance degradation.

Metal contacts in the form of a metal paste are printed on the front and back of the cells at some point of the process and "fired" in a horizontal furnace to provide good ohmic contact with the cell which ensures a low resistivity path for the electrons to the outside circuit. Then special metal bands are soldered to the wafer's top contacts. The other end of the metal bands is soldered to the next wafer's bottom contacts. The "stringing" of cells is done at a different part of the facility, usually during the modules assembly process and is also a critical process because cold (bad) joints may cause high resistance and failure.

High Efficiency Solar Cells

There are other types of solar cells and processes, in addition to the standard, and most use the process sequence discussed above. These fall in the category of high-efficiency solar cells, simply because they are more efficient than the standard lot. Some of these are:

Buried Contact Solar Cells

The buried contact solar cell is a high efficiency commercial solar cell based on a plated metal contact inside a laser-formed groove. The buried contact technology overcomes many of the disadvantages associated with screen-printed contacts and this allows buried contact solar cells to have performance up to 25% better than commercial screen-printed solar cells. A key high efficiency feature of the buried contact solar cell is that the metal is buried in a laser-formed groove inside the silicon solar cell. This allows for a large metal height-to-width aspect ratio.

A large metal contact aspect ratio in turn allows a large volume of metal to be used in the contact finger, without having a wide strip of metal on the top surface. Therefore, a high metal aspect ratio allows a large number of closely spaced metal fingers, while still retaining a high transparency. For example, on a large area device, a screen-printed solar cell may have shading losses as high as 10 to 15%, while in a buried contact structure, the shading losses will be only 2 to 3%. These lower shading losses allow low reflection and therefore higher short-circuit currents. In addition to good reflection properties, the buried contact technology also allows low parasitic resistance losses due to its high metal aspect ratio, its fine finger spacing, and its plated metal for the contacts. The emitter resistance is reduced in a buried contact solar cell since narrower finger spacing dramatically reduces the emitter resistance losses. The metal grid resistance is also low since the finger resistance is reduced by the large volume of metal in the grooves and by the use of copper, which has a lower resistivity than the metal paste used in screen printing.

The contact resistance of a buried contact solar cell is lower than that in screen printed solar cells due to the formation of a nickel silicide at the semiconductor-metal interface and the large metal-silicon contact area. Overall, these reduced resistive losses allow large area solar cells with high FFs.

When compared to a screen-printed cell, the metalization scheme of a buried contact solar cell also improves the cell's emitter. To minimize resistive losses, the emitter region of a screen-printed solar cell is heavily doped and results in a "dead" layer at the surface of the solar cell. Since emitter losses are low in a buried contact structure, the emitter doping can be optimized for high open-circuit voltages and short-circuit currents. Furthermore, a buried contact structure includes a self-aligned, selective emitter, which thereby reduces the contact recombination and also contributes to high open-circuit voltages.

The efficiency advantages of buried contact technology provide significant cost and performance benefits. In terms of $/W, the cost of a buried contact solar cell is the same as a screen-printed solar cell. However, due to the inclusion of certain area-related costs as well as fixed costs in a PV system, a higher efficiency solar cell technology results in lower cost electricity. An additional advantage of buried contact technology is that it can be used for concentrator systems of up to 50x concentration.

PERL and Rear Contact Solar Cells

There are several types of solar cells that can be built on silicon substrates, but which require much more expensive materials, sophisticated equipment, and elaborate processing. Some of these are:

a. The passivated emitter with rear locally diffused (PERL) cell uses micro-electronic techniques to produce cells with efficiencies approaching 25% under the standard AM1.5 spectrum. The passivated emitter refers to the high quality oxide at the front surface that significantly lowers the number of carriers recombining at the surface. The rear is locally diffused only at the metal contacts to minimize recombination at the rear while maintaining good electrical contact.

b. Rear contact solar cells achieve potentially higher efficiency by moving all or part of the front contact grids to the rear of the device. The higher efficiency potentially results from the reduced shading on the front of the cell and is especially useful in high current cells such as concentrators or large areas. There are several configurations of these: interdigitated back contact (IBT), emitter wrap-through (EWT) and metalization wrap-through (MWT).

   Rear contact solar cells eliminate shading losses altogether by putting both contacts on the rear of the cell. By using a thin solar cell made from high quality material, electron-hole pairs generated by light that is absorbed at the front surface can still be collected at the rear of the cell. Such cells are especially useful in concentrator applications where the effect of cell series resistance is greater. An additional benefit is that cells with both contacts on the rear are easier to interconnect and can be placed closer together in the module since there is no need for a space between the cells. These types of solar cells usually cost considerably more to produce than standard silicon cells, and are typically used in specialized applications, such as solar cars or space exploration.

Efficiency Improvement Techniques

So the battle for more efficient solar cell continues. Some of the techniques and design features used to produce the highest possible efficiencies include:

a. Lightly diffused phosphorus to minimize recombination losses and avoid the existence of a "dead layer" at the cell surface;

b. Closely spaced metal lines to minimize emitter lateral resistive power losses;

c. Very fine metal lines, typically less than 20 μm wide, to minimize shading losses;

d. Polished or lapped surfaces to allow top metal grid patterning via photolithography;

e. Small area devices and good metal conductivities to minimize resistive losses in the metal grid;

f. Low metal contact areas and heavy doping at the surface of the silicon, beneath the metal contact to minimize recombination;

g. Use of elaborate metallization schemes, such as titanium/palladium/silver, that would give very low contact resistances;

h. Good rear surface passivation to reduce recombination;

i. Use of anti-reflection coatings which can reduce surface reflection from 10-30%.

Quality Control Concerns

The proverbial, "garbage in, garbage out" principle is in full force where high quality products are made, so looking for and finding the weak links in the process sequence is of utmost importance in any manufacturing process—including that of solar cells and modules. The quality of the supply chain (materials purchased from third parties) is where the quality control cycle starts. Immaculate quality control of all incoming materials is paramount for final product quality. Quality of production equipment is very important as well. Without good, proven, qualified and precisely controllable equipment, even high-quality materials could lose their value. Process design and execution are without a doubt extremely critical as well. One misstep in the solar cell manufacturing process could convert a batch of cells into paper weights.

Finally, and very importantly, are the people! Qualified design, equipment, process and quality control engineers, well trained technicians and operators, as well as experienced managers are all part of the team that must function like a well lubricated machine. One single glitch—be it equipment or people related—can introduce enough havoc in the process to reduce the quality of the final product. The basic rule of thumb is, "The highest quality of the final product is determined by the lowest quality of one the critical steps of the process." One single misprocessed step can ruin an entire batch of otherwise good product.

We cannot afford any quality violations through the entire process sequence, regardless of how many steps it consists of—not when this product is expected to operate 30 years non-stop under hostile climate conditions.

Total quality control and the related procedures, as followed by many well established US companies is needed in all solar products manufacturing operations. The PV industry, however, is not well organized and standardized yet, so we do see a number of variations and deviations in manufacturing operations.

Below we'll review the key product defects and failure mechanisms, in order to know what to focus on when discussing and negotiating quality characteristics and issues of PV modules.

c-Si PV MODULES

A PV module consists of a number of interconnected solar cells (typically 36 to 72 connected in series for battery charging), and many more for large-scale applications. Individual solar cells are soldered in strings and encapsulated into a single, hopefully long-lasting unit simply because PV modules cannot be disassembled for repairs. The main purpose for encapsulating a set of electrically connected solar cells into a module is to protect them from the harsh environment in which they are going to operate.

Solar cells are relatively thin and fragile and are prone to mechanical damage due to vibration or impact unless well protected. In addition, the metal grid on the top and bottom surfaces of the solar cells, the wires interconnecting the individual solar cells, as well as the soldered junctions can be corroded by moisture or water vapor entering the module, if the protecting materials are damaged or absent. So the encapsulation: a) provides a manageable package that can be installed in the field, b) prevents mechanical damage to the solar cells, and c) prevents water or water vapor from penetrating the module and corroding the electrical contacts and junctions.

Many different types of PV modules exist, and the module structure is often different for different types of solar cells or for different applications. For example, amorphous silicon, and other thin film solar cells are often encapsulated in a flexible array, while crystalline silicon solar cells are usually mounted in rigid metal frames with a glass front surface. Module lifetimes and warranties on bulk silicon PV modules are often 20-25 years, which assumes robust and durable encapsulation of the PV modules. Failing encapsulation will cause quick performance degradation and failure with time.

Silicon PV Modules Elements

Most PV bulk silicon PV modules consist of a transparent top surface, an encapsulant, a string of PV cells, rear encapsulant, rear cover and a frame around the outer edge. In most modules, the top surface is glass, the encapsulant is EVA (ethyl vinyl acetate), and the rear layer is usually Tedlar or a number of similar plastic and thermoplastic materials.

Typical module components are:

Cover Glass

The cover of the front surface of a PV module is usually glass with high transmission in the wavelengths which can be used by the solar cells, usually in the range of 350 nm to 1200 nm. In addition, the reflection from the front surface should be low too. While theoretically the reflection could be reduced by applying an anti-reflection coating to the glass surface, in practice these coatings are not robust enough to withstand some of the conditions in which most PV systems are used.

An alternative technique used to reduce reflection is to "roughen," or texture the top glass surface. In this case dust and dirt are very likely to adhere to the top surface and less likely to be dislodged by wind or rain. These glass surfaces are not "self-cleaning" and the advantages of reduced reflection are quickly outweighed by losses incurred due to increased top surface soiling. Texturing the inside of the glass is also practiced by some manufacturers, but there are some disadvantages in doing this as well, so the proper glass has to selected according to the module type and designation.

In addition, the top surface should have good safety properties and good impact resistance, should be stable under prolonged UV exposure and should have a low thermal resistivity. There are several choices for a top surface material including acrylic, polymers and glass. Tem-

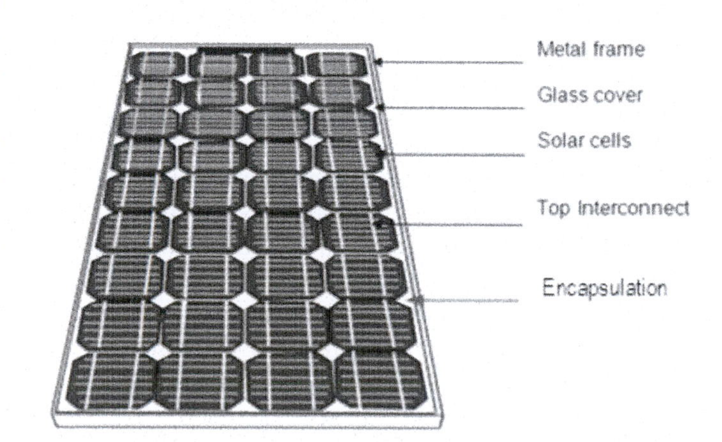

Figure 3-15. Standard silicon PV Module, Top View

pered, low iron-content glass is most commonly used as it is low cost, strong, stable, highly transparent, mostly impervious to water and gases, and has good self-cleaning properties. This type of glass is the most stable and trouble-free component of the entire module assembly. It doesn't deteriorate easily regardless of the harshness of the elements, and unless is broken, it will withstand the test of time for 25-30 years largely unaffected. Once the glass is broken, however, the module must be removed or put on a special maintenance schedule.

Cell Strings

A number of solar cells are interconnected and sealed (laminated) between plastic materials, which insulate them from each other, from the interconnecting wires and from the elements. The cells can be arranged and wired in a number of ways, as shown in Figure 3-16.

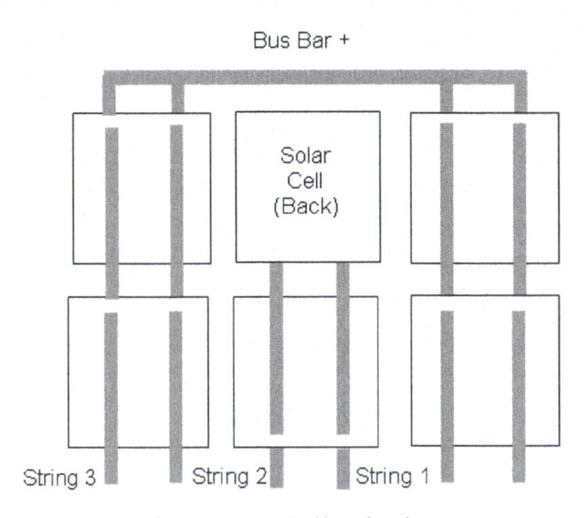

Figure 3-16. Cell stringing

Thus generated DC power is extracted from the module via two wires protruding from the module and routed into a junction (or terminal) box which is fitted with connectors for quick interconnect within the other array components.

Encapsulant (Front Surface)

An encapsulant is used to provide adhesion between the solar cells, the top surface and the rear surface of the PV module. The encapsulant should be stable at elevated temperatures and high UV exposure. It should also be optically transparent and should have a low thermal resistance. EVA (ethyl vinyl acetate) is the most commonly used encapsulant material. EVA comes in thin sheets which are inserted between the solar cells and the top surface and the rear surface. This sandwich is then heated to 150°C to polymerize the EVA and bond the module together.

EVA is responsible for protecting the cells from moisture and reactive species entering the module. Long exposure to UV and IR radiation tends to damage the EVA and it becomes yellow, which reduces its transmittance and reduces the module efficiency. Cracks and pores created in it under long exposure will allow the elements to enter the module and destroy the cells.

Back Cover

Most c-Si modules have a thin sheet of aluminum for back cover, which is screwed into the frame, with the terminal box attached to it. A key characteristic of the back cover is that it must have low thermal resistance and that it must prevent the ingress of water or water vapor. In most modules, a thin polymer sheet, typically Tedlar, is used as the rear surface, which provides electrical and environmental protection for the solar cells. Some PV modules, known as bifacial modules, are designed to accept light from either the front or the rear of the solar cell. In bifacial modules both the front and the rear must be optically transparent, so glass is the preferred material for these.

NOTE: The c-Si module's standard configuration is solid and strong enough to withstand transportation bumps, handling, and high winds during operation. Nevertheless, it is not intended to be used for support, or to be stepped on as some installers do. Even if the glass doesn't break in such case, the weight and impact puts enough stress on the cells to cause micro-cracks and other interruptions, which eventually grow into much bigger problems.

Side Frame

A final structural component of the module is the edging or framing, which provides additional mechanical strength and isolation from the elements. Module edges are sealed by the encapsulant layers in it and by additional adhesive materials for additional protection against the elements. An aluminum frame is then fastened around the edges of the module. The frame structure should be free of projections or pockets which could trap water, dust or other foreign matter.

PV Module Design Considerations

There are a large number of variables to keep in mind when designing or evaluating PV modules. Some of these are listed below: (2)

Cell Packing Density

The packing density of solar cells in a PV module refers to the area of the module that is covered with solar cells compared to that which is blank (between the cells). The packing density affects the output power of the module as well as its operating temperature. The packing density depends on the shape of the solar cells used. For example, single crystalline solar cells are round or semi-square, while multicrystalline silicon wafers are usually square. Therefore, if single-crystalline solar cells are not cut squarely, the packing density of the resulting module will be lower than that of a multicrystalline module, because of excess wasted space between the cells.

Sparsely packed cells in a module with a white rear surface can also provide marginal increases in output via the "zero depth concentrator" effect. Some of the light striking regions of the module between cells and cell contacts is scattered and channeled to active regions of the module.

Power Output

While the voltage from the PV module is determined mostly by the number of solar cells, the current from the module depends primarily on the size of the solar cells and also on their efficiency. At AM1.5 and under optimum tilt conditions, the current density from a commercial solar cell is approximately between 30 mA/cm^2 to 36 mA/cm^2. Single-crystal solar cells are often 100cm^2, giving a total current of about 3.5 A from each cell.

Poly-crystalline silicon modules have comparatively larger size individual solar cells but a lower current density. However, there is a large variation in the size of poly-crystalline silicon solar cells, and therefore the current will vary. Current from a module is not affected by temperature in the same way that voltage is, but instead depends heavily on the tilt angle of the module and the sunlight intensity reaching its surface.

If all the solar cells in a module have identical electrical characteristics, and they all experience the same insolation and temperature, then all the cells will be operating at exactly the same current and voltage. In this case, the IV curve of the PV module has the same shape as that of the individual cells, except that the voltage and current are increased proportionally to the number of cells in the module. If one single cell in the module, however, has different electrical characteristics (i.e. higher resistivity) then the entire module is affected, and will most likely underperform down to the level of the failing cell. The different cell might also overheat and fail under the increased load, thus making the entire module fail. And because not all cells are made equal and do not perform exactly the same, they should be tested and sorted before stringing and encapsulating into finished PV modules.

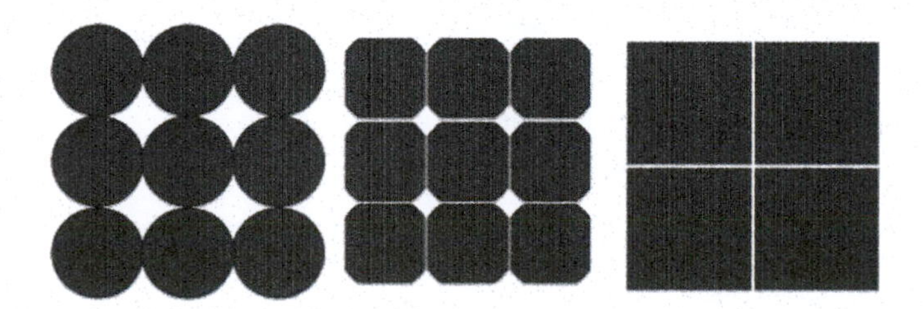

Figure 3-17. Different cell shapes and packing types

This, however, is a tricky operation, because cells tested under "normal" or "standard" test conditions in-house, often perform totally differently in the field. This anomaly could be caused by difference in materials quality and/or improper process execution. In all cases, the established manufacturers who have long experience with solar cells, testing and field applications, have proven ways to sort and assemble the cells in order to eliminate or reduce this type of field problem.

In all cases, proper design calculations and tests must be executed in order to come up with an efficient cell and module design.

The equation for the module power under normal operating conditions used to evaluate the different cells and modules is:

$$I_T = M \bullet I_L - M \bullet I_0 \left[\exp\left(\frac{q\frac{V_T}{N}}{nkT} \right) - 1 \right]$$

where:

- I_T is the total current from the circuit;
- N is the number of cells in series;
- M is the number of cells in parallel;
- V_T is the total voltage from the circuit;
- I_0 is the saturation current from a single solar cell;
- I_L is the short-circuit current from a single solar cell;
- n is the ideality factor of a single solar cell; and
- q, k, and T are respective constants

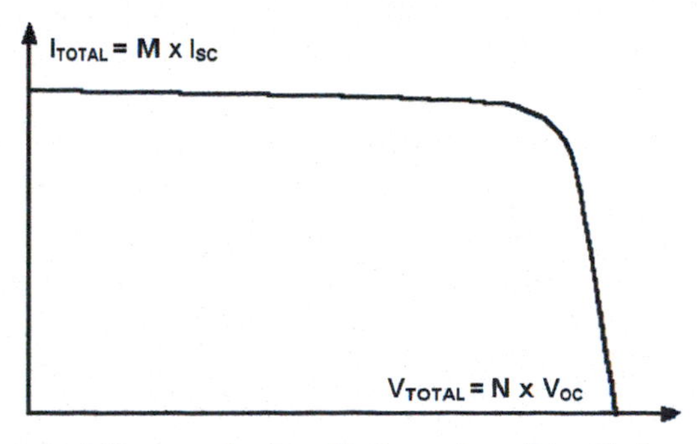

Figure 3-18. I-V curve for N cells in series x M cells in parallel.

This formula can be used also to predict the behavior of different cells in modules in the field. The situation changes drastically under not-so-normal field conditions, such as operation under high desert sunlight. Although there are formulas to calculate the temperature effect (as discussed in more detail in this writing), experience shows that elevated temperatures and humidity create havoc through individual cells affecting them differently and causing them to behave differently over time. This in turn results in unpredictable behaviors of the affected modules, and these anomalies often lead to reduced power output and short lifetime.

Mismatch of Series Connected Cells

As most PV modules are series-connected, series mismatches are the most common type of mismatch encountered. Of the two simplest types of mismatch considered (mismatch in short-circuit current or in open-circuit voltage), a mismatch in the short-circuit current is more common, as it can easily be caused by shading part of the module. This type of mismatch is also the most severe.

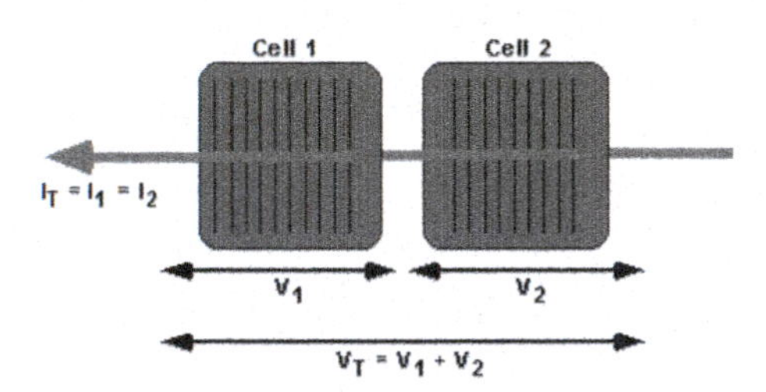

Figure 3-19. Cells in series

For two cells connected in series, the current through the two cells is the same. The total voltage produced is the sum of the individual cell voltages. Since the current must be the same, a mismatch in current means that the total current from the configuration is equal to the lowest current.

a. Open circuit voltage mismatch
A mismatch in the open-circuit voltage of series-connected cells is a relatively benign form of mismatch. So, at short-circuit current, the overall current from the PV module is unaffected. At the maximum power point, the overall power is reduced because the poor cell is generating less power. As the two cells are connected in series, the current through the two solar cells is the same, and the overall voltage is found by adding the two voltages at a particular current.

b. Short-Circuit current mismatch
A mismatch in the short-circuit current of series connected solar cells can, depending on the operating point of the module and the degree of mismatch, have a drastic impact on the PV module. As shown in Figure 3-19, at open-circuit voltage, the impact of a reduced short-circuit current is relatively minor. There is a minor change in the open-circuit voltage due to the logarithmic dependence of open-circuit voltage on short-circuit current. However, as the current through the two cells must be the same, the overall current from the combination cannot exceed that of the poor cell. Therefore, the current from the combination cannot exceed the short-circuit current of the poor cell. At low voltages where this condition is likely to occur, the extra current-generating capability of the good cells is not dissipated in each individual cell (as would normally occur at short circuit), but instead is dissipated in the poor cell.

Overall, in a series-connected configuration with current mismatch, severe power reductions are experienced if the poor cell produces less current than the maximum power current of the good cells. Also, if the combination is operated at short circuit or low voltages, the high power dissipation in the poor cell can cause irreversible damage to the module.

Hot-spot Heating

This condition occurs when there is one low current solar cell in a string of at least several high short-circuit current solar cells, as shown in Figure 3-20.

One shaded cell in a string reduces the current through the good cells, causing the good cells to produce higher voltages that can often reverse bias the bad cell. If the operating current of the overall series string approaches the short-circuit current of the "bad" cell, the overall current becomes limited by the bad cell. The extra current produced by the good cells then forward biases the good solar cells. If the series string is short circuited, then the forward bias across all of these cells reverses the

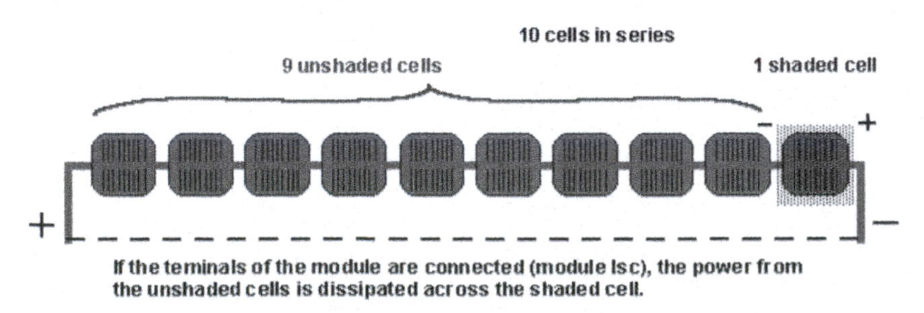

Figure 3-20. Cell shading

bias of the shaded cell. Hot-spot heating occurs when a large number of series connected cells cause a large reverse bias across the shaded cell, leading to large dissipation of power in the poor cell.

Essentially the entire generating capacity of all the good cells is dissipated in the poor cell. The enormous power dissipation occurring in a small area results in local overheating, or "hot-spots," which in turn leads to destructive effects, such as cell or glass cracking, melting of solder or degradation of the solar cell.

Bypass Diodes

The destructive effects of hot-spot heating may be circumvented through the use of a bypass diode. A bypass diode is connected in parallel, but with opposite polarity, to a solar cell. Under normal operation, each solar cell will be forward biased and therefore the bypass diode will be reverse biased and will effectively be an open circuit. However, if a solar cell is reverse biased due to a mismatch in short-circuit current between several series connected cells, then the bypass diode conducts, thereby allowing the current from the good solar cells to flow in the external circuit rather than forward biasing each good cell. The maximum reverse bias across the poor cell is reduced to about a single diode drop, thus limiting the current and preventing hot-spot heating. In practice, however, one bypass diode per solar cell is generally too expensive and instead bypass diodes are usually placed across groups of solar cells. The voltage across the shaded or low-current solar cell is equal to the forward bias voltage of the other series cells which share the same bypass diode plus the voltage of the bypass diode.

The voltage across the unshaded solar cells depends on the degree of shading on the low-current cell. For example, if the cell is completely shaded, then the unshaded solar cells will be forward biased by their short circuit current and the voltage will be about 0.6V. If the poor cell is only partially shaded, some of the current from the good cells can flow through the circuit, and the remainder is used to forward bias each solar cell junction, causing a lower forward bias voltage across each cell. The maximum power dissipation in the shaded cell is approximately equal to the generating capability of all cells in the group. The maximum group size per diode, without causing damage, is about 15 cells/bypass diode, for silicon cells. For a normal 36-cell module, therefore, 2 bypass diodes are used to ensure the module will not be vulnerable to "hot-spot" damage.

Within Module Heat Generation

A PV module exposed to sunlight generates heat as well as electricity. For a typical commercial PV module operating at its maximum power point, only 10 to 15% of the incident sunlight is converted into electricity, with much of the remainder being converted into heat.

Factors which affect the heating of the module are:

a. Reflection from the top surface.
Light reflected from the front surface of the module does not contribute to the electrical power generated. Such light is considered an electrical loss mechanism which needs to be minimized. Neither does reflected light contribute to heating of the PV module. The maximum temperature rise of the module is therefore calculated as the incident power multiplied by the reflection. For typical PV modules with a glass top surface, the reflected light contains about 4% of the incident energy.

b. Electrical operating point.
The operating point and efficiency of the solar cell determine the fraction of the light absorbed by the solar cell that is converted into electricity. If the solar cell is operating at short-circuit current or at open-circuit voltage, then it is generating no electricity and hence all the power absorbed by the solar cell is converted into heat.

c. Absorption of sunlight in areas not covered by solar cells.
The amount of light absorbed by the parts of the module other than the solar cells will also contribute to the heating of the module. How much light is absorbed and how much is reflected is determined by the color and material of the rear backing layer of the module.

d. Absorption of low-energy (infrared) light.
The amount of light absorbed by the parts of the module other than the solar cells will also contribute to the heating of the module. How much light is absorbed and how much is reflected is determined by the color and material of the rear backing layer of the module.

e. Packing density of the solar cells.
Solar cells are specifically designed to be efficient absorbers of solar radiation. The cells will generate significant amounts of heat, usually higher than the module encapsulation and rear backing layer. Therefore, a higher packing factor of solar cells increases the generated heat per unit area.

Nominal Operating Cell Temperature (NOCT)

A PV module will generally be rated at 25°C under 1 kW/m2. However, when operating in the field, they typically operate at higher temperatures and at somewhat lower insolation conditions. In order to determine the power output of the solar cell, it is important to determine the expected operating temperature of the PV module. The nominal operating cell temperature (NOCT) is defined as the temperature reached by open circuited cells in a module under the conditions listed below:

a. Irradiance on cell surface = 800 W/m2
b. Air temperature = 20°C
c. Wind velocity = 1 m/s
d. Mounting = open back side.

The equations for solar radiation and temperature difference between the module and air show that both conduction and convective losses are linear with incident solar insolation for a given wind speed, provided that the thermal resistance and heat transfer coefficient do not vary strongly with temperature. The best case includes aluminum fins at the rear of the module for cooling which reduces the thermal resistance and increases the surface area for convection.

$$T_{Cell} = T_{Air} + \frac{NOCT - 20}{80} S$$

where:

S is the insolation in mW/cm2

Thermal Expansion and Stress

Thermal expansion is another important temperature effect which must be taken into account when modules are designed.

The spacing between cells tries to increase an amount δ given by:

$$\delta = (\alpha_G C - \alpha_c D)\Delta T$$

where:

$\alpha_G \alpha_C$ are the expansion coefficients of the glass and the cell respectively;
D is the cell width; and
C is the cell center to center distance.

Typically, interconnections between cells are looped to minimize cyclic stress. Double interconnects are used to protect against the probability of fatigue failure caused by such stress.

In addition to interconnect stresses, all module interfaces are subject to temperature-related cyclic stress which may eventually lead to delamination.

Other Module Design and Evaluation Considerations

A bulk silicon PV module consists of multiple individual solar cells connected nearly always in series as needed to increase the power and voltage to the desired level. The voltage of a PV module is usually chosen to be compatible with a 12V battery, if used for automotive or battery charging purposes, and any other voltage and current combination as needed for each particular application. An individual silicon solar cell has a voltage around 0.6V under 25°C and AM1.5 illumination.

Take into account the expected reduction in PV module voltage due to temperature and the fact that a battery may require 15V or more to charge the modules containing 36 solar cells in series. This gives an open-circuit voltage of about 18-21V under standard test conditions, and an operating voltage at maximum power and operating temperature of about 17 or 18V.

The remaining excess voltage is included to account for voltage drops caused by other elements of the PV system, including operation away from maximum power point and reductions in sunlight intensity.

The same principle is in effect for modules used in commercial or large-scale PV installations. These, however, contain a much larger number of cells in order to generate higher voltage, in the 150-300Wp range. These modules are also much larger in size in order to accommodate the greater number of cells.

A number of additional factors are considered when designing or evaluating different modules for different applications. The source and quality of the materials and components used in building the PV modules in question must be addressed, although this is easier said than done. Taking a close look at the certification documents may

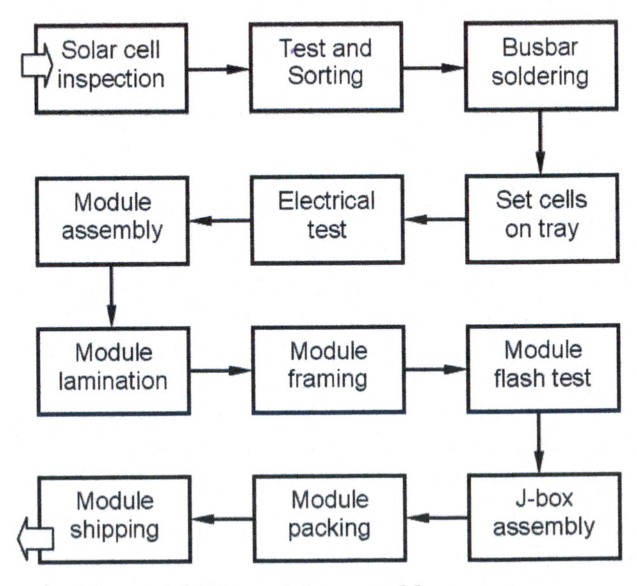

Figure 3-21. PV module assembly sequence

provide a good estimate of the modules' performance and longevity. We also recommend discussing some of the certification test points with the manufacturers.

c-Si PV Module Manufacturing

A number of potential issues are encountered during the cell and module manufacturing processes, all of which must be taken into consideration, if we are to have reliably performing PV cells and modules, lasting 25-30 years.

The major issues to keep in mind when designing or planning to use c-Si PV modules are:

1. Quality of silicon material, chemicals and consumables
2. Cell type and design parameters
3. Quality of the cells' manufacturing equipment and process
4. Module type and design parameters
5. Modules' manufacturing equipment and process
6. Possible cell malfunctions within this type and make of module
7. Possible module malfunctions within the particular array

Once the materials—PV cells, laminates, glass, back cover, wiring etc.—have been received and gathered at the module production site, the module is assembled in the following sequence (see Figure 3-21).

Cell Sorting, Arranging and Soldering

Finished solar cells are flash tested and sorted by their I-V characteristics and power output. Cells that pass the test are placed in bins according to their performance and stored or taken to the module assembly area.

Wiring and Assembly

Cells are connected in a series circuit manually, or by a semi-automated soldering machine using solder coated metal ribbon (usually two in parallel) soldered to the top of one cell and to the bottom of the next cell. This process forms a string of cells which could be as long as desired but usually it is shaped to fit in the respective PV modules tray. Electrical continuity and resistivity tests are performed on some modules to make sure that the bonds are good. "Pull" tests are done sometimes, to check the mechanical strength of the bonds. Thus, connected cells make a complete circuit (string), which is ready for lamination into a completed module.

Lamination

A lay-up for lamination is prepared with clean top glass, and EVA film, and strings of wired cells are placed on it. Sometimes the backing materials (Tedlar and back cover) are placed on top too, forming a complete module. Several lay-ups, each consisting of the above components are lined up in a large cabinet called a laminator. Using silicone vacuum blankets, the batch of lay-ups is heated and vacuum laminated at one time. After cooling, the modules are ready for use. This method of laminating is much cheaper than laminating one or two units at a time. The excess lamination is trimmed and terminal wiring is attached. In most cases, an aluminum extruded frame is assembled around the material and the unit is ready for shipping.

Basically, module laminators consist of a large-area heated metal platen mounted in a cabinet-like vacuum chamber. The top of the cabinet opens for loading and unloading modules. A flexible diaphragm is attached to the top of the chamber, and a set of valves allows the space above the diaphragm to be evacuated during the initial pump step and backfilled with room air during the press step. A pin lift mechanism is sometimes used to lift modules above the heated platen dur-

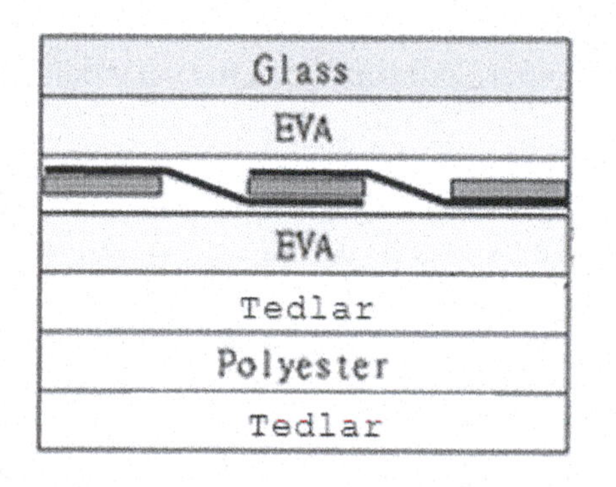

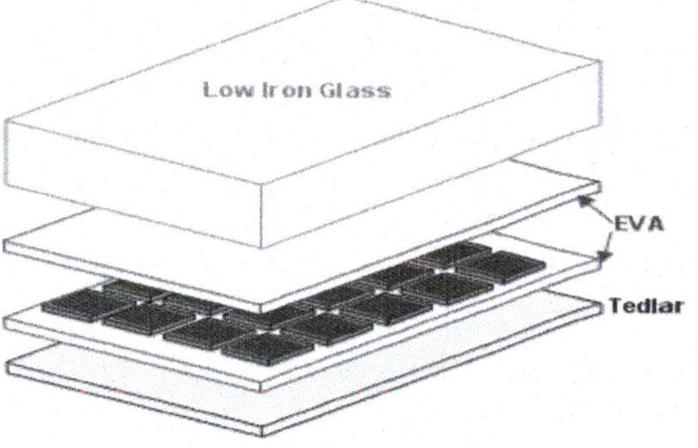

Figure 3-22. PV Modules cross section

ing the initial pump step, but most standard modules don't require it.

Temperature uniformities of ±5°C at the lamination point are sufficient for obtaining good laminations with acceptable gel content and adhesion across the module. While more uniform temperatures are available from some laminator suppliers, there is no real benefit to the module manufacturer.

Laminators are available with two types of cover opening systems: clamshell and vertical post. In the clamshell design, the cover is mounted on a hinge at the back of the laminator, which opens like the hood of a car. This leaves the laminator wide open on three sides, making it easy for an operator to load and unload modules manually. Automated belt-fed laminators, on the other hand, use the vertical post method, which lifts the cover horizontally above the process chamber. Because the cover does not need to travel much for belt loading, the chamber opening and closing times, and resulting process steps (heating and vacuum pump down) are reduced. As a result, most high throughput module lines use belt-fed laminators with vertical cover lifts.

NOTE: Fully automated cell assembly and lamination lines exist today, but most low- to mid-volume assembly operations, especially those in Asia, still prefer manual lay-up and stringing operations, combined with low-throughput clamshell type laminators. This is due mostly to the availability of cheap labor, though automating labor-intensive processes is no guaranty of a high quality product.

Modules Flash Testing and Sorting

After completing the assembly process by adding edge sealers, side frame, terminal box, etc., the modules are placed on a test stand in the solar simulator (flasher) and are illuminated with special type of light that resembles the solar spectrum at STC, for a period of time. The temperature of the modules is kept at 25°C during the test by active or passive cooling. I-V curve is then generated for each module. The output data are used to identify, sort and label the modules according to their output. The modules that pass the test are packed and shipped to the customers.

Certification Tests

All PV modules must be prop-erly tested and certified before they are allowed on the energy market. For this purpose they are sent to test laboratories for official certification. Then they can be used in the destination country.

The testing process consists of a series of visual and electro-mechanical tests as follow:

Standard Test Conditions (STC)

These are the conditions (light characteristics, operating temperature and time) all PV modules are exposed to, in order to establish and certify their nominal performance parameters. PV modules are tested using test conditions accepted as "standard," or "standard test conditions," or STC by the solar industry.

The STC specs are as follow:

a. Vertical irradiance E of 1000 W/m^2;
b. Temperature T of 25°C with a tolerance of ± 2°C;
c. Defined spectral distribution of the solar irradiance at air mass $AM = 1.5$.

I-V Curve

I-V, or current vs. voltage, curve is usually generated and documented during and after these tests, and the efficiency of the solar module is calculated from the test data. Basically, the I–V curve is characterized by the following three points:

a. The maximum power point (MPP). For this point, the power Pmpp, the current Impp and voltage Vmpp are specified. This MPP power is given in units of watt peak, Wp.

b. The short-circuit current Isc is approximately 5-15% higher than the MPP current. With crystalline standard cells (10cm x 10cm) under STC, the short-circuit current Isc is around 3A. Some modules will have higher Isc, and some lower.

c. The open-circuit voltage Voc in silicon cells registers ~0.5V to 0.6V, and for amorphous silicon cells it is ~.6V to 0.9V.

NOTE: Wiring the cells in different configurations will produce different voltage and current combinations, as required by the particular type of module or its application.

Efficiency

The efficiency η is the most commonly used parameter to compare the performance of one solar module to another. It is the property that identifies each module type and determines its performance in the field. Efficiency is defined as the ratio of energy output from the solar cell to input energy from the sun. The more efficient a module is, the more expensive it seems to be.

The efficiency of a solar module is determined as the fraction of incident power which is converted to electricity by the module. It is defined as:

$$\eta = \frac{V_{ocv} I_{sc} FF}{P_{in}}$$

where

V_{oc} is the open-circuit voltage;
I_{sc} is the short-circuit current;
FF is the fill factor
P_{in} is the incident power

Environmental and Mechanical Tests

A number of tests, designed to test the resistance of the modules to environmental effects (rain, snow, heat, freeze, etc.), are usually part of the manufacturers' procedures and the product certification procedure.

These tests include:

a. Thermal stress: cold, hot, wet cycles; humidity tests
b. Electrical rigidity, resistance, temperature
c. Hail launching
d. Mechanical stress tests
e. Salt, fog and dampness tests
f. UV light tests
e. Outdoor tests

Test and Certification Standards

A number of standards for testing and certification of PV modules have been established and accepted for use in different countries, as follow:

a. Most relevant standards for the U.S. and Canada.
   - UL 1703: Standard for flat-plate PV modules and panels
   - UL 790: Standard for standard test methods for fire test of roof coverage
   - AC 365: Standard for building-integrated PV modules (BIPV)

b. Individual U.S. State Requirements
   - California: California Energy Commission (CEC)
   - Florida: Florida State Energy Center (FSEC)

c. Most relevant standards for Europe and parts of Asia
   - IEC 61730: Standard for PV module safety
     — Part 1: requirements for construction
     — Part 2: requirements for testing
   - IEC 61215: Standard for crystalline silicon terrestrial PV modules
   - IEC 61646: Standard for thin-film terrestrial PV modules
   - IEC 60904-X: Standard for PV devices (measurement procedures and requirements)

Note: A full list of international standards accepted and used by the solar industry around the world can be found in Appendix A below.

TYPES OF PV MODULES AND ARRAYS

PV modules come in different makes, types, shapes, sizes and designations, but their ultimate purpose is to produce DC power. The more electricity a PV module generates from a given active surface area, the more efficient it is. Different types of modules, using different solar cells, materials and manufacturing techniques have different efficiencies and longevity, but for most part:

1. The efficiency of the solar cells in each module determines its overall efficiency. One not-so-efficient cell can reduce the efficiency of an entire module. As a matter of fact, one "bad" cell can ruin the performance of an entire array of modules—or at least it can lower the output significantly. The cell's slow degradation, will also affect the module's performance over time.

2. The proper structure of the module, and in particular its ability to keep the elements from penetrating inside and attacking the cell components, determines its durability, degree of accelerated degradation and overall longevity.

This is why product quality and quality control procedures during the entire production process are such an important aspect of PV cells and modules manufacturing and use.

Efficiency

PV modules come in different types and sizes. The efficiency of the different types of modules is one of the most understood and discussed properties (Table 3-2).

Clearly, sc-Si is the most efficient of the single junction PV cells for all practical purposes (III-V semiconductor cells are usually of the multi-junction type). It is, in our opinion, also most reliable for long-term use as well, simply because its structure is simple, its crystal is perfect and not as easily affected by different electro-mechanical abnormalities. Polysilicon, however, is most widely used, mostly because of its pleasant aesthetics, ease of manufacturing and lower cost, even though its efficiency is lower and the imperfections of its crystal bring some unwanted effects in the long run as well.

Generally speaking, we must be aware of the behavior patterns, variables, performance and longevity issues of the different types of PV cells and modules, if we are to design and build efficient and long-lasting PV power plants.

Table 3-2. Performance of different cells and PV modules

Solar cell material	Cell efficiency η_z (laboratory) (%)	Cell efficiency η_z (production) (%)	Module efficiency η_M (series production) (%)
Monocrystalline silicon	24.7	21.5	16.9
Polycrystalline silicon	20.3	16.5	14.2
Ribbon silicon	19.7	14	13.1
Crystalline thin-film silicon	19.2	9.5	7.9
Amorphous silicon[a]	13.0	10.5	7.5
Micromorphous silicon[a]	12.0	10.7	9.1
CIS	19.5	14.0	11.0
Cadmium telluride	16.5	10.0	9.0
III-V semiconductor	39.0[b]	27.4	27.0
Dye-sensitized call	12.0	7.0	5.0[c]
Hybrid HIT solar cell	21	18.5	16.8

PV Arrays

Typically, PV modules are not used alone, but in a combination with a number of similar modules to create an "array." A PV array is simply a group of PV modules installed and wired together as a group. The PV arrays come in many forms, shapes and sizes, and each array can be fixed, sun-tracking with one axis of rotation, or sun-tracking with two axes of rotation. Fixed PV array is the cheapest and most common on the PV energy market. It is not as efficient as the tracking array, but it is the simplest, the cheapest, and the one requiring least maintenance.

The function, advantages and disadvantages of fixed vs. tracking arrays is quite important and has to be considered in any PV project. The technical and financial aspects of different types of modules and arrays are discussed in more detail in following chapters.

A fixed array consists of a number of modules that are mounted permanently on a solid frame, usually steel or aluminum angles, which is cemented, or otherwise solidly fixed into the ground or on a structure. Tracking arrays are also mounted on a frame, but the frame is heavier and is mounted on a pivoting pedestal, that allows it to rotate on top of it, thus following the sun all through the day.

The tracking frame moves with the help of one motor-gear assembly for a one-axis system, or via two motor-gear assemblies for a two-axis system. A controller, which knows exactly where the sun is, sends a signal to the motors, which then activate the gears to move the frame in position. (See Figures 3-23 and 3-24.)

As you can see, trackers generally have higher comparative efficiency, because they are positioned most advantageously towards the sun (constantly at a 90-degree angle), thus receiving the maximum amount of sunlight and generating the largest possible amount of electric power from sunrise to sundown. Of course, multi-junction solar cells which have 2-3 times higher efficiency add to the advantage of tracking systems.

Module/Array Operating Parameters

Several key considerations must be kept in mind when designing and using PV modules and arrays. Some of these are:

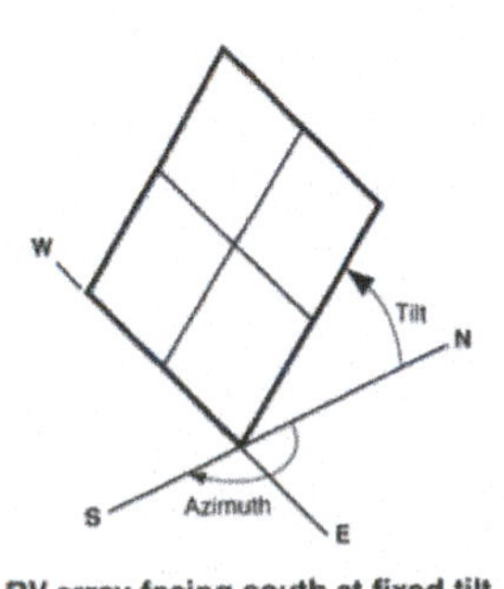

PV array facing south at fixed tilt.

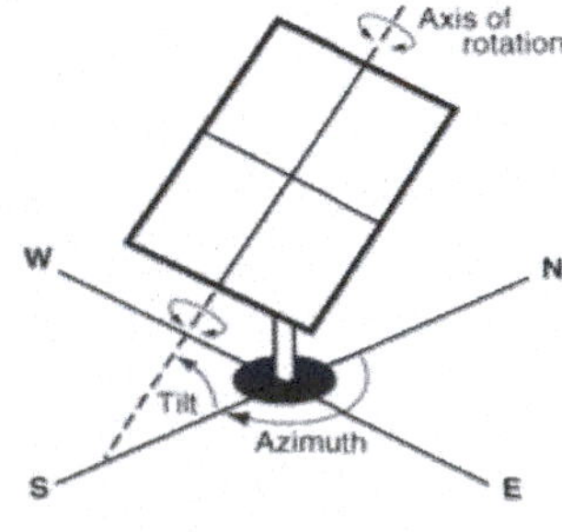

One axis tracking PV array with axis oriented south.

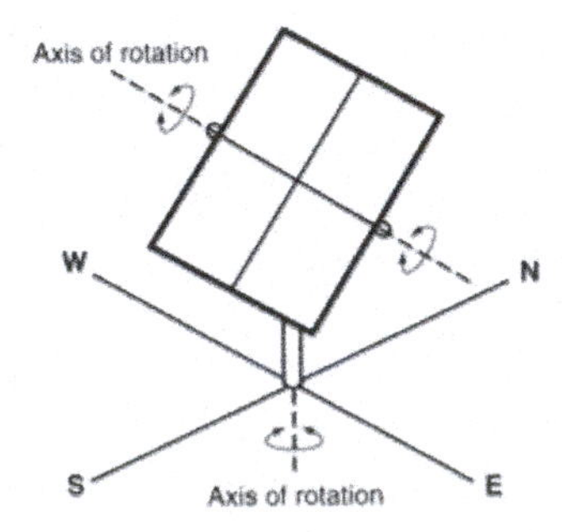

Two-axis tracking PV array

Figure 3-23. Types of PV arrays

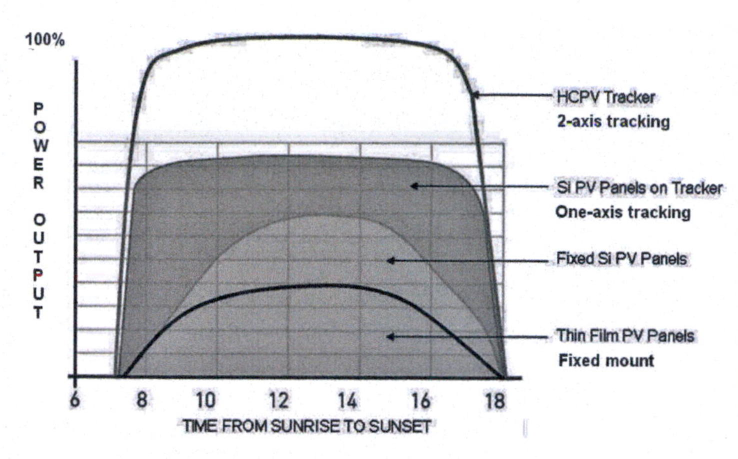

Figure 3-24. Performance of tracking vs. non-tracking PV arrays

Tilt Angle

For a fixed-mounted PV module, the tilt angle is the angle from horizontal of the inclination of the module top surface (0° = horizontal, 90° = vertical). Or, a PV module is at a 0° tilt angle when lying flat on the ground, or mounted parallel to it. For a sun-tracking PV array with one axis of rotation, the tilt angle is the angle from horizontal of the inclination of the tracker axis. The tilt angle is not applicable for sun-tracking PV arrays with two axes of rotation, because they are always oriented at 0° towards the sun, or are always "looking" straight at it.

The default value of a fixed module is a tilt angle equal to the module's latitude. Installers might consider modifying the azimuth angle +/- several degrees according to the particular location or customer's needs. The overall intent is to maximize annual energy production without changing the tilt angle at any time.

Increasing the tilt angle favors energy production in the winter, and decreasing the tilt angle favors energy production in the summer. This is due to the low-in-the-sky position of the sun in winter, and high-in-the-sky position in summer. Of course, a compromise which would increase the generated power is the so-called "seasonal adjustment" of the tilt angle. This consists of changing the angle of all modules in the power plant to match closely the seasonal changes of the sun angle. Ideally, this is done 4 times every year at the beginning of each season at the particular location.

Azimuth Angle

For a fixed-mounted PV array, the azimuth angle is the angle clockwise from true north (zero degree) to the south (180 degrees) that the PV array faces. For a sun-tracking PV array with one axis of rotation, the azimuth angle is the angle clockwise from true north of the axis of rotation. The azimuth angle is not applicable for sun-tracking PV arrays with two axes of rotation.

The default value for a fixed module is an azimuth angle of 180° (south-facing) for locations in the northern hemisphere and 0° (north-facing) for locations in the southern hemisphere. This normally maximizes energy production. For the northern hemisphere, increasing the azimuth angle favors afternoon energy production, and decreasing the azimuth angle favors morning energy production. The opposite is true for the southern hemisphere. Installers might consider modifying the azimuth angle +/- several degrees according to the particular location or customer's needs. Seasonal changes of the azimuth angle are also possible, but not recommended, due to little benefit from a fairly major effort.

Power Rating

The size of a PV array is its DC power rating, or nameplate. This is determined by adding the PV module power listed on the PV modules in watts and then dividing the sum by 1,000 to convert it to kilowatts (kW). PV module power ratings are for standard test conditions (STC) of 1,000 W/m2 solar irradiance and 25°C PV module temperature.

The maximum power generated by a single cell or module is determined by:

$$P_{max} = V_{oc} I_{sc} FF$$

where:

V_{oc} is the open-circuit voltage;
I_{sc} is the short-circuit current; and
FF is the fill factor.

A small commercial type PV system consists of a number of modules, generating 5-500 kW total power. This corresponds to a PV array area (active area) of approximately 30m2 to 3,000m2, but it depends on the efficiency of the modules, and other factors which determine the overall PV system size; i.e., some thin film modules are only 6-8% efficient, so they will need at least 50% more area for installation plus additional mounting frames, wiring etc. Large-scale PV systems consist of thousands of PV modules, depending on the nameplate (maximum power generated at SCT, usually as claimed by the modules manufacturer).

DC-to-AC Derate Factor

The overall DC-to-AC derate factor accounts for losses suffered by the DC nameplate (or the originally estimated and calculated power rating), due to losses from the components in the power plant, while generating, transporting, converting and otherwise transferring the generated DC power into the AC grid. It is basically the mathematical sum of the losses (derate factors) triggered or caused by all components of the PV system. So, we need to multiply the nameplate DC power rating by an overall DC-to-AC derate factor of the system in order to determine its total AC power rating at STC.

Table 3-3. Derating factors for different PV system components (10)

Derate Factors for AC Power Rating at STC

Component Derate Factors	PVWatts Default	Range
PV module nameplate DC rating	0.95	0.80–1.05
Inverter and transformer	0.92	0.88–0.98
Mismatch	0.98	0.97–0.995
Diodes and connections	0.995	0.99–0.997
DC wiring	0.98	0.97–0.99
AC wiring	0.99	0.98–0.993
Soiling	0.95	0.30–0.995
System availability	0.98	0.00–0.995
Shading	1.00	0.00–1.00
Sun-tracking	1.00	0.95–1.00
Age	1.00	0.70–1.00
Overall DC-to-AC derate factor	0.77	0.09999–0.96001

Because STC is just a theoretical value provided by the manufacturer and derived basically under lab test conditions, we need to estimate the power rating during every-day operating conditions; i.e., a spring day in the Arizona desert might produce sunlight close to STC and the modules might even operate under STC conditions for a short while. This, however, is not so during a hot day in July, when the PV modules, inverters and other components are boiling hot and when their efficiency drops significantly.

The overall DC-to-AC derate factor of the system is represented by the sum of all components' derate factors (as used by PVWatts), which is the baseline of the system. In the above case, we have a derate factor of 0.77. So 77% of the available power (generated by the modules) will be converted into DC and sent into the grid, if the module or arrays are operating under STC. This is a 23% loss due to the components' resistance operation, and inefficiency.

As mentioned above, the derate factor would be quite different under extreme conditions—extreme desert heat in particular. This difference must be calculated as well, and important decisions must be made on the basis of the difference between STC and in-sun operation under extreme climates; i.e., the efficiency of a PV array or an inverter rating will drop 20-30% from the STC measurements when operated in the summer desert heat. This drop of efficiency in the field must be taken into consideration when designing, installing and operating a PV system, because the negative effects under these conditions are usually quite significant. Ignoring these changes during the design or evaluation stage will result in serious unpleasant surprises in the field.

Performance Factors

We've looked briefly at some of the key issues and possible degeneration of modules during long-term field operation, always remembering to extrapolate for use in the ever harsh desert environment. Now we will take a closer look at the practical expression of these phenomena, considering worst case scenarios and all possible outcomes:

Temperature Effect

A PV module exposed to sunlight generates heat as well as electricity. For a typical commercial PV module operating at its maximum power point, only 10 to 15% of the incident sunlight is converted into electricity, with much of the remainder being converted into heat.

The key factors which affect the efficiency and thermal behavior of PV modules are:

a. The degree of reflection from the top surface of the module;
b. The electrical operating point of the module components;
c. Absorption of sunlight by the regions which are not covered by solar cells;
d. Absorption of low energy (infrared) light in the module or solar cells;
e. The packing density of the solar cells, and
f. The overall quality of the solar cells and modules.

Each of these factors is seemingly independent in their origin and function, but is inter-related in their long-term effect on the module's efficiency and longevity. We will take a much closer look at the temperature effects below, so it suffices to say that any PV cell or module will lose ~0.5-0.6% of its total output per degree C increase of temperature. This is a significant number, which causes a lot of concern about the auditability of PV modules for long-term operation in extreme heat conditions.

Mechanical and Thermal Effects

There are a number of materials in the PV module that act and interact differently with adjacent materials, depending on temperature, pressure and other factors. The major mechanical forces determining the interaction between the different layers in the module are:

a. Coefficient of friction is the force that resists the lateral motion of solid surfaces in contact with each other and moving in different directions. This includes the types of materials (substrates and layers of materials stacked on top of each other) of which PV modules are made.

b. Coefficient of thermal expansion (linear and volumetric) is the way different materials expand and shrink under different temperature gradients. In other words, all materials tend to change shape and size along the surface and across their volume, with changes in temperature, which ultimately results in mechanical friction if in contact with other materials.

These forces also determine a number of events and phenomena that occur in the modules over time. Hot and cold days, windy conditions, hail, storms, etc. have effect on the modules, and all these events must be taken into consideration in the design and operation of PV power plants. The relations and interactions between the module's components could be described as intimate, ongoing, and relentless. In the Arizona deserts, these materials have to go through blistering 180°F temperatures during the day in summer and sub-freezing temperatures in winter, not to mention the effects of UV light. Also, moisture, air, environmental chemicals, and gasses penetrating the module will cause quick deterioration and even failure.

Since the coefficient of expansion of these materials is very different, they go through never ending shifting, sliding and slipping, expanding and shrinking, constantly rubbing against each other in a never ending dance; a soup of dissimilar molecules. This is where any poor design, or deviation from the materials or manufacturing specs will be revealed, most likely as decreased performance or other failure.

Cracks, chips, pits, crumbling, voids, adhesion failures, delamination, chemical decomposition and reactions within the cells and the modules are expected during the 25-30 years of operation—due to natural elements. Some of these effects, such as yellowing of the EVA, or change of front surface color due to degrading of the AR coating, become visible to the naked eye. Most other internal and some external processes, however, start and continue undetected visually as time goes by.

Moisture and Chemical Ingress into the Module

This is another serious condition which could reduce the modules' efficiency and life. If it occurs soon after the modules are installed, it would very likely be traced to a manufacturing defect, or handling problems during the transport or installation steps. At times, however, moisture in the modules is found months or years after they have been operating in the field. The reasons are usually defects in the laminate layers, caused by defective materials or improper handling or processing. Desert sunlight is especially hard on the organic materials in the modules too, so it is usually only a matter of time before they start breaking down mechanically and chemically, at which point moisture and air could penetrate the inside of the modules.

Moisture ingress is a diffusion process (liquid diffusion that is). Its diffusivity depends on the module type, its frame construction and laminate materials composition and application. Moisture could penetrate into PV modules through the edge seals and from front or back cover defects (cracks, pores and such). Once in the module, moisture could easily penetrate through the encapsulants, causing them to delaminate, discolor, and mechanically disintegrate. In parallel with that, moisture could oxidize and otherwise attack the solar cells' metal contacts, causing them to degrade or fail completely.

The encapsulants and sealants have not changed much during the last 30 years, so the best defense we have in assuring the quality is to make sure that only quality materials are used, and that they are applied according to established manufacturing and QC specs.

Front Surface Reflection

Light reflected from the front surface of the module does not contribute to the electrical power generated. It is considered a loss which needs to be minimized. Reflected light does not contribute to heating the PV module. The maximum temperature rise of the module is therefore calculated as the incident power multiplied by the reflection. For typical PV modules with glass top cover, the reflected light contains about 4% of the incident energy.

Operating Point and Efficiency of the Module

The operating point and efficiency of the solar cell determine the fraction of light absorbed by the cell that is converted into electricity. If the solar cell is operating at short-circuit current or at open-circuit voltage, then it is generating no electricity and hence all the power absorbed by it is converted into heat.

Absorption of Light by the PV Module

The amount of light absorbed by the parts of the module other than the solar cells will also contribute to the heating of the module and does not contribute to generating power. How much light is absorbed and how much is reflected by the non-solar cell areas is determined by the color and materials of the rear backing layer of the module.

Absorption of Infra-red Light

Light which has an energy level below that of the band gap of the solar cells cannot contribute to generating free electrons and electrical power. On the other hand, if it is absorbed by the solar cells or by the module, this light will contribute to heating. The aluminum metallization at the rear of the solar cell tends to absorb this infrared light. In modules without rear aluminum cover, the IR light may pass through the solar cell and exit from the back of the module.

Packing Factor

Solar cells are specifically designed to be efficient absorbers of solar radiation. The cells will generate significant amounts of heat, usually much higher than the module encapsulation and rear backing layer. Therefore, a higher packing factor of solar cells, in addition to increasing the generated power, will also increase the generated (parasitic) heat per unit area. This increase is harmful for long-term operation of modules, especially in areas of extreme heat, so the packing density must always be considered and adjusted accordingly to balance power output and module heating.

Heat Dynamics in PV Modules

Understanding heat behavior in the modules and the interactions within it is paramount for proper PV module design, manufacturing, installation and operation. With mega-fields planned in US deserts, where the temperatures (measured on cells in the modules) exceed 180°F, heat becomes the primary enemy, which we need to understand and overcome, if we are to be successful there.

The operating temperature of a PV module results from an equilibrium between the heat generated by the PV module and the heat loss to the surrounding environment. Heat conduction, heat convection, and heat radiation are the three main mechanisms that we should be fully aware of, since they are major contributors to heat generation and loss in PV modules. Controlling heat loss will help maintain efficiency and extend the lifetime of modules. A more detailed study of these factors follows (2):

Heat Conduction

Conductive heat losses are due to thermal gradients between the PV module and other materials (including the surrounding air) with which the PV module is in contact. The ability of the PV module to transfer heat to its surroundings is characterized by the thermal resistance and configuration of the materials used to encapsulate the solar cells.

Conductive heat flow is analogous to conductive current flow in an electrical circuit. In conductive heat flow, the temperature differential is the driving force behind the conductive flow of heat in a material with a given thermal resistance, while in an electric circuit the voltage differential causes a current flow in a material with a particular electrical resistance. Therefore, the relationship between temperature and heat (i.e., power) is given by an equation similar to that relating voltage and current across a resistor.

Assuming that a material is uniform and in a steady state, the equation between heat transfer and temperature is given by:

$$\Delta T = \Phi P_{heat}$$

where:

- P_{heat} is the heat (power) generated by the PV module
- Φ is the thermal resistance of the emitting surface in °C W^{-1}; and
- ΔT is the temperature difference between the two materials in °C.

The thermal resistance of the module depends on the thickness of the material and its thermal resistivity (or conductivity). Thermal resistance is similar to electrical resistance and the equation for thermal resistance is:

$$\Phi = \frac{1 l}{kA}$$

where:

- A is the area of the surface conducting heat;
- l is the length of the material through which heat must travel; and
- k is the thermal conductivity in units of W m^{-1}°C^{-1}.

To find the thermal resistance of a more complicated structure, the individual thermal resistances may be added in series or in parallel. For example, since both the front and the rear surfaces conduct heat from the module to the ambient, these two mechanisms operate in parallel with one another and the thermal resistance of the front and rear accumulate in parallel. Alternatively, in a module, the thermal resistance of the encapsulant and that of the front glass would add in series.

Convection

Convective heat transfer arises from the transport of heat away from a surface as the result of one material moving across the surface of another. In PV modules, convective heat transfer is due to wind blowing across the surface of the module. The heat which is transferred by this process is given by the equation:

$$P_{heat} = hA\Delta T$$

where:

A is the area of contact between the two materials;

h is the convection heat transfer co-efficient in units of W m^{-2} $°C^{-1}$; and

ΔT is the temperature difference between the two materials in °C.

Unlike the thermal resistance, h is complicated to calculate directly and is often an experimentally determined parameter for a particular system and conditions.

Radiation

A final way in which the PV module may transfer heat to the surrounding environment is through radiation. As discussed before, any object will emit radiation based on its temperature. The power density emitted by a blackbody (a perfect radiation absorber) is given by the equation:

$$P = \sigma T^4$$

where:

P is the power generated as heat by the PV module;

σ is the Stafan-Boltzmann constant, and

T is the temperature of the solar cell in K.

However, a PV module is not an ideal blackbody and to account for non-ideal blackbodies, the blackbody equation is modified by including a parameter called the emissivity, ε, of the material or object. A blackbody, which is a perfect emitter (and absorber) of energy has an emissivity of 1.

An emissivity of an object can often be gauged by its absorption properties, as the two will often be very similar. For example metals, which tend to have reduced absorption, also have a lower emissivity, usually in the range of 0.03. Including the emissivity in the equation for emitted power density from a surface gives:

$$P = \varepsilon\sigma T^4$$

where:

ε is the emissivity of the surface; and the remainder of the parameters are as above.

The net heat or power lost from the module due to radiation is the difference between the heat emitted from the surroundings to the module and the heat emitted from the PV module to the surroundings, or in mathematical format:

$$P = \varepsilon\sigma\left(T_{sc}^4 - T_{amb}^4\right)$$

where:

T_{sc} is the temperature of the solar cell;

T_{amb} is the temperature of the ambient surrounding the solar cell; and the remainder of the parameters are as above.

NOCT (Normal Operating Cell Temperature)

Since we are still on the very important subject of temperature effects on PV module performance and longevity, we need to discuss the NOCT. A PV module will typically be rated at 25°C under 1 kW/m^2. However, when operating in the field, they typically operate at higher temperatures and at somewhat lower insolation conditions. In order to determine the power output, it is important to determine the expected operating temperature of the PV module.

The nominal operating cell temperature (NOCT) is defined as the temperature reached by open circuited cells in a module under the conditions listed below:

- Irradiance on cell surface = 800 W/m^2
- Air temperature = 20°C
- Wind velocity = 1 m/s
- Mounting = open back side.

The equations for solar radiation and temperature difference between the module and air show that both conduction and convective losses are linear with incident solar insolation for a given wind speed, provided that the thermal resistance and heat transfer coefficient do not vary strongly with temperature. The NOCTs for best case, worst case and average PV modules are shown below. The best case includes aluminum fins at the rear of the module for cooling which reduces the thermal resistance and increases the surface area for convection. (See Figure 3-25.)

The best module operated at a NOCT of 33°C, the worst at 58°C and the typical module at 48°C. An approximate expression for calculating the cell temperature is given by:

$$T_{Cell} = T_{Air} + \frac{NOCT - 20}{80} S$$

where:

S = insolation in mW/cm2.

Module temperature will be lower than this when wind velocity is high, but higher under still conditions.

Impact of Module Design on NOCT

Module design, including module materials and packing density, have a major impact on the NOCT. For example, a rear surface with a lower packing density and reduced thermal resistance may make a temperature difference of 5°C or more.

Impact of Mounting Conditions

Both conductive and convective heat transfer are significantly affected by the mounting conditions of the PV module. A rear surface which cannot exchange heat with the ambient (i.e., a covered rear surface such as that directly mounted on a roof with no air gap), will effectively have an infinite rear thermal resistance. Similarly, convection in these conditions is limited to the convection from the front of the module. Roof integrated mounting thus causes higher operating temperature, often increasing the temperature of the modules by 10°C.

Thermal Expansion and Thermal Stress

Thermal expansion is another important temperature effect which must be taken into account when modules are designed. The spacing between cells tries to increase by an amount δ given by:

$$\delta = (\alpha_G C - \alpha_c D)\Delta T$$

where:

$\alpha_G \alpha_C$ are the expansion coefficients of the glass and the cell respectively;

D is the cell width; and

C is the cell center-to-center distance.

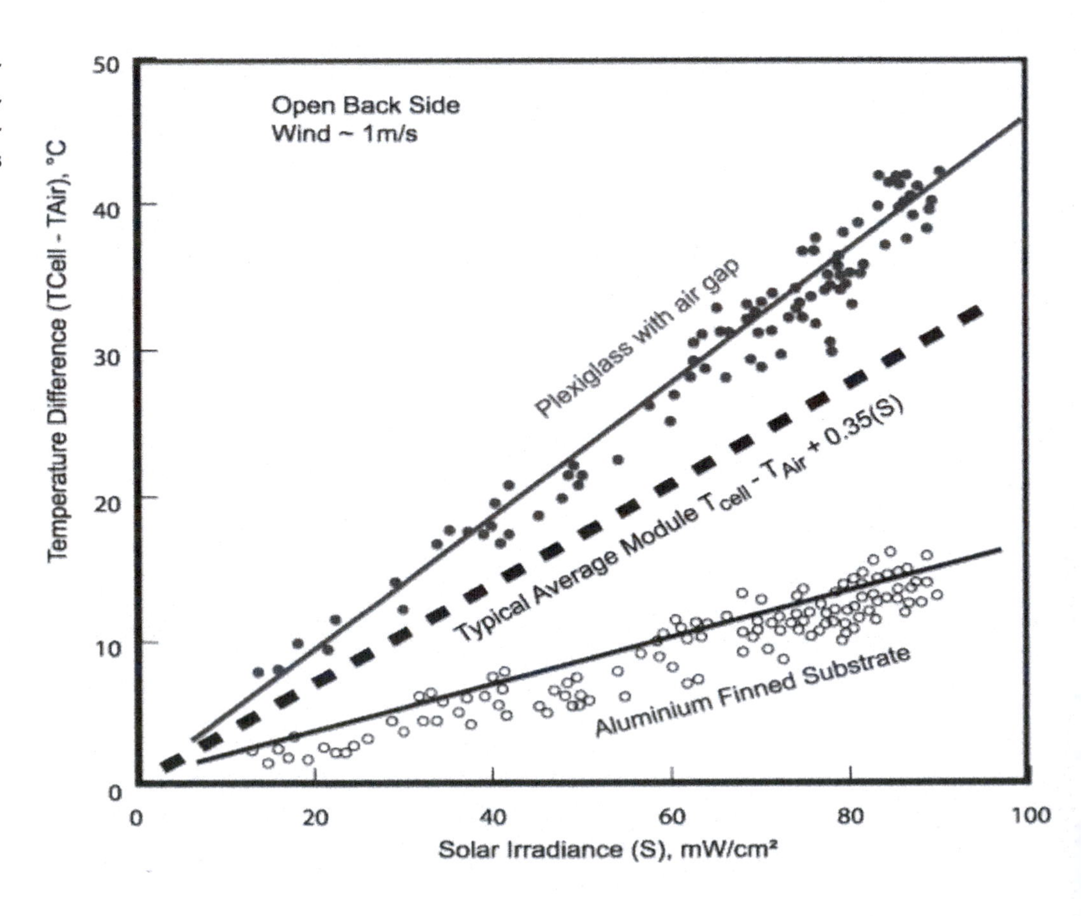

Figure 3-25. Temperature increases above ambient levels, with increasing solar irradiance for different module types

Typically, interconnections between cells are looped to minimize cyclic stress. Double interconnects are used to protect against the probability of fatigue failure caused by such stress. In addition to interconnect stresses, all module interfaces are subject to temperature-related cyclic stress which may eventually lead to delamination.

Electrical and Mechanical Insulation

The encapsulation system has to be able to withstand voltage differences at least as large as the system voltages. Metal frames must also be grounded, as internal and terminal potentials can be well above the earth potential. Any leakage currents to earth must be very low to prevent interference with earth leakage safety devices.

Solar modules must have adequate strength and rigidity to allow normal handling before and during installation. If glass is used for the top surface, it must be tempered, since the central areas of the module become hotter than areas near the frame. This places tension at the edges, and can cause cracking. In an array, the modules must be able to accommodate some degree of twisting in the mounting structure, as well as to withstand wind-induced vibrations and the loads imposed by high winds, snow and ice. Additionally, some installers may habitually, or be forced to walk on the modules during installation. This practice induces internal stress to the modules' frame and the cells proper, which might have detrimental results, so it has to be avoided altogether.

Miscellaneous Issues

There are a number of additional issues that we must be aware of and keep in mind when talking about PV modules. Some of these are:

PV Modules Handling, Packing and Transport Issues.

During their assembly, PV modules go through many processing stations where they are exposed to numerous manual transfers, loading, unloading and other manual operations. During any of these steps the modules could be dropped, hit, scraped, or suffer physical contact which could crack, stress or even break them.

The modules are sold as independent, self-contained "packages." That is, the solar cells are enclosed (encapsulated) in a sandwich of materials, selected, manufactured and installed in the module in such a way as to provide maximum power and protection of the solar cells from impact and from contact with the elements. Any impact, pressure, or similarly careless contact with the modules could be detrimental to the performance and lifetime of the solar cells in them. The resulting damage might be noticeable immediately, but most likely it will result in latent (delayed) loss of performance or failure.

So, if a module is badly damaged—broken glass, bent rear cover or side frame, etc.—it would be detected and separated prior to shipping, or in the worst case before installation. Some of the stressed and cracked modules will not be detected, however, and might even test OK at first, only to fail a short while later. A manufacturer's warranty usually covers modules that fail during the warranty period, but there are cases when the damage does not manifest itself until months or years later. In many such cases, the customer has no recourse and will end up "eating" the damaged product, unless there is special insurance or warranty to cover these exceptions.

PV "packages" are promoted and sold as "hermetic," or nearly hermetic vessels, which means that no water will penetrate through the materials at any time to attack and damage the solar cells. Looking from the scientific point of view most materials—including all materials PV modules are made of (glass and encapsulating materials and such) are permeable to a certain degree. That means that given time and the right conditions, air and moisture will penetrate through these materials and, yes, this includes glass. So the packaged modules should be kept away from water and certainly submersion into water must be avoided at all costs. Any signs of excess moisture must be reported for warranty purposes.

In all cases, each individual module must be carefully unpacked, visually inspected and electrically tested prior to installation. Any and all modules with unusual appearances or performance abnormalities must be separated and held for inspection by the manufacturer's rep. Any unusual drop in power or other abnormal behavior of the modules or strings of modules after installation must be documented and investigated for warranty purposes as well.

Electrical Components

PV modules have a number of components, in addition to the obvious ones. which have different functions and requirements, and which need to be taken into consideration as well, prior to installation. Some of these components are wiring, external bypass diodes, junction box, and terminal. These components are integral parts of the modules' structure and their overall performance, efficiency and longevity.

Any modules with broken or bent wires or parts must be separated for investigation. Abnormalities such as the wrong type and size wiring and poorly designed and assembled junction boxes will cause poor performance or failure if not detected and corrected. Special attention must be paid to, and thorough inspection performed on, each module—especially if these are coming from new manufacturers.

Materials Supply

Although wind and solar technologies are called "renewable," some of the materials their components are made of are far from renewable. On the contrary, some of them are expensive, exotic and rare materials, some of which are in short supply—and some are even on the "endangered species" list.

On top of that, a major part of these exotic materials are found and mined in third-world countries, where quality is often far from being a #1 priority. In all cases, we must know where the materials come from and assess the associated risks, which we've shown elsewhere in this text.

Safety

Generally speaking, all PV modules, including c-Si modules, contain some amount of toxic materials, and must be handled properly. Small amounts of Pb or Sn, for example, could be found inside the modules and sometimes even on their outside surfaces. Some thin film PV modules are especially dangerous, because they contain toxic and carcinogenic compounds of elements like cadmium, tellurium, indium, arsenic, etc. Some trace amounts of these poisons might have remained on the modules' surfaces, which upon contact could trigger allergic reactions in some people.

In all cases, due to the above toxicology problems and to the fact that PV modules have sharp edges, they must be handled with gloves and with utmost caution. Any and all incidents, pains or aches must be immediately reported and investigated.

Key Module Issues, Detailed Technical Discussion

Keeping in mind all possible behavior conditions and problems we discussed above, we will now take a look at the possible problems at each process step of PV cells and modules manufacturing. These issues can be traced to materials, equipment, processes and labor-related problems, and are usually demonstrated during testing or long-term exposure to the elements.

Now we will examine possible defects and failures which can occur during the final module test or, more importantly, during long-term exposure to harsh climatic conditions.

So, the key issues and all possible effects caused by long-term on-sun exposure, following the PV module elements shown in Figure 3-26, are detailed as follows:

ONE—Glass Cover (Top Cover)

The cover glass is basically designed to provide protection to the fragile solar cells and module components from mechanical damage, such as vibration, impact, etc., as well as to prevent the elements (rain, moisture, dust) from entering the module. If the glass itself is stressed, cracked and broken during processing, handling, transport, installation and operation stages, then the insides of the module might already be damaged as well. Subsequent damage could occur in such cases, by means of mechanical stress and chemical destruction of the cells and other active components.

This is even more important for thin film PV modules because the active layers are deposited directly on the front or back covers, thus any stress or breakage of the glass will directly, and immediately, affect the performance and longevity of the damaged modules. The advantage of c-Si PV modules in this area is that the EVA layer is enveloping the active components (solar cells), so that even if the top or bottom covers are damaged, the EVA envelope is given a chance to protect the cells from an invasion of the elements and chemical attack.

Modules with damaged cover glass—be it cracked or with hazy appearance—should not be installed and must be removed from service immediately. Repairing damaged glass is not an option in most cases, so the entire module must be properly disposed of, or sent back to the manufacturer for credit or for replacement. Periodic testing of the voltage/current output of modules with partially cracked cover glass, if acceptable at all, should be part of the PM schedule.

NOTE: In all cases damaged modules should be handled with utmost care.

Cover glass surface soiling is another serious issue during normal operation. Rain and dust will deposit

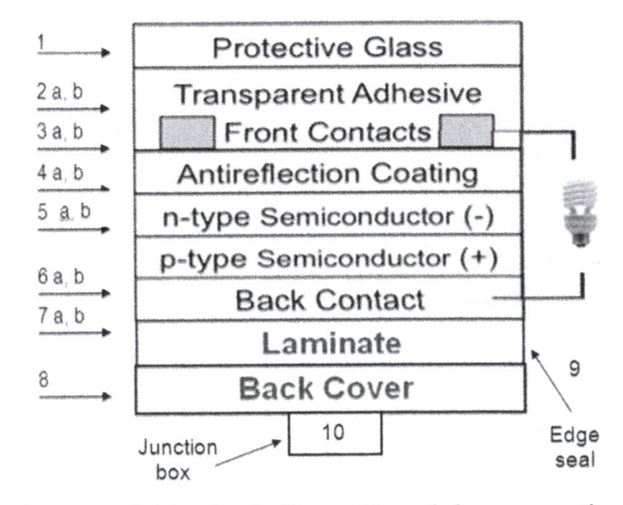

Figure 3-26. Standard silicon PV module cross section

layers of contaminants and water spots on the surface, which will prevent sunlight from reaching the solar cells underneath. With time the layers might grow so thick that very little light goes through them, significantly reducing the output of the modules. Washing the top surface with a soapy water solution, followed by proper rinsing with DI water will restore the modules' efficiency, so it should be part of the PM schedule and carefully executed. This, however, is an expensive undertaking, especially in the deserts. Periodic use of water and chemicals also causes concerns with water table contamination, which has to be kept in mind when deciding on a PV power field location and O&M procedures.

TWO—EVA Encapsulant Deterioration and Delamination

The layer of plastic encapsulation (EVA usually) and the cover glass are in intimate contact and partially bonded. These two, however, have different elasticity, and coefficients of expansion and friction. With long-time exposure to the elements (heating, freezing, excess UV radiation, mechanical stress, fatigue) the EVA plastic will eventually change. Its physical and chemical properties change slowly and we can see it turning a yellowish color. Yellowing of the EVA results in reduction of its transmissivity and causes a decrease of power output.

Yellowing is one of the few changes occurring in the modules that can be observed with the naked eye and which gives us a clear indication of the changes within the module. As the changes continue, air bubbles, cracks, pits and cavities could form in the EVA material, which could cause it to separate from the top glass. This contributes further to increasing optical losses due to reflection and poor transmission, ultimately resulting in serious performance deterioration. Once the EVA material damage and the accompanying delamination processes have started they cannot be stopped. At this point the damaged modules are suspect and must be put under periodic observation as part of the O&M schedule.

EVA Delamination From The AR Layer

Some of the above-mentioned damage and defects in the EVA material will affect also the adhesion of the EVA material to the AR coating to which it is bonded. EVA delamination from the glass and AR coating surface is caused by the never ending UV bombardment, expansion and contraction of the layers and the resulting friction between them. This action might result in an air gap which will cause another (second) optical barrier which will further decrease the power output from the affected cells.

The newly created gaps and cracks might allow moisture, air and reactive gasses to enter in the module which might attack the other modules components.

Changes at the EVA-AR film interphase (air gap, color change, cracking, etc.) are another of the few visually observable changes in PV modules. Depending on the type of cells (mono or poly crystalline) and type of AR coating (many different formulations are available), the visual effects might change the cell's surface color from black to faded black, and from dark blue to light blue (respectively), as well as a number of shades in between.

THREE—Top Contacts and Interfaces

The top contacts (fingers) consist of silver metal with metal ribbons soldered on top of the contacts to provide connection to the other cells and to the outside circuit. This creates several separate contact points (interfaces) as follow:

a. Silicon to silver metal of the top contact,
b. Silver metal to the soldered metal strip, and
c. Soldered metal strip to the EVA encapsulant.

The different interfaces provide good adhesion to their components and are quite stable under normal operating conditions. In abnormal situations, however, such as excess heat, overcurrent conditions, mechanical or thermal stress, and/or exposure to the elements (water vapor, chemicals and reactive gasses), the metals undergo serious changes which are usually initiated in the interfaces. The damage could increase with time and cause significant changes in the contacts' structure. In all cases these events will result in electrical changes or failures (high resistivity or open circuit). Interrupting the contact with the adjacent cells will lead to a failure of the module.

In some cases, one or more of the affected contact layers might burn, chemically disintegrate, or undergo other destructive processes, which could cause the device to fail partially, or completely, in which case the entire solar module might stop generating power. In all cases, there will be noticeable reduction in power from the module, due to mismatch caused by the affected solar cell(s).

Top Contact Adhesion into the Silicon Substrate

Under normal operating conditions, the adhesion between the front contacts (silver metal) and the silicon material which they are fused into is good enough to keep them together and properly conducting electricity for the life of the module. Abnormally, however, mechanical stress, excess heat, and moisture and air penetrating into the module, could affect the front contacts. Mechanical forces could induce stress in the materials and their interfaces, while ingress of water and chemicals might corrode and separate them from the substrate.

These processes would negatively affect the bond at the interface between these key components (silicon substrate and silver contact). In such cases:

a. The diffusion area might be affected and undergo internal changes which might affect the cell's performance, and/or
b. The resistivity at the affected area might increase, which could result in overheating, and eventually a breakdown (open circuit), caused by partial or complete separation of the contact from the substrate.

These phenomena could bring partial or total failure of the solar cell and/or the entire module.

FOUR—AR Coating Surface Damage

A number of distractive mechanisms on the cell's surface, caused by mechanical stress and chemical attacks could affect the AR layer. These changes could occur under different operating conditions and cause the AR coating top surface to change mechanically or chemically. This could cause a change in the AR layer's optical properties, decreasing the cell's efficiency.

In addition, blistering of large areas of the AR coating might also stress the cell and the top contacts, causing further increase of resistance and heat generation. Overheating could damage the contacts and/or their interface. The warped AR surface might also reflect some of the incoming sunlight at a greater rate than designed. In all cases, the final result is reduction in affected cells' efficiency.

These effects (AR layer delamination, blistering etc.) also result in color change of the AR coating which represents another of the few visually observable changes in the solar cell and module. Depending on the type of cells (mono or poly crystalline) and type of AR coating, the visual effects might change the cell's surface appearance to different variations of the original colors.

AR Coating Adhesion to the Substrate

The antireflection coating (AR) is a fairly thin film of inorganic material, i.e., TiO_2, Si_3N_4 and such, which is very thin, semi-transparent and fragile under certain conditions. Very thin and lightly bonded to the cell surface, it is easily damaged by mechanical and chemical means. There are several mechanisms that can contribute to changes of the AR layer and its adhesion to the substrate. One is contamination of the silicon substrate surface prior to AR coating, which would result in poor adhesion over time. The surface contamination could be caused by insufficient cleaning and rinsing, improper handling, or an out-of-spec process during the AR coating deposition. All of these inadequacies would compromise the adhesion integrity of the thin AR film to the silicon substrate.

If the AR layer is not fully adhered to the silicon substrate, there would be physical and optical gaps which would cause excess reflection and/or obstruct sunlight transmission into the cell. Combinations of these conditions will contribute greatly to reduction of the cell's efficiency.

Further, and more seriously, changes in the AR film properties could occur quicker if the damaged area is exposed to the elements via moisture and air leaking into the module (see above). The adhesion forces in the interface between the AR film and the substrate in such cases would weaken even more under the outside attacks, and the AR film might disintegrate or separate from the surface, amplifying the negative effects.

Most changes in the AR coating's adhesion and other changes of its properties are also visible on the module's surface, but equally impossible to correct. When these changes occur, future cell/module behavior is impossible to predict without destructive tests, so the module must be checked periodically as part of the O&M schedule.

NOTE: All visually observable changes mentioned above start, and might even continue, as microscopic imperfections which eventually grow bigger and more visible. A trained eye is needed to observe the changes in many cases.

FIVE—P-N Junction and Diffusion Process Issues

The p-n junction is the most critical area of solar cells' function. It has the greatest impact on cell/module efficiency, performance and longevity. The p-n junction is formed by diffusing (injecting) a specified concentration of foreign atoms in the silicon bulk, which then initiate and drive the photoelectric effect. Unacceptable amounts of impurities in the bulk silicon, such as Fe, Cu, K, Na and any other contaminants in the process chemicals and gasses will affect the diffusion concentration and depth, ultimately changing the behavior of the p-n junction. Changes in the concentration of the foreign atoms in the junction, due to parasitic effects during its field operation would produce similar results.

If the wafer surface was not properly prepared, cleaned and rinsed, or if the diffusion process was not properly executed (time, gas concentration and temperature), the p-n junction might not be stable enough. Thus affected solar cells might operate well in the beginning, but eventually the diffusion layer will start changing (decreased concentration and diffusing further into the substrate) causing a drop in output. These effects sometimes occur months or years after the module has been in operation, and are called "latent" effects. They are most dangerous for the large-scale power plant's long-term success, because if they happen after the warranty period has expired and if the failures are great enough they might cause serious technical and financial difficulties.

The most common negative field effect of the p-n junction is change in the diffusion depth (and the resulting dopant concentration reduction at the junction) over time. While this phenomenon is well understood and the p-n formation parameters are controlled during the manufacturing sequence, certain material and process conditions force greater changes, over time, than allowed by design. In such cases, the cells start losing power quicker than expected and eventually fail altogether.

Diffusion processes techniques vary from manufacturer to manufacturer. Some use the old-fashioned, but most reliable, diffusion furnace process, where the wafers are placed into a glass tube and heated to over 1000°C in a controlled environment. Doping gases are passed through the tube and by precise control of temperature, process time and gas volume, precise deposition concentration and depth are obtained. A high-volume, but less precise, diffusion process has been widely used lately. It consists of spraying or printing the dopant liquid on the wafer surface and then baking it in a conveyor belt type of furnace. This process has more uncontrollable variables and is basically inferior from process control and product quality points of view than the old diffusion furnace method. Nevertheless, it is cheaper and faster, thus we will see it more and more in the future. And due to its less-than-precise and not-so-easy-to-control process parameters, we expect to see more diffusion and p-n junction changes issues in the field as well.

Diffusion layer depth and concentration changes cannot be visually detected, nor can they be measured under field conditions, so at this point we are looking at a black box and relying on the manufacturer's experience and his properly and efficiently executed quality control procedures.

P-N Junction Field Performance

So, what if the diffusion process were less than perfect? First, and most important, if the materials and process specs were even slightly out of control, the solar cells and modules would show low efficiently during the final tests. After sorting, the less efficient would go into the group of "cheaper" cells and modules to be shipped to a customer who hopes they will work well. But if the diffusion process were somewhat out of spec, thus the reason for the lower quality of some cells, then there would be a good chance that the affected cells and modules might deteriorate much quicker than the rest.

Exposure to extreme temperature is especially testy, and is the reason for many malfunctions and failures encountered in c-Si modules. It is possible for the species in the diffusion layer to start migrating into the bulk under abnormal conditions brought upon by extreme temperature regimes and/or any imperfections in the silicon material close to the p-n junction area. These will accelerate the changes in the electrical characteristics of the cells, thus causing additional efficiency decrease.

In summary, as soon as the newly processed solar cells are flash-tested they are ready for encapsulation into a module. Provided that the materials are of high quality and all process steps have been properly executed, the p-n junction and the cells will operate properly for a long time. But if we had materials quality problems or the diffusion process were not properly executed, then we might be looking for a surprise in our power field, over time.

Note: Here we need to mention the great effect of silicon bulk material on the efficiency and other properties of the solar cells that were made from it. We now know that the quality of the solar wafers determines the quality of the solar cells made from them. The silicon wafers are made from 99.9999% pure solar grade silicon material. Even at that purity, however, a slight increase in one of the harmful contaminants (usually metals) might have serious effects on the solar cells' field performance and longevity. Over time, and when operating at extreme temperatures, parasitic metals could start diffusing through the cell and alter p-n junction properties, with the diffusion speed increasing.

SIX—Rear Contact Damage

The rear contact also consists of several interfaces:

a. Silicon bulk to aluminum BSF
b. Silicon to silver back contact, and
c. Silver to soldered metal band

This metal structure and its interfaces are exposed to the same electro-mechanical, thermal and chemical attacks and changes as the front contacts we reviewed above. Although these are somewhat more protected from and less affected by the UV and IR radiation than the front contacts, their long-term quality is as critical. Any elemental impurities, chemical contaminants, mechanical damage, such as cracks, pits and other imperfections in the structure will have profound effect on the cell's efficiency and longevity.

The quality of the metal deposition is of great importance here too, because if it is defective in any way (high resistivity, cracks and pores, poor adhesion, oxidation at the interface, etc.), it will cause overheating, delamination and ultimately reduced efficiency of the cell, and even total failure. Nevertheless, the rear contacts are much less problematic and affect the cells' function less than the front contacts, but failure of the encapsulants and edge sealers could damage them seriously enough to cause performance degradation and failures.

Rear Contact Adhesion

In case of poor contact quality due to process control inadequacies (time, temperature, or metal quality) the metal film can partially separate from the substrate, eventually leading to decreased efficiency due to higher resistivity and overheating.

Excess overheating at the back surface might contribute to further erosion and delamination of the metal film and total failure of the cell and module.

SEVEN—Laminate Issues

Laminate materials (EVA, PET, PVB, Tedlar, etc.) are organic (plastic) materials, which have been in use and remain basically unchanged in composition and use, for the last 30+ years. Also unchanged is their vulnerability to exposure to the elements, where EVA and other organic compounds in the module do undergo mechanical and chemical changes and degradation over time.

Continuous expansion during hot days and shrinking during freezing nights, extreme IR and UV bombardment, and chemicals' ingress will affect the EVA and other organic (plastic) materials. This will eventually result in mechanical changes—discoloration, mechanical stress and disintegration of the lamination materials—creating cracks, bubbles, pits, voids, etc.

Ingress of moisture and gasses into the module via the above formed cracks and voids will contribute further to the degradation process by decomposing the module materials (solar cells, wiring, contacts etc.).

Rear Laminate

The laminate structure on the back of the module usually consists of the EVA envelope with thin Tedlar and/or other plastic materials backing, laid onto the rear metal cover. The rear EVA-on-Tedlar structure is protected from the elements (UV and IR radiation) so it lasts much longer without damage. Its yellowing and even delamination from the back cover don't have such dramatic effects on the module performance.

The adhesion of the EVA envelope to the back of the cells, however, can be affected with time. Excess heating, freezing, and moisture penetrating from the sides might eventually debilitate the adhesion to the solar cells. In that case, moisture and air penetrating the laminate material and reaching the solar cells will cause rapid oxidation of the back surface metallization (aluminum and silver metals), which will also affect the interface between these metals and the back surface of the solar cell. This deterioration of the cells' adhesion at interface might cause increased resistivity and overheating, and eventually delamination of the metal contacts from the substrate, which will cause performance issues and eventually failure of the affected cells and modules.

EIGHT—Rear Cover and Frame

The purpose of the frame and back cover (aluminum sheet or glass plate) is to protect the module's insides from attack by the elements. It is highly unlikely for the frame and back cover to degrade significantly with time under normal use, but a severe mechanical or chemical attack could compromise its integrity, forming cracks and voids which might allow moisture and air to penetrate the module. At that point, any of the above-described events might cause the cells and module to decrease in efficiency or fail.

NINE—Edge Seal

c-Si PV modules usually have side protection, as an extension of their edge protection. Ingress of harmful elements into the module is usually initiated through the sides (the edge seal) and could be slowed by a well-installed and sealed metal frame around the edges. A lot of effort is dedicated to finding better sealing materials, and assembly processes have improved significantly lately as a result. Edge sealing—its effects and weaknesses—are well understood. No matter how good the sealing materials are, however, they are made of plastic (organic) materials, which are affected by the elements—IR and

UV radiation and environmental chemicals. Thirty years under the blistering Arizona desert sun would challenge any organic material or compound.

Edge seals are prone to accelerated changes with time which usually lead to the formation of voids, cracks, pores, and bubbles. These imperfections eventually allow moisture and air to penetrate the modules and cause damage as described time after time above. Therefore, new types of frames and edge seal protection should be developed to provide additional protection.

NOTE: Edge seal quality and protection issues are even more important in case of some thin film PV modules, which are frameless. The modules consist of two glass plates with no sides, so that the edge seal is exposed to the elements. This makes the thin films inside the modules more vulnerable to the elements, which could be detrimental to their performance and longevity.

TEN—Junction Box

The junction box is just that—a small box where the wiring "junction" or connection is made. It is a metal container intended for easy, safe, and reliable electrical connections. It is also intended to conceal them from the elements and prevent tampering. The box is attached to the back cover of the module by means of screws and/or glue and contains connectors for wires coming from the module and those connecting it to the external circuitry.

Corrosion of the contacts in the boxes is the most frequent problem encountered during long-term operation in harsh climates. This might cause increased resistance, reducing output power and eventually resulting in fire or an open circuit. This problem, however, is the only one that can be fixed by replacing the defective parts without tearing the module apart.

Summary

As a summary of issues affecting solar cells and modules performance and longevity, if we represent the actual efficiency and longevity of a solar module using the above described conditions and issues, we get:

$$\eta a = \eta t - (1 + 2a + 2b + 3a + 3b + 4a + 4b + 5a + 5b + 6a + 6b + 7a + 7b + 8 + 9 + 10)$$

$$La = Lt - (1 + 2a + 2b + 3a + 3b + 4a + 4b + 5a + 5b + 6a + 6b + 7a + 7b + 8 + 9 + 10)$$

Where:

ηa is the actual field efficiency of the module
ηt is the theoretical (optimum) efficiency of the module, and
La is the actual longevity of the module
Lt is the theoretical (optimum) longevity of the module

Numbers 1 through 10 are the sequence of conditions and issues discussed above.

It is impossible to predict, let alone put a numerical value on most of the conditions and issues in 1 through 10 above, so these formulas are good only to show roughly the qualitative dependence of the efficiency and longevity of the modules to the members of this long chain of events and how they could affect (usually in a negative way) the cells and modules during their long-term on-sun operation.

As we can clearly see, any discrepancy in the chain of ten events can only reduce the efficiency and/or longevity of the cells and modules, according to the seriousness of the deviation. Or as we mentioned, "The highest quality of the finished device is determined by the lowest quality of any process step," or event in this case.

Applying the related manufacturing process steps, sub-steps and procedures to the each of the members (1-10) of the above formula will result in a long string of variables (literally hundreds of them). Each of these additional sub-routines and variables would have an equally negative effect on the modules' performance and longevity, if not properly executed. This is why quality of materials, quality of design, and process control, combined with know-how and experience, are so important in ensuring acceptable quality of the final product. This is not impossible, but we fear that very few people are fully aware of the complexity of these matters. Our hope is that now, with all the issues on the table, we will be able to start an open discussion about their resolution. Standardization of materials quality and process controls is the ultimate solution to the issues at hand, so until that occurs we must use great care when designing and evaluating solar cells and modules for large-scale power generation, especially in harsh climates.

PV MODULES CERTIFICATION AND FIELD TESTS

Testing PV modules before installation in the field for long-term operation is a necessary step, but the results are not always indicative of actual lifetime field performance. Many modules qualified and certified as required, and operating for several years in the field, have been found malfunctioning after the abuse of Mother Nature.

Once the solar modules are manufactured, one of the best batches goes to US or European labs for testing and final certification. This is a very important test, which determines the future of this set of modules, and possibly that of the company that manufactures them.

Thus, selected modules undergo standard test procedures. Some pass the testing program, but start deteriorating prematurely and at times even fail completely. Sometimes they fail at the pre-test and pre-certification stages—even before the testing procedure has started.

Pre-test Failures

So let's take a look at an unusual trend of failures, which were discovered even before the actual testing was began. These failures tell us a lot about the manufacturer and don't give us too much confidence in their products. See for yourself:

According to Intertek, a certified test facility in California, a number of test modules fail this pre-test screening. Here are five typical reasons why PV modules coming for testing at their facility fail initial inspection—even before the testing can be started. The reasons are shocking, for a supposedly hi-tech business, such as PV cells and module manufacturing operations (7):

1. Inappropriate/incomplete installation instructions
2. Models provided for testing do not accurately represent the entire production model scheme being listed (largest module must be submitted for test)
3. Testing requested without prior construction evaluation being performed
4. Complete bill of materials with ratings and certification information not provided prior to start of the project
5. Lack of back-sheet panel RTI rating

According to the author's extensive experience with engineering and quality control operations in the world's semiconductor and solar industries, this phenomena is quite unusual from a quality control point of view. It also points to weaknesses in the overall management systems of the manufacturers in question.

Keep in mind also that the above listed failures are among the very few that can be readily recognized—before the modules are packed, shipped and installed in the field. Then how are we going to catch the possible failures of the thousands of modules we need for our 100MWp power plant, during and after installation in the field?

This is a question that we must answer in the daily quest for PV products needed for large-scale installations. How do we make sure that the modules we are ordering don't fall in this category? The quick answer is:

1. Take your time to thoroughly check all engineering and quality control procedures used during the processing of your batch of PV modules. The manufacturer must provide some of this information. If not, then go to someone who will.
2. Find out as much as you can about the "sand-to-module" history of the product, before placing a large order.
3. Even better, have a team of qualified engineers visit the manufacturer and inspect the process line your modules go through. This is a difficult task, but short of that, you are working with a black box that may or may not work properly. The choice is yours.

Performance Tests and Failure Rates

The PV modules that pass the pre-test inspection at the certification lab are then put through a real test and undergo procedures such as damp heat, UV exposure, heat and freeze, mechanical strength and other tests that require 3-6 months or more for completion. To make sure the modules will pass the certification tests and to save time and money, manufacturers run similar tests at their facilities before sending the modules for certification; i.e., current standards require approximately 1000 hours of exposure to damp heat, but some manufacturers extend this procedure 2-3 times in their in-house tests, to ensure that the modules perform adequately.

Long test procedures result in the delayed introduction of materials to the PV market; however, so many companies, in an attempt to "accelerate" the aging process, utilize alternative test procedures. For example, HAST (highly accelerated stress test) has become popular among PV modules and components manufacturers. The HAST procedure provides some useful, albeit superficial test results. It needs to be standardized and correlated with damp heat and outdoor performance, before it can be relied upon for providing meaningful and acceptable test results.

So, modules that pass manufacturers' in-house tests are sent to a qualified test and certification lab (usually in the US or EU) for final testing and certification. At the lab they are put through a series of tests to determine their efficiency, overall performance, and longevity.

During 1997-2009, TUV Rheinland ptl in Tempe, AZ, recorded the results from a series of such tests with several thousand PV modules from a number of different major PV module manufacturers. These results were published in *Photovoltaics International*, May 2010, clearly outlining the test failures and related issues, as follows (5):

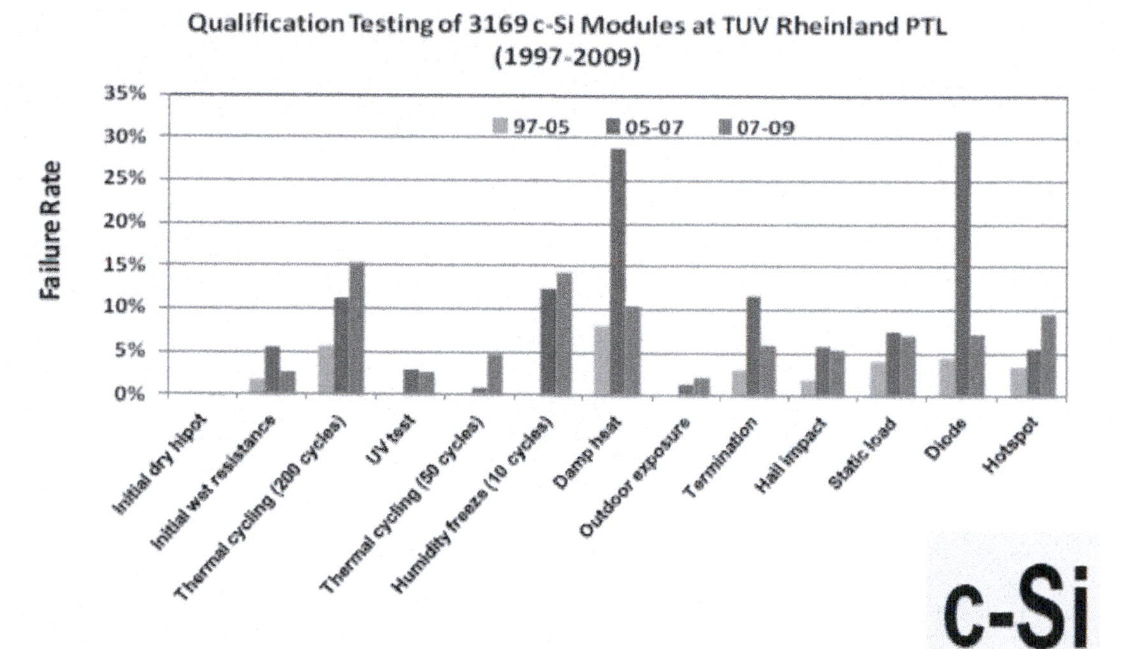

Figure 3-27. c-Si modules tests 1997-2009 (5)

A. *Crystalline silicon PV modules (extrapolated from Figure 3-27)*

3% failed at the first wet resistance test (down from 5% in 2007)
16% failed after 200 thermal cycles* (up from 12% in 2007)
3% failed UV test
5% failed after 50 thermal cycles (up from 1% in 2007)
14% failed after 10 humidity freeze cycles (up from 12% in 2007)
11% failed damp test (down from 28% in 2007)
2% failed outdoor exposure test.
6% failed termination test (down from 2% in 2007)
5% failed hail impact test.
7% failed static load test.
7% failed diode test (down from 31% in 2007)
9% failed hotspot test (up from 6% in 2007)

*NOTE: 200 thermal cycles represent 200 days on-sun operation, while 30 years of thermal cycling = ~10,000 cycles.
**NOTE: 10 humidity freeze cycles represent only part of an Arizona winter season worth of freeze cycles, while 30 years would consist of over 1,000 cycles, and many more in northern states.

B. *Thin film PV modules (extrapolated from Figure 3-28)*

1% failed at the first wet resistance test (down from 20% in 2007)
12% failed after 200 thermal cycles* (down from 20% in 2007)
12% failed after 10 humidity freeze cycles** (down from 16 in 2007)
31% failed damp test (down from 70% in 2007). There is big problem here...
5% failed outdoor exposure test.
12% failed termination test.
6% failed hail impact test.
12% failed static load test.
10% failed hotspot test.

*NOTE: 200 thermal cycles represent 200 days on-sun operation, while 30 years of thermal cycling = ~10,000 cycles.
**NOTE: 10 humidity freeze cycles represent only part of an Arizona winter season worth of freeze cycles, while 30 years would consist of over 1,000 cycles, and many more in northern states.

In general, the tested PV modules' quality during 2005-2007 was absolutely unacceptable. Quality has improved in some areas during the last several years, but it is still much lower than the best possible and profitable for use. Even higher degradation and failure rates are possible in extreme climate areas, which might lead to lower power output and profits.

An added uncertainty here is the fact that there were significantly less failures (30% less in some cases) during 1997-2005 which number increased during 2005-2007 and then decreased again later. This titter-tatter of failures can be contributed to a number of factors, such as changes in the quality of raw materials and the influx of solar cells and modules manufacturing facilities around the world. It also might be related to the world demand for PV modules, which forced manufacturers to speed up

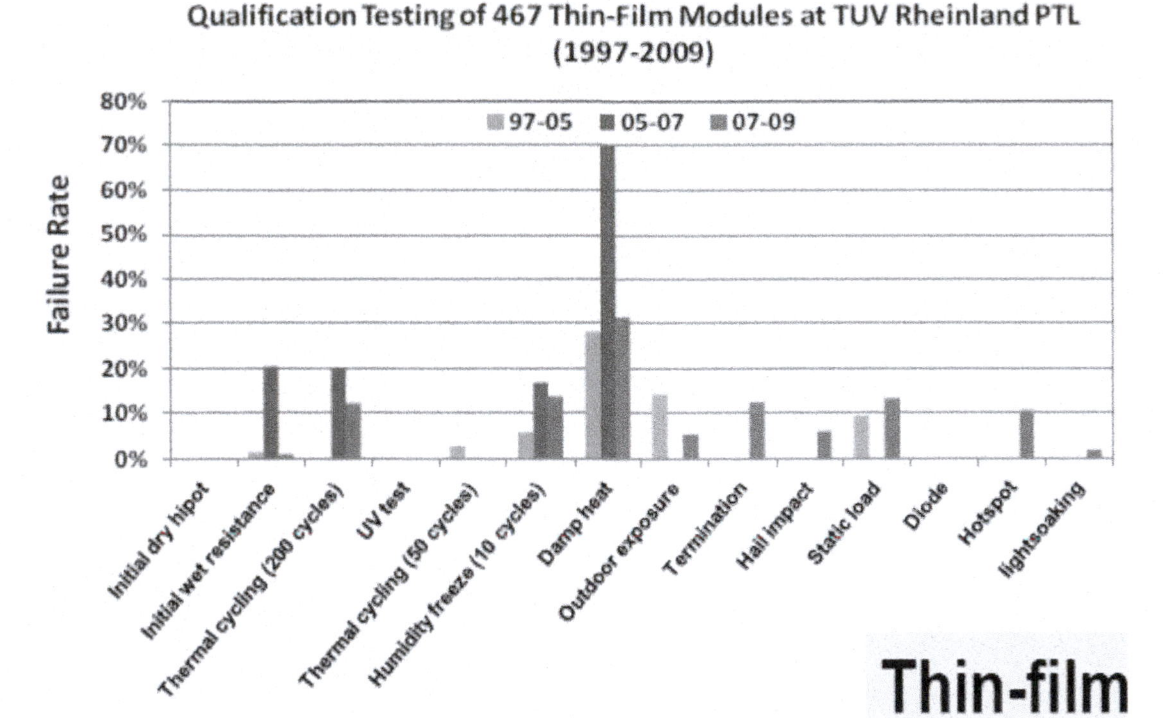

Figure 3-28. Thin film modules tests 1997-2009 (5)

the production process beyond their ability to maintain and control quality. The overall conclusion here is that the PV industry is immature at best and has not found its "legs" as yet. This also means that the quality issues need to be brought out in the open, and elevated to the highest possible levels of engineering. We know how to do this, so it is just a matter of time.

Conclusions and Recommendations

Some of the conclusions and recommendations of the TUV Rheinland test center team, in regard to these tests and their results are as follow:

- Considering 10-25 years of warranty provided for PV modules, failure rates in qualification testing are still unacceptably high.
- The top 4 failure rates for c-Si modules were related to damp heat, thermal cycling, humidity freeze and diode tests.
- The top 4 failure rates for thin-film modules were related to damp heat, thermal cycling, humidity freeze and static load tests.
- New manufacturers have higher influence on the failure rates of c-Si modules, whereas they have lower influence on the failure rates of thin-film modules. Encouragingly, overall failure rates for both technologies have decreased for the 2007-2009 period as compared to the 2005-2007 period.
- To pass full qualification testing, these top 4 tests are recommended as a minimum, before initiating full qualification testing by the module manufacturers.
- To differentiate among manufacturers who all have qualification certificates, these top 4 tests may be recommended as a minimum, before purchase decisions are made by consumers/system integrators.
- To initiate long-term lifetime reliability or test-to-failure testing, these top 4 tests may be considered as minimum key tests by researchers.

We need to remember always that the above tests, and the conclusions thereof, reflect only a fraction of the time and stress these modules will experience during 20-30 years of non-stop exposure to the sun and the elements. The most critical, but not so stable component of the modules—the EVA encapsulant—is easily affected by heat, freeze, UV radiation, etc. Once it is damaged, cells are in danger of degradation and premature failure.

Dr. Dauskardt, Stanford University (13) concludes:

1. Delamination can occur between EVA and the front surface of the solar cells.
2. More frequent in hot and humid climates.
3. Exposure to atmospheric water and/or ultraviolet radiation leads to EVA decomposition to produce acetic acid, lowering the pH and increasing corrosion.

4. EVA, Tg ~ -15°C so lower temperatures may result in "ductile-to-brittle" transition in adhesive/cohesive properties.

This is self-explanatory: Poor performance and failures cannot be avoided, but need to be calculated, and plans for their elimination (or reduction) need to be made in any plant design. We'd only dare suggest bringing the issues—especially when large-scale PV installations are concerned—to the attention of cell and module manufacturers to look for solutions with their help. A partnership between manufacturers, installers, investors and customers is the best way to ensure their long-term efficiency, longevity and profitability of the large PV projects.

IEC, UL and ASTM Test Results

There are several different internationally recognized test methods that are used for testing and qualification/certification of PV modules. These are IEC, UL and ASTM. They are quite different in the test setup and interpretation, and the reader can get details from the references or from the web.

The conclusions below are a good summary of the 3 methods, derived from tests done at TUV Rheihland test center at ASU, in Tempe, AZ, provided in the article by Dr. Mani, named, "Hot Spot Evaluation of Photovoltaic Modules," as follow (6).

"CONCLUSIONS: In this work, all three (IEC, UL and ASTM) test methods were investigated with 18 modules composed of low and high shunt resistance cells. Out of 18 (9 mono-Si and 9 poly-Si) modules tested, 9 poly-Si and 7 mono-Si modules passed the hotspot tests of all three standards. Both failures (mono-Si with voltage limited and current limited cells) occurred in the UL method. One failure occurred within the first few minutes (current limited) and the other failure occurred about 70 hours of stress (voltage limited).

1. The IEC 61215 method was found to be the most economical method in terms of test duration but the issue with this method is that it selects only the cell with the highest shunt resistance for the stress testing. The highest shunt resistance cell has the characteristics of uniform heating and this uniform heat distribution minimizes the hot spot failure during the stress testing duration warranted by the IEC 61215 standard. Another reason for the absence of hot spot failure in this test method could be due to the current sharing by both the stressed cell and bypass diode attached to the cell string.

2. The UL 1703 standard calls for a good cell screening method on limited (only ten) randomly selected cells (there is a high likelihood of not selecting the worst cell) to identify the low, median and high shunt resistance cells; however, this method requires the construction of a special test module and also the periodic testing of production modules is not practical. The other issue with this method is completely neglecting the current sharing task of the bypass diodes. In actual production modules, there are always bypass diodes connected in the circuit. This bypass diode shares the current/heat of the stressed cell of the string and reduces the hot spot failure of the cell. Since the UL 1703 method stresses the individual cells without any bypass diode, it may be viewed that this test method causes undue failure or the worst case failure if the diodes are assumed to fail in the field under open-circuit modes.

3. The ASTM E2481-06 method appears to be the best method for accurate cell selection and stress testing procedures. The results obtained after performing tests on 18 modules indicate that none of the modules failed in this test method. The reason for the absence of hot spot failure in the ASTM method but the presence of hot spot failure in the UL method could be potentially attributed to the presence of bypass diode in the test module used in the ASTM method, which bypasses the current and partly shares the heat dissipated in the cell. The stress testing duration of the ASTM method may be extended beyond 50 hours to match with 100 hours of the UL method. In this work, only a limited number of modules (18) have been tested. It is intended to continue this work with a large number of modules to ensure that the results and conclusions obtained in this short study are statistically valid."

From the above test data we see PV modules passing one test but failing another, and that the failures could occur at any time—from immediately to weeks, or months later. Clearly, certification tests are vitally important for getting a good idea of the modules performance, but they give us only part of the picture. The longevity of the modules—even the best made—still remains a black box. This also means that we need to know a lot about the modules, the source and quality of the process materials, manufacturing and quality control procedures, in-house tests and the manufacturer's overall quality control system.

So what is the best way to ensure high quality of the modules in future applications? The answers are few, but we have to start somewhere, so:

1. Standardization of materials, equipment, processes and quality control procedures at manufacturer and supply chain facilities is the ultimate solution. This

is for the most part nonexistent right now in PV manufacturing operations and field installations, so while we wait, we need to be:

2. Establishing two-way dialogs with PV device manufacturers and their supply chain. That would help build confidence in their products. This is not done presently, or at least it can only be done with great difficulty, due to the accepted "black box" nature of the products sold on the energy markets, and
3. Opening wide the communication lines between customers, utilities, regulators, installers, investors and plant operators. This will help clarify the issues at hand and bring the PV energy generation to the high level of attention they need and deserve. This effort is far from defined, let alone standardized, although some communication lines are beginning to open slowly, with California leading the effort.

Field Tests

There is a difference between results from laboratory tests, as above, and what happens under actual field conditions. Results depend on the types of modules used and weather conditions in the test field, so keep in mind that:

1. The results from tests under the extreme heat in the Arizona desert sun or the high humidity in Southern Florida will be drastically different from those under the cloudy skies of Portland, ME, or Berlin, Germany, and that,

2. In all cases, failures and degradations increase after 15-30 years of non-stop operation under climate conditions of extreme heat or humidity.

The field tests in Table 3-4 show the average annual degradation of different PV modules from different manufacturers. Most of the results from c-Si PV modules are actually quite encouraging, but notice also that most of the manufacturers whose products show good results are reputable, well-established companies who have been in the solar business for a while now and have figured out how to make a high-quality product. In contrast, most new manufacturers have not had a chance to work out the process bugs and/or conduct long-term field tests, to prove the reliability of their products.

The field test results prove that c-Si technology, if properly made and used, can be a viable and reliable energy source and that some of the major and well-established manufacturers have achieved reliability. Keep in mind, however, that most of the above tests (except one) have been conducted 10-11 years max), so the conclusions reflect only a fraction of the time, stress and abuse all PV modules will experience during 30 years of non-stop ex-

Table 3-4. Field test results (26)

Manufacturer	Module Type	Exposure (years)	Degradation Rate (% per year)	Measured at System Level?	Ref.
ARCO Solar	ASI 16-2300 (x-Si)	23	-0.4	N	2
ARCO Solar	M-75 (x-Si)	11	-0.4	N	3
[not given]	[not given] (a-Si)	4	-1.5	Y	4
Eurosolare	M-SI 36 MS (poly-Si)	11	-0.4	Y	5
AEG	PQ40 (poly-Si)	12	-5.0	N	6
BP Solar	BP555 (x-Si)	1	+0.2	N	7
Siemens Solar	SM50H (x-Si)	1	+0.2	N	7
Atersa	A60 (x-Si)	1	-0.8	N	7
Isofoton	I110 (x-Si)	1	-0.8	N	7
Kyocera	KC70 (poly-Si)	1	-0.2	N	7
Atersa	APX90 (poly-Si)	1	-0.3	N	7
Photowatt	PW750 (poly-Si)	1	-1.1	N	7
BP Solar	MSX64 (poly-Si)	1	0.0	N	7
Shell Solar	RSM70 (poly-Si)	1	-0.3	N	7
Würth Solar	WS11007 (CIS)	1	-2.9	N	7
USSC	SHR-17 (a-Si)	6	-1.0	Y	8
Siemens Solar	M55 (x-Si)	10	-1.2	Y	9
[not given]	[not given] (CdTe)	8	-1.3	Y	9
Siemens Solar	M10 (x-Si)	5	-0.9	N	10
Siemens Solar	Pro 1 JF (x-Si)	5	-0.8	N	10
Solarex	MSX10 (poly-Si)	5	-0.7	N	10
Solarex	MSX20 (poly-Si)	5	-0.5	N	10

posure to the sun and the elements. Also keep in mind that age and extreme climate conditions have detrimental effects. We need to always assume that power degradation after 20-30 years will increase somewhat, and might progress catastrophically in some cases.

PV TECHNOLOGIES PRODUCTION STATUS, 2010

Price vs. efficiency vs. longevity is a war that is just now beginning. PV modules' price goal is set at $1.0/Wp or less, but is not expected to go much lower for now. Some thin film manufacturers have achieved a remarkable reduction of their production costs, but when everything else is taken into consideration, by the time their modules get in the field, the final price is close to that of the competing c-Si technologies.

The efficiency is approaching the theoretical efficiency of c-Si, but it gets harder to increase as we get closer to the top limits. Longevity, however, is the skeleton in the closet nobody wants to talk about. It is because it is the weak point of most PV technologies, and because it is hard to test and prove.

Several PV manufacturers are racing toward high efficiency and longevity records. Most of these are well-established, proven manufacturers, supported by large industrial or financial entities, who offer reliable products. Some, however, are newcomers, struggling with quality and throughput issues, with no idea about the longevity of their product—none! This only confirms that the PV industry is in its embryonic days of development, and that the race is just beginning. Bringing the issues into the open is one way to sort them out and to discuss and resolve them.

Major PV cells and modules manufacturers are putting a lot of money, effort, and even their reputations on the line, trying to win the race. Some examples follow: (11)

SunPower

Technology: All-back contact monocrystalline

— High-Efficiency Product Status: Volume production (2009: 398 MW)

— Commercialized Cell Efficiency: 22%

SunPower has been the heavyweight champion of the world when it comes to commercialized cell and module efficiencies for the last half-decade. The company's back-contact cell design, in commercial production since 2005, moves the metal contacts to the back of the wafer, maximizes the working cell area, and eliminates redundant wires. Impressively, SunPower has been able to achieve consistent improvements in efficiency with each successive generation of commercialized cells, and this has translated to gains in the module arena as well. Its Gen 2 cells, currently in high-volume production, have an efficiency of 22%. Further improvements are on the way: Gen 3 cells, which reportedly have already started shipping, have efficiencies in excess of 24%.

The Verdict: As Gen 3 rolls out and exceeds efficiencies of 24% (something the company has already achieved in low volume), SunPower is likely to be the efficiency leader when it comes to high-volume PV cells and modules for the foreseeable future.

The problem, as Michael Kanellos (12) points out, is that 24% is awfully close to the realistic ceiling, meaning there may not be much further to go from there. As the other firms on this list start to narrow the difference, the company's price premium will erode, and its high cost structure will come under increased scrutiny.

SunPower has already recognized this, and has aimed at what seems to be a realistic target of $1/W by 2014. Whether this will be enough to survive in a commoditized world of low-cost Chinese manufacturing remains to be seen. Fortunately for the firm, though, its downstream business does afford it some measure of insulation.

Sanyo

Technology: Heterojunction with Intrinsic Thin Film [HIT]

— High-Efficiency Product Status: Volume production (2009: 255 MW)

— Commercialized Cell Efficiency: 19.8%

Ahead of the rest, but a distant second behind SunPower, Sanyo's high-efficiency product has been in volume production 1997. Its proprietary HIT cell is a hybrid of monocrystalline silicon surrounded by ultra-thin amorphous silicon layers (see here for details). The amorphous silicon layer enables superior temperature characteristics and low light performance compared to standard crystalline silicon technology. Continuous improvements have led to best commercialized cell efficiencies of 19.8% (launched this year), compared to 18.4% six years ago.

The Verdict: Sanyo has the same basic problem as SunPower: HIT costs considerably more to manufacture than standard c-Si. At the same time, its cells are about two percent less efficient than SunPower's, which means the cost pressure is significantly more. Sanyo should continue to hold the number-two spot as regards to commercial efficiency over the next three years, but unless it can start driving step-function improvements in either cost or efficiency, this will matter less and less in the commoditized global market.

The company will, however, enjoy a competitive advantage in its home country of Japan, where residential systems dominate and space constraints mean there will always be a preference for higher efficiency products. Additionally, the company is banking on the success of specialty products (e.g., BIPV modules, combined module-battery packs) in less price-sensitive markets going forward to ensure demand.

Suniva

Technology: ARTisun monocrystalline

- High-Efficiency Product Status: Volume production (2009: 16 MW)
- Commercialized Cell Efficiency: 18.3%

The brainchild of PV pioneer Dr. Ajit Rohatgi, a Georgia Tech scientist, Suniva began commercial production of its monocrystalline cells in late 2008. Unlike many struggling PV startups that entered the market around that time, this company has remained strong over the last 18 months. It has exhibited one of the quickest production ramps of any Western PV company, going from an initial 32-96 MW to a current 170 MW of cell capacity, and is sold out for 2010.

By its own admission, Suniva's technology does not represent a radical innovation; rather, the company has its own paste and texture recipes, is able to customize and optimize every layer of the cell design to its own specifications, and has leveraged its considerable R&D experience to optimize each processing step.

The Verdict: While Suniva is clearly not going to overtake SunPower or Sanyo anytime soon, reports suggest that the company has a much better cost structure compared to those two players, one that is more in line with low-cost manufacturers. That, combined with its current efficiency advantage over other firms, makes it competitively positioned. A 19% efficiency cell is in the works and should maintain competitiveness in the near future as well. The key question is whether the company can maintain this advantage going forward, given that major Chinese players are hell-bent on catching up (see below). Moreover, the company does not really have a differentiated technology that can guarantee this.

Suntech Power (STP)

Technology: Pluto monocrystalline

- High-efficiency Product Status: Low volume (2010 run rate of 4 MW per month)
- Commercialized Cell Efficiency: 19%

The Chinese cell/module behemoth threw its hat into the next-gen c-Si ring in spring 2009, when it announced the development of its proprietary "Pluto" technology, which can be used to retrofit existing cell lines. The Pluto design is based on the PERL (passivated emitter with rear locally diffused) technology developed at Australia's University of New South Wales, where efficiencies of 25 percent have been achieved in the laboratory.

Unique texturing technology with lower reflectivity ensures more sunlight can be absorbed throughout the day even without direct solar radiation, and thinner metal lines on the top surface reduce shading loss. Average cell efficiencies in low-volume production were 19%, with plans to hit 20% in two years. The company aimed to reach 450 MW of Pluto capacity by mid-2010, and envisioned that Pluto would eventually become its core product.

The Verdict: At 19%, Pluto would place Suntech behind only SunPower and Sanyo in the efficiency stakes. Importantly, Pluto offers higher efficiency with the potential to simultaneously lower costs; the cells are made with copper, rather than more expensive silver paste contacts. Pluto thus holds the key to global domination for Suntech.

Unfortunately, the company has had trouble ramping production beyond its current levels of 4 MW per month, which it describes as "glitches" with the process flow (see this article for a detailed explanation). Although it is too early to be certain, one is inclined to think the snags will eventually be resolved; the question is more 'when' than 'if'. Too long, and Suntech runs the risk of lagging behind its Chinese brethren (Yingli [YGE] and Trina [TSL], see below) on both cost (which it already does) as well as efficiency, and facing heated competition from less differentiated Chinese manufacturers (Eging PV, Ningbo, Neo Solar).

Trina Solar (TSL)

Technology: Quad Max square monocrystalline

- High-efficiency Product Status: Development (first shipments expected Q3 2010)
- Commercialized Cell Efficiency: 18.1% (pilot)

Trina's new cell tries to avoid cutting corners, quite literally—Quad Max's square shape allows it to harvest more sunlight by avoiding surface area loss typical with traditional monocrystalline cells, which are octagonal-shaped (also known as "pseudo-square"). In a 72-cell module, the additional active surface area translates into a power output advantage of eight percent.

The company has developed a new process for the technology, which involves two high-temperature thermal processes, an additional printing and cleaning step, and use of special paste for the cell surface. Initial shipments are expected in the third quarter of 2010, but don't expect meaningful megawatts until 2011.

The Verdict: "True" square mono has been a talking point in the industry for a number of years without anything to show for it. Trina's move is therefore a much-needed step in the right direction. At 18.1% efficiency, though, it places Trina at the bottom of the pack as far as high-efficiency initiatives are concerned. This will matter less as long as Quad Max does not represent a meaningful increase in manufacturing costs, since Trina is currently the second cheapest manufacturer of c-Si PV in the world, and Quad would drive a 0.6% increase in module efficiency, which would boost product margins. It is still the early days of the technology, though; as Suntech's example shows, there is potential for problems galore when going from low- to high-volume production.

Yingli Solar (YGE)

Technology: PANDA N-type monocrystalline

— High-efficiency Product Status: Pilot (commercial launch in Q3 2010)

— Best Commercialized Cell Efficiency: 18.5% (pilot)

Yingli's foray into the world of high-efficiency cell technology has come courtesy of a three-way research collaboration with the Energy Research Center of the Netherlands [ECN] and process tool maker Amtech Systems (announced in June 2009). PANDA uses ECN's design, the solar diffusion technology and dry phosphosilicate glass [PSG] removal expertise of Amtech's Tempress Systems subsidiary, and Yingli's process technology.

The PANDA cell is N-type (for more on that, see here), which means it has greater impurity tolerance and does not suffer from the light-induced degradation that conventional P-type cells do. Yingli claims the corresponding module will also have better performance under high-temperature and low-light conditions. Plans for PANDA are aggressive: in March 2010, the company announced it would construct 300 MW of ingots, wafers, cells and module capacity by the end of the year, and first shipments expected by the end of October.

The Verdict: As with Trina, Yingli has a way to go as far as commercial ramp-up of PANDA is concerned, but average cell efficiencies of 18.5% in pilot production are comfortably above Quad Max's 18.1%, although comfortably behind Pluto's 19%. Given N-type's higher impurity tolerance, PANDA also gives Yingli the option of using lower quality (and thus cheaper) polysilicon for its cells, which confers a direct cost advantage. This would further cement the firm's position as the lowest-cost c-Si manufacturer in the world and make life very difficult for its competitors. With a $5.3 billion loan in hand, the company has some cash to burn before it gets the recipe right.

A newly developed solar module with 24% efficiency is a good thing and might last forever, if it is kept in an air conditioned room. c-Si PV modules, however, will be installed in the deserts, where temperatures approach 200°F or more. Keeping in mind that all PV modules experience problems in the field, the question of the longevity of the newly developed c-Si PV technologies remains unanswered. Time will tell who's got it right, but until then, we'd advise PV power plant designers, integrators and investors to hold back their enthusiasm and stay away from the "high efficiency, but of unproven longevity PV technologies," and instead make sure the PV technology they choose has been proven reliable, and will be able to handle the 30+ years in harsh desert or excessively humid areas. This is the key to success in the long run!

CONCLUSIONS

Crystalline silicon technologies have a long track record of success. The encouraging aspect of their performance is that there are documented tests and long-term installations with a flawless record. The overwhelming conclusion from all available documentation, including our own tests through the years, is that the quality, performance and longevity of c-Si PV modules and systems is directly related to the quality of materials, equipment, processes in their construction, as well as the performance of the people involved in their manufacture.

Basically speaking, well-established, reputable companies do have the ability and experience to produce a world-class product, which functions per the manufacturer's specs. At the same time, a number of newcomers and low-cost manufacturers are working hard at getting their manufacturing and customer support processes streamlined, and we expect that many of them will be successful.

c-Si PV technology could be successfully used in many areas of the world, but we'd like to caution the reader that it is not the ultimate solution to the energy crisis, as some portray it. It is useful, but only in certain geographic areas and special situations. Due to their large temperature coefficient and humidity dependence, we'd suggest that c-Si PV modules could be used without limitations or restrictions in areas with moderate temperatures and humidity. Their use in large-scale power plants in extreme climate areas, such as deserts and humid areas of the world, however, is not well proven. More research is needed to understand the phenomena at hand, accompanied by appropriate long-term tests which must be conducted to verify the c-Si modules' performance and longevity under these conditions.

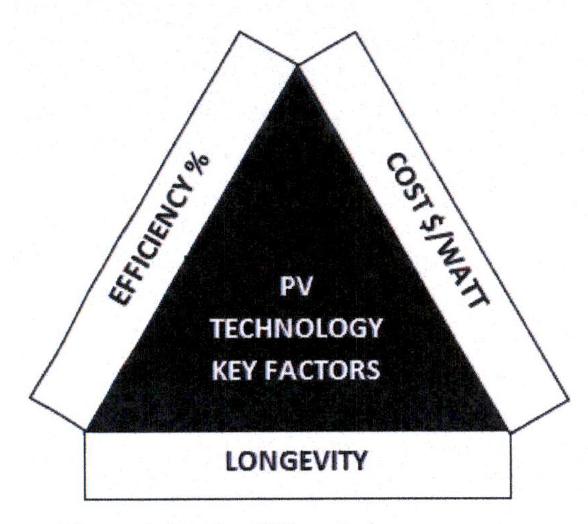

Figure 3-29. Key PV modules parameters

So, cost and efficiency goals are almost reached, and the work in these areas continues to set new records. Durability and longevity, however, are not so clearly defined. Some tests show good results, including field tests done under normal conditions. The above-mentioned extreme climates do and will present a serious problem to all PV technologies—c-Si included. The emphasis should be on longevity, and must be resolved well before we embark on deploying mega plants in US deserts and humid areas. Since these areas are the most promising for large-scale PV installations, this must be a priority for manufacturers, customers, installers, investors and operators alike.

Notes and References

1. 2009 global PV cell and module production analysis, Shyam Mehta | GTM research
2. "PVCDROM," C.B. Honsberg and S. Bowden, www.pveducation.org, 2010.
3. Mora Associates, Internal papers, http://www.moraassociates.com/
4. Optical Degradation Of C-Si Photovoltaic Modules, A. Parretta," G. Graditi
5. Testing the reliability and safety of photovoltaic modules: failure rates and temperature effects," Dr. Govindasamy TamizhMani 2010, "Photovoltaics International, Vol. 8, pp. 146-152.
6. Hot Spot Evaluation of Photovoltaic Modules, Govindasamy (Mani) TamizhMani and Samir Sharma, Photovoltaic Testing Laboratory (ASU-PTL), Arizona State University, Mesa, AZ, USA 85212
7. Five Reasons PV Modules Fail Product Certification Testing the First Time, Intertek,
8. Photovoltaic array performance model, D.L. King, W.E. Boyson, J.A. KratochvilSandia National Laboratories, Albuquerque, New Mexico 87185-0752
9. Wikipedia, http://www.wikipedia.org/
10. PVWATT. Reprinted with permission from the National Renewable Energy Laboratory, http://www.nrel.gov/rredc/pvwatts/changingparameters.html#dc2ac(July, 2010)
11. 7 Firms at the Forefront of Crystalline Silicon Efficiency, by Shyam Mehta, June 28, 2010.
12. SunPower Announces Efficiency Record: Is the End Near?, Michael Kanellos, 2010
13. Adhesion and Thermomechanical Reliability of PV Devices and Modules, by Reinhold H. Dauskardt, Stanford University
14. GTM Research, internal papers, http://www.gtmresearch.com/
15. Temperature Coefficients for PV Modules and Arrays: Measurement Methods, Difficulties, and Results, David L. King, Jay A. Kratochvil, and William E. Boyson
16. Operational Behavior Of Commercial Solar Cells Under Reverse Biased Conditions W. Herrmann, M. Adrian, W. Wiesner TUV Rheinland
17. IEC 61215, Crystalline silicon terrestrial photovoltaic (PV) modules-design qualification and type approval,1993
18. L.D. Partain (Ed.), Solar Cells and Their Applications, Wiley, 1995.
19. S.M. Sze, Semiconductor Devices: Physics and Technology, Wiley, 2004, 2nd ed.
20. Live Science.com: http://www.livescience.com/technology/080702-pf-solar-tower.html
21. Renewable Energy UK: http://www.reuk.co.uk/First-European-Solar-Power-Tower.htm
22. Environment News Service: http://www.ens-newswire.com/ens/mar2007/2007-03-30-02.asp
23. International Electrotechnical Commission (IEC) 61215, "Crystalline silicon terrestrial photovoltaic (PV) modules—design qualification and type approval," (2005)
24. Underwriters Laboratory (UL) 1703, "Standard for Flat-Plate Photovoltaic Modules and Panels," (2004)
25. American Standards of Testing and Measurements (ASTM) E2481, "Standard Test Method for Hot Spot Protection Testing of Photovoltaic Modules" (2006)
26. Comparison of Degradation Rates of Individual Modules Held at Maximum Power, C.R. Osterwald, J. Adelstein, J.A. del Cueto, B. Kroposki, D. Trudell, and T. Moriarty

APPENDIX A

CERTIFICATION STANDARDS

ASTM E44.09 PV Standards

- E 927-91 Specification for Solar Simulation for Terrestrial PV Testing
- E 948-95 Test Method for Electrical Performance of PV Cells using Reference Cells under Simulated Sunlight
- E 973-91 Test Method for Determination of the Spectral Mismatch Parameter Between a PV Device and a PV Reference Cell
- E 1021-95 Test Methods for Measuring Spectral Response of PV Cells
- E 1036-96 Test Methods Electrical Performance of Nonconcentrator Terrestrial PV Modules and Arrays using Reference Cells

Table 3-5. Advantages and disadvantages of c-Si cells and modules

CRYSTALLINE SILICON BASED PV CELLS AND MODULES
Advantages: + c-Si PV technology is the oldest, most mature and best understood PV technology as of this writing. It has been around for the last 50 years with thousands of engineers and PhDs working feverishly on improving the materials, equipment and processes, and the different aspects of the c-Si PV devices design, performance and operation. + Silicon materials are abundant and could be produced cheaply and in very high quality. No shortage is expected any time soon, and the prices are not expected to rise too much in the near future. + The efficiency of c-Si cells and modules is in the 16-19% range at the module level, which is the highest of any single-junction PV technology on the energy market. + c-Si modules are the most reliable PV products available now days, with millions of them installed, operated and tested around the world.
Disadvantages: - Each step of the PV cells and modules manufacturing process—starting with the raw materials and up to shipping—can create efficiency and longevity problems if not properly planned, designed and executed. - c-Si cells and modules manufacturing process is complex, consisting of a long chain of events with many links. Each link has its own peculiarities and requirements. Missing or messing up one link will break the chain, thus producing defective product. - c-Si cells and modules degrade with time. ~1.0% power output decrease every year is expected by each manufacturer and is built in their long term warranty. - c-Si cells and modules lose efficiency with temperature increase. ~0.5% power is lost with each degree C increase of the cell / module temperature above STC. This can amount to up to 30% output drop in some extreme cases. - c-Si cells and modules suffer most when exposed to harsh climatic conditions, such as these in the deserts and in very humid regions. Excess heat, moisture and UV radiation cause accelerated degradation and premature failure of the modules. - Most c-Si cells and modules are imported which makes us dependent—AGAIN--on other countries for our energy supply. What is stopping us from manufacturing US made PV products?

- E 1038-93 Test Method for Determining Resistance of PV Modules to Hail by Impact with Propelled Ice Balls
- E 1039-94 Test Method for Calibration of Silicon Non-Concentrator PV Primary Reference Cells Under Global Irradiation
- E 1040-93 Specification for Physical Characteristics of Non-Concentrator Terrestrial PV Reference Cells
- E 1125-94 Test Method for Calibration of Primary Non-Concentrator Terrestrial PV Reference Cells using a Tabular Spectrum
- E 1143-94 Test Method for Determining the Linearity of a PV Device Parameter with Respect to a Test Parameter
- E 1171-93 Test Method for PV Modules in Cyclic Temperature and Humidity Environments
- E 1328-94 Terminology Relating to PV Solar Energy Conversion
- E 1362-95 Test Method for the Calibration of Non-Concentrator Terrestrial PV Secondary Reference Cells
- E 1462-95 Test Methods for Insulation Integrity and Ground Path Continuity of PV Modules
- E 1524-93 Test Methods for Saltwater Immersion and Corrosion Testing of PV Modules for Marine Environments
- E 1596-94 Test Methods for Solar Radiation Weathering of PV Modules
- E 1597-94 Test Method for Saltwater Pressure Immersion and Temperature Testing of PV Modules for Marine Environments
- E 1799-96 Practice for Visual Inspection of PV Modules
- E 1802-96 Test Methods for Wet Insulation Integrity Testing of PV Modules

IEEE PV Standards

- 928 IEEE Recommended Criteria for Terrestrial PV Power Systems
- 929 IEEE Recommended Practice for Utility Interface of Residential and Intermediate PV Systems
- 937 IEEE Recommended Practice for Installation and Maintenance of Lead-Acid Batteries for PV Systems
- 1013 IEEE Recommended Practice for Sizing Lead-Acid Batteries for PV Systems
- 1144 Sizing of Industrial Nickel-Cadmium Batteries for PV Systems
- 1145 IEEE Recommended Practice for Installation and Maintenance of Nickel-Cadmium Batteries for PV Systems

- P1262 Recommended Practice for Qualification of PV Modules
- P1361 Recommended Practice for Determining Performance Characteristics of Batteries in PV Systems
- P1373 Recommended Practice for Field Test Methods and Procedures for Grid-Connected PV Systems
- P1374 Guide for Terrestrial PV Power System Safety

IEC PV Standards

- IEC-891 Correction of Temperature and Irradiance to Measured I-V Character of Crystalline Silicon PV Devices
- IEC-904-1 Measurement of PV I-V Characteristics
- IEC-904-2 Requirements for Reference Solar Cells
- IEC-904-3 Measurement Principles for Terrestrial PV Solar Devices with Reference Spectral Irradiance Data
- IEC-904-4 On-Site Measurements of Crystalline Silicon PV Array I-V Characteristics
- IEC-904-5 Determination of Equivalent Cell Temperature (ECT) of PV Devices by Open-Circuit Voltage Method
- IEC-904-6 Requirements for Reference Solar Modules
- IEC-904-7 Computation of Spectral Measurement of a PV Device
- IEC-904-8 Guidance for Spectral Measurement of a PV Device
- IEC-904-9 Solar Simulator Performance Requirements
- IEC-1173 Overvoltage Protection for PV Power Generating Systems
- IEC-1194 Characteristic Parameters of Stand-Alone PV Systems
- IEC-1215 Design and Type Approval of Crystalline Silicon Terrestrial PV Modules
- IEC-1277 Guide-General Description of PV Power Generating System
- IEC-1701 Salt Mist Corrosion Testing of PV Modules
- IEC-1702 Rating of Direct-Coupled PV Pumping Systems
- IEC-1721 Susceptibility of a Module to Accidental Impact Damage (Resistance to Impact Test)
- IEC-1727 PV-Characteristics of the Utility Interface
- IEC-1829 Crystalline Silicon PV Array—On-Site Measurement of I-V Characteristics

UL PV Standards

- UL-1703 Flat-Plate PV Modules and Modules
- Subject: UL-1741 Proposed Draft of the Standard for Power Conditioner Units for Use in Residential PV Power Systems (Work in Progress)

NEC Standards

- ANSI/NFPA 70-1996 National Electrical Code, Article 690, Solar PV Systems

Chapter 4

Thin Film Photovoltaic Technologies

"The whole of science is nothing more than a refinement of everyday thinking."

Albert Einstein

There are a number of PV technologies in addition to, and different from, the c-Si PV technologies described in the previous chapter. These are the so-called: thin film PV (TFPV) technologies and related products. We will take a look at their properties, focusing on their suitability for use in commercial and large-scale PV installations.

THIN FILM PV TECHNOLOGIES

Thin film PV (TFPV) technologies are a relatively new branch of the solar industry, which has grown much faster in popularity and size than the other PV technologies during the last several years. Since the active layers in TFPV cells and modules are deposited in the form of thin films, via thin film deposition methods, we refer to them as "thin film" PV products. Thin films of special photovoltaic materials can produce solar cells or modules with relatively high conversion efficiencies, while at the same time using much less semiconductor material than c-Si cells. In addition, thin film equipment and manufacturing methods allow efficient, cheap, fully automated mass production which is the reason for their success lately.

This, however, comes at the expense of reduced efficiency (average 6-9%), which is not expected to increase much in the future (in mass production mode). On the other end of the efficiency spectrum, multi-layer thin film CPV cells have reached efficiencies over 40% and are getting higher by the day with theoretical efficiency limits in the 80% range.

Due to their versatility, TFPV products have become very popular for use in a number of applications. Recently they have gained a share in large-scale installations as well. Some TFPV modules also show better efficiency under reduced solar radiation than the c-Si competition. This is very useful in many regions with cloudy climates, and could account to their quick rise in European and other world energy markets.

The major types of TFPV technologies considered for commercial and large-scale installations are:

1. Cadmium telluride thin films
2. CIGS thin films
3. Amorphous silicon thin films
4. Silicon ribbon
5. Epitaxial silicon thin films
6. Light absorbing dyes thin films
7. Organic/polymer thin films
8. Ink thin films
9. Nano-crystalline cells
10. Indium phosphide
11. Single-junction III-V cells
12. Multi-junction cells
    a. Gallium arsenide based cells
    b. Germanium based cells
    c. CPV solar cells

Below we see each of these technologies and their specific structure and function, focusing on their use in large-scale PV power generation and related issues.

The Major TFPV Technologies

We classify these PV technologies as "major" for the purposes of this text, because they are presently considered for large-scale PV power generation projects.

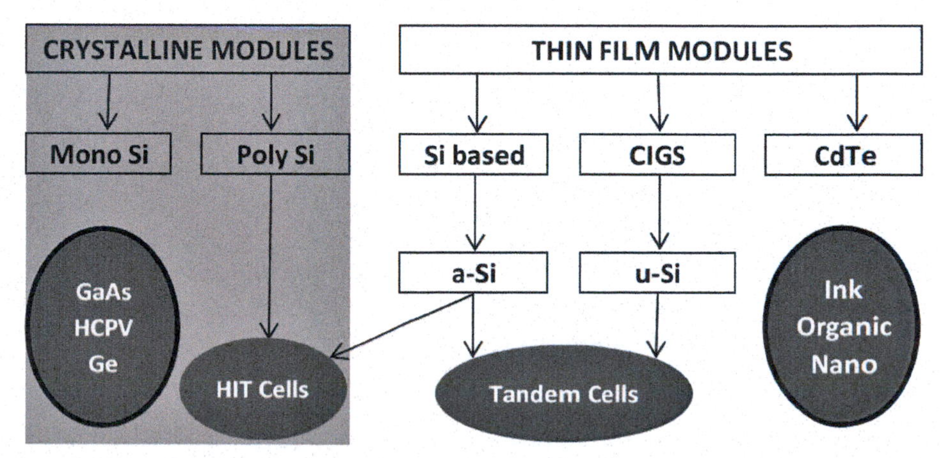

Figure 4-1. Key PV technologies 2010

These are the cadmium telluride, CIGS, and amorphous silicon thin films:

Cadmium Telluride (CdTe)

Cadmium telluride (CdTe) is a type of solar cell and module based on thin films of the heavy metal cadmium and its compounds, cadmium telluride (CdTe) and cadmium sulfide (CdS). CdTe is an efficient light-absorbing material, quite adaptable for the manufacture of thin-film solar cells and modules. Compared to other thin-film materials, CdTe is easier to deposit in mass production environments and more suitable for large-scale production.

CdTe bandgap is 1.48 eV, which makes it almost perfect for PV conversion purposes. At 16.5% demonstrated efficiency in the lab, it is a candidate for a major role in the energy future. Mass production modules are sold with 8-9% efficiency. No significant increase is expected with the present production materials and methods, although manufacturing costs are down—at or below $1.0/Wp.

With a direct optical energy bandgap of 1.48 eV and high optical absorption coefficient for photons with energies greater than 1.5 eV, only a few microns of CdTe are needed to absorb most of the incident light. Because only very thin layers are needed, material costs are minimized, and because a short minority diffusion length (a few microns) is adequate, expensive materials processing time and costs can be avoided.

The structure, as shown in Figure 4-2, consists of a front contact, usually a transparent conductive oxide (TCO), deposited onto a glass substrate. The TCO layer has a high optical transparency in the visible and near-infrared regions and high n-type conductivity. This is followed by the deposition of a CdS window layer, the CdTe absorber layer, and finally the back contact.

For high-volume devices, the CdS layer is usually deposited using either closed-space sublimation (CSS) or chemical bath deposition, although other methods have been used to investigate the fundamental properties of devices in the research laboratory. In all cases, mass production and automation is possible, which is the greatest advantage of this technology.

The CdTe p-type absorber layer, 3-10 μm thick, can be deposited using a variety of techniques including physical vapor deposition (PVD), CSS, electrodeposition, and spray pyrolysis. To produce the most efficient devices, an activation process is required in the presence of CdCl2 regardless of the deposition technique. This treatment is known to recrystallize the CdTe layer, passivating grain boundaries in the process, and promoting inter-diffusion of the CdS and CdTe at the interface.

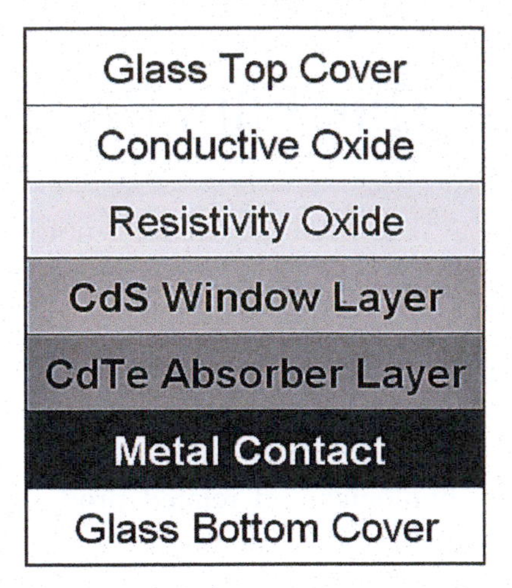

Figure 4-2. CdTe/CdS thin-film solar cell

Forming an ohmic contact to CdTe, however, is difficult because the work function of CdTe is higher than all metals. This can be overcome by creating a thin p+ layer by etching the surface in bromine methanol or HNO3/H3PO4 acid solution and depositing Cu-Au alloy or ZnTe:Cu. This creates a thin, highly doped region that carriers can tunnel through. However, Cu is a strong diffuser in CdTe and causes performance to degrade with time. Another approach is to use a very low bandgap material, e.g. Sb2Te3, followed by Mo or W. This technique does not require a surface etch and the device performance does not degrade with time.

CdTe PV modules manufacturing is a sophisticated process; much more sophisticated than that of the conventional c-Si modules process, which uses simple 1970s manufacturing equipment, materials and processes. CdTe TFPV modules are manufactured with the help of modern, complex and expensive semiconductor type equipment and processes. Because of that, the precision and accuracy of the resulting process steps, and ergo the quality of the final product, are limited only by the quality of the materials and supplies, and the capabilities of the engineers, technicians and operators on the production lines.

CdTe thin-film solar modules are now being mass produced very cheaply, and it is expected with economies of scale that they will achieve the cost reduction needed to compete directly with other forms of energy production in the near future. Since CdTe thin film PV devices still have a long way to go to achieve maximum efficiencies, it will be interesting to see which materials and methods are most successful.

The most efficient CdTe/CdS solar cells (efficiencies of up to 16.5%) have been produced using a Cd2SnO4 TCO layer which is more transmissive and conductive than the classical SnO2-based TCOs, and including a Zn2SnO4 buffer layer which improves the quality of the device interface.

CdTe PV research, done by manufacturers, universities and R&D labs focuses on some of these challenges:

a. Boosting efficiencies by exploring innovative transparent conducting oxides that allow more light into the cell to be absorbed and at the same time more efficiently collect the electrical current generated by the cell.

b. Studying mechanisms such as grain boundaries that can limit voltage.

c. Understanding the degradation some CdTe devices exhibit at the contacts and redesigning the devices to minimize this phenomenon.

d. Designing module packages that minimize any outdoor exposure to moisture.

e. Engaging aggressively in both indoor and outdoor cell and module stress testing.

These efforts are geared to address the main problems with CdTe PV modules: a) the relatively low efficiency which contributes to using more land and mounting hardware, b) temperature power degradation, c) annual power degradation, and other negative long -term effects.

Availability of the rare metals used in CdTe TFPV technology and their toxicity are other serious issues which manufacturers and regulators have put on the back burner. Let's hope that we won't have to wait for a serious accident before bringing these issues out in the open and discussing possible solutions.

CIGS

Early solar cells of this type were based on the use of CuInSe2 (CIS). However, it was rapidly realized that incorporating Ga to produce Cu(In,Ga)Se2 (CIGS) structure, results in widening the energy bandgap to 1.3 eV and an improvement in material quality, producing solar cells with enhanced efficiencies. CIGS have a direct energy bandgap and high optical absorption coefficient for photons with energies greater than the bandgap, such that only a few microns of material are needed to absorb most of the incident light, with consequent reductions in material and production costs.

The best performing CIGS solar cells are deposited on soda lime glass in the sequence—back contact, absorber layer, window layer, buffer layer, TCO, and then the top contact grid. The back contact is a thin film of Mo deposited by magnetron sputtering, typically 500–1000 nm thick.

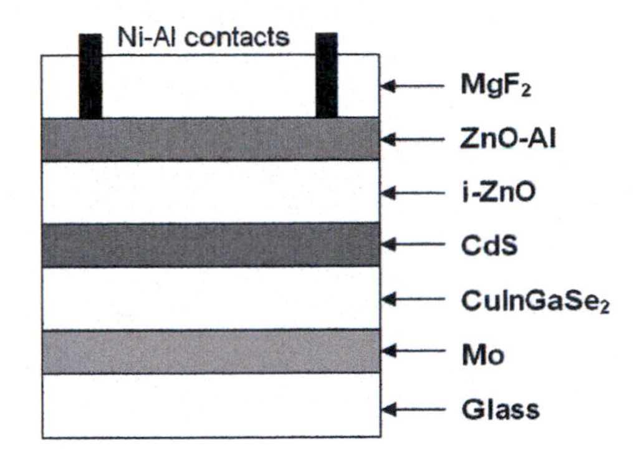

Figure 4-3. CIGS cell cross section

The CIGS absorber layer is formed mainly by the co-evaporation of the elements either uniformly deposited, or using the so-called three-stage process, or the deposition of the metallic precursor layers followed by selenization and/or sulfidization. Co-evaporation yields devices with the highest performance while the latter deposition process is preferred for large-scale production.

Both techniques require a processing temperature >500°C to enhance grain growth and recrystallization. Another requirement is the presence of Na, either directly from the glass substrate or introduced chemically by evaporation of a Na compound. The primary effects of Na introduction are grain growth, passivation of grain boundaries, and a decrease in absorber layer resistivity.

The junction is usually formed by the chemical bath deposition of a thin (50–80 nm) window layer. CdS has been found to be the best material, but alternatives such as ZnS, ZnSe, In2S3, (Zn,In)Se, Zn(O,S), and Mg-ZnO can also be used.

The buffer layer can be deposited by chemical bath deposition, sputtering, chemical vapor deposition, or evaporation, but the highest efficiencies have been achieved using a wet process as a result of the presence of Cd2+ ions. A 50 nm intrinsic ZnO buffer layer is then deposited and prevents any shunts. The TCO layer is usually ZnO:Al 0.5–1.5 μm. The cell is completed by depositing a metal grid contact Ni/Al for current collection, then encapsulated.

CIGS solar cells have been produced under lab conditions with efficiencies of 19.5%, and lately modules with efficiencies of 15.7% were verified as well. Commercial, mass produced, CIGS PV module efficiency, however, is still lower than CdTe PV modules—and this will have a major impact on their future unless ways to increase their efficiency and reduce their costs are found soon.

CIGS TFPV modules have similar problems as those plaguing CdTe TFPV technologies. They have low efficiency, require larger mounting infrastructure, exhibit power loss under excess heat and have a significant annual degradation rate. Scarcity of materials and related toxicity issues are, as in the CdTe PV case, on the back burner for now. These issues must be evaluated from the point of view of large-scale installations, where thousands and millions of these modules will be installed. In such cases, minute amount of toxic materials in each module are multiplied mega times and become a substantial threat to the environment. Also, special measures must be taken for proper disposal or recycling of these modules.

CIGS research is focused on several of today's challenges of this promising technology:

a. Pushing efficiencies even higher by exploring the chemistry and physics of the junction formation and by examining concepts to allow more of the high-energy part of the solar spectrum to reach the absorber layer.

b. Dropping costs and facilitating the transition to a commercial stage by increasing the yield of CIS modules—which means increasing the percentage of modules and cells that make it intact through the manufacturing process.

c. Decreasing manufacturing complexity and cost, and improving module packaging.

NOTE: At a meeting of PV specialists in February 2011, called PV Module Reliability Workshop (PVMRW), the degradation and longevity of PV technologies and products were discussed by representatives of se veral manufacturing companies. The susceptibility to moisture of SIGS modules was addressed as one of the major concerns, and packaging solutions were presented. Location-specific reliability tests and evaluations was also one of the topics, which is a step in the right direction. We are glad that such open discussions are underway, since this is the fastest way to resolve the issues and put the promising technologies on the energy market.

Amorphous Silicon

Amorphous silicon (a-Si) is produced via thin film processes, based on depositing thin layers of silicon films on different substrates. Silicon thin-film cells are mainly deposited by chemical vapor deposition (CVD), typically plasma-enhanced (PE-CVD), using silane and hydrogen reactive and carrier gasses for the actual deposition. Depending on the deposition parameters and the stoichiometry of the process, this reaction can yield different types of thin film structures, such as amorphous silicon (a-Si, or a-Si:H), protocrystalline silicon or nanocrystalline silicon (nc-Si or nc-Si:H), also called microcrystalline silicon.

These types of silicon feature dangling and twisted bonds, which result in deep defects (energy levels in the bandgap) as well as deformation of the valence and conduction bands (band tails), which lead to reduced efficiency. Proto-crystalline silicon mixed with nano-crystalline silicon is optimal for high, open-circuit voltage.

Solar cells and modules made from these materials tend to have lower energy conversion efficiency than those made from bulk silicon, but have some operating advantages (such as lower temperature degradation). They are also less expensive to produce, although the capital equipment expense is greater, due to equipment complexity.

a-Si has a somewhat higher bandgap (1.7 eV) than crystalline silicon (c-Si) (1.1 eV), which means that it absorbs the visible part of the solar spectrum more efficiently than the infrared portion. nc-Si has about the same bandgap as c-Si, so nc-Si and a-Si can advantageously be combined in thin layers, creating a layered cell called a "tandem cell," where the top a-Si cell absorbs the visible light and leaves the infrared part of the spectrum for the bottom cell in nc-Si.

The biggest problem with a-Si TFPV technology and a barrier to its success, however, is its low efficiency. Today's best cell efficiencies are about 12% in the lab, which is almost 50% lower than other PV technologies. Mass produced a-Si cells and modules are in the 8% efficiency range today.

A second problem with a-Si is its manufacturing cost as related to initial capital investment, which is quite high as compared with the competing PV technologies. Two proposed solutions to this problem are higher manufacturing rates, and batch (simultaneous) pro-

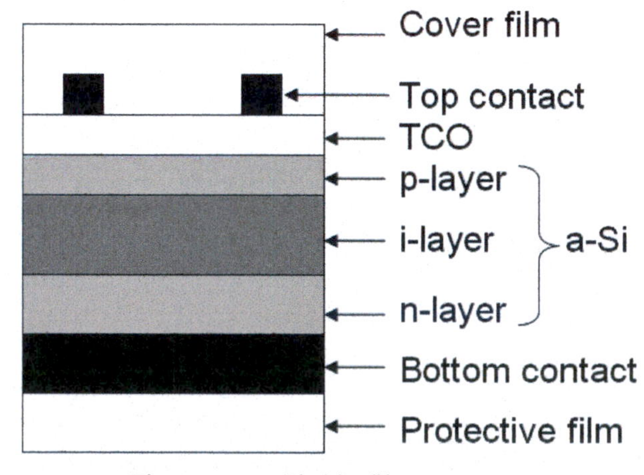

Figure 4-4. a-Si thin film structure

cessing of multiple modules. Good progress has been made in rates that are 3-10 times higher than those being used in production, but all this is still on a lab scale and is to be proven in reality and on a large scale.

On the positive side, while some of the more efficient cells and modules lose about 20-30% of their output in the field, due to excess heat exposure, a-Si loses only about 5-10%, due to its lower temperature coefficient. Also, the active thin film structure is composed mainly of silicon films which have inert and homogeneous natures that show better chemical and mechanical stability than some of the competing thin films—in case of an encapsulation failure. a-Si modules are also more resistive to the negative effects of shading in the field. Of equal importance, a-Si modules do not contain any hazardous materials, which is paramount where large-scale PV installations are concerned. These qualities put a-Si on the top of the list of PV technologies suitable for large-scale power generation in deserts and other inhospitable areas.

Even with low efficiency (well under 10%), a-Si thin film technology is being successfully developed for building-integrated photovoltaics (BIPV) in the form of semi-transparent solar cells which can be applied as window glazing. These cells function as window tinting while generating electricity. It remains to be seen if the amount of generated electricity covers initial and operating expenses.

A triple-junction a-Si TFPV power system has been operating near Bakersfield, CA, for several years, and is providing proof of the excellent performance of this technology. The 500 kW grid-connects system has been performing well, meeting or exceeding its design goals. Performance data from this larger-scale installation confirms data obtained from smaller a-Si systems and proves that this thin film PV technology can be successfully used in large-scale power plants, if the low efficiency can be justified.

Great research effort is underway at universities and R&D labs around the world, geared towards solving efficiency and cost issues and obtaining a-Si that is truly competitive in the energy market. a-Si manufacturers need to work on understanding the key areas of this technology and the processes, and focus on their optimization by:

a. Improving the light-stabilized electronic quality of a-Si and low-gap a-Si:H cells to achieve broader spectrum conversion, and increased and stable overall efficiency.

b. Increasing the growth rates of a-Si, a-SiGe, etc. layers while maintaining high electronic quality, to obtain increased throughput and reduced capital cost.

c. Developing high-growth-rate methods for nanocrystalline silicon while maintaining high electronic quality as needed for increased efficiency, stability and reduced cost.

d. Understanding and controlling light-induced degradation in a-Si:H as needed for increased efficiency and understanding of the intrinsic limits of the efficiency.

e. Developing in-situ in-line process monitoring for increased yield.

f. Improving light-management to obtain maximum efficiency.

g. Improving stability and conversion efficiency of a-Si modules in actual use by addressing the Staebler-Wronski negative effects, where the conversion efficiency of the a-Si module decreases when it is first exposed to sunlight.

h. Reducing capital equipment costs for manufacturing a-Si panels by improved manufacturing processes that include increasing the deposition rates.

i. Improving module-packaging designs to make them more resilient to outdoor environments and less susceptible to glass breakage or moisture ingress.

j. Developing new module designs for building-integrated applications.

These and other similarly complex issues must be addressed, because the future of a-Si PV products depends on their proper and timely resolution.

NOTE: The untimely exit of Applied Materials from the a-Si equipment manufacturing business in the summer of 2010 cast a shadow of doubt over the a-Si technology. Applied Materials is an example of a new-comer in the PV field with great potential and ambition, but inexperienced with the energy market's wants, needs, and overall peculiarities. Applied Materials had its own reasons to leave the field, but a-Si TFPV is here to stay. It is a promising technology that has already found niche markets, and will become even more popular with time, until it finds its place in the large-scale energy markets.

The Developing TFPV Technologies

The TFPV technologies we classify as "developing" for the purposes of this text are silicon ribbon, epitaxial silicon, light absorbing dyes, organic thin films, ink thin films, nano-crystalline, and indium phosphide thin film.

Silicon Ribbon

Called EFG ("edge defined film fed growth"), this method, is not exactly a "thin film" process, as we know it, but the resulting material is thin enough, so it belongs in this category. Here, a graphite dye is immersed into molten silicon, making it rise into the dye by capillary action. It is then pulled as a self-supporting very thin sheet of silicon which hardens in the air above the dye. It can then be cut in different shapes and sizes for processing into solar cells and modules. This method is more efficient than conventional c-Si ingot and wafer processes in terms of producing c-Si substrates of exact thickness and avoiding slicing it into wafers. Conventional processes waste 20-40% of the silicon material, use a lot of energy, and produce tons of hazardous waste materials.

Another similar process we need to mention here is called "dendritic web growth process." It consists of two dendrites, which are placed into molten silicon and withdrawn quickly, causing the silicon to exit and solidify as a thin sheet. A modification of this method now in use is called the "string ribbon method," where two graphite strings are used (instead of the dendrites) to draw the silicon sheet, which makes process control much easier. Again, the silicon sheet can be cut into different shapes and sizes.

In all cases the silicon produced by these methods is multi-crystalline with a quality approaching that of the directionally solidified material. Although lab tests show efficiencies in the 17-18% range, solar modules made using silicon ribbons and produced via these methods, generally have efficiencies in the 10-12% range.

After the initial hoopla, that silicon ribbon technologies would dominate the market, their share is quite small—less than 1% of total sales today—and does not seem to be growing. This is mostly due to the fact that the process is not easy to control, and the wafers' surface is not uniform enough, thus resulting in breakage, processing defects, and performance inefficiencies. The silicon sheets forming process is also complex and uses a lot of energy, which makes it comparably more expensive than some of the other mass produced TFPV technologies.

Epitaxial Thin Film Silicon

The high cost of silicon material accounts for about half of the production cost of current conventional, industrial-type silicon solar cells. In order to reduce the amount of consumed silicon, the photovoltaics (PV) industry is counting on a number of options presently being developed. The most obvious is to move to thinner silicon substrates by producing thinner Si wafers, or shaving the thicker wafers, but this is proving hard to do for a number of reasons. A more feasible approach is the so-called epitaxial deposition of a thin film of silicon on a cheap substrate, thus creating efficient but cheap solar cells.

There are several approaches that can be used to create such a thin film cell:

a. Epitaxial single crystal sc-Si

To create an epitaxial thin-film solar cell on a cheap substrate we start with highly doped sc-Si wafers (e.g., from low-grade silicon or scrap Si material), and deposit an epi layer of Si by chemical vapor deposition (CVD). The resulting mix of a high quality epi layer and a cheap substrate is a compromise between high cost and efficiency, and yet offers a solution to gradual transition from a wafer-based (heavy material dependence) to a thin-film technology (less material and more sophisticated processing). This process is easier to implement than most other thin-film technologies today, but it remains to be seen if its efficiency and cost will be able to compete in the energy market.

b. Epitaxial Polysilicon thin film

To produce thin-film polysilicon solar cells, a thin layer (only a few microns) of polysilicon Si is deposited on a cheap foreign substrate, such as ceramic or high-temperature glass. These seed layers are then epitaxially thickened into absorber layers several microns thick using high-temperature CVD with a deposition rate exceeding 1 μm/min. Polycrystalline silicon films with grain sizes between 1-100 μm appear to be particularly good candidates.

Good polycrystalline silicon solar cells can be obtained using aluminum-induced crystallization of amorphous silicon. This process leads to very thin layers with an average grain size around 5 μm. This technology is still in R&D stages, but shows high cost-reduction potential and might become very important, especially in case of silicon shortage, or very high prices in the future.

Light-absorbing Dyes (DSSC)

These are special types of dye-sensitized solar cells, where a ruthenium metalorganic dye (Ru-centered) is used as a monolayer of light-absorbing material. The dye-sensitized solar cell depends on a mesoporous layer of nanoparticulate titanium dioxide to greatly amplify the surface area (200-300 m2/g TiO2, as compared to approximately 10 m2/g of flat single crystal).

Photogenerated electrons from the light-absorbing dye are passed on to the n-type TiO2, and the holes are passed to an electrolyte on the other side of the dye. The circuit is completed by a redox couple in the electrolyte, which can be liquid or solid. This type of cell allows a more flexible use of materials, and is typically manufactured by screen printing and/or use of ultrasonic nozzles, with the potential for lower processing costs than those used for bulk solar cells.

However, the dyes in these cells also suffer from degradation under heat and UV light, and the cell casing is difficult to seal due to the solvents used in assembly. In spite of these problems, this is a popular emerging technology with special applications and significant commercial impact forecast within this decade.

The first commercial shipment of DSSC solar modules was recorded in July 2009 from G24i Innovations.

Organic/Polymer Solar Cells

Organic and polymer solar cells are built from thin films (typically 100 nm) of organic semiconductors such as small-molecule compounds like poly-phenylene vinylene, copper phthalo-cyanine (a blue or green organic pigment), and carbon fullerenes and fullerene derivatives, such as PCBM. Energy conversion efficiencies achieved to date using conductive polymers are low compared to inorganic materials. However, they were improved in the last few years and the highest NREL certified efficiency has reached 6.77%. In addition, these cells could be beneficial for some applications where mechanical flexibility and disposability are important.

These devices differ from inorganic semiconductor solar cells in that they do not rely on the large built-in electric field of a p-n junction to separate the electrons and holes created when photons are absorbed. Instead, the active region of an organic device consists of two materials, one which acts as an electron donor and the other as an acceptor. When a photon is converted into an electron hole pair, typically in the donor material, the charges tend to remain bound in the form of an exciton, and are separated when the exciton diffuses to the donor-acceptor interface. The short exciton diffusion lengths of most polymer systems tend to limit the efficiency of such devices. Nanostructured interfaces, sometimes in the form of bulk hetero-junctions, can improve performance. Instability of the films, especially under harsh environmental effects is a major problem, which needs to be resolved, before full-scale implementation. Even with its advantages, this technology still has far to go to full market acceptance and serious deployment.

Ink PV Cells

A fairly new development, this light-activated power generating product is based on a unique and patented solvent-based silicon nanomaterial platform that can be applied like ink on any substrate. Developers claim that this approach has cost savings over traditional silicon products by using less silicon and having a more efficient manufacturing process as well as unique optical advantages.

This new technology consists of processing the quantum dots in the silicon "ink" in a way that makes it possible to use the old "roll-to-roll" printing technology used for printing on paper or film. Applying ink directly on any substrate (including a flexible one) allows applications such as tagless printing for clothing labels and portable chargers for consumer and military customers.

By controlling the sizes of the dots from 2 to 10 nm, the absorption or emission spectra of the resulting film can be controlled. This allows capture of everything from infrared to ultraviolet and the visible spectrum in between which is not possible with conventional technology.

The technology is also used as an efficient light source. By controlling particle size, you can produce light of any color or a combination of particle sizes that will give off white light. This application might provide additional, and possibly larger markets for this technology in the near future—at least in some specialized areas.

Nano-crystalline Solar Cells

These structures make use of some of the usual thin-film light absorbing materials, but are deposited as a very thin absorber on a substrate (supporting matrix) of conductive polymer or mesoporous metal oxide having a very high surface area to increase internal reflections. Hence, the probability of light absorption increases.

Using nanocrystals allows one to design architectures on the length scale of nanometers, the typical exciton diffusion length. In particular, single-nanocrystal ('channel') devices, an array of single p-n junctions between the electrodes and separated by a period of about a diffusion length, represent a new architecture for solar cells and potentially high efficiency.

We envision the development of this type of photo-conversion to be in the R&D labs for a while yet, but it opens new possibilities in areas where other technologies simply cannot compete, thus opening promising niche markets for these cells.

III-V Materials and Devices

These are PV technologies based on the deposition of thin films of the III-V materials. These devices are divided into single- and multi-junction devices as follow:

Single-junction III-V Devices

A number of compounds such as gallium arsenide (GaAs), indium phosphide (InP), and gallium antimonide (GaSb) have adequate energy band gaps, high optical absorption coefficients, and good values of minority carrier lifetimes and mobility, making them excellent materials for making high efficiency solar cells. These materials are usually produced by the Czockralski or Bridgmann methods, which provide high quality materials with increased efficiency and reliability, but at a higher price.

After silicon, GaAs and InP (III-V compounds) are the most widely used materials for single-junction (SJ) solar cells manufacturing. These materials have optimum band gap values (1.4 and 1.3 respectively) for SJ conversion of sunlight. The construction of solar cells made of these materials is similar to the regular single-junction c-Si solar cells we discussed in Chapter 3.

The major disadvantage of using III-V compounds for PV devices is the high cost of producing the materials they are made of and the related manufacturing processes. Also, crystal imperfections, including bulk impurities, severely reduce their efficiencies, so that only very high quality materials could be considered. Too, they are heavier than silicon, which requires the use of thinner cells, but they are weaker mechanically, so their design requires a delicate balance of thickness vs. weight.

The combination of high efficiency, high price, crystal imperfections intolerance, and mechanical weakness makes these devices useful for limited applications, where efficiency and overall behavior is more important than price. Thus they are not widely used in the general PV market, but still can be found in some important niche markets.

Multi-junction Cells and Devices

Single-junction PV devices convert only a portion of the sunlight (with photons just above the band-gap level of the semiconductor material). The problem is that photons with lower or higher energy do not generate electron-hole pairs and are lost as heat, which is also detrimental to the cells, reducing their efficiency and deteriorating them over time.

One way to solve this problem and to increase the efficiency of the PV devices is to add more junctions, thus creating multi-junction (MJ) solar cells. By selecting materials, properties, number and types of junctions, and manufacturing processes designed to capture the majority of sunlight photons, we can reach very high efficiencies.

Thus far, multi-junction solar cells made primarily using the III-V compounds have clearly proven that by minimizing thermalization and transmission losses, large improvements in efficiency can be made over those of single-junction cells. These devices find use in generating power for space applications and in concentrator systems. They show great promise for high efficiency and reliability under harsh climate conditions, such as those in the deserts.

Future development of multi-junction devices using low-cost thin-film technologies is especially promising for producing more efficient and yet inexpensive devices. Cost reductions will also be significant when thin-film technologies are directly produced on building materials other than glass, because many materials such as tiles and bricks can be substantially cheaper than glass and have much lower energy contents.

The devices in this group are gallium arsenide-based, germanium-based, and CPV solar cells.

Gallium Arsenide Based Multi-Junction Cells

High-efficiency multi-junction cells were originally developed for special applications such as satellites and space exploration, but at present, their use in terrestrial concentrators might be the lowest cost alternative in terms of $/kWh and $/W. These multi-junction cells consist of multiple thin films produced via metal-organic vapor phase epitaxy. A triple-junction cell, for example, may consist of the semiconductors GaAs, Ge, and GaInP2.

Each type of semiconductor will have a characteristic band gap energy which, loosely speaking, causes it to absorb light most efficiently at a certain color, or more precisely, to absorb electromagnetic radiation over a portion of the spectrum. Semiconductors are carefully chosen to absorb nearly all of the solar spectrum, thus generating electricity from as much of the available solar energy as possible.

GaAs-based multi-junction devices are some of the most efficient solar cells to date, reaching a record high of 40.7% efficiency under "500-sun" solar concentration and laboratory conditions.

This technology is currently being utilized mostly in powering spacecrafts. Demand for tandem solar cells based on monolithic, series-connected, gallium indium phosphide (GaInP), gallium arsenide GaAs, and germanium Ge p-n junctions is rapidly rising. Prices are rising dramatically as well.

Twin-junction cells with indium gallium phosphide and gallium arsenide can be made on gallium arsenide wafers. Alloys of In.5Ga.5P through In.53Ga.47P may be used as the high band gap alloy. This alloy range allows band gaps in the range of 1.92eV to 1.87eV. The lower GaAs junction has a band gap of 1.42eV.

In spacecraft applications, cells have a poor current match due to a greater flux of photons above 1.87eV vs. those between 1.87eV and 1.42eV. This results in too little current in the GaAs junction, and hampers the overall efficiency since the InGaP junction operates below MPP current and the GaAs junction operates above MPP current. To improve current match, the InGaP layer is intentionally thinned to allow additional photons to penetrate to the lower GaAs layer.

In terrestrial concentrating applications, the scatter of blue light by the atmosphere reduces photon flux above 1.87eV, better balancing junction currents. GaAs was the material of the highest-efficiency solar cell, until recently, when Germanium-based MJ cells capped the world record at 41.4% efficiency.

Indium Phosphide Based Cells

Indium phosphide is used as a substrate to fabricate cells with band gaps between 1.35eV and 0.74eV. Indium phosphide has a band gap of 1.35eV. Indium gallium arsenide (In0.53Ga0.47As) is lattice matched to indium phosphide with a band gap of 0.74eV. A quaternary alloy of indium gallium arsenide phosphide can be lattice matched for any band gap in between the two.

Indium phosphide-based cells are being researched as a possible companion to gallium arsenide cells. The two differing cells may be either optically connected in series (with the InP cell below the GaAs cell), or through the use of spectra splitting using a dichroic filter.

The presence of varying quantities of toxic materials in these devices must be considered when planning their use in large quantities.

Germanium-based Single- and Multi-junction Cells

Germanium (0.86eV band gap) is a semiconductor material, with properties far superior to other substrate materials used for PV cells and modules. It is ~40-50% more efficient than silicon and has a much lower temperature coefficient. It is several times more expensive than silicon, too, but with new superior slicing techniques, it can be cut into very thin wafers, saving a lot of material. This, combined with its higher efficiency and less degradation than silicon, could put it on the competitors' list within the next few years.

Germanium-based solar cells have been used mostly for space applications, but a number of manufacturers have geared up for mass producing them for high concentration HCPV and other high efficiency applications (the record is currently 42.3% efficiency) .

CPV Solar Cells

Concentrating photovoltaics (CPV) is a branch of the PV industry, using special cells, optics and tracking mechanisms, developed in the 1970s by several companies under contracts and financing from U.S. Departments of Energy. Early CPV systems used silicon-based CPV cells, which had a problem with elevated temperatures.

These were later replaced by GaAs-based multijunction cells, which have much higher efficiency, but still suffer from the effects of high temperatures. At first GaAs CPV cells were made by using straight gallium arsenide in the middle junction. Later cells have utilized In0.015Ga0.985As, due to the better lattice match to Ge, resulting in a lower defect density.

As you can see in Figure 4-5, CPV cells are complex structures, consisting of many layers (some deposited, some diffused) in, or piled on top of, Germanium semiconductor material, which has more superior process and performance characteristics than silicon.

Current efficiencies for InGaP/GaAs cells are in the 40% range and constantly increasing. Research into methods to produce band gaps in the range between the Ge and GaAs is ongoing. Lab cells using additional junctions between the GaAs and Ge junction have demonstrated efficiencies above 41%. InGaP/GaAs CPV cells on GaAs substrate have demonstrated 42.3% efficiency.

CPV cells are mounted under lenses, which concentrate sunlight falling on the cells 100 to 1000 times. This allows high efficiency and reliability, better land utilization, and other benefits. Cell-lens assemblies are mounted on trackers, which track the sun precisely through the day, providing the most power possible. Efficiencies of 30-32%, measured in the grid are obtainable with these devices. We foresee CPV technologies as the primary choice for installation and use in large-scale power plants, especially those in desert regions.

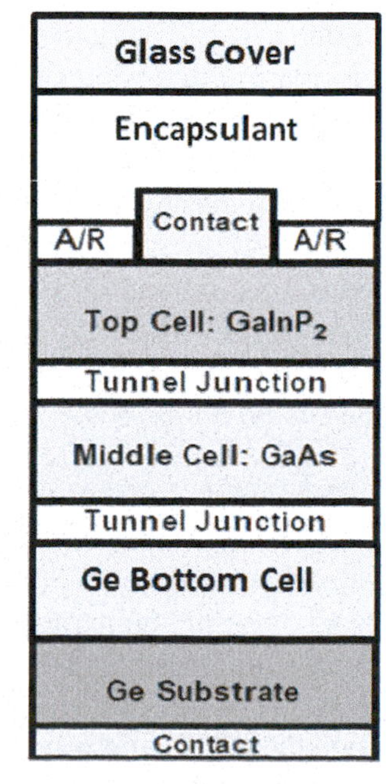

Figure 4-5. CPV cell (10)

Research Trends

Current research on high-performance multi-junction thin film devices focuses on several major challenges:

1. Determining high-bandgap alloys based on I-III-VI and II-VI compounds and other novel materials for the top cell.
2. Considering low-bandgap CIS and its alloys, thin-film silicon, and other novel approaches for the bottom cell.
3. Studying the difficult task of integrating the thin-film tunnel junction (interconnect) with the top cell. This work includes understanding the role of defects, how they affect the transport properties of this junction, and the diffusion of impurities into the bulk material.
4. Fabricating a monolithic, two-terminal tandem cell based on polycrystalline thin-film materials that requires low-temperature deposition for several layers.
5. Avoiding deterioration of the top cell when fabricating the bottom cell if a low-bandgap cell is fabricated after a high-bandgap cell with a superstrate structure, such as CdZnTe.
6. Avoiding temperatures and processes that could damage the CIS bottom cell if a high-bandgap cell is fabricated on top.

THIN FILM PV CELLS AND MODULES ANALYSIS

Thin film PV (TFPV) technologies are a newer branch of the PV industry, where new sophisticated vacuum/plasma deposition equipment and processes similar to those used in the semiconductor industry are used to deposit very thin (microns in most cases) light-sensitive films onto substrates such as glass, plastics or metals.

TFPV in general provide a good alternative for generating power from the sun in a simple and cheap manner. Nevertheless, as with any new and unproven technology, they are faced with some challenges which need to be addressed.

Background

The physical and chemical properties of TFPV technologies are well understood because they are similar to those of semiconductor devices, which have been around for decades. Thin films have been used in their present configuration in the semiconductor industry since the 1970s, and while the processes are basically the same, the equipment has gone through many changes through the years. The TFPV industry benefits tremendously from using such well understood processes and sophisticated equipment, which lends itself to automation and mass production.

There are, however, several differences in the way semiconductor and TFPV films and products are made and used.

1. The semiconductor industry uses small process chambers, where usually only one or a few wafers are placed and processed as a batch under the strictest of controls. Deviations or exceptions are simply not tolerated, so the quality of raw materials, consumables and procedures is second to none. The product is thoroughly checked for electro-mechanical defects—100% product inspections, several times during the process sequence. Sometimes it is inspected after each step, and even several times after some steps.

   Most high volume thin film PV deposition processes, on the other hand, are executed in large in-line, conveyor type tools, consisting of a series of deposition chambers. The conveyors move slowly along the line, transferring the glass substrates from chamber to chamber where the different materials of the thin film (layers) are deposited under varying conditions. The emphasis of the process design and execution is on fast deposition times, resulting in high volume production and cheap final product with acceptable, but not superb quality.

2. Finished TFPV products are indeed very cheap—with a market value of ~\$1/Wp, or ~\$70/m2. Compared with a market value over \$250,000/m2 for some finished semiconductor devices, this is a big difference, so only very high volume and the cheapest starting materials can ensure decent profit for TFPV module manufacturers. There are of course limits to how fast and how cheaply one can push the production line before starting to produce junk, but TFPV manufacturers have been quite successful in keeping the delicate balance between speed, safety, cost and quality thus far.

3. Both semiconductor and TFPV devices contain a small quantity of toxic materials in their structure, such as Cd, Ga, As, In etc., but this is where the similarities end. Semiconductor devices are spread around so that only very small quantities of toxic materials can be found in any specific area. TFPV modules, on the other hand, are making a quick entry into the large-scale energy

market, where millions of them containing toxic materials cover thousands of acres of virgin desert lands. This unprecedented massive deployment of a fairly new product that is unproven for long-term exposure to a hostile climate elicits a number of questions, the answers to which are unavailable. We can only hope that these and other unresolved issues revealed in this text will be addressed properly by the US scientific community, manufacturers, regulators and product users in the near future.

Thin Films Structure

As the name implies, "thin films" are just that; very thin films (layers) of organic or inorganic materials. Thin film PV (TFPV) modules consist of several very thin layers (thin films) of different materials piled up on top of each other, to form a structure that is suitable for trapping and converting sunlight into electricity. The thickness of each film is usually several microns (1 micron is 0.001mm, or 0.0004 inch). Visualize the thickness of a human hair (avg. 100 microns) and you'll get a good idea of what 100 microns is—50-100 times the thickness of an individual thin film.

Now visualize layering these super-thin films until there are 8 or 10. This is how TFPV structures are made and it is what they look like. A better visualization might be using a piece of Scotch tape as an example. Standard Scotch tape is ~0.060mm thick, or 60 microns, which is at least 10-20 times thicker than most thin films. Yes, the entire thin film structure—all different layers combined in your TFPV modules is many times thinner than a Scotch tape strip. And the different layers (different chemical compounds) are even thinner. They are stacked on top of each other, held together by weak electro-mechanical forces of complex nature and behavior. We'll take a closer look at these forces and the related interactions in this text.

Thin films of any type and size are affected by chemical, mechanical, thermodynamic and electric forces and changes in, between, and around them. TFPV structures also depend heavily on the surrounding materials, glass, laminates, etc., and components in the PV module for protection. Complex picture, yes, but well understood by design and process engineers and research scientists, thanks to the broad experience we've gained from the semiconductor industry, which is based on thin films and processes.

NOTE: Remember this picture, because we will revisit it later on in this chapter, to explain the behavior of thin films in TFPV modules under different environmental conditions.

Thin Film Manufacturing Process

So let's see how these thin films and TFPV modules are made today:

Thin Film Deposition Processes

Thin film PV cells and modules manufacturing processes, similar to thin film processes used in the semiconductor industry, are well controlled. The level of process control depends only on the quality requirements and budget restrictions. The actual thin films deposition is usually done on a substrate (glass, plastic or such) which has been thoroughly inspected, cleaned and otherwise properly prepared for the deposition step.

a. Substrate preparation is a key factor in maintaining process control, and determines the overall quality, performance and longevity of the final product. Basically speaking, dirty substrate will not only produce defective product, but will also contaminate the equipment, forcing lengthy and expensive clean-up procedures. Pre-deposition cleaning of large glass substrates (panes) is done in automated washers, where brushes, soap and/or high pressure water solutions remove all particles and organic material from both surfaces. The glass panes are then rinsed with DI water and dried with forced air and heat applied to both surfaces. This step is also critical, because moisture retained on the large surfaces is a great enemy of vacuum/plasma processes and could seriously affect product quality.

b. The substrate is then placed in a large vacuum chamber (usually on a horizontal or vertical conveyor belt that moves the material along the process path), where powerful vacuum pumps suck out all the air, to clear any mechanical (dust) and chemical contaminants (reactive gasses and water). The substrates are then usually heated, to remove any residual moisture and to heat the surface close to the temperature of the deposited species, reducing potential thermal disequilibrium and stress of the thin films to be deposited on it. Argon gas-based DC or RF plasma is ignited and maintained at a proper pressure and power density during the process, facilitating the deposition process and the related reactions.

c. The material to be deposited as thin film is then evaporated (by actually melting the material and directing the resulting vapor clouds onto the deposit surface), or it is sputtered (by dislodging small clusters of the material via high voltage-generated ion bombardment) onto the substrate. These processes are also called chemical vapor deposition (CVD) and physical vapor deposition (PVD), respectively.

In both cases thin film material particles impinge on and adhere to the heated substrate surface on impact. The strength of the adhesion between film and substrate, or between the individual films, depends on the design and execution quality of the entire process sequence—quality of all materials, cleanliness of substrate and chamber interior, vacuum integrity, actual process temperature, forward and reverse plasma and substrate bias power levels, time duration, partial and total gas pressures, speed, and other process variables.

d. Coated substrate with the films deposited on top of it is taken out of the chamber and exposed to a number of additional operations, such as wet chemistry treatment, rinsing, drying, annealing, and wire attachment. The above CVD, PVD and wet chem processes are repeated several times for some devices, following strict process and quality control procedures all through the sequence. Upon completion of the PV structure creation, the substrate with the deposited thin films is joined to similar substrate (glass usually) with the help of adhesives and encapsulants. Thus, TFPV modules are tested, sorted and packaged for shipment to lucky customers.

Process and Device Issues

No process is perfect, and a defect could be generated at any step of the manufacturing process, including the tightly controlled thin film deposition processes in the semiconductor fabs. Starting with the basics, materials selection is paramount for obtaining quality product. "Garbage in, garbage out," goes the saying. So the quality of the substrates, process chemicals and gasses, and all consumables must meet and exceed specifications—not a simple task. The different pieces of process equipment must be well taken care of, tuned and qualified periodically, to make sure they are in good shape and operating within spec.

a. Step one of the thin film process is cleaning substrate materials. Contaminated cleaning materials or equipment or improper procedures will render the glass surface unsuitable for proper adhesion of deposited films. Cleaning the substrate with poorly selected chemicals, or shoddy procedures, might even introduce fatal impurities, such as Cu, K, Na, Fe metal ions, and phosphates. These impurities would eventually start their own demolition processes from within the thin film structure enclosed in the PV module, reducing its efficiency and longevity. Dirt particles, fingerprints, residual moisture, and strong electrostatic charge on the glass surface entering the vacuum chamber will have a profound, usually negative, effect on process integrity and final product quality. These are only few of the things to watch for at this stage of the process.

b. The plasma deposition process could also introduce impurities and defects in film structures if improper parameters or dirty hardware are used. Impurities in carrier gasses, contaminated vacuum chamber walls, back-streaming vacuum pumps, unplanned drops or increases of total or partial pressure, air leaks, process time or belt speed discrepancies are process abnormalities which could introduce other impurities or create defects with immediate or latent problems.

c. Out-of-spec process materials and consumables could be blamed for many process failures. Poor quality substrates, contaminated gasses and chemicals (overlooked during incoming quality control procedures) could cause slight or very great defects in the finished product. But most often, it is the process itself that creates problems in thin film manufacturing.

Generally speaking, there are a number of known, unknown, controlled and uncontrolled manufacturing process factors and variables in most thin film deposition sequences:

a. Human error is the #1 variable in high-volume thin film manufacturing. People tend to cut corners, improvise, push the wrong buttons, rush through operation, maintenance and inspection steps, etc. These kinds of workplace behaviors cause defects of unpredictable proportions and consequences. Handling substrate materials and consumables, as well as operating the complex process equipment requires highly trained engineers, technicians and operators, who in many cases have bad habits that are hard to break even with the best of training.

b. Equipment quality and malfunctions are #2 on the list of process-related variables in sophisticated thin film manufacturing operations. Thin film process equipment is complex in its design, operation and maintenance. Most production equipment made now is of good quality, but even the best equipment can malfunction and create headaches. Key components and process control instrumentation slowly drift out of spec over time, and product quality could vary from batch to batch. One serious drift in the multi-step process could cause serious defects in a batch. Anomalies during processing also occur unexpectedly and quickly; if not handled promptly and properly they could cause quality problems.

c. Poor quality supplies, materials and chemicals purchased from third-party manufacturers are another serious problem that we encounter in the thin film process. Since high volume operations use a lot of outsourced materials, the incoming quality is hard to verify 100% of the time, so any problem at the third-party vendor's plant will have negative effects in the final PV product quality.

A number of additional factors and variables affect the quality of thin film PV modules, but it suffices to say that the TFPV manufacturing process consists of dozens of different materials and complex process steps, each of which must be immaculately designed, planned, executed and controlled. Basically, the lowest quality of any process step (or material) in the process sequence determines the highest quality of the final product. One mishandled step, or one bottle of contaminated chemical or gas, could lead to rejection of a batch of modules, or worse—to their failure in the field, where things get very expensive...and embarrassing.

TFPV Functional Considerations

Even though the electro-mechanical and chemical properties of PV thin films are thoroughly studied and well understood, improper design and manufacturing procedures are still encountered at times. We already discussed the TFPV manufacturing process and the related issues, so now we'll take a closer look at the behavior of thin films in TFPV modules. (Refer to Figure 4-6.)

Remember our discussion at the beginning of this chapter that thin films are 1 micron thin, or thinner, and that this is less than 1/100 the thickness of human hair. Recall, too, that the entire thin film structure in TFPV modules is several times thinner than a piece of Scotch tape.

Some of these thin films are so thin in places that you could actually count the molecules across the film thickness, if you had a way to do that. And as you look at thin film very closely, you'll see all kinds of imperfections as well; interruptions, distortions, breaks, cracks, splits, slips, pits, voids, bubbles, lose particles, and all kind of other sub-structures in/on the thin film structure. In other words, the films, although they look perfectly smooth and uniform to the naked eye, are anything but.

And you will also notice in Figure 4-6 a clear delineation between the films—the boundary (interface) between each pair. These are very special areas, which play a huge role in the performance of the PV devices made out of these films. They are also the weakest points in most thin film structures, because it is where the electrical resistivity, mechanical stress and chemical degradation are usually initiated, stored, executed and amplified. These boundaries represent the time and place where the process was interrupted and switched from one material deposition to another; and/or from one deposition step to another. This involves moving the substrate, which could be accompanied by longer than specified delays, switching gasses, changing vacuum and plasma power levels, and a number of additional inter-step process modifications.

The inter-step changes do cause momentary out-of-control conditions which could cause a number of process and product abnormalities, such as contamination of the surface by carrier gases or pump oil, and temporary process destabilization (total and partial pressures, gas mixing, power fluctuations, temperature imbalance, etc.). All the combinations and permutations of possible process abnormalities and extraordinary scenarios is too much to discuss here, but you get the idea; this is a most complex set of parameters and conditions that must be executed perfectly all the time at all process steps, to obtain the best quality of final product. For the purposes of this writing, however, we'll just agree that the inter-layer boundaries (interfaces) are extremely critical areas, which have different properties than the parent materials. These areas are

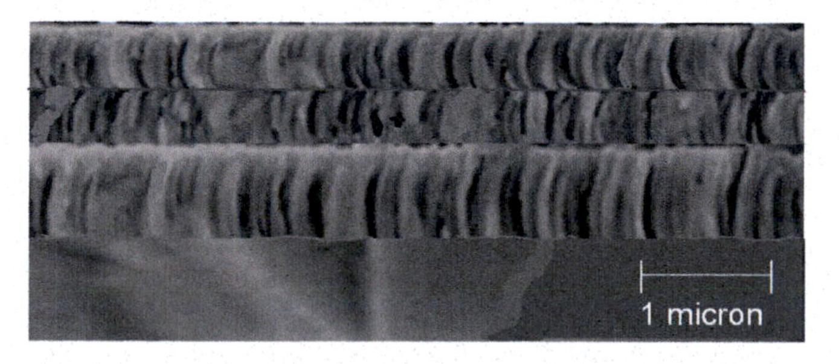

Figure 4-6. Cross section of a typical TFPV structure

more fragile, so there is a limit as to how much abuse they can handle during long-term field operation.

Each material pair and its interface has well defined mechanical, chemical, electric and thermal stress limits, depending on the critical de-bonding energy and chemical inertness levels specific for each interface between two different materials. How much mechanical stress, heat, electrical charge, chemical reactivity or combination of these will be needed to cause delamination, adhesion problems, mechanical, electrical or chemical changes and/or disintegration of the bond between the materials, and eventually the materials themselves?

Amazingly enough, the energy levels and forces acting in these areas are very small, relative to the mass of the materials and strength of the films; i.e., Van der Wall (VDW) =/~0.001-0.4 eV/bond, H-bonds =/~0.11-0.44 eV/bond, SiO2 cracking in water E=/~1.39 eV/bond. So, without getting into much detail, just looking at such extremely small numbers one can deduct that it doesn't take much effort to damage an interface which will affect the entire thin film structure. Additionally, in most cases, many of these forces are acting together—on and off many times daily for the duration (30 years or more) of the module's on-sun operation. Worst of all, once these destructive processes have started, they cannot be stopped. As a matter of fact, things usually only get worse with every daily cycle, and can accelerate quickly.

Taking a closer look at the major destructive processes acting upon thin film PV structures, we see:

Mechanical Stress; Thin Films and Boundaries Disintegration

Thin film modules are exposed to mechanical stress from the moment they leave the production line. Never ending vibration, hitting, rubbing, pushing, pressing and squeezing of the modules, affect the layers in the thin film structure. Stress is induced during handling, packing, transport (long truck and train rides are the worst), installation (careless handling during installation is a major problem) and operation (high winds, hail and storms are some of the worst enemies). Each of these actions and interactions increases the mechanical stress in the film structure. (See Figure 4-7.)

Because of the films' non-uniformity, and the different coefficients of expansion of the different adjacent films, there is a lot of stress and friction within each film and among the adjacent films. As with other materials stuck together, they will be slipping and sliding, expanding and contracting , thus creating never ending bending, pulling, pushing and tugging against each other and the materials around them, be it thin films or encapsulants, glass etc.

In most cases these activities will produce some usually unpredictable changes. The question is what and how bad will these changes be? Stress, cracks, voids and general weakening of the thin film system is expected with time, followed at times by partial or full disintegration of the films, depending on operating conditions. Results from these changes would be expressed as gradual loss of power, intermittent power, and finally complete failure.

Electrical and Heat Stress

Generation of electric energy is the primary purpose of any PV module. The photoelectric process and the accompanying extraction and transmission of electrons (electric current) through the different thin film layers, their interfaces and contacts is usually accompanied by heat generation. Parasitic (excess) heat is generated when the resistivity of the materials increases, which inevitably happens when the internal temperature of the module goes up. And it starts going up from the second the sun hits the module. The higher the sun, the more electricity is generated, and the higher the resistivity goes.

On a bright sunny day in the desert, the module interior could see temperatures as high as 180°F, or more. The heat build-up is a combination of the simultaneous increase of air temperatures and internal resistivity of the thin films and their interfaces. This heat is not enough to damage or destroy the films, since each one can withstand much higher temperatures. The excess heat, however, forces the materials to expand and shift in one direction. They then shrink at night and move in the opposite direction, which creates friction at the interfaces and stresses the materials in each layer. The more temperature differential the layers are exposed to, the more they stress and shift.

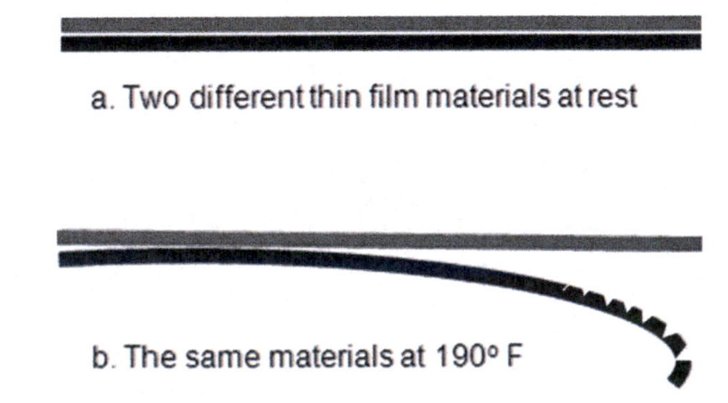

Figure 4-7. Thin film behavior under elevated temperature and humidity

Now imagine these films, packed tightly in the module, getting very hot during the day in the desert (measured inside module temperatures exceed 190°F in Arizona deserts), and freezing at below −20°F at night in winter. This process goes non-stop, 365 times a year. This translates into ~3.650 min-max temperature cycles in 10 years, and close to 11,000 cycles in 30 years. Can the thin films withstand this constant push and pull, up and down, left and right? Would one of them give up and break down mechanically? Would that cause an untimely power drop, or complete failure?

The results from one, two, or even hundreds, of these cycles would go unnoticed, but the non-stop (30 years and over 10,000 cycles) effect of the never ending push and pull of the layers will fatigue thin film structure, shifting it into a different performance mode. As the module gets older, the resistivity of the films and their interfaces increases proportionally, according to the internal heat build-up and dissipation within the module. Electrical output will drop by ~1.0% every year, partially due to the above effects, or combinations of them.

There is no getting away from the harmful effects of excess heat and moisture in field operations. They are variables in the PV generation equation, which cannot be ignored, but are hard to control.

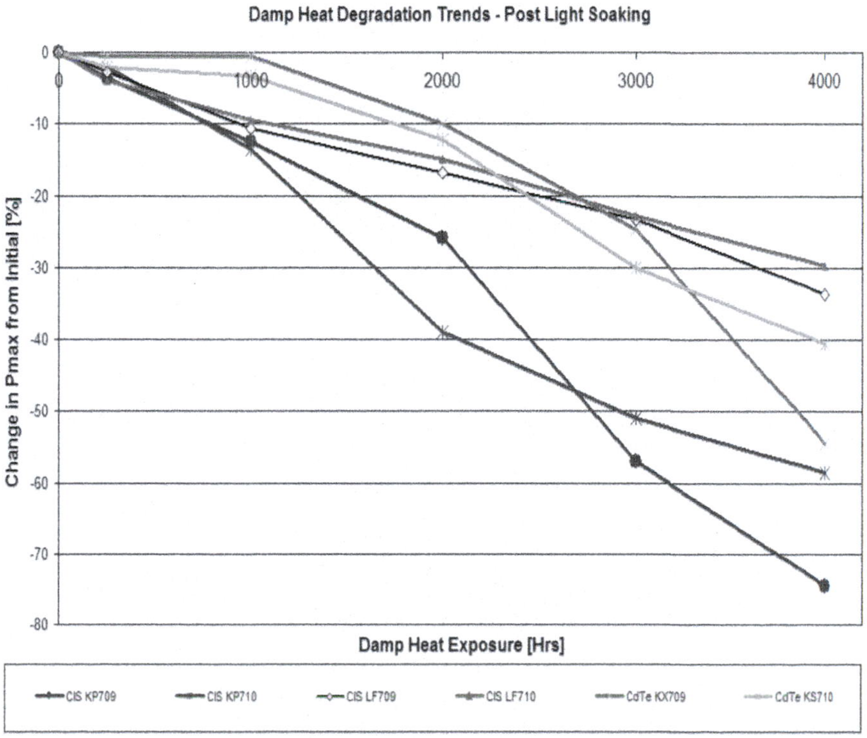

Chemical Reactions and Thin Films Decomposition

All unprotected thin films—without exception—react with many chemicals. It is usually the interface (the boundary between the films) that is affected first, and is where problems can be observed. Some films and interfaces will disintegrate instantly with a simple touch (human sweat contains salts), while others will withstand weak chemicals, and some will require strong acids or bases in order to dissolve or decompose them or their interface. Nevertheless, all thin films are subject to mechanical and chemical changes under certain conditions.

CdTe, CIGS, a-Si and other thin film structures, as well as interfaces between layers, are affected in a similar fashion. Some of them are more chemically resistive than others, but prolonged exposure to weak acids (brought on by moist air, or contained in rain water) penetrating the module lamination (or entering through cracked glass) will eventually cause the films to react, delaminate and decompose. The type and speed of the decomposition process, and the newly created chemical species, will be determined by the types of thin films, the active chemicals species and the types of reactions these invoke.

Again, we count on the encapsulants and glass or metal frame to keep moisture, air and related chemicals out of the module for 30 years. How many modules will survive the test of time? What would be the total failure rate? These are questions to which we simply have no answers because there are no precedents.

Moisture ingress is the culprit of pronounced power output degradation of CIS and CdTe modules in Figure 4-8, reducing their output significantly within 6 months of exposure. This example also leads to the conclusion that good encapsulation is paramount, but since it is never perfect, harsh climate conditions will force moisture to penetrate thin film structure and change its composition and behavior, resulting in power loss and, eventually, failure.

Environmental effects initiate and accelerate the evolution of defects in TFPV modules. Damaged encapsulation allows moisture to penetrate modules and attack layers, resulting in their delamination or separation from the substrate, as shown in Figure 4-9.

Figure 4-8. Moisture caused degradation of CIS and CdTe modules (8)

Just as in tooth decay, once moisture or environmental acids and gasses find a gap in the encapsulation they start the decomposition process which cannot be stopped without outside intervention. A dentist has the option to mechanically separate (drill) decay out of the tooth surface, but drilling decomposition damage out of a TFPV module is impossible. This means that modules will start losing power at a rate proportional to the inflicted damage, usually much quicker than the standard 1.0% power loss per annum. Eventually—when the internal decay has affected large parts of the critical areas of the thin film structure—the affected modules will fail completely and must be replaced. The process is accelerated when affected modules are exposed to extreme heat and humidity (see Figure 4-10).

The effects depicted in Figure 4-11 play a significant role in compromising thin film structure integrity. Some of these are vicious and fast acting, while some are slow and cause less damage. The actions of each effect are hard to predict, because different operating conditions have their own peculiarities. It is even harder to predict the combined effect of these processes during non-stop, 30-year, on-sun operation.

The inevitable exposure to excess UV and IR radiation, thermal cycling, mechanical stress, moisture ingress, and chemically active environmental species leads to unpredictable degradation, and unreliable kinetics and reliability models, because of the ever changing types and numbers of forces acting on the materials in inhospitable regions.

In summary, we don't know what to expect of fragile thin film layers in TFPV modules exposed to unending mechanical, chemical, electrical and thermal action and on-sun operation for 30 years or more. Still, there is enough evidence to conclude that PV modules made with quality materials via proper processing would have a greater chance to survive over time than those made of low-quality materials, and/or using poor design and flawed manufacturing processes.

Efficiency, Costs and Longevity

TFPV modules have several serious inherited issues and exhibit a number of abnormal behaviors (discussed above) that need to be brought out. This is even more important now, since there is a lot of misunderstanding, misinterpretation and general confusion on the subject of efficiency vs. cost vs. longevity of TFPV

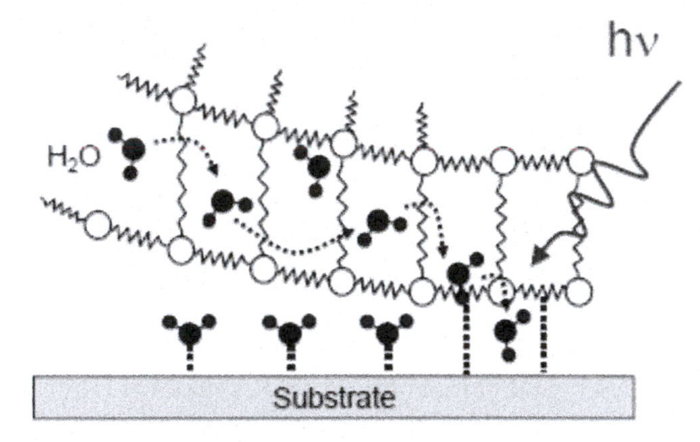

Figure 4-9. The chemical invasion process on molecular level (11)

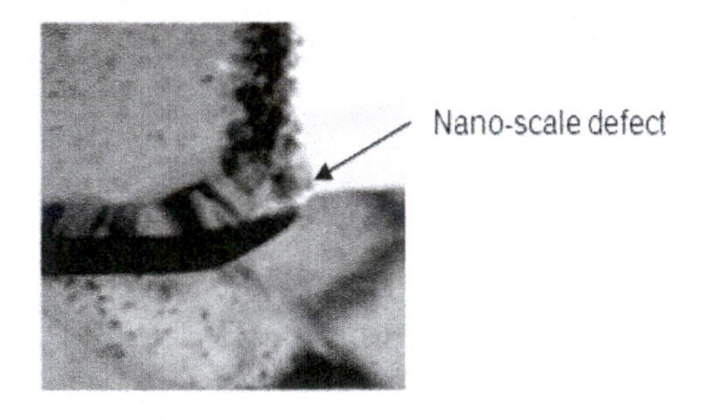

Figure 4-10. The invasion process has started (11)

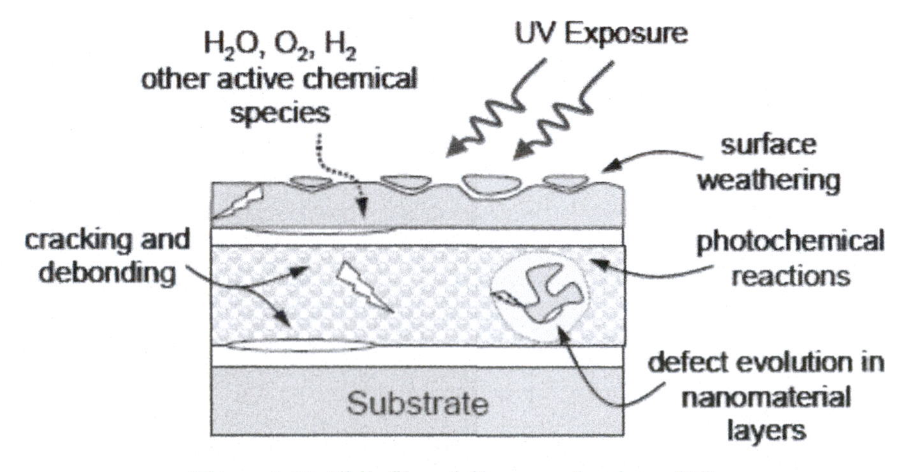

Figure 4-11. Thin films failure mechanisms (11)

modules.

First and most importantly, we need to keep in mind that TFPV module manufacturers are relative newcomers to the solar field. Most have been in existence only a few years, with even shorter field experience and insufficient on-sun exposure test data. Therefore, they have little experience with long-term field behavior of their TFPV modules, especially under extreme climate conditions.

The other obvious drawback of TFPV modules is their low efficiency, which is even lower under the hot desert sun (due to temperature degradation). Also, they demand use of larger than normal structures and land areas, and there is an annual 1% or more power loss, which further decreases power output over time. Most of the available test data are from "standard" tests done under "normal" conditions, with no mention of performance issues under the extreme conditions of the US deserts where many large TFPV plants are installed, or where future installations are planned. So let's take a closer look at the parameters affecting present-day TFPV technologies:

Efficiency

A major concern with commercially available TFPV modules is their relatively low efficiency, in the range of 8-9% measured at the module level. When the modules are connected in an array, plugged into the grid, and all related losses are taken into consideration, we end up with 4-6% efficiency. We have seen record low performance of TFPV power plants operating at 2-3% efficiency in the grid. Just think...2% efficiency. It takes ~10 acres to generate 1.0 MWp power with present-day 8% efficiency TFPV modules. At 2% in-the-grid conversion efficiency, we will need more than 40 acres to produce 1.0 MW AC power—not an exciting proposition.

Land Use

The land area needed to generate electric power at 8% DC conversion efficiency (at the module level) averages 10-12 acres per each MWp installed, or more than 1,000 acres would be needed to install one 100 MWp TFPV power plant. Obviously, at this efficiency, land-related requirements and expenses are huge, and not easily justified no matter how you look at it—unless the land is free, as is the case with that belonging to BLM (US Bureau of Land Management), who can and does give it away as they see fit.

In any case, TFPV modules require 2-3 times the land needed to install c-Si modules for the same power output. Mounting TFPV modules on trackers to boost output might be one solution, but the disproportionately large numbers of trackers needed (due to low efficiency) increase initial costs that usually cannot be justified by slim returns.

BOS

Balance of system (BOS) consists of support frames, controllers, inverters, wiring, cement, hardware, and everything else needed to install and operate a TFPV system. Labor for installing BOS components is also part of these calculations.

Table 4-1. Cost breakdown in % of total system

Installation	c-Si PV %	TFPV %
PV Efficiency	16	10
Land purchase	10	14
Land preparation*	3	5
Fencing & security	1	2
PV Modules	58	38
Inverters & controls	8	12
Substructure	12	16
EPC & management	5	7
Wiring & electrical	3	6

*Including land permits, PPA, leveling, etc.

Operation		
Maintenance	2	4
Recycling & disposal	2	6

This shows also that we should expect a $500-750/kWp increase in materials and installation costs—a penalty for using low efficient TFPV modules which require more land, inverters, support structures and interconnect materials. With all fairness, the TFPV BOS cost can be reduced with some work, but it is still higher than c-Si power arrays, so it should be considered in all power field designs and installation calculations.

BOS costs are related in a way to the PV market's ups and downs, but follow the commodities and components supply and demand laws as well. All indications are that overall costs of materials (metals and plastics) will keep increasing, so the difference will grow and must be addressed and resolved soon, lest it becomes another stumbling block for the less efficient TFPV technologies deployment.

Cost of Capital Equipment

Thin film PV manufacturing sequence is similar to that of semiconductor devices, including equipment and processes, so it faces a number of similar hurdles as they advance toward higher-volume and lower-cost goals. There are also substantial similarities among all thin films, which allow for simplifying the analysis of thin film module manufacturing costs, since most thin film cells and modules have at least some of the following characteristics:

a. Transparent front layer (or glass) that also protects from the environment;
b. Transparent and conductive top layer or grid that carries away current;
c. Thin (1-4 micron), central sandwich of semiconductors that form one or more junctions to separate charge;
d. Back contact that is often a metal film;
e. Back sheet that protects from the environment and that could be supportive (rigid or flexible);
f. Various intermediate processing steps: scribes and depositions to interconnect strip cells, annealing steps to activate or complete certain components;
g. Lamination to attach encapsulation;
h. Buss bar attachment to carry off power;
i. Including a number of peripheral operations, such as isolation scribes at the borders, glass or other substrate handling, cleaning and heating.

Thin film modules differ in performance, and that determines their cost per Wp before installation. To get a true picture of manufacturing costs, our analysis must be based on cost per square meter. Performance can be folded in later to get $/Wp comparisons.

The major categories of capital cost are: 1) the equipment used to make the active semiconductor layers; 2) the equipment used to make the rest of the module (e.g., metal contacts, lamination, scribing, annealing, etc.) and 3) all other support equipment, infrastructure and labor involved in the process.

The most critical aspect of capital cost, and the biggest deference, is in the equipment used to deposit semiconductor thin films. The rest of the capital costs will be similar for all different thin film technologies. In all cases thin film equipment is more complex, more expensive than that used for making c-Si cells and modules, and requires qualified personnel to run it.

All these costs and expenses affect the price of the final product, but thin film equipment can be automated and run very efficiently, contributing to the lower cost per watt of some TFPV modules. As a matter of fact, some TFPV module manufacturers have broken the $1.0/Watt price barrier and are aiming to drop the price even lower. The overall quality of mass-produced modules is also increasing, as manufacturers gain experience.

Maintenance Cost

Many industries estimate maintenance costs using a fixed percentage (about 4%) of initial capital cost as the expected annual maintenance cost of equipment. This may appear small, but since it is not reduced by dividing by a depreciation period, it is actually quite similar to the amortized capital cost. This increases the impact of high capital costs and emphasizes the difficulty of some approaches.

Thin film operations require especially complex and expensive labor-intensive procedures. Vacuum pumps and instrumentation equipment must be immaculately maintained, calibrated and tuned at all times. The cost of maintenance is included in the product cost, and because the TFPV industry has the lowest cost/Wp, we must assume that maintenance costs are not exuberant.

Raw Materials Availability and Costs

A major expense in the production of TFPV modules is the cost of the materials they are made of. As an example, CdTe TFPV modules require 10-12 grams cadmium for a finished PV module, 10-12 grams tellurium, and 2-4 grams S. Large quantities of process materials, chemicals, and gasses are needed also.

Many of the key materials used by the TFPV industry are by-products from mining operations and their cost varies with availability and with the world's commodity prices. Materials for CdTe TFPV modules cost about $48/m2 (2009 prices). From this only about 10% is the cost of active materials (semiconductor type thin films). The rest is the cost of encapsulation, substrates, and other materials and components. Therefore, the largest portion, or ~$43/m2 of the TFPV module cost will be the same across TFPV technologies. Only the cost of the active materials may vary for the different thin film deposition processes—the cost of specific materials and their thicknesses (quantity).

R&D Cost

Like other semiconductor technologies, thin film PV is scientifically driven. For a company to maintain its profitability, it will have to be a technical leader. Staying ahead in PV device design and module manufacturing requires this technical leadership. The baseline estimate of R&D cost for thin film manufacturing companies is ~$4/m2. This is about $1 M for a 20 MW plant. It is also close to the cost of labor for the manufacturing plant.

However, in the future, as volumes increase, R&D costs will remain proportionally smaller and should become a manageable part of the total cost. For example, if R&D costs maintained the same percentage of module price in dollars per watt, at a price of $1/W, R&D costs would be an affordable $1-$2/m2. Clearly, as volumes increase, R&D costs will not rise at the same pace.

Warranty

The warranty we'd expect to get for any size PV project is a limited warranty for the particular PV modules we use. This warranty is issued by the manufacturer and is usually divided into sub-sections, as follow:

a. Workmanship warranty
   This is usually a 6-months, 1- or 2-year repair, replacement or refund (RRR) warranty, in which the manufacturer promises/warrants its PV modules to be free from defects in materials and workmanship, if used under "normal conditions, including installation and O&M."

b. Twenty-year limited peak power warranty
   This is warranty that most PV module manufacturers provide to cover the expected and unavoidable short-term 1% annual power output loss which is characteristic for most c-Si and thin film PV modules.

c. Exclusions and limitations
   There are a number of conditions to be taken in account in case of a claim, and usually go along the lines of:
   i. Warranty claims must be filled within the warranty period.
   ii. Limited warranties do not apply to any PV-modules which in the manufacturers' discretion have been subjected to:
      1. Misuse, abuse, neglect or accident;
      2. Alteration, improper installation or application;
      3. Disregard of manufacturer's installation and maintenance instructions;
      4. Repair by un-approved service technician
      5. Power failures, surges, lighting, flood, fire, accidental breakage or other natural or man caused events or accidents.
   iii. The limited warranty usually does not cover transportation, return, or removal and re-installation costs.
   iv. Warranty claims usually stipulate that if the manufacturing label or any other identification has been altered, removed, or made illegible, the warranty is void.

Recycling

Many thin film modules contain toxic materials and are required by law to be removed, packed, picked up, transported and recycled as hazardous waste at the end of their useful life. This requires the manufacturers of such products to include, in the initial price, the cost of removal, packing, transport, recycling and disposal of the product; or as an alternative, to have a special fund or insurance that will ensure proper and efficient execution of the end-of-life steps outlined above.

40,000 MT PV components have been forecast for decommissioning and recycling by 2020. This number will double or triple during the following decade. This includes silicon and thin film based PV components recycling. The thin film component will be ~20% by then. 8,000 MT of CdTe thin film PV modules will be the majority of the recycled product in this category. This number will grow exponentially with the unprecedented growth of the large-scale TFPV power plants.

EU already has directives for voluntary and extended manufacturer responsibility, where the decommissioning, transport, storage, processing and disposal of the modules is the ultimate responsibility of the original manufacturer. The directive has been integrated into the legal system of several member states, and we hope that a similar initiative will be undertaken in the US, before we end up with a huge Hazmat nightmare in the California and Arizona deserts.

Production Cycle Safety

Thin film PV modules are usually manufactured in semiconductor type equipment, using specifications and procedures similar to those used in semiconductor fabs. These procedures—especially in the US and western countries—are quite strict and include health and safety clauses and precautions which have been proven to ensure safe environment under all working conditions and with any materials and components.

This is one of the best points, when comparing thin film PV modules with other technologies, including c-Si. Thin film manufacturing cannot be done in a dirty place, as is sometimes the case with c-Si manufacturing facilities. Nor can it be done sloppily, because the equipment and its sophisticated controls simply will not allow it. c-Si cells and modules could be, and often are, manufactured sloppily, since many of the operations are done by hand, by workers who might be ill trained or asleep on the job.

Yes, human error can be found anywhere, but it is much less frequent in thin film manufacturing facilities than in c-Si facilities.

Field Operation and Environmental Safety

All PV modules have a propensity for causing environmental damages. Some of the most dangerous, however, are TFPV modules which contain significant amounts of toxic materials (CdTe, CdS, GaAs etc.). The dangers during manufacturing these products have been well researched and understood. Their long-term operational safety in the field and especially in hot desert and humid areas, however, have not been thoroughly investigated, and we know practically nothing about their behavior during 30 years under extreme heat or elevated humidity in some of the proposed mega-installations.

The manufacturers and proponents of toxic thin film products have been successful in offering incomplete explanations and solutions until now, which, in light of the impressive success of some TFPV manufacturers, have been accepted by the general public. These issues have not been reviewed in detail by the scientific community and the regulators, so we hope that an open discussion will follow soon.

In summary, we are supportive of the responsible expansion of all TFPV technologies which can be used in residential and small commercial installations and even bigger installations in moderate climate areas. However, we caution against massive expansion of some TFPV technologies into large-scale power fields in US deserts. Efficiency, performance, longevity and safety issues must first be addressed and resolved to give us reasonable expectations for long-term safety and success of those yet unproven for use in harsh environments.

TEST RESULTS, ISSUES AND DEVELOPMENTS

To ensure the quality and proper function of TFPV modules, a number of tests must be conducted during and the manufacturing process. Modules that pass these tests are then sent to EU and US labs for certification, as needed to market these in the respective countries. Other sets of modules are sent for field testing to these or different facilities. While certification tests are standardized and mandatory (i.e., no modules can be sold without a UL label in the US), there are presently no official standardized field test requirements. So for now, the official certification tests are all we have to go by when evaluating different TFPV modules made by different manufacturers.

Certification Test Results

We've been talking about potential problems with thin film PV modules and unintended consequences. Now we will take a close look at what official certification tests of all kinds of PV modules show. Keep in mind the above-mentioned issues as you review and evaluate the test results, and imagine what could happen if some of these modules were installed and operating in a power field.

PV Modules Tests 2007-2009

The results from a series of tests with several hundred PV modules from a dozen major PV module manufacturers, done by TUV Rheinland during 2007-2009 and published in *Photovoltaics International*, May 2010 (1), outline clearly the issues at hand with both, c-Si and thin film modules, as follow:

a. Crystalline silicon PV modules test results

- 3% failed at the very first wet resistance test (down from 5% in 2007)
- 16% failed after 200 thermal cycles* (up from 12% in 2007)
- 3% failed UV test
- 5% failed after 50 thermal cycles (up from 1% in 2007)
- 14% failed after 10 humidity freeze cycles (up from 12% in 2007)
- 11% failed damp test (down from 28% in 2007)
- 2% failed outdoor exposure test
- 6% failed termination test (down from 2% in 2007)
- 5% failed hail impact test
- 7% failed static load test
- 7% failed diode test (down from 31% in 2007)
- 9% failed hotspot test (up from 6% in 2007)

*NOTE: 200 thermal cycles represents 200 days on-sun operation, while 30 years of thermal cycling = ~10,000 cycles.

**NOTE: 10 humidity freeze cycles represent only part of an Arizona winter's worth of freeze cycles, while 30 years would consist of over 1,000 cycles, and many more in northern states.

b. Thin film PV modules test results:

- 1% failed at the very first wet resistance test (down from 20% in 2007)
- 12% failed after 200 thermal cycles* (down from 20% in 2007)
- 12% failed after 10 humidity freeze cycles** (down from 16 in 2007)
- 31% failed damp test (down from 70% in 2007). There is big problem here…
- 5% failed outdoor exposure test

12% failed termination test

6% failed hail impact test

12% failed static load test

10% failed hotspot test

*NOTE: 200 thermal cycles represents 200 days on-sun operation, while 30 years of thermal cycling = ~10,000 cycles.
**NOTE: 10 humidity freeze cycles represent only part of an Arizona winter's worth of freeze cycles, while 30 years would consist of over 1,000 cycles, and many more in northern states.

Test Criteria

A module design shall be judged to have passed the qualification tests, and therefore to be certified, if each sample meets the following criteria:

a. Degradation of the maximum power output at standard test conditions (STC) does not exceed 5% after each test or 8% after each test sequence;
b. The requirements of tests 10.3 (and 10.20) are met;
c. There is no major visible damage (broken, cracked, torn, bent or misaligned external surfaces, cracks in a solar cell which could remove a portion larger than 10% of its area, bubbles or delamination, loss of mechanical integrity;
d. No sample has exhibited any open circuit or ground fault during the tests;
e. For IEC 61646 only: the measured maximum output power after final light-soaking shall not be less than 90% of the minimum value specified by the manufacturer (2).

Qualification test conditions are simple but stringent, and the quality of both c-Si and thin film modules tested until 2007 was absolutely unacceptable. Quality has improved in some areas during the last several years, according to the tests, but is still lower than practicable and profitable for use in large-scale PV power generating installations. Large failure rates would bankrupt any such installation.

An obvious conclusion from the above tests is that some modules perform better than others. Is this due to different quality materials, or because of different processes used? The answer is less than obvious, so this means that the issues at hand need to be brought out in the open, discussed and resolved ASAP. The PV industry has a bright future, but is still learning how to walk, and we need to help it get on its feet sooner by maximizing product quality and reducing failures.

Damp Tests

This test seems to be the weakest point of TFPV technology, so a closer look is justified. Damp heat studies of CIGS and CdTe cells have been conducted by subjecting cells and min-modules to an environment of 60°C/90% RH.

Two key conclusions can be made:

(1) Both CIGSS and CdTe cells degrade rapidly under 60°C/90% RH…

(2) Damp heat stress will cause changes in junction transport properties and minority carrier transport characteristics of the cell absorber.

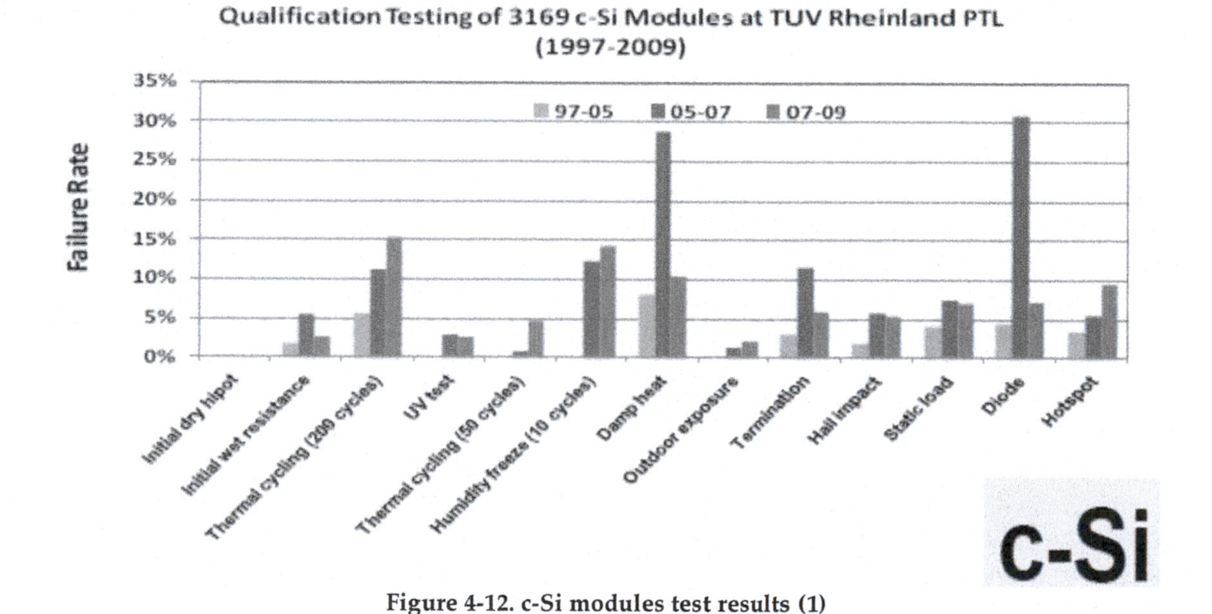

Figure 4-12. c-Si modules test results (1)

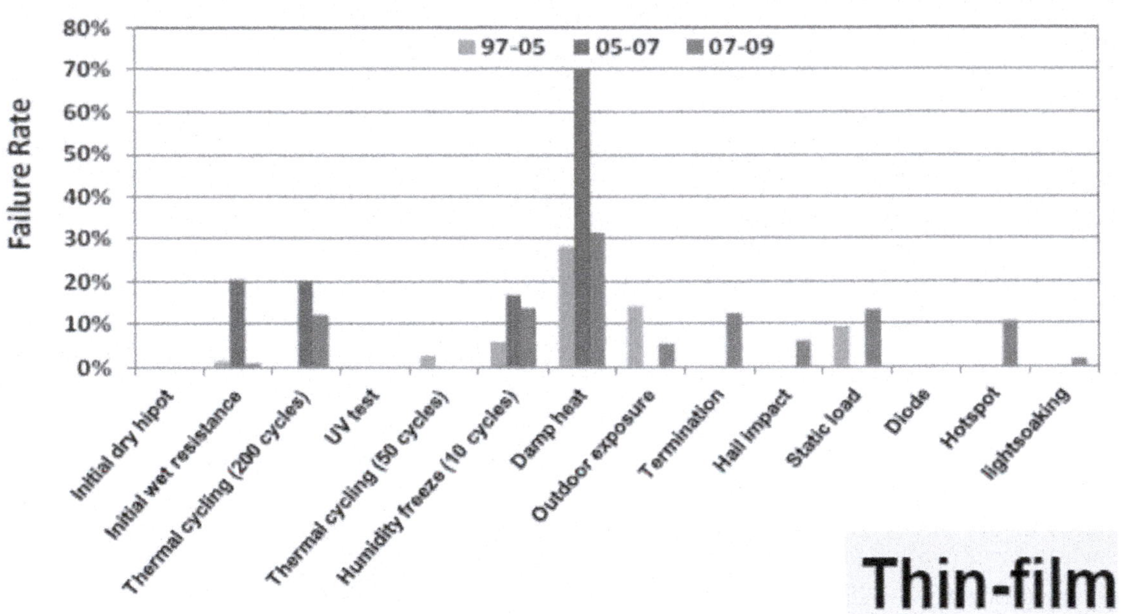

Figure 4-13. Thin film modules test results (1)

The damp tests in Figure 4-14 show fatal degradation for the un-encapsulated thin film's structure, which confirms that:

1. Thin film structures will be attacked, disintegrated and decomposed if and when exposed to the elements, and
2. Thin film structures depend exclusively on encapsulation for protection from the elements.

A series of tests performed by the author with different metal and non-metal TF structures exposed to hot and humid conditions ("sweat box" tests under different temperature and humidity regimes) show clearly that bare thin film structures (any type and combination thereof) will degrade if exposed to damp heat. See Figure 4-14. The destruction process usually starts slowly at the films' interfaces, and then accelerates quickly, causing delamination and/or destruction of the entire TF structure. TF structures enveloped in impermeable materials have a better chance to survive this test. An important conclusion here is that the modules' encapsulation is the only thing separating them from destruction and total failure.

This raises a serious question, "What and how much protection is needed to ensure proper operation of TFPV modules, for 30 years in excessively hot and humid areas?" It is obvious that encapsulation determines performance and longevity of modules, because once moisture and other environmental chemicals and gasses reach the thin film structure, it undergos unpredictable changes. These changes depend on the thin film composition and stability, and also on the nature of the attacking species, but most often the attacks will result in reduced power output and eventually lead to total failure.

NOTE: Thin-film PV modules are different from conventional c-Si modules, as follow:

1. The thin film structure is deposited directly on the front or rear glass cover, thus it is exposed to outside elements in case of delamination from the glass surface.
   Note: In c-Si modules, if the EVA delaminates from the glass, the cells are still enveloped in it, thus they are still protected from the outside elements.

2. Most TFPV modules usually have no side frame*, which leaves them fully dependent for protection by the edge sealers, which are fully exposed to the elements' and could be easily damaged or destroyed.

These are important differences, because leakage via un-framed module edges can easily reach the active thin film structure which is otherwise unprotected, thus damaging it and causing gradual power decrease and eventually total failure.

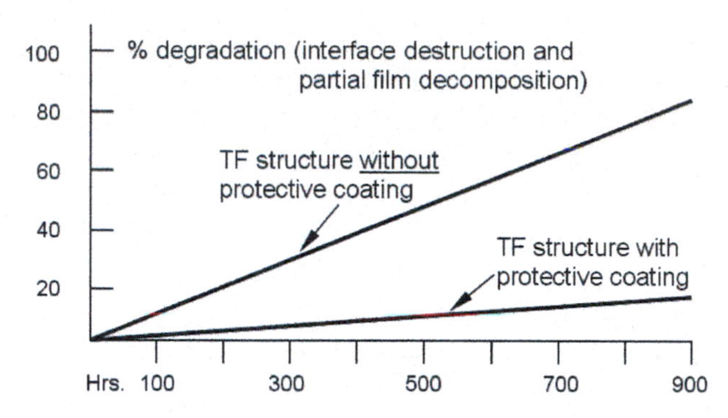

Figure 4-14. Damp test of thin film structure with and without protective coating

Field Test Results

There is a difference between the test results from tests done under lab conditions, and those performed under actual field conditions. The final results will depend on the type of modules used and weather conditions in the field, so keep in mind that results from tests under extreme heat in the Arizona desert sun or high humidity in southern Florida will be drastically different and probably much worst after 5-20 years of non-stop operation than what we see in the lab tests above.

Some partial, and far from complete, test data are shown below in support of our arguments of the efficiency and longevity of TFPV modules:

Field Test Data

Several modules were tested under field conditions (several years of outdoor operation), with the results below reflecting the degradation rate of different types of modules from different manufacturers:

The conclusions of the authors of these tests are:

"First, module degradation rate determinations should be made from performance data over periods of at least three years. Shorter time spans are likely to give inaccurate degradation rate (RD) values because of seasonal variations and initial module performance stabilization.

"Second, many (but not all) crystalline Si modules degrade at rates slower than the 1% per year rule of thumb. A more reasonable rule of thumb is probably 0.5% per year. Conversely, many (but not all) thin-film modules appear to have RD values somewhat higher than 1% per year.

"Third, RD appears to vary over a fairly wide range, from values as high as several percent per year, down to zero (no measurable degradation). It would therefore seem important for system designers to have accurate degradation rate information available."

The PV industry has accepted 1% annual degradation as "normal," but based on our and others' experience with PV modules, we question this acceptance as an unreasonable compromise of performance caused by mediocre materials quality and inadequate process design and execution. We would suggest that zero, or near-zero, annual degradation is possible (as we can see in the BP modules in these tests) and badly needed, if we are to rely on any PV modules to provide efficient and reliable power for 30 years, regardless of the location. Annual degradation rates close to zero would be a good indication of the excellence of materials and processes used in the modules which should be our goal for the near future.

Zero degradation rates have been achieved and maintained by the semiconductor industry for a long time. Can you imagine your Pentium processor which is made of similar materials and operates at comparably extreme temperatures and humidity conditions dropping 1% computing power every year? Since the materials, equipment and processes are similar, near-zero degradation rates should be expected from the TFPV industry too.

A major concern with quickly degrading PV modules, including TFPV, is that since we don't know why they are degrading, there is reason to suspect a serious, if not terminal problem within. It should be noted, however, that although degradation rates are sometimes related to failure rates, there is no proven direct correlation between these two, which emphasizes the complexity of the technology and the uncertainties we are faced with.

Until materials and manufacturing processes are standardized to achieve zero annual degradation, we must keep in mind that if PV modules continue to operate even when the output power is gradually decreasing, the slow degradation is neither infant mortality nor normal life expectancy decrease. Instead, at least for now, it should be regarded as one of the factors contributing to the gradual "wear-out" mode, which should've been properly investigated and calculated during the design and test stages of any PV project. If many modules degrade at an unexpectedly fast rate, at some point they will shut-down entire arrays which will no longer be able to meet the minimum input voltage window of the inverters, and eventually the entire system would shut down. This is a catastrophic failure that will render the power plant useless and out of commission, as we have witnessed on several occasions since the early 1980s.

*60C/90% RH approximates a monsoon condition in July-September in Arizona and California deserts and parts of SE USA.

Actual Field Failure Case

According to an Auriga USA, LLC, May 2010 report, "We are aware of a disruption in the manufacturing process that occurred in late 2009 that resulted in some TFPV modules having significant degradation after being installed for just one month. Although the modules passed the initial flash inspection at the factory, they later degraded by as much as 20% (from 75Wp to 60Wp) once installed in the field. We discussed the potential problem with several of [manufacturer's name deleted by author] prominent installers, and we were satisfied that the problem is not an ongoing concern. We estimate that of the roughly 1.3GW of 2010 production, only ~10MW has been identified as problematic."

So, 10MW out of 1.3GW production volume is less than 1% defective modules coming from the production line. This is actually not a bad loss of yield for any manufacturer. 10MW, however, represents ~133,000 modules manufactured with an unidentified defect, sold, packed, transported, unpacked and installed in a power field as "good." Then shortly after that they failed. This is a lot of modules when they fail on our field. At a minimum sale price of $1.0/watt, this is over $10 million—a lot of money, capable of bankrupting any PV project, if not taken care of properly and on time. Fortunately, in this case the failure was caught within the warranty period.

So how did this happen? Keep in mind that the manufacturer is one of the largest and most reliable TFPV manufacturing companies in the world. We will never know what happened exactly, because the details were not disclosed. Since the TFPV process sequence is similar to that of semiconductor devices, we can deduct right off the bat that the in-process inspection and test procedures of the 133,000 defective PV modules batch were not effective enough, or were simply not executed properly. Proper QC/QA procedures and inspection would have identified the bad modules as different from the norm and stopped the batch from leaving the facility.

Now the company would tighten its QC/QA procedures and such failure not happen again. But what if everything is done right—including in-process and final tests and inspections—and the batch still failed? Then we must look deeper into the materials quality and the entire process sequence, suspecting and looking for the worst enemy of thin films—"latent" failures. These failures, which occur early on or long after PV modules are installed in the field, are usually caused by materials problems and/or process-induced defects, which are not detectable during routine inspection and test procedures. Latent failures will cause excessive power loss, and complete failure with time, depending on the nature and gravity of the defects and their location within the devices.

The defects causing these problems could be due to anything from poor quality incoming materials to contaminated substrate surfaces, or dirty vacuum chambers in one of the process steps. They could be due to improper process parameters—temperature, process time, gas pressure etc. Not much can be done once the defect has been introduced in the process. The module's fate is sealed and only time will tell how bad the defect and its consequences are.

If failure happens within the warranty period, modules will be replaced, as in the above case. But what if it happens after the warranty expires? What if 133,000 TFPV modules in our 100MWp power plant fail and must be replaced long after the warranty expires? Negotiating with or suing the manufacturer could be alternatives, but not a good options for a number of reasons. "Eating" the loss in order to keep good relations with the manufacturer and the solar community could be another choice, with a compromise of sorts being the most likely. In all cases, this is lost time, lost output, and lots of wasted effort, which is hard to express in numbers.

Again, the reasons for TFPV modules' degradation and failures are as many as the number of materials and steps in the manufacturing process. It is important for system designers and integrators to have accurate information available for the particular modules used in the projects under their watch. This information (usually in form of results from qualification tests must be correlated with long-term (3 years minimum) field tests at the location of the planned installation. Only then would we have some confidence in the reliability of our PV modules.

*Note: Research into side frame protection has been going on for a long time, and experience shows that a properly designed and installed side frame on any PV module plays a critical role in helping it retain its performance characteristics longer and/or will extend its useful life significantly. The fact that millions upon millions of unframed TFPV modules will be installed and operated for 30 years in US deserts makes us wonder about their longevity. And some of these contain toxic materials which begs a question of potential environmental contamination upon attack by the elements. We hope this issue will be discussed openly and resolved in the near future.

Current Research on PV Materials, Processes and Devices

Obviously, the TFPV industry is in its infancy and a lot of work remains to be done before it becomes a proven, mature technology. There are many research groups active in the field of photovoltaics in universities and research institutions around the world. This research can be divided into three areas: a) making solar

Table 4-2. c-Si and TFPV modules degradation rates under field operating conditions (5)

Manufacturer	Module Type	Exposure (years)	Degradation Rate (% per year)	No. of Modules
BP Solar	BP 585F (x-Si)	7	-0.30	2
BP Solar	BP 270F (x-Si)	8	-0.32	2
Kyocera	KC40 (poly-Si)	4.5	-0.91	2
Solarex	SX40U (poly-Si)	5.6	-0.01	2
Siemens	PC-4-JF (x-Si)	9.5	-0.51	1
Photowatt	PWX500 (poly-Si)	6	-0.13	1
Sanyo	H124 (a-Si/x-Si HIT)	2.6	-1.59	1
ECD Sovonix	[none] (a-Si) †	12	-1.17	1
Solarex	SA5 (a-Si)	12	-0.69	1
Uni-Solar	UPM-880 (a-Si)	12	-0.62	2
APS	EP55 (a-Si)	9.5	-1.62	2
Solarex	MST-22ES (a-Si)	6	-0.86	1
Uni-Solar	US-32 (a-Si)	8.5	-0.39	1
EPV	EPV40 (a-Si) †	6.5	-1.40	2
BP Solarex	MST-50 MV (a-Si)	4	-2.47	2
Siemens	ST40 (CIS) †	7	-1.63	1
Solar Cells Inc.	[none] (CdTe) †	10	-1.84	1

cells and modules cheaper and/or more efficient to effectively compete with other energy sources; b) developing new technologies based on new solar cell architectural designs; and c) developing new materials to serve as light absorbers and charge carriers.

Thin-film photovoltaic cells and modules use less than 1% of the expensive raw material (silicon or other light absorbers) compared to silicon wafer-based solar cells, leading to a significant price drop per Watt peak capacity. Many research groups around the world actively research different thin-film processes; however, it remains to be seen if these solutions can achieve a similar market penetration as traditional bulk silicon solar modules.

An interesting aspect of thin-film solar cells is the possibility of depositing cells on all kinds of materials, including flexible substrates (PET for example), which opens a new dimension for applications. The future will surely bring even more good news to the thin film industry.

Below is a list of new and exotic thin film materials and processes and a short description of ongoing R&D efforts:

1. Metamorphic multi-junction solar cell

This is an ultra-light and flexible multi-junction (MJ) cell, similar to the Germanium based CPV cell that converts solar energy with record efficiency. It represents a new class of solar cells with clear advantages in performance, engineering design, operation and cost. For decades, conventional cells have featured wafers of semiconducting materials with similar crystalline structure. Their performance and cost effectiveness is constrained by growing the cells in an upright configuration.

Present-day MJ cells are rigid, heavy, and thick with a bottom layer made of Ge semiconductor material. In the new method, the cell is grown upside down. These layers use high-energy materials with extremely high quality crystals, especially in the upper layers where most of the power is produced. Not all of the layers follow the lattice pattern of even atomic spacing. Instead, the cell includes a full range of atomic spacing, which allows for greater absorption and use of sunlight.

The thick, rigid Germanium layer is not used, thus reducing the cell's cost and 94% of its weight. By turning the conventional approach to cells on its head, the result is an ultra-light and flexible cell that also converts solar energy with record efficiency of 40.8% under 326 suns concentration.

2. Polymer Lens

The invention of conductive polymers may lead to the development of much cheaper cells that are based on inexpensive plastics. However, organic solar cells generally suffer from degradation upon exposure to UV light, and hence have lifetimes which are far too short to be viable, at least at this stage of their development.

The bonds in the polymers are always susceptible to breaking up when radiated with shorter wavelengths. Additionally, the conjugated double bond systems in the polymers which carry the charge react more readily with light and oxygen.

So, most conductive polymers, being highly unsaturated and reactive, are highly sensitive to atmospheric moisture and oxidation, making commercial applications difficult.

3. Nanoparticle Lens

Experimental non-silicon solar panels can be made of quantum hetero-structures, e.g. C or quantum dots, embedded in conductive polymers or meso-porous metal oxides. Also, thin films of many of these materials on conventional silicon solar cells can increase the optical coupling efficiency into the silicon cell, boosting overall efficiency.

By varying the size of the quantum dots, the cells can be tuned to absorb different wavelengths. Although the research is still in its infancy, quantum dot-modified photovoltaics may be able to achieve up to 42% energy conversion efficiency due to multiple exciton generation.

Solar cell efficiency could theoretically be raised to more than 60% using quantum dots. Electrons can be transferred from photo-excited crystals to an adjacent electronic conductor. Researchers have demonstrated the effects in quantum dots made of PbSe, but the technique could work for quantum dots made from other materials too.

4. Photonic Crystals Lens

These are nanostructured materials in which repeated variations in the refractive index on the length scale of visible light produces a photonic band gap. This gap affects how photons travel through the material and is akin to the way in which a periodic potential in semiconductors influences electron flow. In the case of photonic crystals, light of certain wavelength ranges passes through the photonic band gap while other wavelength ranges are reflected. The photonic crystal layer could be attached to the back of a solar cell.

5. Transparent Conductors

Many new solar cells use transparent thin films that are also conductors of electrical charge. The dominant conductive thin films used in research now are transparent conductive oxides (TCO), and include fluorine-doped tin oxide (SnO2:F, or FTO), doped zinc oxide (e.g.: ZnO:Al), and indium tin oxide (ITO). These conductive films are also used in the LCD industry for flat panel displays.

The dual function of a TCO allows light to pass through a substrate window to the active light-absorbing material beneath, and also serves as an ohmic contact to transport the photogenerated charge carriers away from that light-absorbing material. Present TCO materials are effective for research, but perhaps are not yet optimized for large-scale photovoltaic production. They require special deposition conditions at high vacuum, and can sometimes suffer from poor mechanical strength, while most have poor transmittance in the infrared portion of the spectrum (e.g., ITO thin films can also be used as infrared filters in airplane windows). These factors make large-scale manufacturing more costly.

A relatively new area has emerged using carbon nanotube networks as a transparent conductor for organic solar cells. Nanotube networks are flexible and can be deposited on surfaces a variety of ways. With some treatment, nanotube films can be highly transparent in the infrared, possibly enabling efficient low-bandgap solar cells. Nanotube networks are p-type conductors, whereas traditional transparent conductors are exclusively n-type. The availability of a p-type transparent conductor could lead to new cell designs that simplify manufacturing and improve efficiency.

6. Infrared Solar Cells

Researchers have devised an inexpensive way to produce plastic sheets containing billions of nano-antennas that collect heat energy generated by the sun and other sources. The technology is the first step toward a solar energy collector that could be mass-produced on flexible materials. While methods to convert the energy into usable electricity still need to be developed, the sheets could one day be manufactured as lightweight "skins" that power everything from hybrid cars to computers and iPods with higher efficiency than traditional solar cells.

The nano-antennas target mid-infrared rays, which the Earth continuously radiates as heat after absorbing energy from the sun during the day. Also, double-sided nano-antenna sheets can harvest energy from different parts of the Sun's spectrum. In contrast, traditional solar cells can only use visible light, rendering them idle after dark.

7. UV Solar Cells

Researchers have succeeded in developing a transparent solar cell that uses ultraviolet (UV) light to generate electricity but allows visible light to pass through it. Most conventional solar cells use visible and infrared light to generate electricity. Used to replace conventional window glass, the installation surface area could be large, leading to potential uses that take advantage of the combined functions of power generation, lighting and temperature control Also, easily fabricated PEDOT:PSS photovoltaic cells are ultraviolet light selective and sensitive.

8. 3-D Solar Cells

These are truly three-dimensional solar cells that capture nearly all of the light that strikes them and could boost the efficiency of photovoltaic systems while reducing their size, weight and mechanical complexity. The new 3D solar cells capture photons from sunlight using an array of miniature "tower" structures that resemble high-rise buildings in a city street grid.

9. Meta-materials

Meta-materials are heterogeneous materials employing the juxtaposition of many microscopic elements, giving rise to properties not seen in ordinary solids. Using these, it is possible to fashion solar cells that are perfect absorbers over a narrow range of wavelengths. This is still pure research.

10. Photovoltaic Thermal Hybrid

These systems combine photovoltaic with thermal solar. The advantage is that the thermal solar part carries heat away and cools the photovoltaic cells; keeping temperature down lowers the resistance and improves the cell efficiency. Modified CPV systems have been tested, where the CPV cells are cooled by active flow of liquid, thus generating both heat and electricity.

11. HIT Solar Cell

A novel device developed by Sanyo is the HIT cell. In this device thin film layers of amorphous silicon are deposited onto both faces of a textured wafer of single crystal silicon. This improves the efficiency and reliability of the solar cell by avoiding the usual problems related to doped and fired silicon solar cells.

This type of solar cell would be more expensive to produce, but might provide the much needed solution to the problems c-Si and TFPV modules encounter when exposed to extreme desert temperatures. Such cells would exhibiting less heat sensitivity and power degradation, but, we have not seen any proof of that happening as yet.

12. Organic Solar Cells

Organic materials have been subjects of intense research lately, with the most interesting being the molecular and polymeric semiconductors, Fullerene (C60) and its derivatives. These materials are used in organic LEDs and thin film transistors for use in smart cards. The advantage of using these materials is their manufacturing simplicity and the possibility of use on flexible materials.

The basic structure consists of glass or other such substrate, coated with indium tin oxide (ITO) which is a transparent conductor and acts as the top electrode. Several layers of light-absorbing organic materials could be deposited on the ITO and then completed by depositing a metal back contact onto the organic materials.

Cell efficiencies of 5% have been achieved, but low efficiency and problems with large-scale manufacturing processes must be resolved before we see these technologies on the market.

13. CTZSS Solar Cells

This is a new type of solar cell, developed by IBM (summer 2010), who claims that it could potentially lower the price of solar power in the future. The new CTZSS solar cell is made from copper, tin, zinc, sulfur and selenium, all of which are somewhat earth-abundant, according to IBM.

The new process does not use vacuum chambers and related processes (the sort of equipment used for chip-making). Instead, it deploys a solution-based approach, which is different and quite interesting, but it brings us back to the days when we used to make c-Si cells via solution-based processes. So we've come a full circle, it seems.

In mass production, that could mean producing cells through printing or dip and spray coating. Bringing such a cell to market, however, won't be easy. Investors have become nervous about championing new types of solar cells. Various CIGS companies also employ printing, and other roll-to-roll manufacturing processes to produce thin film solar cells. It has taken most quite a while to iron out the kinks in the manufacturing process. So, will IBM make solar cells? We'd guess not, but some other manufacturers might benefit from their research and experience. We will see...

14. Thermovoltaics and Thermophotovoltaics

Researchers are working on ways to combine "quantum and thermal mechanisms into a single physical process" to generate electricity and make solar power production twice as efficient as existing technologies. It's called PETE photon-enhanced thermionic emission. This process does require some exotic materials (cesium-coated gallium nitride) and works best at high temperatures more likely with concentrators or parabolic dishes than flat solar panels.

15. Intermediate Band

This is intermediate band nitride thin film semiconductor material, which might be an alternative to the multijunction designs for improving power conversion efficiency of solar cells. The intermediate-band solar cell is a thin-film technology based on highly mismatched alloys. The three-bandgap, one-junction device has the potential of improved solar light absorption and higher power output than III-V triple-junction compound semiconductor devices.

16. Plasmonics

Plasmonic technology uses engineered metal structures to guide light at distances less than the scale of the wavelength of light in free space. Plasmonics can improve absorption in photovoltaics and broaden the range of usable absorber materials to include more earth-abundant, non-toxic substances, as well as to reduce the amount of material necessary.

Materials and Consumables Supply

Another area of the TFPV industry which might benefit from vigorous standards and regulations is the supply chain and control of the quality of raw and process materials coming from third-party vendors. A consistently reliable supply of raw materials needed to manufacture some components and entire PV modules is a major issue which we need remember when discussing alternative energy use and its attributes—"renewable," "green" and "clean."

Although wind and solar technologies are called "renewable," some of their material components are far from renewable. On the contrary, some are expensive exotic and rare materials, and in very short supply. Silicon is an exemption, because it is one of the most abundant materials on the face of the Earth. Its cost is expected to vary—mostly in upward direction—due to the price of energy fluctuations, which it heavily depends on, but it is abundant, and there is no danger of depletion anytime soon.

Most TFPV technologies, however, depend on rare and hard-to-find elements such as Cd, Te, As, In, Ga, and Se. Mining and refining such elements is expensive and accompanied by hazardous operations and waste that harm the environment and people's health.

On top of that, a significant part of these exotic materials are located in Third World countries, which makes control of the price structure and quality even harder, so putting our future in their hands would be like jumping from the frying pan into the fire.

CONCENTRATING PV (CPV) TECHNOLOGY

Concentrating PV (CPV) technology has been around for 30 years. It was developed by several US companies in the 1970s and 1980s with the help of DOE and its satellite national labs. HCPV is the most efficient PV technology in the world, but its commercialization has been bogged down by a number of factors unrelated to its efficiency.

CPV equipment operation is somewhat different and more complex than other PV technologies. In contrast with 'flat-plate' PV modules, where a large area of photovoltaic material (usually crystalline silicon) is exposed to the maximum sunlight, CPV systems, as the name suggests, use lenses or mirrors to focus (or concentrate) sunlight onto a small amount of non-silicon photovoltaic material.

General Description

HCPV systems operate by concentrating sunlight through Fresnel lenses onto specially designed, and very efficient (over 40%) CPV solar cells. Additional components, such as secondary optics, insulators, heat sinks, and trackers are needed to keep the light focused precisely onto the solar cells, to dissipate the generated heat and conduct the generated electricity to the electric grid, or user's site.

CPV solar cells mounted into CPV assemblies are installed in modules, which are mounted on a steel frame—a frame that pivots on a pedestal (Figure 4-15). GPS controllers send a signal to the x-y drive motors which drive actuators to position the frame and modules exactly perpendicular to the sun all day long with a .01% degree margin of error.

Basic Operation

HCPV systems (trackers) consist of several key components, as follow:

1. CPV solar cells
2. Heat sinks

Figure 4-15. HCPV tracker

Table 4-3. Thin film PV devices advantages and disadvantages

THIN FILM CELLS AND MODULES

Advantages

+ Thin films, due to their extensive use in semiconductor devices manufacturing, are the most understood structures and best developed processes in the PV industry.
+ Thin film PV modules are adaptable to mass production methods of fabrication and so they are the easiest and cheapest to manufacture of all PV devices,
+ Thin film PV modules can be made in different sizes and shapes, thus they can be installed in areas where other technologies cannot function.
+ The price of thin film PV devices will continue falling, as material properties and efficiencies improve, and as the mass production equipment and processes are optimized.

Disadvantages

- Low efficiency is the major issue plaguing the TFPV technology. It is due to reduced number of collected charge carriers per incident photon and not much can be done to change that, so the efficiencies are not expected to increase significantly.
- Thin film deposition processes require complex equipment and sophisticated processes, which require precise process control.
- Thin films are fairly fragile structures, affected by mechanical and chemical changes brought upon them by different environmental and operational factors.
- Different thin films have different coefficient of expansion, and coefficient of friction, which create stress in the films during long term field operation.
- If moisture enters the modules, as it often does in even the best designed and best constructed modules*, then the heat and water will cause the thin film layers to degrade by virtue of chemical decomposition of the thin film layers.

  * Most thin film PV modules consist of two pieces of glass, held together by organic encapsulants, which degrade with time, allowing air and moisture to penetrate into the thin film structure and damage it slowly
- Some thin film structures require use of exotic and rare metals, which are not abundant enough and which might create material shortages in the near future.
- Some thin film PV modules contain hazardous materials, such as Cadmium, Arsenic etc., so there is a potential danger of environmental contamination.
- Thin film modules containing such hazardous materials must be handled with utmost care and recycled as hazardous waste at the end of their useful life.

3. Bypass diodes
4. Secondary optical elements
5. CPV assembly (containing the above components)
5. Fresnel lenses
6. CPV modules (housing the above components)
7. Steel frame
8. x-y drive gear-motors
9. GPS based controller

The CPV solar cells are made out of GaAs, Ge or other more efficient and more durable semiconductor materials with thin films deposited on them to create a multi-junction device of very high efficiency—over 42.3% presently. See Figure 4-16.

CPV cells are sophisticated semiconductor devices that have followed Moore's law to an extent, as far as increase of efficiency per active area is concerned. At the very least, they are much closer to it than any competing PV technology to date.

The CPV cells are arranged into a special assembly, called CPV cell assembly, which ensures efficient collection of sunlight and cooling of cells during operation, as well as the proper conduction of electric power into the wiring harness. Each cell is configured with a blocking diode, which protects it from reverse bias currents and surges, and also isolates it from the circuit in case of total failure. The cell and diode are usually mounted on a heat sink, which extracts heat from the cell during operation and dissipates it in the surrounding environment outside the module.

A Fresnel lens is used to capture a large area of sunlight and focus it onto the small CPV cell. The Fresnel lens is basically a flat (or slightly curved) plastic or glass lens that uses a miniature saw-tooth design on its bottom to redirect and focus the incoming light onto the small area cell, several inches away from the lens. When the saw-tooth teeth on the lens are arranged in concentric circles, light is focused at a central point, and

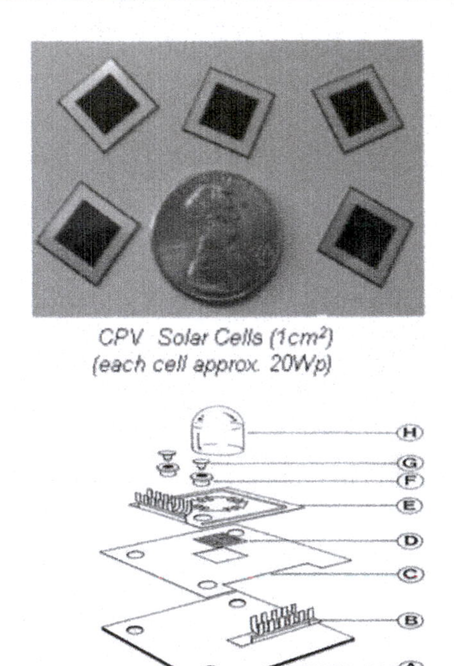

Figure 4-16. CPV cells and cell assembly

the equipment with this design is called the point-focus CPV system. When the teeth run in straight rows, the lenses act as line-focusing concentrators, and the resulting equipment is called the linear-focus CPV system.

The concentration ratio of each CPV device can vary as well, depending on lens and cell design. For example, if sunlight falling onto 100 cm2 is focused onto 1 cm2 of PV material, the ratio is considered as 100 suns, or 100 x. If the light from a 1000-cm2 Fresnel lens is focused onto a 1-cm2 CPV cell, then the ratio is 1000 suns, or 1000 x. If thus concentrated sunlight light falls onto a well designed CPV cell, it will produce 100, or 1000 times the electricity respectively that a normal c-Si PV cell of the same size would produce under the same operating conditions. Commercial concentration ratios are between 200 and 500 suns, and as much as 1000 suns, but there are theoretical and practical limits of min.-max. sun concentration levels, with the mid-range 400-600 considered the most efficient and practical for now, in the author's opinion.

Most CPV systems use only direct solar radiation, so these installations operate best under direct sunlight (such as in desert areas) and always involve trackers, forcing the modules to rotate and follow the sun all day, which keeps the lenses and CPV cells looking right at the sun at all times, thus generating maximum possible electric energy while the sun is shining.

The cells are placed in modules, which are mounted on a large frame, which rotates around its two axes to follow sun movement all day. Tracking is achieved with one or two gear-motor assemblies, which get a signal from a photo-detector or GPS controller, which knows where the sun is at all times and runs the motors to move the frame accordingly. Precise tracking is needed, to keep the sun at 90 degrees with respect to the Fresnel lens and CPV cell, so that sunlight is focused onto the solar cell at all times. If the sunlight comes at an angle, or is diffused (as on a cloudy or foggy day), some, if not most, sunlight will get de-focused, reflected, or otherwise fall away from the cells, resulting in very low light-to-power conversion efficiency.

Tracking systems add cost in terms of motor and controller maintenance, but this cost is relatively small compared with other O&M cost savings that trackers provide. For example, the tracker's motor requires only annual lubrication and a single motor controls more than 50 kWp of PV. Also, tracker bearings require no lubrication and are designed for more than 25 years of use.

The operating and maintenance (O&M) cost of a utility-scale tracking system ends up being less than US$0.001/kWh more than that of a fixed configuration. And this calculation does not factor in the O&M savings from increased energy production for the same power output, which is a different subject all together.

HCPV Efficiency

It is easy to see from Figure 4-17 that HCPV trackers are twice as efficient as the regular c-Si modules in fixed mount, which is the most widely used technology. The best part is that while c-Si PV technology is hitting the top of its physical limits of 28% efficiency, HCPV cells have long way to go before reaching the upper limits of their efficiency, which is ~80%.

So, we are looking at 60% efficiency by 2020 and 80% by 2040. Just imagine 80% of the sunlight reaching the CPV cells being converted into useful electricity. With all other factors considered, we are looking at 65-75% efficient HCPV tracking systems in the future. We are fully confident that that future will arrive, and if we are smart enough, even sooner than predicted.

The power balance sheet of a HCPV tracker looks good, but is not perfect.

Grid efficiency = Pt – Ol – Rd – Ci = 29.5%* (estimate of SolarTech, Inc. USA).

Pt Total power generated by the CPV cell (summer of 2010)	= 41.0%
Ol Optics loss and misalignment	= 4.0%
Rd Cell temperature	

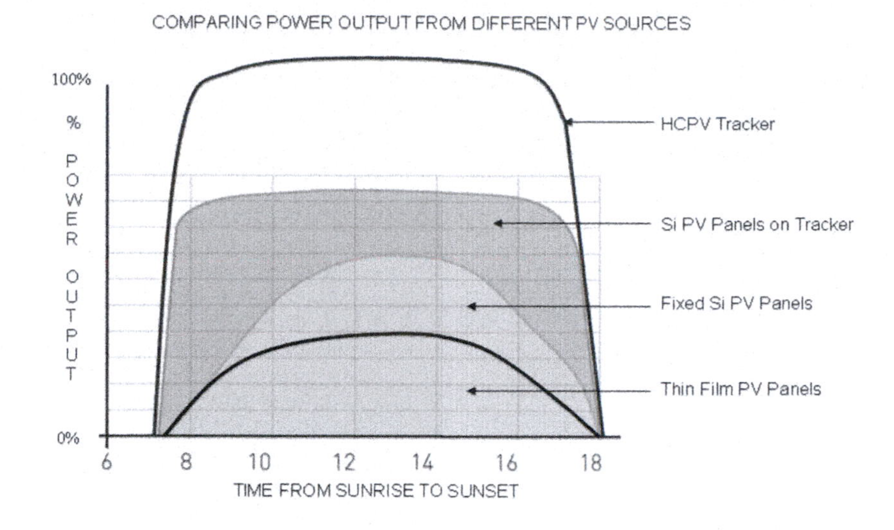

Figure 4-17. Power output from different PV technologies

degradation (at 190°F)	=	2.5%
Ci Inverter and electric circuits loss	=	5.0%
Total DC power converted into AC	=	29.5%

NOTE: HCPV trackers stop generating power as the air haziness (clouds, fog, etc.) increase, while some PV technologies continue generating proportionally. Because of that, HCPV trackers can only be used in hot desert areas with maximum direct radiation.

So, our 41% efficient CPV cells mounted on a tracker will deliver only part of the generated power into the grid. And according to the above table, it will average ~29.5% efficiency under full power on a bright sunny day. Good, but not spectacular. There are other losses too, related to the electro-mechanical and optical systems' misalignment, and poor maintenance, which must be taken into consideration, in a full and precise power plant design calculation.

Improvements in materials and the various key components (cells, lens materials, optics, heat sinks, GPS controls, drive gears) will account for further increases in overall efficiency in the near future. Based on R&D work done in the past and combined with that done recently, we are confident that HCPV trackers (point-focus type) will soon be able to obtain 35% grid efficiency with the present state-of-the-art, and up to 45-50% when their efficiency is increased to ~60% by 2015-2020.

The future of this technology looks good, and only hard work separates HCPV technologies companies from broadly deploying their technology on the energy markets of the world. But a lot of work must be done until then then. Many questions must be answered and many issues resolved.

HCPV Cost Estimates

The cost of the HCPV technology is based on:

1. Location. HCPV operates most efficiently in desert areas with hot and bright sunshine;
2. Price of materials. HCPV trackers consist of 90% steel, aluminum and glass, so the price of these commodities determines the total cost of the installation; and
3. Size of the installation. The larger the installation, the lower the cost, due to bulk orders of materials.

For a large size installation—over 5.0 MWp—installed in the Arizona desert, at 2009 commodity prices, the cost of manufacturing HCPV tracking systems would be ~$1.20/Wp.

CONCLUSIONS

The cost of PV technologies is getting close to the LCOE and the efficiency is getting close to the upper theoretical limits. Efficiencies are still going up, albeit slowly, while costs are going down consistently, and hopefully permanently. This is the good news. What about the reliability and longevity of PV technologies?

It is apparent from what we have seen in this text that reliability and safety are the frontiers which need to be addressed, with related issues resolved, before today's PV technologies can be considered truly reliable and safe for long-term operation worldwide.

Reliability issues have been addressed by a number of companies and research institutions, and we get more positive information on these developments every day. A lot of effort and money is spent on ensuring long-term reliable operation of PV modules under all conditions. The missing link here is the insufficient proof that any of existing PV technologies could last 30 years, operating reliably under any climate conditions. We do believe that the developments have successfully addressed the issues of some PV technologies and we are hopeful that there will be no serious reliability setbacks with most of them in the future.

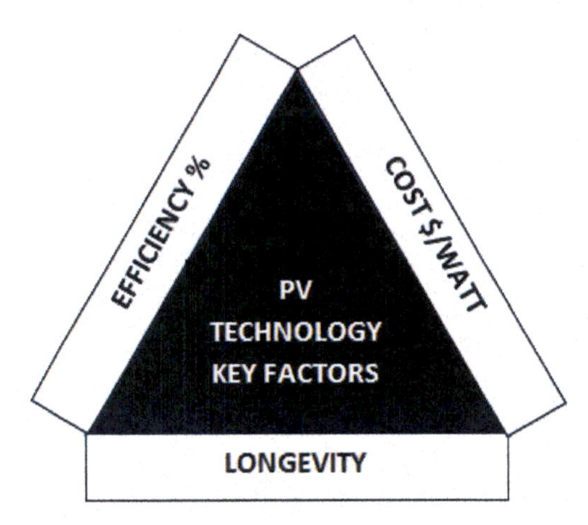

Figure 4-18. What's next?

Safety is another important and equally complicated subject. It goes without saying that any solar (green and clean) technology must be safe—to manufacture and to use. Thus far we are not convinced that all PV technologies, due to the presence of extremely toxic materials in some, are safe for all applications. A lot of additional product development, testing and proof are needed to ensure 30+ years unquestionably safe operation of some thin film PV technologies in the planned desert mega fields. So far, this is the largest skeleton in the PV industry's closet, which we tried to bring out in this text, hoping to clarify and put relevant issues on the table for discussion by the scientific community, regulators and the public. These discussions are absolutely necessary to ensure the safety of the US deserts and the inhabitants.

Notes and References

1. Testing the reliability and safety of photovoltaic modules: failure rates and temperature effects," Dr. Govindasamy TamizhMani 2010, "Photovoltaics International, Vol. 8, pp. 146-152.
2. IEC 61646, Thin film terrestrial photovoltaic (PV) modules-design qualification and type approval,1998.
3. Performance Analysis Of Large Scale, Amorphous Silicon,Photovoltaic Power Systems, Alan Gregg.
4. Damp Heat Effects on CIGSS and CdTe Cells, Larry Olsen, Sambhu Kundu, Mark Gross, and Alan Joly Pacific Northwest National Laboratory, Richland, WA.
5. Comparison of degradation rates of individual modules held at maximum power, C.R. Osterwald, J. Adelstein, J.A. del Cueto, B. Kroposki, D. Trudell, and T. Moriarty.
6. SolarTech, Inc. HCPV trackers references, www.sst-usa.net
7. Weather Durability of PV Modules; Developing a Common Language for Talking about PV Reliability By: Kurt P. Scott Atlas Material Testing Technology LLC.
8. Common Failure Modes for Thin-Film Modules and Considerations Toward Hardening CIGS Cells to Moisture A "Suggested" Topic Kent Whitfield, MiaSole.
9. BOS cost savings needs and potential for large scale ground based pv systems until 2010, Manfred Bächler, Phönix SonnenStrom AG, Hirschbergstr. 8, D—85254 Sulzemoos, Germany.
10. Spectrolab http://www.spectrolab.com/DataSheets/PV/pv_tech/msce.pdf
11. Adhesion and Thermomechanical Reliability for PVD and Moddules, Reinhold H. Dauskardt.
12. IEC 61215: Crystalline Silicon terrestrial photovoltaic (PV) modules- design qualification and type approval, 2005.
13. IEC 61646: Thin-film terrestrial photovoltaic (PV) modules—Design qualification and type approval, 2008.
14. C. Osterwald: Terrestrial Photovoltaic Module Accelerated Test-to-failure Protocol, Technical Report, NREL/TP-520-42893, 2008.
15. ASTM E2481: Standard Test Method for Hot spot Protection Testing of Photovoltaic Modules, 2006.
16. B. Damiani, M. Hilali, and A. Rohatgi: Light induced degradation in manufacturable multi-crystalline silicon solar cells, Technical Report, Sandia National Laboratories #A0-6062 and NREL #XAF8-17607-05, 2005.
17. Thin Film Solar Cells: An Overview, K.L. Chopra, P.D. Paulson and V. Dutta. http://159.226.64.60/fckeditor/UserFiles/File/tyndc/reference/19938777866147.pdf

Chapter 5

PV Power Generation

"Imagination is more important than knowledge."
Albert Einstein

Our objective in this chapter is to take a close look at photovoltaic (PV) power generation by reviewing the tools we have to work with: available sunlight, PV technologies and the way we can integrate these into PV power generating plants. We will use the term "large-scale" which includes any commercial and utility type installations that are designed for grid-connected power generation. Many of these are planned for installation in desert regions and because of that we pay special attention to the characteristics and behavior of the different technologies destined for long-term operation in harsh climates.

World energy markets are expected to grow significantly in the future, but due to the quickly changing political and economic environment, the growth is unpredictable. Nevertheless, two things are certain: the PV industry as a whole will grow steadily, and crystalline silicone (c-Si) will remain a leader in residential, commercial, and large-scale energy generation market segments.

Below we will review some of the prerequisites for creating and operating a successful large-scale power plant, the most important of which are:

1. Availability of solar energy (preferably unobstructed direct sunlight),
2. Proper location (favorable climate and available land),
3. Proper technology selection (PV equipment, inverters etc.),
4. Availability of interconnection points (transmission lines, substation etc.),
5. Suitable local, regulatory, and utilities conditions, and
6. Last, but not least, a team of capable and experienced professionals which is absolutely necessary to properly plan, design, setup and operate a PV power plant.

These are only few of the critical factors that we must tackle most often when planning a PV power plant. So, let's take a closer look at the key prerequisites and qualifications for a successfully designed, installed and operated large-scale PV project.

SOLAR ENERGY USE IN PV PROJECTS

So, we are planning to design an efficient and profitable PV power installation. A number of things must be considered in this process, the first of which is figuring out how much solar energy is available and the most efficient ways to use it.

Before we go to our project, let's see how much energy we have to work with in general, and how it is used in our daily activities. The total solar energy reaching the Earth's atmosphere is estimated at 175PW (PW is 1,000,000,000MW). 30% is reflected back into space and some is absorbed by the clouds. Whatever sunlight reaches the Earth's surface is absorbed by the oceans and land mass. A major portion of the sunlight reaching the Earth falls on the world's deserts, because they are mostly void of cloud cover. Most large-scale solar power plants are located in the desert areas, and that is also where most of the solar power will be generated in the future. This is why our emphasis in this text is on the operation of PV technologies in the world's deserts and the challenges they present.

Annual world-wide energy consumption is approximately 1/8000 of the total sunlight energy reaching the Earth' surface, so theoretically only a small part of the deserts could provide the entire world's energy demand. Putting this in practice, however, is quite complicated, because capturing and converting sunlight into useful energy, and delivering it to the point of use is not easy. So let's tackle the issues.

Sunlight for PV Generation

We mentioned above that sunlight coming from the sun has to be captured and converted into its two forms that are most useful for human consumption, thermal and electric energy. In order to do this right, we need to know the properties of sunlight as well as the properties and behavior of everything related to the process of converting sunlight into useful energy.

We know that full unobstructed sunlight, like what we see in the deserts, is called "beam" or "direct" radiation, while when obstructed by clouds, dust and other particles in the atmosphere it is called "diffused" radiation. The direct and diffused radiations, as well as the albedo (light reflection from ground surfaces) are useful forms of energy. They have different properties under different atmospheric conditions and seasons, so we do need to know how they affect solar power generation in order to design a most efficient PV power plant.

Another factor to consider is the path that sunlight travels to reach the location of our PV plant. For this we need to take a close look at the Earth' path around the sun, and around its axis, and understand the behavior of sunlight at the location we are considering. The Earth makes a full circle around the sun every 364.99 days, while at the same time the Earth makes one full revolution around its axis every 24 hours. The rotation of the Earth around the sun determines the seasons, while the rotation around its axis determines time of day. The combination of these two factors is very important for the proper design and operation of solar generating systems.

Calculating these two parameters, we can predict how much sunlight we can get at anytime of year and adjust the solar power collecting devices for maximum performance.

Figure 5-1 shows that on June 21 of every year the northern hemisphere is inclined slightly towards the sun. This "slight" inclination is enough to provide a shorter path for the sunlight, more intense direct radiation and generally warmer and sunnier summertime. On December 21 of every year, the Earth is inclined in the opposite direction, taking the northern hemisphere further from the sun, where the sunlight travels a longer distance to reach us, which translates in less direct radiation and shorter, colder and darker days.

Figure 5-2 shows the amount of power produced by a solar power generating system at different seasons and different hours of the day. Basically it tells us that we will get maximum power at noon on June 21st and much less at noon on December 20th of the same year, due to the Earth's position, as explained above. Anytime in between, we will get solar energy somewhere in between the above maximum and minimum levels. Again, it all depends on local weather condition, because a cloudy and rainy day in December would produce almost the same amount of sunlight as such a day in June.

We also see from Figure 5-2 that PV modules will generate more energy if they are pointed directly at the sun all day (in effect tracking it), instead of just being anchored in fixed position. The difference is exaggerated during the winter season, because of the sharper angle of the sunlight falling on fixed tilt modules. We will expand on these details, including tracking vs. fixed solar power collection in this text, so for now we just have to understand that the most important parameters for efficient solar power generation are the geographic location, the seasons, local weather, time of day, the technology we use, and if we use tracking or fixed PV modules.

The darker areas of the map in Figure 5-3 show

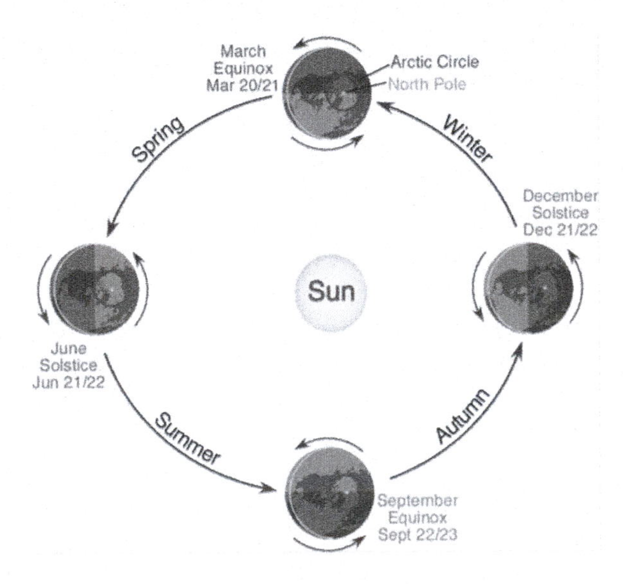

Figure 5-1. Earth path and cycles (seasons)

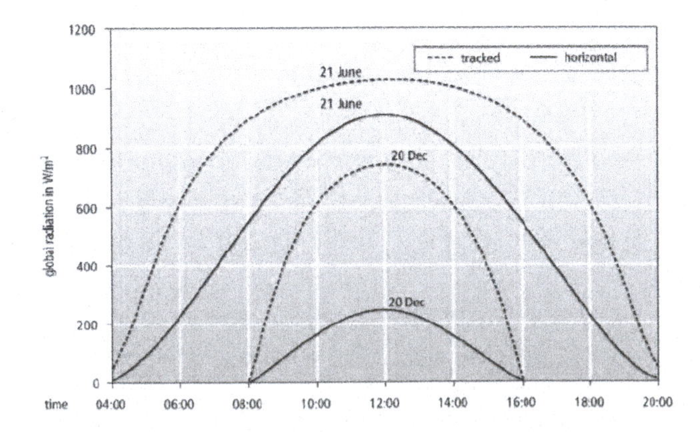

Figure 5-2. Seasonal global radiation

the highest solar radiation areas of the world, which are most suitable for solar power generation. Of course any area will generate some power, but the darker areas will generate many times the power generated by the lighter areas. So again, location, location, location.

Similarly, the darker areas on the US map in Figure 5-4 show where the sunlight is more intense and where the power output of any type of solar generating device will be greatest. Obviously, the deserts of the southwest USA seem to be the most appropriate for this purpose, and that is why there are such a great number of solar (thermal and PV) power plants installed and planned for installation in those areas.

These areas are the focal point of this book as well, as we have discussed in the previous chapters, because we believe that the energy future is related to the proper and efficient use of PV technologies in US and world deserts. Unfortunately, the inhospitable climatic conditions in the deserts are problematic for most PV technologies, which is also something we are addressing and explaining in this book, with the hope of shedding more light on the issues and bringing them out in the open for discussion and solution.

Insolation Levels and Variations

We can see now that the weather (cloud, fog and smog cover mostly) plays a great role in the performance of any PV installation. Inconsistent cloud covers cause variability issues (power fluctuations) which are undesirable and even harmful. There are some variability problems on mostly sunny days and fully cloudy days that will present some problems, mostly expressed in reduced output, but the variability is greatest during partially cloudy days, when the sun is randomly going in and out of the clouds. During this time the power output is going up and down uncontrollably, creating serious fluctuations of power being transmitted into the power grid.

Power output increases gradually from zero in early morning to full (peak) power at noon and then gradually back to zero by late afternoon. This is another, albeit more predictable and controllable, variable. On a clear sunny day that ascent and descent will be a smooth bell curve, but on a cloudy day it would be a mess of ups and downs of solar intensity and the related power fluctuations. Some fluctuations would be gradual, some fast, and some slow, but basically with no way to predict and compensate for them in advance. See Figure 5-5.

There are a number of effects that influence the behavior of the PV system on partially cloudy days, the major of which are related to a number of effects: special-temporary, incident angle, inverter, area affected, and temperature effects, as follow in more detail:

Spatial-temporary Effects

These are short-time variations in cloud cover—the sun going in and out randomly. To summarize the effect of this variable on the power output of a solar plant, we

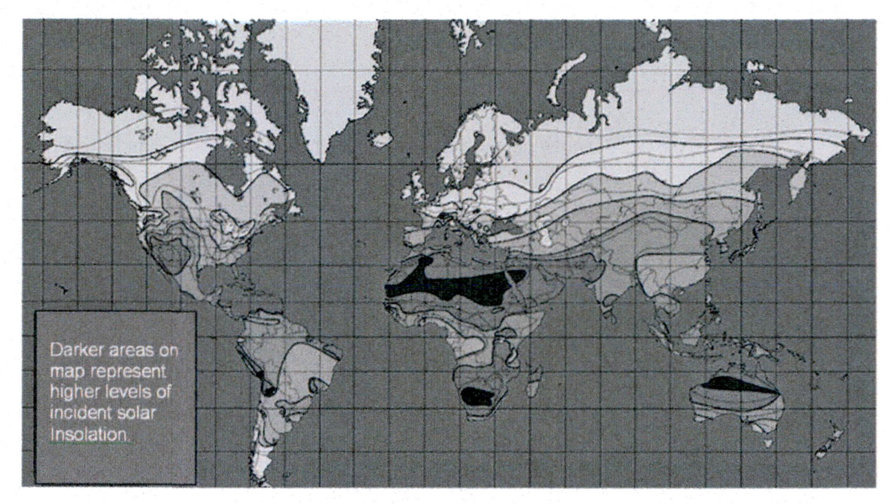

Figure 5-3. World's solar radiation.

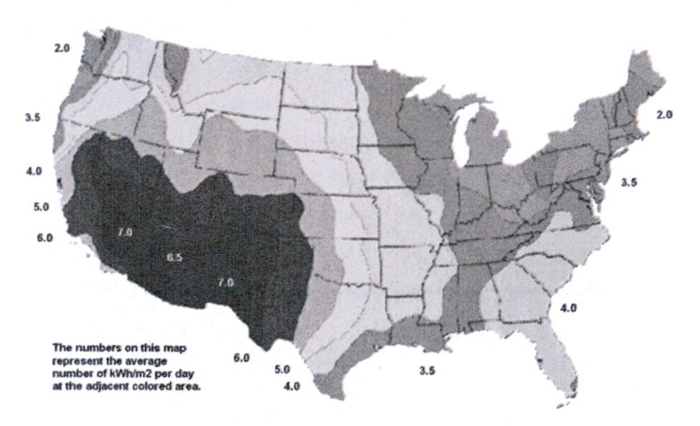

Figure 5-4. US solar radiation.

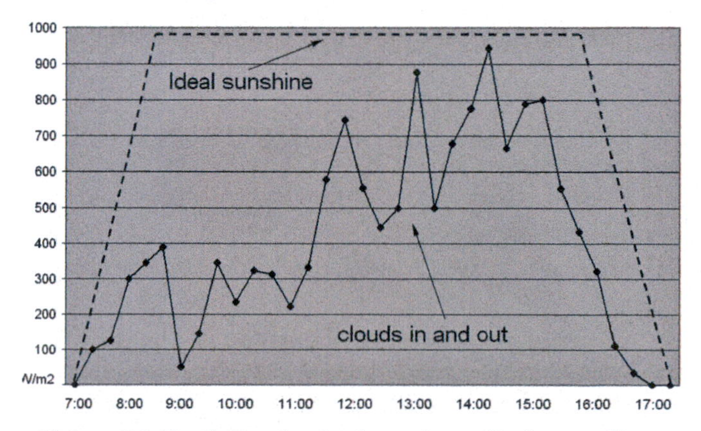

Figure 5-5. Partially cloudy day solar radiation readings

need to take into consideration all other factors (see below) but basically speaking, the larger the plant, the larger the output power fluctuations will be. Also, the shorter the duration of the sunlight intensity (insolation) fluctuations, the larger the power output effect. In other words, if the sun comes in and out of the clouds every second or two, the PV system (collectors-transmission-inverters) will not have enough time to process the signal and convert it properly into useable energy. This is one of the major reasons we always recommend that large solar power plants should be installed in areas with high solar insolation and minimal weather fluctuations.

Incident Angle Effects

Indicent angle effects are caused by improper positioning of the PV modules in relation to the sun. This anomaly could be caused by a number of factors, the most important of which is the awkward position of the fixed PV modules systems during early morning and late afternoon. During these times, the fixed systems collect only a small percent of the overall PV power reaching their surface.

Inverter Effects

Inverter effects are caused by the inverters' inability to properly follow and adapt to the random ramp-up and ramp-down patterns of generated power during variable conditions. Because of that, inverters are not able to convert 100% of the incoming DC power, which causes additional power losses. Basically, when a cloud covers the sky partially, the inverters sense the change and start regulating the power-down cycle, in less than efficient manner in most cases. At a certain point—which varies from inverter to inverter—the cloudy sky will reduce the power output from the modules to such a low level that the inverters cannot handle it and simply shut down. So even if the modules are generating some power, the inverters do not "see" it, and it is wasted. These abnormal conditions are caused by and/or identified as: MTTP issues, IEEE 1547 dropouts, inverter "clipping" abnormalities etc.

Tracking Systems

Trackers are much better in collecting the maximum sun energy, due to the fact that their x-y drive controllers position the PV modules to face the sun properly at all times. Trackers, however, are affected by partial cloud cover even more than their fixed PV module cousins, because tracker controls might get confused by the sun hiding behind erratic cloud cover, thus not providing the best angle for sunlight collection.

Cloud Cover

The effects of cloud cover vary with the size of the power plant. Large plants cover large areas of land, where partially cloudy days might complicate things even further by throwing a shadow over one side of the field, while the other side is clear and gets full sun. This situation will make attempts to get a complete handle on the incoming vs. outgoing power almost impossible. This condition is especially critical in fields using trackers because the shaded side will act unpredictably adding to the complexity of the variability factor.

Temperature Effects

Temperature effects are abnormalities caused by high temperature on solar cells and panels. Silicon and thin film PV modules drop 0.5% efficiency with each degree C temperature increase. Temperature increases are implicated in several failure or degradation modes of PV cells and modules, as elevated temperatures increase the stresses associated with thermal expansion and significantly increase degradation rates.

These factors are hard to keep track of, let alone control as needed for producing consistent maximum power output. Because of that, we need to take them into consideration during the design of the power field and then do whatever is necessary to control their effect during operation.

Derating Factors

Once we have captured solar energy via PV devices (solar cells or modules) we need to convert it into electricity. Some of the energy is lost during the conversion process; i.e., only 15% is converted by a PV module, and the losses continue down the line before the energy can be used. In a PV power plant, the loses are from PV equipment, wires, connectors, controllers, inverters, transformers etc.—equipment that is needed to capture and convert sunlight into AC power.

The losses from each piece of equipment determine its efficiency, and the corresponding power loss amount is the derating factor. The derating factors of all plant components must be used to calculate the PV system's ability to produce DC and AC power. Some of these factors follow.

DC Rating (Nameplate)

The size of a PV system is its nameplate or DC power rating, which is the amount of DC power it is supposed to produce according to the manufacturer at STC (1,000 W/m^2 solar irradiance and 25°C module temperature). The DC rating of a PV system is determined by adding the total PV modules' power listed on each PV module in watts and then dividing the sum by 1,000 to convert it to kilowatts (kW); our 100 MWp PV system consists of 500,000 PV modules, each measuring 200Watts at STC, and is expected to generate 100 MWp of power at STC.

This is a theoretical number, simply because PV modules seldom operate at STC in the field, and for sure not in the fields we are most interested in—harsh desert areas with the highest temperature extremes. Nevertheless, the DC rating is important for establishing a baseline of the generation power and is used in many consequent calculations and considerations.

DC Derating Factor

This is basically the difference between the DC rating (nameplate) as provided by the manufacturer and the actual power generated at location. We perform a series of field tests after installation to verify and baseline the performance at this particular location. The tests could be done with each string of modules independently, which will provide an average DC power produced for each string. The different strings could be added to obtain the actual total DC power produced by the entire installation. When compared with the nameplate, the difference will give us the actual DC derating factor of the entire power plant under full power. Field measurements are usually lower than the nameplate rating for a number of reasons.

DC derating factor =
Actual generated DC power/Nameplate

A derate factor of 0.95, for example, indicates that post-install field testing power measurements at the location were 5% less than the manufacturer's nameplate rating measured at STC. Or, our power plant with 100 MWp nameplate produced only 95 MWp during a certain time period under actual operating conditions.

In this case:

DC derating factor =
95MWp/100MWp = 0.95, or 95% of nameplate

During this test and after consecutive time-lapse tests, we can find a number of variables and discrepancies which must be taken into consideration when estimating the initial overall plant efficiency, its performance levels and the related DC derating.

DC to AC Derating Factor

Another critical variable of the overall performance estimate of a PV power-generating system is its DC to AC derating factor (or DC to AC conversion efficiency), which is simply the multiple of the derate factors of the different power plant components.

DC to AC Derating factor =
multiple of components' derate factors

In this case, DC to AC derating factor = 0.95 x 0.92 x 0.98 x 0.995 x 0.98 x 0.99 x 0.95 x 0.98 x 1.00 x 1.00 x 1.00 = 0.77. Or, 77% of the nameplate DC power will be converted into AC power at this particular location with this particular equipment. In other words, our 100 MWp power plant will generate 77 MWp of AC electricity and send it into the grid.

Table 5-1. Average derating factors as used in PVWATT (NREL) (3)

Component Derate Factors	PVWatts Default	Range
PV module nameplate DC rating	0.95	0.80–1.05
Inverter and transformer	0.92	0.88–0.98
Mismatch	0.98	0.97–0.995
Diodes and connections	0.995	0.99–0.997
DC wiring	0.98	0.97–0.99
AC wiring	0.99	0.98–0.993
Soiling	0.95	0.30–0.995
System availability	0.98	0.00–0.995
Shading	1.00	0.00–1.00
Sun-tracking	1.00	0.95–1.00
Age	1.00	0.70–1.00
Overall DC-to-AC derate factor	0.77	0.09999–0.96001

- **PV module nameplate DC rating**
  This accounts for the accuracy of the manufacturer's nameplate rating. Field measurements of PV modules may show that they are different from their nameplate rating or that they experience light-induced degradation upon exposure. A derate factor of 0.95 indicates that testing yielded power measurements at STC that were 5% less than the manufacturer's nameplate rating.

- **Inverter and transformer**
  This reflects the inverters and transformer's combined efficiency in converting DC power to AC power. A list of HYPERLINK "http://www.gosolarcalifornia.ca.gov/equipment/inverters.php" inverter efficiencies by manufacturer is available from the Consumer Energy Center. The inverter efficiencies include transformer-related losses when a transformer is used or required by the manufacturer.

- **Mismatch**
  The derate factor for PV module mismatch accounts for manufacturing tolerances that yield PV modules with slightly different current-voltage characteristics. Consequently, when connected together electrically, they do not operate at their peak efficiencies. The default value of 0.98 represents a loss of 2% because of mismatch.

- **Diodes and connections**
  This derate factor accounts for losses from voltage drops across diodes used to block the reverse flow of current and from resistive losses in electrical connections.

- **DC wiring**
  The derate factor for DC wiring accounts for resistive losses in the wiring between modules and the wiring connecting the PV array to the inverter.

- **AC wiring**
  The derate factor for AC wiring accounts for resistive losses in the wiring between the inverter and the connection to the local utility service.

- **Soiling**
  The derate factor for soiling accounts for dirt, snow, and other foreign matter on the surface of the PV module that prevent solar radiation from reaching the solar cells. Dirt accumulation is location- and weather-dependent. There are greater soiling losses (up to 25% for some California locations) in high-traffic, high-pollution areas with infrequent rain. For northern locations, snow reduces the energy produced, and the severity is a function of the amount of snow and how long it remains on the PV modules. Snow remains longest when sub-freezing temperatures prevail, small PV array tilt angles prevent snow from sliding off, the PV array is closely integrated into the roof, and the roof or another structure in the vicinity facilitates snow drift onto the modules. For a roof-mounted PV system in Minnesota with a tilt angle of 23°, snow reduced the energy production during winter by 70%; a nearby roof-mounted PV system with a tilt angle of 40° experienced a 40% reduction.

- **System availability**
  The derate factor for system availability accounts for times when the system is off because of maintenance or inverter or utility outages. The default value of 0.98 represents the system being off 2% of the year.

- **Shading**
  The derate factor for shading accounts for situations in which PV modules are shaded by nearby buildings, objects, or other PV modules and arrays. For the default value of 1.00, the PVWatts calculator assumes the PV modules are not shaded. Tools such as HYPERLINK "http://www.solarpathfinder.com" Solar Pathfinder can determine a derate factor for shading by buildings and objects. For PV arrays that consist of multiple rows of PV modules and array structures, the shading derate factor should account for losses that occur when one row shades an adjacent row.

- **Sun-tracking**
  The derate factor for sun-tracking accounts for losses for one- and two-axis tracking systems when the tracking mechanisms do not keep the PV arrays at the optimum orientation. For the default value of 1.00, the PVWatts calculator assumes that the PV arrays of tracking systems are always positioned at their optimum orientation, and performance is not adversely affected.

- **Age**
  The derate factor for age accounts for performance losses over time because of weathering of the PV modules. The loss in performance is typically 1% per year. For the default value of 1.00, the PVWatts calculator assumes that the PV system is in its first year of operation. For the 11th year of operation, a derate factor of 0.90 is appropriate.

Note: Because the PVWatts overall DC-to-AC derate factor is determined for STC, a component derate factor for temperature is not part of its determination. Power corrections for PV module operating temperature are performed for each hour of the year as the PVWatts calculator reads the meteorological data for the location and computes performance. A power correction of -0.5% per degree Celsius for crystalline silicon PV modules is used.

PV MODULES FUNCTION

PV modules have a number of mechanical, chemical and electric characteristics, with which design engineers, customers, installers, and investors alike must be familiar. These specs and characteristics describe the history and the state of the module. They are the first thing to pay close attention to when considering certain types, brands or sizes of PV cells and modules, and when designing, installing and operating a large-scale power plant. Let's take a closer look at some of these:

PV Modules, Technical Specifications

PV modules are complex electro-mechanical assemblies with a number of components which determine their function. Modules usually come with a set of specifications and installation and operation instructions which describe their type, structure, use and expected behavior. Some of the key concepts of PV modules' function and on-the-field operation are described below (7).

Rated Power at STC (in Watts)

This is the module power rating at STC, or 1,000 watts per square meter of solar irradiance on the module surface at 25°C (77°F) cell temperature. Because module power output depends on environmental conditions, such as irradiance and temperature, each module is tested at STC so that modules can be compared and rated on a level playing field. When less sunshine hits the module, less power is produced. Likewise, the hotter it gets, the less power your modules will produce due to the cells' temperature degradation phenomena.

STC references the actual cell temperature, not ambient air temperature. As dark PV cells absorb radiant energy, their temperature increases and will be significantly higher than the ambient air temperature. For example, at an ambient air temperature of about 23°F, a PV cell's temperature will measure about 77°F—the temperature at which its power is rated.

If the ambient air temperature is 77°F (and irradiance is about 1,000 W/m2), module cell temperature will be about 131°F and power output will be reduced by about 15%. If the air temperature is 120° F, as it is at noon on a clear summer day in the Sonora desert, the power output might drop by another 15-20%.

Rated Power Tolerance (%)

This is the specified range within which a module will either over perform or underperform its rated power at STC. Power tolerance is a much-debated module specification. Depending on the module, this specification can vary greatly, from as much as +10% to –9%. A 100 W module with a –9% power tolerance rating may produce only 91 W straight out of the box. With potential losses from high temperatures, it will likely produce even less than that.

Because modules are often rated in small increments, it is not uncommon for those that fall under the lower power tolerance of the next model to be rated as a higher wattage module. Case in point: A module with a +/–5% power tolerance rating that produces 181 W during the factory testing process could be classified as a 190 W module, as opposed to a 180 W module. For maximum production, look for modules with a small negative (or positive only) power tolerance.

Rated Power Per Square Foot (Watts)

This is the power output at STC, per square foot of module (not cell) area. It is calculated by dividing module rated power by the module's area in square feet and is also known as "power density." The higher the power density, the less space is needed to produce a certain amount of energy. With some of the newer-generation modules, power density values are higher due to increased cell and module efficiency.

The greatest variation in this specification is in comparing crystalline PV modules to thin-film modules. If space is tight for array placement, consider choosing modules with higher power densities, though more efficient modules can be more expensive. Choose modules with lower power densities, and you'll need more modules for the same amount of energy. That means more infrastructure (module mounts, hardware, etc.) and more installation time.

Module Efficiency (%)

This is the ratio of output power to input power, or how efficiently a PV module uses the photons in sunlight to generate DC electricity. If 1,000 W of sunlight hit 1 square meter of solar module, and that solar module produces 100 W of power from that square meter, then it has an efficiency of 10%. Similar to power density, the higher the efficiency value, the more electricity generated in a given space.

Series Fuse Rating (Amps)

This is the current rating of a series fuse used to protect a module from overcurrent, under fault conditions. Each module is rated to withstand a certain number of amps. Too many amps flowing through the module—perhaps backfed amps from paralleled modules or paralleled strings of modules—could damage the module if it's not protected by an overcurrent device rated at this specification. Backfeeding from other strings is most likely to exist if one series string of modules stops producing power due to shading or a damaged circuit. Because PV modules are current-limited, there are some cases where series fusing may not be needed.

When there is only one module or string, there is nothing that can backfeed, and no series string fuse is needed. In the case of two series strings, if one string stops producing power and the other string backfeeds through it, still no fuse is needed because each module is designed to handle the current from one string. Some PV systems even allow for three strings or more with no series fuses. This is due to 690.9 Exception B of the *NEC* and is possible when the series fuse specification is substantially higher than the module's short circuit current (Isc). When required, series fuses are located in either a combiner box or in some batteryless inverters.

Connector Type

This is the way the module output terminal or cable/connector is configured. Most modules come with "plug and play" weather-tight connectors to reduce installation time in the field. These are connectors such as Solarlok (manufactured by Tyco Electronics), and MC and MC4 (manufactured by Multi-Contact USA). Solarlok and MC4 are lockable connectors that require a tool for opening.

Because so many PV systems installed today operate at high DC voltages, lockable connectors are being used on modules in accessible locations to prevent untrained persons from "unplugging" the modules, per *2008 NEC* Article 690.33(C). Due to this new code requirement, most PV manufacturers are updating their connectors to the locking type. Depending on how fast this change is reflected in the supply chain, connectors on a particular module may be of an older style or lockable—so be sure to check.

Some manufacturers still offer modules with junction boxes (J-boxes). J-boxes allow the use of conduit in between modules, as raceways are required for PV source and output circuits (with a maximum system voltage greater than 30 volts) installed in readily accessible locations per *2008 NEC* Article 690.31(A). This approach is used to prevent an unqualified person from accessing array wiring.

Materials Warranty (Years)

This is a limited warranty on module materials and quality under normal application, installation, use, and service conditions. For most modules material warranties vary from 1 to 5 years. Most manufacturers offer full replacement or free servicing of defective modules.

Read the warranty conditions carefully prior to purchasing the modules, because they vary from vendor to vendor. The problem that is most often encountered and one that is most disputed is shipping charges and replacement (labor) cost. In other words, the expense of shipping a batch of modules from abroad, uninstalling the defective ones and reinstalling the new batch might be higher than the cost of the modules themselves, thus the vendor might object to absorbing these charges, unless all these conditions and exemptions are clearly identified and addressed in the warranty documents.

Power Warranty (Long-term)

This is a limited warranty for module power output based on the minimum peak power rating (STC rating minus power tolerance percentage) of a given module. The manufacturer guarantees that the module will provide a certain level of power for a period of time—at least 20 years. Most warranties are structured as a percentage of minimum peak power output within two different time frames—90% over the first 10 years and 80% for the next 10 years.

For example, a 100 W module with a power tolerance of +/–5% will carry a manufacturer guarantee that the module should produce at least 85.5 W (100 W x 0.95 power tolerance x 0.9) under STC for the first 10 years. For the next 10 years, the module should produce at least 76 W (100 W x 0.95 power tolerance x 0.8).

Module replacement value provided by most power warranties is generally prorated according to how long the module has been in the field. Again, the cost of shipping and replacement of the non-conforming modules could be high, and replacement conditions must be well understood and agreed upon by both parties prior of purchasing the modules.

Cell Type

This describes the type of silicon that comprises a specific cell, based on the cell manufacturing process. There are four basic types of modules for the non-commercial market—monocrystalline, multicrystalline, ribbon, and amorphous silicon (a-Si). Each cell type has pros and cons. Monocrystalline PV cells are the most expensive and energy intensive to produce but usually yield the highest efficiencies. Though multicrystalline and ribbon silicon cells are slightly less energy intensive and less expensive to produce, these *cells* are slightly less efficient than monocrystalline cells.

However, because both multi- and ribbon-silicon modules leave fewer gaps on the module surface (due to square or rectangular cell shapes), these *modules* can often offer about the same power density as monocrystalline modules per unit area. Thin-film modules, such as those made from amorphous silicon cells, are the least expensive to produce and require the least amount of energy and raw materials, but they are the least efficient of the cell types. They require about twice as much space to produce the same power as mono-, multi-, or ribbon-silicon modules. Thin-film modules do have better shade tolerance and high-temperature performance but are often more expensive to install because of their lower power density and structural requirements.

Some manufacturers now offer a cell with a combination of cell types—Sanyo's "bifacial" HIT modules are composed of a monocrystalline cell and a thin layer of amorphous silicon material. In addition to generating power from the direct rays of the sun on the module face, this hybrid module can produce power from reflected light on its underside, increasing overall module efficiency.

Cells in Series

This is the number of individual PV cells wired in series, which determines the module design voltage. Crystalline PV cells each operate at about 0.5 V. When cells are wired in series, the voltage of each cell is additive. For example, a module that has 36 cells in series has a maximum power voltage (Vmp) of about 18 V. Why 36? Historically, these modules—known as 12-V modules—were designed to push power into 12-V batteries. But to deliver the 12 V, they needed to have enough excess voltage (electrical pressure) to compensate for the voltage loss due to high temperature conditions. Modules with 36 ("12-V") or 72 ("24-V") cells are designed for battery-charging applications.

Modules with other numbers of cells in series are intended for use in batteryless grid-tied systems. Grid-tied modules now combine a certain number of cells for the goal of maximizing power with grid-tied inverters and their maximum power point tracking (MPPT) capabilities. Due to the increased availability of step-down/MPPT battery charge controllers, grid-tied PV modules can also be used for battery charging, as long as they stay within the voltage limitations of the charge controller.

Maximum Power Voltage

Maximum power voltage (Vmp) is the voltage generated by a PV module or array when exposed to sunlight and connected to a load—typically an inverter or a charge controller and/or a battery. Batteryless grid-tied inverters and MPPT charge controllers are built to track maximum power throughout the day, and the Vmp of each module and array, as well as array operating temperatures, must be considered when sizing an array to a particular inverter or controller.

Increasing temperatures cause voltage to decrease; decreasing temperatures cause voltage to increase. Fortunately, series string-sizing programs for grid-tied inverters will allow you to input both the high and low temperatures at your installation site, and calculate the correct number of modules in series to maximize system performance.

Maximum Power Current (Imp)

This is the maximum amperage produced by a module or array (under STC) when exposed to sunlight and connected to a load. This specification is most commonly used in performing calculations for PV array disconnect labeling required by *NEC* Article 690.53(1), as the rated maximum power-point current for the array must be listed. Maximum power current is also used in array and charge controller sizing calculations for battery-based PV systems.

Open-circuit Voltage (Voc)

This is the maximum voltage generated by a PV module or array when exposed to sunlight with no load connected. All major PV system components (modules, wiring, inverters, charge controllers, etc.) are rated to handle a maximum voltage. Maximum system voltage must be calculated in the design process to ensure all components are designed to handle the highest voltage that may be present. Under certain low-light conditions (dawn/dusk), it's possible for a PV array to operate close to open-circuit voltage.

PV voltage will increase with decreasing air temperature, so Voc is used in conjunction with historic low temperature data to calculate the absolute highest maximum system voltage. Maximum system voltage must be shown on the PV array disconnect label.

Short-circuit Current (Isc)

This is the amperage generated by a PV module or array when exposed to sunlight and with the output terminals shorted. The PV circuit's wire size and overcurrent protection (fuses and circuit breakers) calculations per *NEC* Article 690.8 are based on module/array short-circuit current. The PV system disconnect(s) must list array short-circuit current (per *NEC* 690.53).

Maximum Power Temperature Coefficient (% per degree C)

This is the change in module output power in percent of change per degree Celsius for temperatures other than 25°C (STC temperature rating). This specification allows us to calculate how much module power will be lost or gained due to temperature shifts. In hot climates, cell temperatures can reach in excess of 70°C (158°F). Consider a module maximum power rating of 200 W at STC, with a temperature coefficient of –0.5% per degree C. At 70°C, the actual output of this module would be approximately 155 W.

Modules with lower power temperature coefficients will fare better in higher-temperature conditions. Notice the relatively low values listed for thin film modules. This specification reflects their usually better high-temperature performance.

Open-circuit Voltage Temperature Coefficient (mV per degree C)

This is the change in module open-circuit voltage in millivolts per degree Celsius at temperatures other than 25°C (STC temperature rating). Expressed as millivolts per degree Celsius, it can be shown as percentage per degree Celsius, volts per degree Celsius, or volts per degree Kelvin. If given, this specification is most commonly used in conjunction with open-circuit voltage to calculate maximum system voltage (per *NEC* Article 690.7) for system design and labeling purposes.

For example, consider a series string of ten 43.6 V (Voc) modules installed at a site with a record low of –10°C. Given a Voc temperature coefficient of –160mV per degree Celsius, the rise in voltage per module will be 5,600 mV [–160 mV per degree Celsius x (–10°C – 25°C)], making an overall maximum system voltage of 492 V [10 x (5.6 V + 43.6 V)]—under the 600 VDC limit for PV system equipment.

Nominal Operating Cell Temperature (NOCT)

This is the temperature of each module, given an irradiance of 800 W/m2 and an ambient air temperature of 20°C. This specification can be used with the maximum power temperature coefficient to get a better real-world estimate of power loss due to temperature. The difference in cell temperature and ambient temperature is dependent on sunlight's intensity (W/m2). Less-than-ideal sky conditions are common in many areas, so a standard of 800 W/m2 is the basis for this specification, rather than 1,000 W/m2, which is considered full sun. The construction and coloring of each module is slightly different, so actual cell temperature under these conditions will vary per module.

For example, if a particular module has an NOCT of 40°C and a maximum power temperature coefficient of –0.5% per degree Celsius, power losses due to temperature can be estimated at about 7.5% [0.5% x (40°C – 25°C)].

Supply Chain

A large number of materials are needed during the many steps of PV cells and modules manufacturing, most of which are provided by third-party manufacturers or vendors. In all cases, the supply chain plays an integral role in ensuring the quality of the final product. If just one product is out of spec, or of lower quality, then the entire batch of PV cells and modules will be of equally low quality. Keeping track of, checking and controlling the quality of all incoming products at all times is an enormous job, which only a few companies can claim is under complete control.

Some key components of the PV industry's supply chain are:

Silicon Materials Supply Chain

The c-Si PV industry uses very large quantities of silicon material, from which the c-Si wafers and solar cells are manufactured. Silicon production capacities have increased several-fold during recent years, due to increased demand for PV cells and modules around the world. The rapid expansion of silicon material capacities, however, is not a guaranty that the quality of the material has risen at the same pace.

The quality of the silicon material is not something that can be seen with the naked eye, nor is it easy or quick to test. Since the quantities coming into a wafer and cell plant for processing are so vast, it is impossible to test and verify the quality of every silicon chunk used in the manufacturing process. The silicon PV industry also uses different types of "off-grade" silicon materials, remaining (surplus or rejects) from ingots and wafer manufacturing operations, which are sometimes mixed with the virgin silicon. In addition, rejected silicon wafers from the semiconductor and solar production operations are also mixed in the melts, which adds another variable to the quality of the raw materials, and which affects the final products' quality.

Another silicon source is from recycling (stripping and surface grinding) of fallouts from solar and semiconductor processing operations. This is tricky, however, because while the top and bottom surfaces of the wafers could be shaved off, the bulk material might still contain impurities remaining from the previous processes. These impurities are hard to detect and if undetected could affect the quality and performance of the solar cells processed with these wafers.

Taking all this into consideration, we could attribute a significant portion of the responsibility for the quality of the final product to the silicon material manufacturers. Nevertheless, the PV cells and modules manufacturers are ultimately responsible for the overall quality of their product and must be well aware of the impact of all incoming materials—including the quality of silicon coming from all sources.

Thin film PV modules, on the other hand, use very little semiconductor material which is usually expensive and justifies precise control of its quality. The glass plates are the major materials costs for manufacturing these products. Nevertheless, the quality of the supply chain (incoming materials, components, supplies and consumables) is equally important and critical as with the c-Si products described above. And again, the manufacturers bear the ultimate responsibility for the quality of the final product—regardless of the source of materials.

Consumables Supply Chain

The PV industry supply chain covers the ordering, manufacturing and transport of all materials, including consumables, needed during all steps of the manufacturing of PV cells and modules.

The key consumables are:

1. Module materials, such as wafers, thin film materials and targets, chemicals, glass covers, aluminum frames, back covers, laminates, wiring, junction boxes, diodes, hardware etc.
2. Process chemicals and gasses, such as cleaning chemicals, acids and bases, metal pastes, AR coating chemicals and gasses, DI water system supplies, etc.
3. Process supplies, such as gloves, masks, product handling and transport hardware (baskets, cassettes, packaging), wipes, etc.

Ensuring the highest quality of the process materials and chemicals is paramount for obtaining high quality PV cells and modules. Here again, the highest quality final product will be determined by the lowest quality of any material in the supply chain.

The consumables supply chain (third-party suppliers), facilities, equipment and processes must be inspected, verified and qualified periodically to ensure the quality of incoming materials and products.

General Supply Chain Concerns

The major aspects of a well run, quality oriented, supply chain are a) proper selection of process materials, b) proper selection of suppliers, c) proper qualification of the key suppliers, and d) periodic auditing of suppliers' operations and quality control systems. Such qualification and auditing procedures are marginal or nonexistent in some countries.

In our experience, the Asian PV industry supply chain is based on "static approach," using long-term contracts and the cheapest deal around to keep production going. Central distribution of supply chain deliveries is widely practiced in countries where government subsidies and policies drive the key industrial functions—including those of PV cells and manufacturing operations. Very little, if any, direct quality responsibility is transferred to the suppliers, nor is there much room for qualifications or audits under such management (or lack thereof).

Competition among supply chain vendors is another vehicle for assuring excellent quality and lower price from the supply chain members. This system is also not in force most Asian countries. Supply chain integration, data exchange within the chain, JOT orders and quality improvement initiatives are discussed in that part of the world, but their full implementation in real life is incomplete.

NOTE: Exempt of this generalization is Japan and some other more advanced Asian countries.

Supply Chain Management of Change

Management and control of the supply chain (the quality and flow of materials, chemicals and components from third-party vendors) as needed for the manufacturing process is usually complicated by suppliers' unforeseen changes in materials, equipment, processes and personnel. Constant changes of the product design and manufacturing procedures are the norm these days in the battle for lower prices. In other words, if we order 100 MWp PV modules today, and 100,000 of the 500,000 modules are shipped tomorrow, the rest of the shipments, manufactured at a later date, might be processed with different materials or components. Different types or qualities of materials might have been received from the supply chain vendors while completing the order, thus the quality of the first shipment might be different from the following. Even if the final quality is better, it is still NOT what we've ordered and we at least need to know about the changes, keep track of them and follow-up with appropriate evaluations, tests, documenting every detail. This is not easily done in most cases.

In conclusion: both the supply chain quality management and the management of change systems are complex, cumbersome, expensive, hard to implement and difficult to audit. There is, however, no other way to ensure the quality of the final product. Some compromises might be useful in simplifying the documentation required by these procedures; nevertheless, a thorough investigation and audits are the only way to ensure excellent quality of PV products—quality we can rely on during 30 years of continuous operation.

Production (Capital) Equipment

Production equipment, similar to that used to manufacture semiconductor devices, is also used to manufacture solar cells and modules (c-Si and thin films) in fully or semi-automated production lines. Equipment type and size varies according to the respective PV technology, but is basically made and sold by several large international companies. The major players in the production equipment manufacturing and sales during 2009-2010 were ALD Vacuum Tech., Applied Materials (now out of the solar business), CentroTerm, GT Solar, Manz Automation, Meyer Burger, NPC, Oerlikon, Roth and Rau AG, Schmidt Group, Ulvac Japan. Most are large companies, or part of a large international conglomerate.

There are many variables in the solar production equipment business, but the most important and controversial one is the lack of equilibrium between manufacturing capacities and solar products consumption. This uncertainty is what keeps most equipment manufacturers awake at night. Averaged from several sources, we see that the ratio of PV cells and modules production capacity vs. consumption in 2008, 2009 and 2010 was approx. 2:1 (12,000 MW total capacity vs. 5,500 MW consumed in 2008, 19,000 MW vs. 7,000 MW in 2009 and 28,000 MW vs. 14,000 MW in 2010). Inventory averages (product in storage ready to ship) for that period were in the 70-80 days range, all of which translates into low equipment and resources utilization rates averaging 40-50%. Production facilities were also idle 40-50% of the time which means that no additional production equipment was needed during 2009-2010.

The overall picture has been changing lately, where demand is catching up with supply, so that we now see forecasts of over 30% increase of production equipment sales in 2011-2012. This, of course, depends on many factors, and may or may not happen, but the trend points toward the positive side of the equation. This might mean that equipment prices will increase accordingly, and with that final product costs, but we will just have to wait and see.

In conclusion, we'd like to emphasize that the solar wafers', cells' and modules' processing equipment (c-Si and TFPV) is the most reliable part of the solar industry. Most equipment manufacturers have had many years of experience, accompanied by a lot of trial and error, so what we have now is the best state-of-the-art production equipment ever.

Major concerns with production equipment today:

1. Using production equipment made by newly established, or converted from other industry, equipment manufacturers, many of whom are popping up in Asia. The quality of their equipment is questionable since it has not been proven reliable for this use,
2. Production equipment and processes are not standardized; different manufacturers have different ways to produce and use the equipment.

Because of these reasons, variability of the final product is inevitable. The sheer number of cells and modules that must be made and assembled around the world suggests that standardized methods for manufacturing and assembly of PV modules and products are needed, in order to ensure 100% adherence to production efficiency and product quality norms. This, however, is not going to happen anytime soon, so we need to take measures to ensure the quality of the PV products we are designing, making, or buying.

Needed Manufacturing Standards

SUPPLY CHAIN	MFG. PROCESS	QC/QA
Raw materials	Equipment	Initial inspection
Process chemicals	Process specs	In-process inspect
Consumables	O&M procedures	Final inspection
Waste treatment	Enforcement	Long-term tests

PV Modules Test Procedures

Every PV module type and size destined for residential and large-scale projects in the US and EU is supposed to undergo an official standardized testing sequence of its electro-mechanical and esthetic properties. See the list in Table 5-2 of different inspection methods, test procedures, and conditions used to determine the quality of PV cells and modules.

See Appendix A below for a complete list of test procedures, used during PV modules qualification and certification testing programs.

Table 5-2. IEC 61215/IEC 61646 test procedures

Code	Qualification Test	Test Conditions
10.1	Visual Inspection	according defined inspection list
10.2	Maximum Power Determination	measurement according to IEC 60904
10.3	Insulation Test	1000 VDC + twice the open circuit voltage of the system at STC for 1 min, isolation resistance * module area > 40 MΩ·m² at 500 VDC
10.4	Measurement of Temperature Coefficients	Determination of the temperature coefficients of short circuit current, open circuit voltage and maximum power in a 30°C interval
10.5	Measurement of NOCT	total solar irradiance = 800 W/m² wind speed = 1 m/s
10.6	Performance at STC and NOCT	cell temperature = NOCT / 25°C irradiance = 800 W/m² / 1000 E/m² measurement according to IEC 60904
10.7	Performance at low Irradiance	cell temperature = 25°C irradiance = 200 W/m² measurement according to IEC 60904
10.8	Outdoor Exposure Test	60 kWh/m² solar irradiation
10.9	Hot-Spot Endurance Test	5 hour exposure to > 700 W/m² irradiance in worst-case hot-spot condition
10.10	UV-preconditioning test	15 kWh/m² UV-radiation (280 - 385 nm) with 5 kWh/m² UV-radiation (280 - 320 nm) at 60°C module temperature
10.10*	UV-Exposure according IEC 61345	Min.15 kWh/m² UV-radiation (280 - 400 nm) with 7.5 kWh/m² UV-radiation (280 - 320 nm) at 60°C module temperature
10.11	Thermal Cycling	50 and 200 cycles -40°C to +85°C
10.12	Humidity Freeze Test	10 cycles -40°C to +85°C, 85% RH
10.13	Damp Heat	1000 h at +85°C, 85% RH
10.14	Robustness of Terminations	As in IEC 60068-2-21
10.15	Wet Leakage Test	Evaluation of insulation of the module under wet conditions
10.16	Mechanical Load Test	Three cycles of 2400 Pa uniform load, applied for 1 h to front and back surfaces in turn
10.17	Hail Test	25 mm diameter ice ball at 23 m/s, directed at 11 impact locations
10.18	Bypass diode thermal test	Asses adequacy of thermal design of by-pass diodes at a current of 1.25 x Isc running through the diodes at module temperature of 75°C
10.19**	Light soaking	Light exposure of cycles of at least 43 kWh/m² and module temperature of 50°C ± 10 °C, until Pmax is stable within 2 %

* Tests can alternatively be used

** Tests only relevant for IEC 61646 qualification

Basically speaking, a PV module shall be judged to have passed the qualification tests, and therefore to be UL or IEC type approved, if each sample meets the following criteria:

1. The degradation of the maximum power output at standard test conditions (STC) does not exceed 5% after each test nor 8% after each test sequence;
2. The requirements of IEC 61646, tests 10.3 (insulation) and 10.20 (wet leakage current) are met;
3. No major visible damage (broken, cracked, torn, bent or misaligned external surfaces; cracks in a solar cell which could remove a portion larger than 10% of its area; bubbles or delamination; loss of mechanical integrity;
4. No sample has exhibited any open circuit or ground fault during the tests;
5. For IEC 61646 only: the measured maximum output power after final light-soaking shall not be less than 90% of the minimum value specified by the manufacturer.

These tests and guidelines are quite helpful for use in any PV cells, modules and systems design, installation and operation, but we need to point out that no matter how good the qualification or production test procedures are, PV cells and modules must withstand the test of time—30 years of non-stop operation under harsh environmental conditions of hot deserts, freezing mountains, humid climates, rain, snow and everything else that comes their way. The tests show a glimpse of what to expect, but are far from predicting the 30-year performance and longevity of the modules.

In Figure 5-6, PV module (a) fails quickly during testing or soon after installation in the field, while module (b) fails slowly during a long time period. This time period might be days, months or even years. And this is where things get complicated, because while reliability is measured in a lab setup under improvised field conditions, the actual field behavior which determines the module's durability and longevity is more often than not different.

Reliability is a very important characteristic of PV devices performance and overall behavior since it measures their discrete and absolute failure modes. Durability, however, is expressed in rate of change of the devices—their degree of robustness. Durability is not always expressed in total failures, but rather in gradual decline of efficiency, and/or shortened service lifetime. These phenomena are generally harder to test and measure, but are extremely important to the long run.

Durability is equally important, but even harder to measure and often neglected when estimating PV system's performance and costs. The general assumption is that once the PV modules have passed the qualifications and production tests they will deliver power as specified for the duration. This, however, is far from the truth. The durability which determines the longevity is very important and yet hard to express in numbers. Because of its complexity, manufacturers do not provide enough information and data for estimating the longevity of the products, and customers are in most cases ignorant enough to overlook this important variable, relying on the manufacturer's warranty and promises. This leaves large gaps in the performance and warranty contracts and leads to confusion and surprises.

In summary, test procedures are intentionally quite thorough and will address some of the obvious quality questions by showing a mechanical or electric defect during the testing. Even the best and most thorough of testing procedures, however, cannot be held responsible for latent defects and failures (these that take a long time to develop) in the future. Also, and very importantly, the manufacturers usually send hand picked, crop-of-the-crop, PV modules for testing, which in most cases is representative of only small part of the entire volume of their process and final product. This way the well made, hand-picked modules pass the qualification tests, while the bulk of the production volume might have questionable quality.

NOTE: Keep in mind that our major concern is PV modules and BOS components destined for operation in large-scale installations under extreme climate conditions (deserts and humid areas) where these have not been proven reliable as yet. For the proper execution of these tasks we recommend taking a very close look at the PV modules manufacturing and quality procedures, tests data and warranties. Then make a list of detailed questions and enter into intense negotiations with the manufacturers in order to assess the average product quality, extrapolate the "best-worst" scenarios and build them into the technical and financial models of the installation. This is the only way to gain some confidence in the product and produce reliable technical and finance models.

Solar Cells and Modules Sorting Issues

Table 5-3 shows a typical outcome from sorting a batch of solar cells which is needed to provide homogeneity of each module. So using this example we will end up with PV modules varying in efficiencies from 12 to 18.5% which will be sold at different prices but with the same warranty and life expectations.

This is an accepted practice of the PV industry, but the question here is, "Why did some cells in the batch turned out 18.0% efficient, while others are only 12%?" This is a huge difference—1/3 difference between the most and least efficient cells, to be exact. In the author's experience, this can be explained by a number of variations in the quality of materials and the different steps of the manufacturing process.

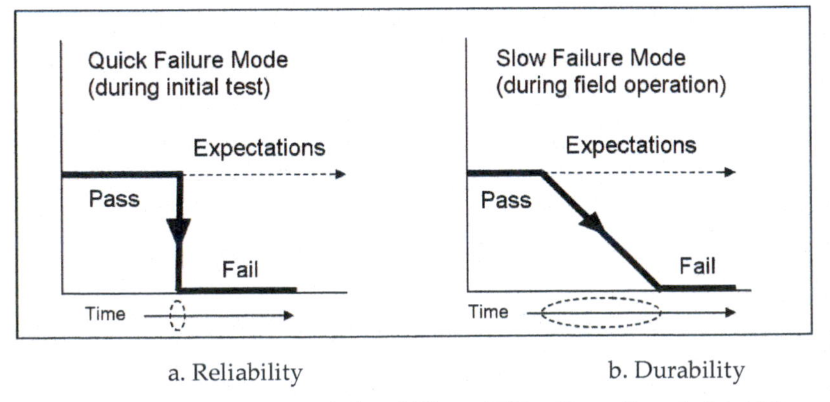

Figure 5-6. Reliability and durability of PV cells and modules (15)

Table 5-3. Different solar cells sorted by efficiency

Eff(%)	Pm(Wp)	Vap(V)	Iap(A)	Voc*(V)	Isc*(A)
18.50	2.86	0.515	5.56	0.63	5.86
18.25	2.83	0.515	5.49	0.63	5.80
18.00	2.79	0.515	5.41	0.63	5.71
17.75	2.75	0.515	5.34	0.62	5.72
17.50	2.71	0.515	5.26	0.62	5.64
17.25	2.67	0.515	5.19	0.62	5.56
17.00	2.63	0.515	5.11	0.62	5.47
16.75	2.59	0.515	5.04	0.62	5.39
16.50	2.55	0.515	4.96	0.62	5.31
16.25	2.52	0.515	4.89	0.61	5.33
16.00	2.48	0.515	4.81	0.61	5.25
15.75	2.44	0.515	4.74	0.61	5.25
15.50	2.40	0.515	4.66	0.61	5.08
15.25	2.36	0.515	4.58	0.61	4.99
15.00	2.32	0.515	4.51	0.61	4.91
14.75	2.28	0.515	4.53	0.61	4.82
14.50	2.25	0.515	4.36	0.60	4.84
13.50	2.09	0.515	4.06	0.60	4.49
12.00	1.86	0.515	3.61	0.60	4.00

To get a better idea of where the batch-to-batch and within batch variations could come from we'll look again at the problematic areas of the process.

Some of the variations in quality from cell to cell could be caused by:

1. The starting silicon material impurities and defect content vary from ingot to ingot.
2. The position of the different wafers in the silicon ingot determines their quality. Wafers sliced from the center of the ingots often have different properties than those sliced from the top and bottom sections of the ingot. The bottom section of the ingot is usually heavily contaminated and doped, and if it is not cropped properly could result in poor performing and eventually failing cells and modules.
3. Different consumable quality or contamination levels, used from batch to batch determine the proper surface etch and cleanliness which determine the final quality of the cells.
4. Mishandling of wafers is a huge problem in operations where large number of process steps necessitate handling of wafers (be it manually or by robots), loading and unloading in/from cassettes, conveyor belts etc. At each step there is great possibility of introducing air-born or contact (touching with dirty gloves) contamination. Attention to detail and proper quality control are most needed here.
5. Chemical processes (etch, clean etc.) produce variable results at different times (chemical baths increase in contamination with time) and even within the wafer batch (wafers in the center of the cassettes etch and clean at different rates vs. those at the ends.)
6. Insufficient or improper rinse-dry procedures could leave contaminants or spots on the wafer surfaces on some batches and not on others, Surface contaminants are potential latent problems.
7. Thermal processes, such as diffusion, AR coating and metallization do produce different quality product at different times, due to shift changes, start-up, shut-down and maintenance procedures, chamber or gas contaminations, humidity variations, in-process malfunctions, etc.

This is only a small sample of what can go wrong during the hundreds of steps of the manufacturing cycle, which could produce different quality cells and modules from batch-to-batch and within batches.

PV Modules Issues

PV modules are complex electro-mechanical devices. Cell structure, wiring, electronics and many other factors determine the proper and trouble-free operation of the modules for their long non-stop operation.

Listed below are some of the major issues encountered in PV modules during testing and/or field operations:

Mismatch or Shading

Serious reduction of output and/or damage to the solar cells in the PV modules might occur if:

a. The cells in the module are mismatched (different output), be it from manufacturing negligence, or subsequent degradation of some cells, and

b. Total or partial shading of the modules might lead to thermal overload and overheating (hot-spots) which are often misunderstood or ignored.

One of the quality characteristics that is verified within UL and IEC qualification testing is the ability of PV modules to withstand the effects of periodic hot-spot heating that occurs when cells are operated under reverse biased conditions. Testing, however, provides partial results, and this should be taken into consideration during design, evaluation, and operation of PV systems. As environmental conditions and the modules' behavior change with time, hot spot issues could appear without warning. Once these problems are present, the affected modules must be inspected, evaluated and removed from operation, or checked periodically to make sure the problem doesn't grow or expand.

Solar Cell Degradation:

The overall cell degradation phenomenon ranging from 0.5% to 1.5% of total output loss per year, and the mechanisms thereof, are variable, very complex, and impossible to control. It is assumed that it is primarily due to series resistance (R_S) increases over time, most likely due to inadequate metallization, corrosion of the contacts (caused by moisture in the module); and/or deterioration of the adhesion between the metal contacts and the substrate material, due to manufacturing abnormalities, and/or decrease in shunt resistance (R_{SH}) due to metal migration through the p-n junction; and/or Antireflection (AR) coating deterioration or delamination.

There are also a number of other reasons as well, but since we cannot control any of them after the PV modules have been installed, we must find a way to account for, and control them during the long manufacturing sequence—all the way from sand to module.

Short Circuits

Short circuiting at cell interconnections, as illustrated in Figure 5-7, is also a common failure mode, especially for very thin cells, since top and rear contacts are much closer together and stand a greater chance of being shorted together by pin-holes or regions of corroded contacts or damaged cell material.

Temperature Coefficient

The temperature coefficient (Tc) is generic terminology that can be applied to several different PV performance parameters, such as voltage, and current and power fluctuations under temperature variations. Most often, however, it is referred to as the total power output loss upon exposure to excess temperatures. The procedures for measuring the temperature coefficient for modules and arrays are not yet standardized, so it is quite common to see different Tc expressions and test methods. Systematic influences and errors are common in the test methods used, as well, but the worst part is misconceptions regarding their application under field operation conditions.

Tc basically is the percent of change (usually decrease) of power output from a solar cell or module per each degree C above standard test conditions, STC (set at 25°C). Tc average of 0.5-0.6% power loss per °C is accepted by the PV industry for c-Si cells and modules. It is slightly lower for some types of thin film PV modules. So, if our c-Si PV module were rated at 100 Wp at STC, it would drop 0.5% output with every degree C above 25°CSTC. Or when exposed to full sunlight at noon in the Arizona desert, the PV module would heat up to 85°C, which is 60°C above STC, so that its power output will drop 30% (60 x 0.5%). Our 100 Wp PV module would be generating less than 70 W while at this temperature.

This is a significant loss, so Tc is an important parameter in determining PV power system design and sizing, since often the worst case operating condition dictates the array type and size. In other words, PV mod-

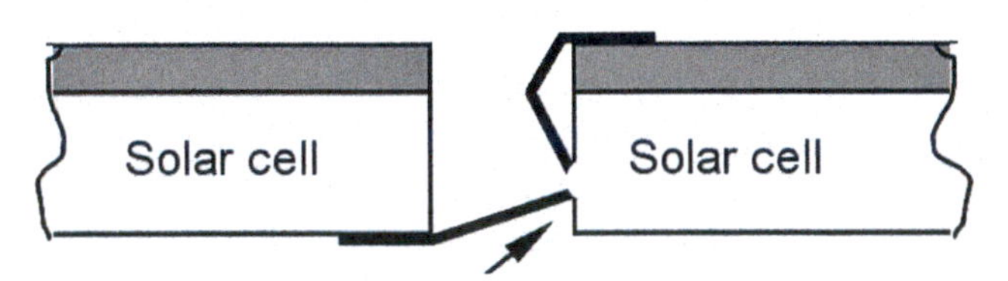

Figure 5-7. Severe solar cell shunting mechanism

ules and arrays are supposed to operate at maximum output around the noon hour, when the sun is highest. But if the output drops 30% during that time due to extreme heat, we must change the calculation methods in arriving at the peak power output of the array. Or— if our 100 MWp (as measured under STC) power array produces only 70 MWp at high noon in the desert, when we need the power most, we may have to add some more PV modules to compensate for the 30% power loss during that critical time. Because of that anomaly Tc is a most important parameter when considering installations in desert areas, where extreme heat conditions will have a great influence on it and the overall system performance.

Open Circuits (Cells)

This is a common failure mode, although redundant contact points plus "interconnect-busbars" allow the cell to continue functioning. Cell cracking can be caused by: thermal stress; hail; or damage during processing and assembly, resulting in "latent cracks," which are not detectable on manufacturing inspection, but appear sometime later and cause partial or total failure.

Cracked solar cell

Figure 5-8. Severe open circuit failure

Interconnect open circuit could be caused by fatigue due to cyclic thermal stress, mechanical damage at a previous stage of the process, or wind loading.

Module Open Circuit

Open circuit failures also occur in the module structure, typically in the bus wiring or junction box.

Module Short Circuit

Although each module is tested before sale, module short circuits are often the result of manufacturing defects. They occur due to insulation degradation with weathering, resulting in delamination, cracking or electrochemical corrosion.

Module Hot-spot Heating

Module hot-spot heating occurs when there is one low current solar cell in a string of at least several high short-circuit current solar cells. If the operating current of the overall series string approaches the short-circuit current of the "bad" cell, the overall current becomes limited by the bad cell. Essentially the entire generating capacity of all the good cells is reduced by dissipation in the poor cell. The enormous power dissipation occurring in a small area results in local overheating, or "hot-spots," which in turn leads to destructive effects, such as cell or glass cracking, melting of solder, or degradation of the solar cell.

Module Glass Breakage

Shattering of the top glass surface can occur due to vandalism, thermal stress, handling, wind or hail. Once the glass is broken, the interior of the module—including the cells—becomes exposed and vulnerable to the effects of the elements. Moisture penetrating the module could destroy the cells and shunt the entire unit within weeks.

Module Delamination

It is usually caused by reductions in bond strength between the front glass and the EVA or other encapsulant. It is either environmentally induced by moisture or photo-thermal aging and stress which is induced by differential thermal and humidity expansion.

Bypass Diode Failure

By-pass diodes, used to overcome cell mismatching problems, can themselves fail, usually due to overheating, often due to undersizing, or over-capacity use. The problem is minimized if junction temperatures are kept below 128°C.

Encapsulation Failure

UV absorbers and other encapsulant stabilizers ensure a long life for module encapsulating materials. However, slow depletion, by leaching and diffusion does occur and, once concentrations fall below a critical level, rapid degradation of the encapsulant materials occurs. In particular, browning of the EVA layer, accompanied by a build-up of acetic acid, has caused gradual reductions in the output of some arrays, especially those exposed to higher temperatures, such as in desert or humid environment.

Optical Losses

Optical losses chiefly effect the power from a solar cell by lowering the short-circuit current. Optical losses consist of light which could have generated an electron-hole pair, but does not, because the light is reflected from the front surface, or because it is not absorbed in the solar cell. For the most common semiconductor solar cells, the entire visible spectrum has enough energy to create electron-hole pairs and therefore all visible light would ideally be absorbed.

There are a number of ways to reduce the optical losses:

a. Top contact coverage of the cell surface can be minimized (although this may result in increased series resistance).

b. Anti-reflection coatings can be used on the top surface of the cell.

c. Reflection can be reduced by surface texturing.

d. The solar cell can be made thicker to increase absorption (although any light which is absorbed more than a diffusion length away from the junction will not typically contribute to short-circuit current since the carriers recombine).

e. The optical path length in the solar cell may be increased by a combination of surface texturing and light trapping.

Material (Silicon Wafer) Thickness

While the reduction of reflection is an essential part of achieving a high efficiency solar cell, it is also essential to absorb all the light in the silicon solar cell. The amount of light absorbed depends on the optical path length and the absorption coefficient of the material and its thickness.

Thermal Expansion and Thermal Stress

Thermal expansion is another important temperature effect which must be taken into account when modules are designed. Typically, interconnections between cells are looped, as shown in Figure 5-7, to minimize cyclic stress. Double interconnects are used to protect against the probability of fatigue failure caused by such stress. In addition to interconnect stresses, all module interfaces are subject to temperature-related cyclic stress which may eventually lead to delamination.

Mismatch Effects in Arrays

In a larger PV array, individual PV modules are connected in both series and parallel. A series-connected set of solar cells or modules is called a "string." The combination of series and parallel connections may lead to several problems in PV arrays.

One potential problem arises from an open-circuit in one of the series strings. The current from the parallel connected string (often called a "block") will then have a lower current than the remaining blocks in the module. This is electrically identical to the case of one shaded solar cell in series with several good cells, and the power from the entire block of solar cells or modules is lost.

Module Failures—Industry Experience

The following is the experience of the industry leader, BP Solar, as presented during the PV workshop at NREL, February, 2010. (9) Extensive tests and inspections of modules from different manufacturers show the following most common failures and defects. These are followed by suggestions for their elimination or correction:

Broken Interconnects

a. Interconnects break due to stress caused by thermal expansion and contraction or due to repeated mechanical stress.
b. Early modules suffered open circuits due to broken interconnects.
c. What makes it worse:
   — Substrates with high thermal expansion coefficients
   — Larger cells
   — Thicker ribbon
   — Kinks in ribbon

Solutions to Broken Interconnects

a. Substrates with lower thermal expansion coefficients—that is one of the reasons why glass superstrate modules are so popular.
b. Built in stress relief (But not kinks because they concentrate stress)
c. Built in redundancy—if one fails the module continues to operate
d. Thinner ribbon
e. Softer more pliable material
f. Discrete bonds versus continuous attachment

Broken Cells

a. Crystalline Si cells can (and will) break due to mechanical stresses.
b. Early modules suffered open circuits due to broken cells since there was only one attachment point for each polarity.
c. What makes it worse
   — thinner cells

— single crystal especially if cleave plane is oriented along bus bar
— pre-stressed or chipped cells
— larger cells in large modules

Solutions to Broken Cells

a. Build in crack tolerance using redundant interconnects and multiple solder bonds on each cell.
b. Do not orient cleave planes along tabbing ribbons.
c. Presorting of cells to remove those with cracks or chips.
e. IR and NIR inspection.
f. Dynamic mechanical load testing of new designs to determine the potential for cell breakage and whether the breakage leads to power loss.

Corrosion

a. Moisture induced corrosion of cell metallization.
b. Key to survival is to minimize the ionic conductivity in the package, especially the encapsulant.
c. Field failures of PVB encapsulated modules in 1980's was due to high ionic conductivity in moist PVB.
d. What makes it worse
   — Metallization sensitivity to moisture
   — Encapsulant with humidity dependent conductivity
   — Encapsulant that absorbs a lot of moisture.

e. For crystalline Si corrosion of front contacts is dependent on both the metallization system and the encapsulation system.
f. Experience—with same cell metallization system, one EVA passed 1000 hours of damp heat (< 5% power loss) while modules with 2ndtype of EVA degraded in power by close to 50% after 1000 hours.
g. Performed testing of competitor's (IEC 61215 certified) crystalline Si modules through 1250 hours of damp heat testing (versus 1000 hours from IEC 61215). 8 out of 10 failed due to power loss in excess of 5%.

Solutions to Corrosion

a. Utilize an encapsulant that does not increase in conductivity when it absorbs water vapor.
b. Incorporate moisture barriers in superstrate and substrate.
c. Utilize a metallization system that is compatible with the encapsulation system chosen.
d. Do damp heat testing beyond 1000 hour.

Delamination

a. Delamination observed to varying degrees in a small percentage of PV module types.
b. Delamination can be between superstrate (i.e. glass), substrate (i.e. Backsheet) and encapsulant or between encapsulant and cells. Usually the result of an adhesive bond that is sensitive to UV, humidity, or contamination from the material (Excess Na in glass or dopant glass left on cell)

Solutions to Delamination

a. Careful selection of adhesives and primers—Stable to UV and moisture.
b. Control of raw materials and processes.
c. Module testing to detect and eliminate any changes in materials or processes.

Encapsulant Discoloration

a. Will result in some loss of transmission and therefore reduced power.
b. Worst reported case was in slow cure EVA caused by low concentration system at Carissa Plains.
c. Standard cure EVA formulation A9918 does discolor.
   — Caused by heat and UV.
   — Bleached by oxygen
   — So with breathable backsheet center of cells discolor while outside ring remains clear.
   — Without concentration it takes 5 to 10 years to see discoloration and longer to start appreciably reducing output power. It was not EVA itself that discolored, but additives in the formulation

Eliminating Encapsulant Discoloration

a. Make sure actual encapsulant package is tested to UV exposure at high temperature. BP Solar does 6 months of UV testing for all encapsulants, backsheets and even labels.
b. Original EVA yellowing alleviated via changes in EVA additive package and adding UV absorber (Cerium Oxide) to low iron glass.
c. Most glass manufacturers have now removed Cerium Oxide from low iron glass so it is important to verify that the encapsulant being utilized is not sensitive to UV induced discoloration.

Solder Bond Failures

a. Solder bonds can fail due to stresses induced by thermal cycling.
b. Solder can creep when loads are applied at elevated temperatures.

c. Solder fatigue can be caused by cyclic loading, e.g. thermal, mechanical or electrical repetitive stress.
d. Early modules typically only had one solder bond per interconnect per cell, so failure of this solder bond resulted in an open circuit failure of the whole module.
e. Even today non-cell solder bonds often have no redundancy, so failure of one of these bonds can lead to dropout of a cell string, a whole module or even a whole string of modules.

Alleviating Solder Bond Failures

a. Utilize multiple solder bonds on each tabbing ribbon.
b. Utilize softer ribbon and leave stress relief.
c. Perform periodic pull tests to assure quality of solder bonds being made.
d. Perform thermal cycle tests well beyond the 200 cycles from IEC 61215.
e. Implement training and QA inspections to assure that non-cell solder bonds are being fabricated correctly.
f. In critical areas (like termination wires) use both solder and mechanical connections.
g. Do not rely on pottants as second attachment for termination wires. This can lead to arcing danger.

Broken Glass

a. Type of glass breakage is dependent on the type of glass used. (tempered, heat strengthened, or annealed)
b. High impacts like hail, rock or bullet will break glass. Can almost always identify spot where object hit.
c. Mechanical loading from snow and/or wind can break glass.
d. Failure of or misuse of support structure can lead to glass breakage
e. High temperature (hot spot or arc) can break glass.
f. Annealed glass can also break due to:
   — Stress built into the package during manufacture.
   — Stress applied by the framing/mounting system.
   — a temperature difference of as little as 25°C from center to edge.

How to Keep Glass from Breaking

a. Use tempered glass wherever possible.
b. Pay attention to mounting method and mounting points.
c. Test mounting system for snow load per IEC 61215.
d. In high traffic areas try to protect glass (or any other superstrate) from direct impacts.
e. Minimize hot spots and arcs.

Hot Spots

a. Hot-spot heating occurs in a module when its operating current exceeds the reduced short-circuit current (Isc) of a shadowed or faulty cell or group of cells.
b. When such a condition occurs, the affected cell or group of cells is forced into reverse bias and must dissipate power.
c. If the power dissipation is high enough or localized enough, the reverse biased cell can overheat resulting in melting of solder and/or silicon and deterioration of the encapsulant and backsheet.

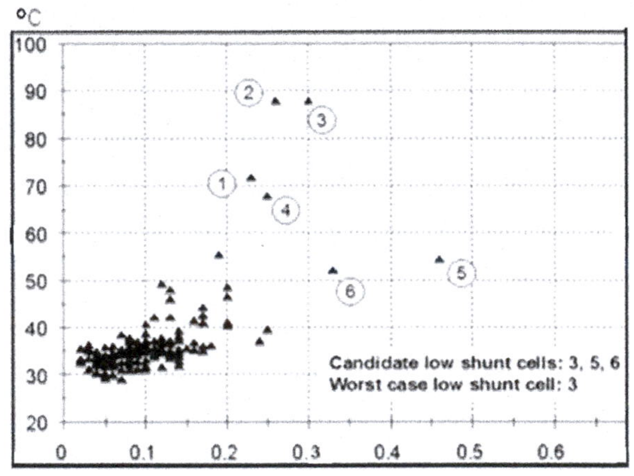

Figure 5-9. Shunting dependence on temperature and reverse current. (Reverse current in Amps, measured at 10 Volts)

How to Avoid Hot Spots

a. Most cells can be adequately protected by use of bypass diodes (say 1 diode every 20 cells).
b. Still, there may be some cells with localized shunts that will heat excessively at the reverse bias level allowed by the bypass diodes.
c. Solution is to screen out low shunt cells.

Ground Faults

a. PV modules are supposed to have high resistance stand-off between the electric circuit and the ground plane.
b. Occasionally this protection is compromised.
c. The consequences can be serious as there is nothing to stop the PV current from flowing in ground loops until one component gets so hot it melts or burns.
d. Many ground faults are the result of poor installation practices, such as mounting modules with

clips that penetrate the module insulation and contact the solar cells at numerous locations.

Alleviating Ground Faults

a. Mounting modules within the cell area should be avoided or done with extreme caution (and probably the addition of significant electrical insulating material like use of a glass substrate).
b. Grounding of the array circuit actually increases the potential for this type of fault.
   — In a grounded array it only takes 1 ground fault to cause current flow in a ground loop.
   — For an ungrounded circuit it takes 2 ground faults to cause a problem, giving the system operator a chance to detect the first one and fix it before the second one occurs.

Junction Box Failures

a. Single point for potential failure that can often be attributed to poor workmanship.
b. Water ingress and subsequent corrosion can be a problem.
c. How well is the J-box attached to the module back sheet?
d. Some adhesive systems are good for short-term pull but poor at maintaining long-term adhesion.

Addressing Junction Box Issues

a. Use only qualified materials and boxes.
b. Make sure boxes pass same set of tests that modules pass—damp heat, humidity freeze, thermal cycle, robustness of termination and wet high pot.
c. Test adhesion well beyond qualification requirements.
   — At BP Solar we perform a boiling water test to verify adhesion under worst case conditions.
d. Evaluate worst case failures—what happens if a wire becomes detached?
e. Quality control during manufacture.

Structural Failures

a. Often it is the way the module is mounted that determines whether it can survive a particular load.
b. You want me to follow the manufacturers' installation instructions?
c. Snow load and ice can deform the frame and break the glass.
d. Sometimes the entire array structure is not capable of surviving high winds.

Avoiding Structural Failures

a. Follow the module manufacturer's installation instructions.
b. Follow local building codes where available as they are usually based on local weather history.
c. Test new approaches (for example in a wind tunnel) before using them in the field.
d. The best designed system built using the highest quality components will not work well and may be unsafe if it is installed improperly.
e. Installer training and certification programs like the one run by BP Solar are critical to achieving highly reliable and safe PV systems.

Summary

a. Crystalline Si Module reliability and performance
   — Very good but still with room for improvement
   — Not all modules are created equal; poor material selection, and improper assembly and processing will yield different degradation rates and lifetimes.
b. New Module Technology
   — Can't test for 25 years before releasing commercial products.
   — Strong, accelerated test programs required.
   — Process controls required to assure production modules perform as well as test modules.

A number of additional tests and R&D activities (as described in ref. 10, 11, 12, 13 below) could be used to support the fact that PV modules are capable of functioning efficiently and without major problems for many years, IF (**note the big IF**) high quality materials, and proper manufacturing processes and techniques have been used. This is a very big IF, however, because there are so many steps in the cradle-to-grave process of PV cells and modules materials production, manufacturing sequence, installation and operation processes. If just one step is out of spec, the quality of the final product will be affected—usually negatively.

Although we have no doubt that there are a number of high quality manufacturers, the majority of PV modules today are produced by new, low-cost manufacturers in third-world countries who lack quality as we know it. Long-term efficient and trouble-free operation of their products is questionable. This issue is well known, but has not yet gained the visibility it deserves, especially when discussing quality issues of products that will be baking 30 years in the Arizona deserts. We do believe that with time customers, engineers and investors will become increasingly aware of these issues, and re-evaluate their options, using greater care as to which technologies and manufacturers they use.

We also believe that most of the issues will be resolved by standardizing the manufacturing processes at which point large-scale installations with c-Si and TFPV technologies will be much more reliable. Until then, we must pay attention to the quality of the materials, the manufacturing processes and subsequent installation, operation and maintenance procedures of PV products and projects we design or evaluate.

Field Issues

PV modules start their real life at the factory, where a careless operator pushes and shoves them hurriedly into the package and then into the awaiting container. Then a careless crane operator at the seaport loading terminal drops the container onto the pile of other containers. In the field, modules get piled on top of each other and a third careless guy might finish the job started at the factory. But if he doesn't, then the guy standing on top of the modules in Figure 5-10 will certainly shorten their life. What he doesn't know is that even if the glass succeeds in protecting the cells from immediate breakage, 150 lbs. spread over a ½-foot surface area will make the glass give in some and put tremendous stress on the cells and the interconnects underneath. In this case it is the responsibility of design engineers to provide a more appropriate installation procedure in order to avoid stressing the modules.

The pressure on the top surface glass concentrated in the small contact areas (where the shoes touch the module glass) is significant, and is directly transmitted

Figure 5-10. PV modules installation

to the EVA and the cells underneath it. A low level stress, a small crack in the cell, or a small tear in the metallization is all that's needed to cause a cell, and its module, to start losing power and eventually fail. And worse, if it doesn't fail immediately (while still under warranty), the small damage will grow over time under the additional stress of the elements and will cause a failure after the parts and labor warranty has expired.

Table 5-4 represents a number of modules from different manufacturers which were tested by NREL engineers under field conditions. Note the varying test results in the table—from 0.0% to 5.0% annual degradation. Products from the same manufacturer show different results, and a number of similarly puzzling data points deserve a close look. We could draw a number of conclusions from all this, but the overwhelming one is

Table 5-4. PV modules field degradation rates (18)

Manufacturer	Module Type	Exposure (years)	Degradation Rate (% per year)	Measured at System Level?	Ref.
ARCO Solar	ASI 16-2300 (x-Si)	23	-0.4	N	2
ARCO Solar	M-75 (x-Si)	11	-0.4	N	3
[not given]	[not given] (a-Si)	4	-1.5	Y	4
Eurosolare	M-SI 36 MS (poly-Si)	11	-0.4	Y	5
AEG	PQ40 (poly-Si)	12	-5.0	N	6
BP Solar	BP555 (x-Si)	1	+0.2	N	7
Siemens Solar	SM50H (x-Si)	1	+0.2	N	7
Atersa	A60 (x-Si)	1	-0.8	N	7
Isofoton	I110 (x-Si)	1	-0.8	N	7
Kyocera	KC70 (poly-Si)	1	-0.2	N	7
Atersa	APX90 (poly-Si)	1	-0.3	N	7
Photowatt	PW750 (poly-Si)	1	-1.1	N	7
BP Solar	MSX64 (poly-Si)	1	0.0	N	7
Shell Solar	RSM70 (poly-Si)	1	-0.3	N	7
Würth Solar	WS11007 (CIS)	1	-2.9	N	7
USSC	SHR-17 (a-Si)	6	-1.0	Y	8
Siemens Solar	M55 (x-Si)	10	-1.2	Y	9
[not given]	[not given] (CdTe)	8	-1.3	Y	9
Siemens Solar	M10 (x-Si)	5	-0.9	N	10
Siemens Solar	Pro 1 JF (x-Si)	5	-0.8	N	10
Solarex	MSX10 (poly-Si)	5	-0.7	N	10
Solarex	MSX20 (poly-Si)	5	-0.5	N	10

caution. Be very careful when buying PV modules. Make sure you know the vendor, take a close look at their history and discuss potential issues. Make sure that the PV modules in your power plant operate per spec from day one, and follow their performance daily, noting any variations because they all mean something worth analyzing. The first several months of module operation usually provide significant information for their future behavior—provided you know what to look for.

PV Module Choice

We now know all there is to know about all existing, and some future, PV technologies—their advantages and issues. Keeping all this in mind we need to choose a PV technology for our new large-scale project—a 100 MWp PV power plant to be installed in the Arizona desert. Based on the available information, Table 5-5 shows some of the major performance and longevity characteristics of c-Si and TFPV modules.

Remember: a) our emphasis is on the application of PV technologies in large-scale fields in harsh climates (US deserts and humid areas), and b) most of the variables cannot be quantified. Thus, our general deduction is that c-Si cells and modules are more durable and reliable for this particular application.

NOTE: The author, nor anyone involved in the writing of this text (to our knowledge) has any connection with, or interests in, c-Si or TFPV technologies, or their manufacturers, investors, etc.

Obviously, there are some differences between these most promising PV technologies—thin films and c-Si PV modules. So what do we need to do, to pick the most suitable for our project and ensure their efficient long-term operation in the field? A lot of know-how, experience, diligence, and hard work are needed to evaluate the land and climate conditions, the available technologies and their proper application. A lot of data are needed for accurate estimates which must be verified with tests to ensure the best possible technology choice of power field setup.

PV is not a new technology, nor is its application, but the energy field has changed dramatically lately, and now PV technologies are taking a more serious place in the energy market. For us to ensure their progress, we need to make sure that the PV modules we use—especially in large-scale power fields—will produce energy efficiently and will last for the duration for which they are designed.

The issues are summarized best in Figure 5-11. Basically speaking; cost is to be minimized, performance maximized and durability (or longevity) demonstrated...still. The durability issue pops up immediately as something that we need to work on, but it doesn't seem to be a priority now. Manufacturers are feverishly working on the cost and performance of their products, because it is what relates directly to the bottom line, while durability and longevity are somewhat on the back burner shrouded in veil of secrecy. Durability and longevity are factors that will make or break the long-term efficient operation of large-scale power fields, especially those installed in deserts and other extreme climate ar-

Table 5-5. Comparison of c-Si and thin film modules

	Description	TFPV Modules	Si PV Modules	Therefore Si modules:
A.		Process specifics and cost		
1	Raw materials	Exotic	Abundant	No shortages expected
2	Process chemicals	High purity	Standard	Lower cost
3	Equipment	Complex	Simpler	Lower capital cost
4	Mfg. process	Hi-tech	Low tech	Lower labor cost
5	Process quality	Excellent	Standard	Lower QC/QA cost
6	New Product has	Excellen quality	Standard	Lower initial quality
B.		Electro-mechanical properties		
1	Power output	4-8 W/ft2	10-15W/ft^2	~2 times more power
2	OC voltage	45V-100V	20V-50V	2 times lower
3	String power	200-750W	200-3000W	4 times higher
4	Voltage isolation	Intermittent	Excellent	Overall superiority
5	Test failure rates	Higher	Lower	Less failures[1]
6	Module design	Poor	Excellent	Overall superiority
7	Blocking diode	Yes	No	Lower initial & O&M cost
8	Glass strength	Annealed	Tempered	5-6 times stronger
9	Encapsulation	Poor	Standard	More reliable[2]
10	Opaque to UV	No	Yes	Superior longevity
C.		Field behavior and cost		
1	Land use	10-12 acres/MWp	4-6 acres/MWp	2-3 times less land use
2	Cost $/Wp (2011)	$1.80/Wp	$1.80/Wp	Comparable[3]
3	Cost tendency	Uncertain	Decreasing	Better price stability
4	BOS components	Increased qty.	Standard	Lower cost overall
5	Temp. coefficient	0.3-0.4/C^0	0.5-0.6/C^0	Somewhat higher
6	Power degrading	1.0-1.5%/year	0.5-1.0%/year	Lower degrading rate
7	Field failure rates	Higher	Lower	Less failures[4]
8	Field Stability	Poor	Good	Overall superiority
9	Hazards	Cd, Te, As, etc.	Pb, solvents	Much less toxicity
10	EOL recycling	100% hazmat	Partial hazmat	Lower EOL cost

Notes: 1. Damp heat test failures
2. Some TFPV modules have no side frames
3. Cradle-to-grave, including BOS and EOL recycling, costs
4. Extrapolated from damp heat and other test results

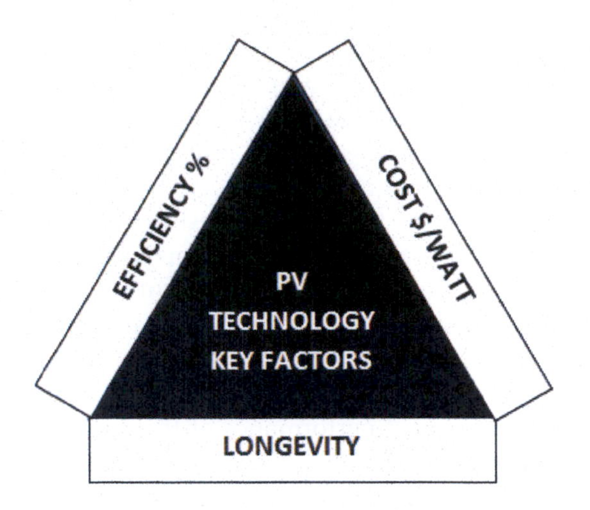

Figure 5-11. Work to be done

eas, so we must put the unresolved issues on the table, to discuss the issues and correct the problems at hand while there is still time. In more detail:

Minimize Cost

This can be easily done by carefully managing the supply chain and manufacturing operations. More often, though, it is not done properly, because manufacturers are in a hurry to cash in by making as many PV modules as they can. Where the supplies come from, how they are processed, and their quality are secondary to the quantity produced. This might be the way to make a quick buck, but it is far from providing high quality PV modules to last 30 years in a 100 MWp project in the Arizona deserts.

Maximize Performance

Several world PV modules manufacturing companies have maximized the efficiency of their products, while most others are interested solely in increasing the quantity of the existing product. We would not dare install such products in a 100 MWp project in Death Valley, unless and until we get all assurances of proven efficiency and reliable performance and the manufacturers' direct participation in the projects at hand.

Demonstrate Durability (Longevity).

And once more (but not the last) time: without proven durability and longevity there will be no c-Si or thin film 100 MWp power plants in the desert anytime in the near future… or ever, unless the manufacturer is the power plant owner and is taking all the risks.

So how do we demonstrate the durability of our product? To start with, the quality of the PV products, including their supply chain and cradle-to-grave quality must be properly documented, disclosed and discussed between the manufacturer and the customer.

Cost and performance factors have been manufacturers' and users' focus for a long time. Durability and longevity, however, are the hardest to address and solve because they are a conglomeration of a many complex phenomena which are hard to understand, let alone put in numeric form and solve on paper. Nevertheless, this is what is needed if we are to install Gigawatts of PV modules in the deserts of the US and the world. Avoiding the issues or postponing them is the wrong thing to do and is something that will prove damaging to the industry and energy markets in the long run.

PV Cells and Modules Manufacturers

The number of solar cell and module manufacturers rose several fold in the last few years. This unprecedented growth is due to the rising demand for PV energy and is expected to continue at this pace. With it, however, arise a number of questions related to the experience of manufacturers and the quality of their products.

Note: Keep in mind that most PV modules and other components are expected to operate efficiently and trouble-free for 30 years or more no matter where they are installed. This is a tall order, and something that we don't expect from almost any other product. To make things even worse, we simply have no experience with such long-time exposures, nor do we have any evidence that these products will last that long while performing as specified. So we rely on the quality of the manufacturers' materials and processes to ensure product quality. But how much do we know about the manufacturers and their processes, and how do they assure us of the long-term, flawless performance of their products?

Note in Table 5-6 the rapid rise of some manufacturers, who did not have much PV experience and capacity until just a few years ago. Now they are making, and mostly exporting, millions of cheap solar cells and modules. The cheaper the better! But where did the expertise come from? How did they manage to obtain the experience and understanding necessary to set up facilities and establish a quality process in a matter of months? What type of quality control system do they use for the supply chain and manufacturing processes? Does the final product quality go up proportionally with the production volume?

More importantly, when did they test the performance and longevity of their modules? If they ramped up within a year or two, they barely had a chance to take care of the basic equipment and process issues, let alone testing and optimizing the long-term performance and

Table 5-6. PV modules manufacturers (% share)

	MARKET SHARE OF PV MODULE MANUFACTURERS, 2008-2010									
Company	**03/08**	**06/08**	**09/08**	**12/08**	**03/09**	**06/09**	**09/09**	**12/09**	**03/10**	**06/10**
Del Solar	0	0	0	0	0	0	0	0	9	2
Sun Power	34	29	18	26	16	14	15	18	10	3
Kyocera	9	6	6	3	2	4	4	4	11	4
Yingli	0	0	0	0	2	2	2	22	16	2
Sharp	16	16	15	20	18	24	27	14	10	15
Trina	0	0	0	0	1	0	0	1	2	28
Suntech	1	4	12	7	23	6	11	11	15	23
Others	41	45	50	44	38	49	40	31	28	24

longevity.

US and EU customers and investors bought, and continue to buy, large quantities of PV modules made by the newcomers, with price as the primary consideration. "Cheap is good," the thinking went. Cost and performance were discussed in detail, but the longevity issue was somewhat hazy and the answers were few and far between.

So what do we expect from these products? What guarantees do we have that they will survive 20-30 years of desert heat? What will happen if these millions of cheaply made modules start failing 10-15 years down the road? There are no answers to these questions, because we simply have no precedent, nor experience with these issues. We hope that everything will be OK, but hope is not a good business driver. We really must look for the answers and solve the issues first.

We do hope, that this text adequately addresses some of the key issues related to cells and modules performance and longevity, to allow a more educated approach to module selection and power field design in the near future. At the very least we expect a more open discussion of the issues we bring forward here.

INVERTERS

DC-AC inverters are an integral part of any large-scale PV plant operation, since they receive and handle the output of the PV field and determine the final level and quality of power sent into the power grid. They are also a major portion of the capital equipment cost and maintenance expense of the plant, so we will dedicate this section to their function, operation, and related issues.

Inverters Basics

Solar (PV) inverters are a major and critical part of PV electric power generating systems. Inverters are designed to properly and efficiently convert the variable DC output from PV modules in the field, into a clean sinusoidal 50 or 60 Hz AC electric current at specified voltage, which is then sent directly to the electrical grid, or to an off-grid electrical network.

Power control, data transfer and communications capability are usually included in a good inverter design, in order for the operators to monitor the system's performance and to take action as needed. Depending on the grid infrastructure, wired (RS-485, CAN, Power Line Communication, Ethernet) or wireless (Bluetooth, ZigBee/IEEE802.15.4, 6loWPAN) networking options can be used.

The brain of the inverter is a real-time microcontroller, which executes the very precise algorithms required to invert the DC voltage generated by the solar module into clean AC. The controller is programmed to perform the control loops necessary for all the power management functions, including DC/DC and DC/AC. A good controller has the ability to maximize the power output from the PV system through complex algorithms called maximum power point tracking (MPPT). The PV system's maximum output power is dependent on operating conditions and varies from moment to moment due to temperature, shading, dirt, cloud cover, and time of day, so tracking and adjusting for this maximum power point is a continuous process. For systems with battery energy storage, the controller manages the charging operations. It also keeps uninterrupted power supply to the load by switching over to battery power once the sun sets or when a cloud cover reduces the PV output power.

The controller contains advanced peripherals like high precision pulse-width-modulated (PWM) outputs and analog-to-digital-converter (ADC) for implementing control loops. The ADC measures variables such as PV output voltage and current, and then adjusts the DC/DC or DC/AC converter by changing the PWM duty cycle. Most units are designed to read the ADC and adjust the PWM within a single clock cycle, so real time control is possible.

Communications on a simple system can be handled by a single processor, while more elaborate systems with complex displays and reporting on consumption and feed-in-tariff payback, may require a secondary processor, potentially with Ethernet capability. For safety reasons, isolation between the processor and the current and voltage is also required, as well as on the communications bus to the outside world.

In summary, the inverter is an all-encompassing receive-convert-send-control power manipulator, which makes continuous decisions for each situation and adjusts the power variables of the DC to AC transformation for maximum conversion efficiency, power output and safety. We could conclude that the electrical dynamic performance of a PV plant depends on, and is even dominated to a large extent by, the inverters. That makes them an integral and critical part of the proper, efficient and profitable operation of any PV plant. And there is a strong link between inverters' cost, efficiency, quality and size on one hand, and the cost of the generated electric power per kWh, on the other. These are the particulars we look at first, when evaluating or designing large-scale power plants—regardless of what PV generating equipment is involved.

Usually, higher quality and larger size inverters assure a better return on the dollar, or such has been the belief and the trend thus far. So present-day inverters of 1 MW and even 2 MW are used in some large PV plants. The trend toward higher quality inverters is justified to a degree, but the price at times is so exuberant, that it does not justify the extra efficiency or years of additional life we can get from these high-end inverters. The trend towards larger inverters, however, is more complicated, and its improper application might contribute to higher capital and O&M cost, which forces us to look into alternative ways to use inverters when we design, build and operate large-scale plants.

String Inverters

String inverters are basically smaller inverters that can be mounted in different (and space restricted) areas of the power field. They can handle the high DC voltages and three-phase output required by large-scale installations; and, by having a large number of maximum power point trackers, they can convert power more efficiently. String combiners and external string monitoring systems are not required which saves on wiring expenses and reduces wiring losses.

Compact transformer stations can be used to connect to medium-voltage grid points, and because of their small size the transformers can also be placed at different locations around the field. The 630-kVA transformer station is among the most commonly used and usually has short lead times. As the height of the transformer station is limited, it is possible to place it behind the modules. The opposite substructure of modules is only slightly more shaded if distances between them is unchanged.

Low loss transformers reduce the nightly power consumption to below 0.4% of total production, and short circuit losses in the transformer have little effect on overall yield. Outgoing feeder panels with HH-fuses can be inserted in the medium-voltage area of transformers of this size, instead of the more expensive power-switches.

String inverters and transformers are relatively small, easy to install, and have short lead times because they are widely used. Special training is not required to install, maintain or replace string inverters, so service contracts (as needed for central inverters) are not needed. There are additional advantages of using string inverter systems, and they should be considered in some commercial and utility type power plants.

Clustered Inverters

Several inverters are usually used in each commercial and utilities power plant. The inverters can be installed and wired in a number of ways, depending on the field location, size and power use characteristics. Large-scale PV installations are usually wired in strings, with several strings feeding one inverter. The strings and inverters could be sized and configured in several different ways to fit the particular installation, usually feeding centralized inverters at large-scale installations. One approach for optimizing the efficiency and cost of a large-scale installation is to subdivide the power field into 1.0-MWp clusters, each consisting of a 1.0-MWp string of PV modules fed into one integrated central inverter cluster (1.0 MW total), comprised of 1, 2, or 4 inverters per individual cluster. Larger inverters are available now, so the modular approach could be expanded to 2- to 4-MW and larger clusters.

As an example, a 100-MWp power field could consist of a 100 x 1.0-MWp inverter cluster. If each 1.0-MW cluster consists of 2 pc. 500-kW inverters, then there would be 200 inverters total in the field. If each 1.0-MW unit contains 4 pcs. 250-kW inverters, there would be 400 inverters total. Each inverter cluster is fed by several PV module strings—each consisting of several sub-strings. The power generated by each 1.0-MWp PV string is fed into an inverter cluster, where the DC power is converted into AC power and sent into the grid.

Inverters of larger size—1.0- to 2.0-MW and larger—are recommendable for use in large power fields, but they are the newcomers and have not been proven reliable on the PV market as yet. Since inverters are quite expensive, and are even considered as the most expensive part of the power field (when maintenance is included), this is a clear disadvantage, so we'd advise caution, careful consideration, proper design calculations, and thorough negotiations with manufacturers before making a final decision on using 1- to 2-MW or larger inverters.

One fairly new and quite beneficial feature offered by some inverter manufacturers lately is their "overload" capacity. This is an extra design feature that allows a power load of 120% in some cases, or 20% more than the designed 100% power-handling capability. In a 1.0 MW inverter this means that it will be able to handle 20%, or 200 kW power in addition to the designed 1.0 MW capacity. This feature is exceptionally useful in ensuring that any extraordinary events, forcing higher than design output (solar explosions, field power variations, etc.), or any work done at the site that changes the strings' output and forces the power injected into the inverters to go over the design level of 100%, won't cause any damage to the inverters. Not only that, but the extra power is efficiently used in the DC-to-AC power conversion. This also allows installation of additional modules to compensate for power loss in the future and under special circumstances.

In large-scale power fields, a modular setup (group of 1.0 MWp PV modules and inverters) might be most effective from technical and financial points of view. This modular design allows a greater degree of freedom at the large generating capacity and is recommended for any power farm over 10 MWp DC power generating capacity. The PV modules and inverters in this case could be grouped in 250-kW, 500-kW, or 1.0-MWp modules, and wired in a number of different ways, to optimize the conversion efficiency and minimize the cost of installation and operation.

Most inverters cannot operate in the deserts without proper cooling because their efficiency drops significantly when the air temperature reaches 110°F. Special enclosures equipped with A/C units are used in such case. Some power is wasted for the cooling, but this will allow the inverter to operate at maximum efficiency and will add years to useful life.

See Appendix B below for specs of inverters used in large-scale fields in the US.

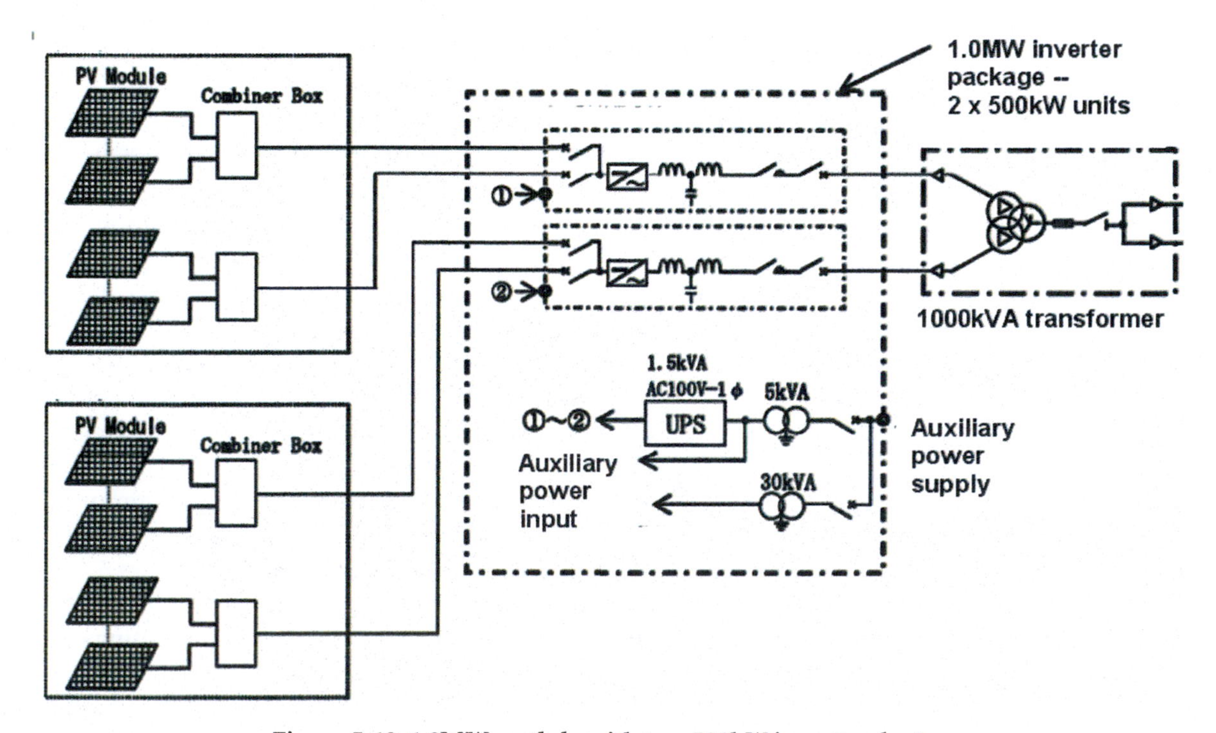

Figure 5-12. 1.0MW module with two 500kW inverter cluster

Related Components Cost Estimates

Although the availability and cost of inverters and related components depend on the manufacturer, type and size of equipment, there are some guidelines, which we could use for ball-park estimates of a 1.0 MWp modular installation, as follow:

Table 5-7. Estimate of 1 MW PV field electric components.

Qty.	Equipment Description	US$
2 pcs.	500kW Inverters	280,000
2 pcs.	internal transformer (included)	
	Switches & 4-Way Triad	180,000
	1MVA Cooper Transformers	40,000
	Square D low voltage Switchgear	12,000
	Combiner boxes (additional)	?
	Wire, Connectors & Conduit	4,000
3,000ft	1/0 Al Primary cable	6,000
600ft	Copper cable	6,000
	Pole & Feeder upgrades	10,000
	Construction materials and labor	8,000
1.0MW	DC-AC module total cost	$548,000

NOTE: This is an estimate only. The actual cost will vary.

Obviously, inverters and related electric and electronic components needed for proper and safe conversion of DC to AC power would add over $0.50/Wp to the cost of the installation. Inverter costs have been going down lately, while their efficiency has been increasing, so this trend will help with reaching a balance in the near future.

The inverters' reliability, however, is not as well defined or predictable. The industry is fairly new (especially larger-size inverters manufacturing) with few years of field operation experience. The industry is loosely standardized, so manufacturers have varying quality control standards and procedures. Because of that, some of them have the reputation of "quality" providers, while the rest are either in the lower quality bracket, or are unproven. In any case, due to some bitter lessons from the past, thorough knowledge of the manufacturer's modus operandi—including manufacturing and QC/QA procedures and supply chain details—are paramount for the successful choice of reliable equipment.

Inverter Issues and Failures

Inverters are electro-mechanical devices, consisting of complex semiconductor components, transformers, gauges, relays, electronics, connectors and wiring which operate efficiently and reliably under "normal" or "standard" conditions—the famous, accepted and totally non-controversial STC (25°C). These "standard" conditions for inverters testing also include incoming DC power with known quality and characteristics. Things, however, are seldom "normal" or "standard" in field operations, and inverters seldom see operation at 25°C and constant power input for long periods of time. They are, in fact, exposed to a number of environmental attacks and operational mishaps.

This is especially true when inverters are exposed to the world's harsh climate regions—like deserts—where mechanical and chemical attacks are the norm. Inverters, like any other electro-mechanical device, will deteriorate and fail with time under these conditions, and we must be aware of this, in order to properly design and efficiently operate the PV power plant during its long exposure to the elements.

A closer look at potential major inverter issues in field operations follows.

Power Loss

A major problem with inverters is that they tend to have a window of operation—lower and upper voltage limits—so unless the power is kept within these limits something bad will happen. Because of the fluctuations of the solar insolation and weather conditions, no matter where the fields are (even deserts have partially and fully cloudy days and ferocious monsoon seasons), the inverters' power output fluctuates accordingly. The field design should anticipate and avoid excursions above the upper limit (set at 600 in the US), so the inverters should never "see" voltage exceeding 600V, even if manufacturers provide some over-power flexibility. If this happens, then it is simply a design issue that must be corrected.

The bigger problem in most cases is the lower operating limit, also called "starting voltage," which must be met for the inverter to start operation. Inverters also tend to "drop" power when the voltage of the feeder string approaches their lower limit. So, if the lower limit is 200V, and the string is not providing it, or it goes below that level, then the power output will be interrupted. This can happen when the sun goes under a cloud and the incoming string voltage drops below 200V.

The power fluctuates during the day with the sun going in and out of the clouds. Optimum (best possible) performance of an inverter operating under variable solar insolation is shown in Figure 5-13b. Note how the voltage is kept almost level while the sun is behind clouds and the feeder string is generating different power output, as that in Figure 5-13a. And while the total power (P = V x A) from the inverter is fluctuating, the voltage is kept at a level that allows the inverter to op-

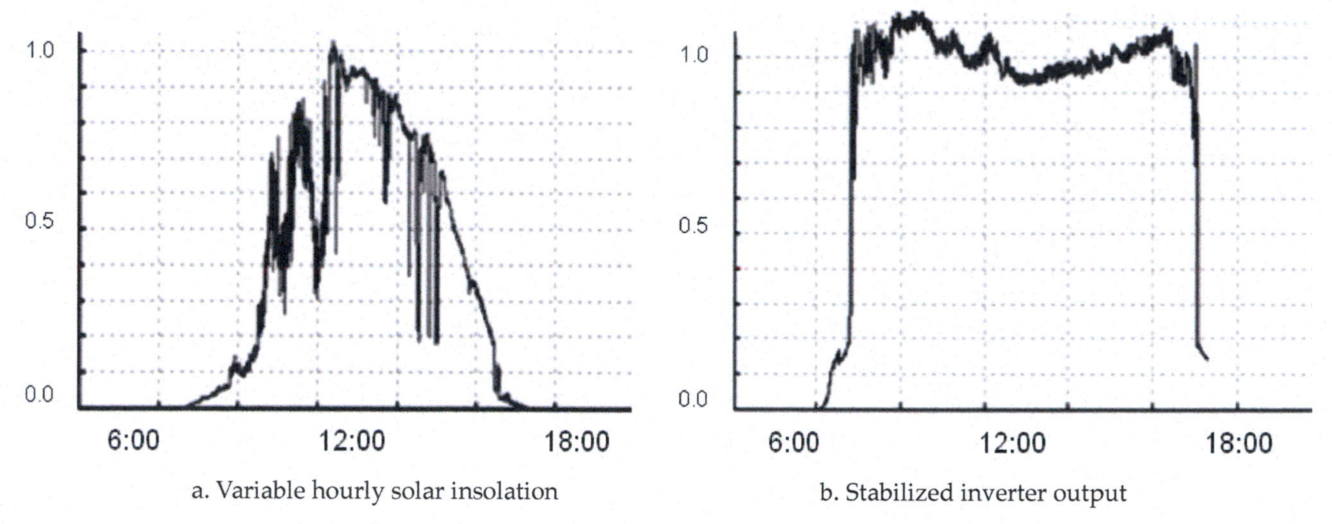

a. Variable hourly solar insolation

b. Stabilized inverter output

Figure 5-13. Optimum inverter operation at variable solar insolation

erate uninterrupted, thus converting any and all incoming DC power into AC and feeding it into the grid (albeit with varying power level). According to the manufacturer, only heavy clouds would force the inverter into a stand-by mode.

The good news for large-scale applications is that the power fields, and the inverters in them, are very likely to be installed in desert areas, where the sun is more constant than any other areas in the world. This provides a more consistent sunlight, with less fluctuations and power interruptions. But even then, there could be some great weather surprises during certain periods of the year.

Temperature Derating

Another major variable that needs special attention—especially in large-scale installations in the deserts—is the quick and serious temperature derating of most inverters. Even 98% efficient inverters lose efficiency quickly at elevated temperatures; i.e., most manufacturer specs call for high efficiency rating at temperatures much lower than the temperatures we measure in US deserts (and which could reach 190°F in the metal enclosures—inverter cabinets, containers etc.) while the units are in operation. Without cooling, these enclosures become an oven in which the inverters' guts bake, and as the temperature climbs, their output drops quickly. Different inverters have different derating factors, but we dare say that no matter what the derating factor is, all inverters operating in desert areas will reduce output and even malfunction if operated without active cooling. A/C units are therefore needed to keep inverters operating at or below 100°F. If their internal temperature is allowed to increase enough over the 100°C mark, the internal electronics circuits will simply shut down and the inverters will stop generating power. Even worse, the electronics might just fry out, thus potentially causing major failures and even disasters.

There is no good news in the case of large-scale applications here, because the power fields, and the respective inverters, are most likely to be installed in desert areas, where the heat built up in an operating inverter exceeds the temperature tolerance.

Humidity Effects

Humidity also impedes the inverters' efficient operation, so that on a monsoon day in the Arizona desert, where temperatures in the inverter cabinet could reach 180°F, and with humidity over 80%, we will see inverter efficiency drop drastically. Humidity is one of the greatest enemies of all electronic and electric circuits and devices. There are a number of harmful effects that can occur within a circuit when humidity increases above manufacturers' specified limits. Damage to some sensitive components might occur at these conditions as well.

The inverters' efficiency would be improved, and their life extended if they were housed in an air-conditioned room, where the temperature and humidity could be kept within spec. This is especially true for utility type projects, which are most likely to be installed in desert areas, where heat and humidity levels exceed the inverters' design tolerances, and where special precautions must be taken, to ensure long-term reliability.

It has been determined that inverter failures are the most frequent, and most serious incidents in a PV power plant operation. This is often caused by their manufacturers' lack of experience in early production stages, so newly designed and/or manufactured inverters might have some serious problems. These problems could be compounded by poor power field design, improper installation or negligent operation. Some inverter failures might be caused by weather disturbance, grid glitches, and interconnecting issues, which could cause serious problems and damage, if the inverters are not designed and built to handle such abnormalities.

Large-scale PV power generation is an emerging industry that has the potential to offer improvements in power system efficiency, reliability and diversity, and to help contribute to increasing the PV energy's percentage in the power generation mix. While a great amount of knowledge has been gained through past experience, the practical implementation of distributed and central PV generation has proved to be more challenging than perhaps originally anticipated—especially for installations in very hot and/or humid areas.

Future Efforts

Here are several future steps to be undertaken by equipment manufacturers and power field design engineers that might assist in assuring inverters' reliability:

1. Research and develop standards and regulation concepts to be embedded in inverters, controllers, and dedicated voltage conditioner technologies that integrate with power system voltage regulation, providing fast voltage regulation to mitigate flicker and faster voltage fluctuations caused by local PV fluctuations.
2. Investigate DC power distribution architectures as an into-the-future method to improve overall reliability (especially with micro-grids), power quality, local system cost, and very high penetration PV distributed generation.
3. Develop advanced communications and control concepts that are integrated with solar energy grid integration systems. These are the key to providing very sophisticated microgrid operation that maximizes efficiency, power quality, and reliability.
4. Identify inverter-tied storage systems that will integrate with distributed or multi-string PV generation to allow intentional islanding and system optimization functions (ancillary services) to increase the economic competitiveness of distributed generation.

These and other similar initiatives will bring the inverters to the desired level or efficiency and reliability, as demanded by today's PV technologies and utilities power systems.

Inverter Maintenance

Inverters are an integral part of the PV power field structure and major contributors for its efficient and safe performance. The effect of one inverter malfunctioning will reverberate through the entire installation, affecting (usually negatively) the operation of other plant components, and the output will suffer. Inverters come with 5-10 years or more limited warranty, which covers the different parts and components, as well as with a guaranty for certain level of performance. More often than not, inverters will malfunction without notice, due to a failed electrical component. To take quick action in such cases, most manufacturers have a stand-by maintenance team to offer diagnostics and repairs by phone or in person. Figure 5-14 is a schematic of one such support center of a major inverter manufacturer.

Provided that a maintenance contract is signed, this manufacturer promises 24/7 technical phone support. If the problem cannot be diagnosed and fixed by phone, then a technical support team will be dispatched. The documented MTTR (mean time to repair) for this manufacturer is 4 hrs. This means that most problems are diagnosed and fixed within 4 hours from notice. This is very good performance, achieved best when the service center

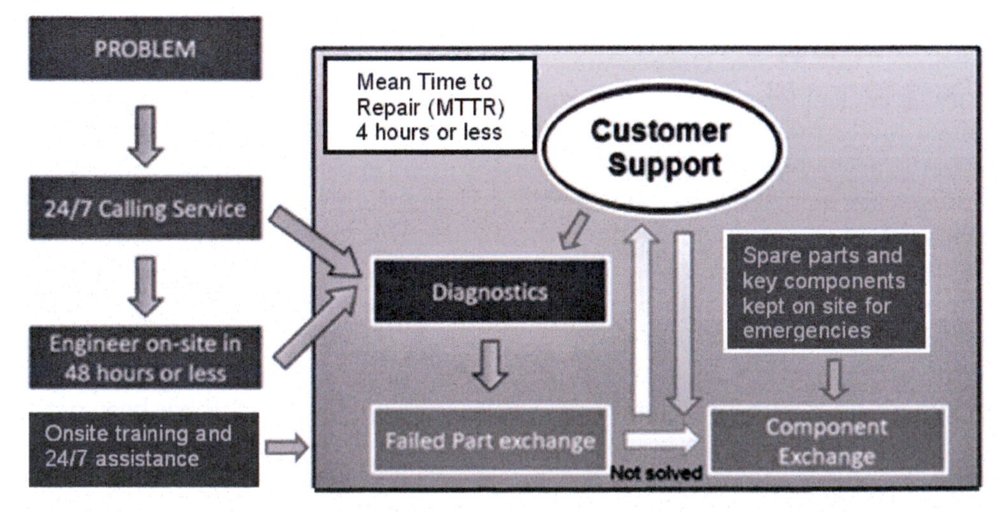

Figure 5-14. Customer support scheme

is located close to the power plant.

Another major issue is that with time the inverters, just like the PV modules, lose efficiency. So, a 95% efficient inverter at time of installation, will go down to let's say 93% by year #5 just by virtue of normal wear and tear. Degradation is more complicated and pronounced in extreme climate regions. This variable must be well understood, discussed and negotiated with the manufacturer at the beginning of the project. It must be also included in the PV power plant design calculations, because it will inevitably affect the PV installation's total power output.

Inverters also have a limited life expectancy, which is reduced further when operated in climates of extreme heat and elevated humidity. Because of that, periodic overhauls (every 5 or 10 years) or replacement of the inverters must be allowed for in the plant's O&M budget. This is a major expense, so its proper estimate and budgeting is of utmost importance to the project's financial success.

Finally, there is a long list of maintenance tasks that must be planned and expected from the manufacturer, or service contractor, in assuring the inverters' proper long-term operation.

The recommended inspection, test, cleaning, repair, tuning and replacement procedures are:

1. *Monthly inspection*
   - Presence of dust and gas in cabinet
   - Excess vibrations and noise
   - Abnormal heating
   - LCD abnormal operation
   - LED abnormal operation
   - LED failures
   - Cleaning of air filters (if necessary)

2. *Annual inspection:*
   - Visual inspection of all components
   - Internal parts inspection for corrosion and discoloration
   - Check for corrosive gases in the cabinet
   - Electrical tests; total output and power variations
   - Cleaning of internal parts in the cabinet
   - Cleaning of air filters (if necessary)

3. *Five-year inspection*
   - Discharge resistor
   - Electrolytic dapacitors
   - Filter capacitor
   - Reactor
   - Semiconductors (diodes)
   - MCCBs
   - Contactors
   - PWBs
   - Fuses
   - Fans
   - A/C units
   - Wiring
   - Bolts, nuts, screws
   - Surge protection circuit
   - Electrolytic capacitors in control circuit
   - Insulation resistance
   - Protection circuit check
   - Operation time of timers
   - Main circuit waveforms
   - Timers operation and accuracy
   - Output voltage waveforms
   - LCD display
   - AC output phase check
   - Conversion efficiency verification
   - Utility interface control parameters
   - Full power operation check (if possible)
   - Cleaning of the internal parts in the cabinet
   - Consumable parts replacement (per contract)

PV POWER PLANT CONSIDERATIONS

Large-scale PV power plants are a fairly new development in the world's energy markets. Their location related to the available solar insolation and socio-political factors is of great importance and usually determines the initial design, installation and operation parameters. In the previous chapters we reviewed the available PV technologies most suitable for large-scale PV installations, and some of the related problems, so now we will take a look at their actual application in PV energy generation.

PV power plants for commercial and utilities power generation are basically larger size installations, using photovoltaic (PV) technologies for generating electric power from the sun. They vary in type, size and function according to their make-up and designation. Some are designed for stand-alone operation (not connected to the grid), while others are grid-tied (plugged into the grid), and some are designed for both applications. Commercial power plants could be as small as 5.0 kWp, and as large as 1.0 MWp or more, and are designed for providing power to larger commercial enterprises or the grid. Utility type applications require larger capacities and utility PV power plants range in size from 1.0 MWp to 1.0 GWp or larger. We will use the term "large-scale" when referring to commercial and utility installations in this text.

Our goal is to examine the possibility of installing and operating large-scale PV power plants in the desert regions of the US and abroad. The principal role of these installations is to:

1. Supply clean, renewable and cheap energy,
2. Contribute to cleaning the environment, and
3. Assist in socio-economic development

There are several tasks to be accomplished and some problems to be resolved in achieving these goals, so the major considerations we tackle in this text are:

1. PV technology suitability and readiness for large-scale PV use,
2. Land availability and acquisition,
3. Permitting, regulatory, political and socio-economic considerations
4. Transmission lines and interconnection points
5. Power purchase conditions, and
6. Financing barriers and opportunities

PV Technologies and Their Use

The PV systems on the energy market today can be divided into several categories:

1. Residential PV systems are basically 1-100 kWp PV arrays mounted on a rooftop, or ground-mounted in the back yard. These usually consist of c-Si or TFPV modules, fixed-mounted in arrays, connected to an inverter which feeds the house appliances, and supplies excess energy to the grid.

2. Commercial type PV systems are 5 kWp to 1.0 MWp roof- or ground-mounted systems which also consist of fixed-mounted c-Si or TFPV modules connected to an inverter feeding the facilities or sending excess power into the grid. In some instances, the PV modules are mounted on trackers to optimize land use and provide more power.

3. Utilities power plants are usually over 1.0 MWp, ground-mounted installations, preferably located in high-intensity, solar-insolation (desert) type areas. c-Si or TFPV modules are fixed-mounted in strings, connected to inverters and feed the grid. Trackers are used in some instances as well for optimizing land use and providing maximum amount of power.

The key PV technologies, suitable for commercial and/or utility type solar power generating plants today are c-Si PV modules (mono or poly crystalline silicon cells) and thin film PV modules (CdTe, CIGS, a-Si etc.), and HCPV trackers. There are some emerging PV technologies too, but their use is still limited, so we will postpone reviewing their practical use until later.

Each of the above PV technologies has a subset of devices and equipment, which are, to one degree or another, suitable for commercial and/or utilities type solar power generating applications. We will review each of these technologies and equipment subsets.

c-Si modules—mono and polycrystalline modules—are shaping up as the most used PV technologies by far. This is mostly due to the quickly developed production capacities in Asia, which were able to flood world markets with cheap PV modules. Thin film PV modules have been a worthy competitor in residential markets, and are now quickly increasing their presence in the commercial and utilities energy markets too. HCPV is a promising PV technology with very high efficiency and reliability, but it has moving parts which make it hard to operate and maintain. It is most suitable for large-scale power generation in the deserts, so we see it as a worthy competitor in the near future.

Utilities are the key to successful use of solar energy, and most of them are making an honest effort to include solar power to their energy portfolios. SCE and PG&E are leading the pack for now and for the foreseeable future. See Figure 5-15. Looking beyond 2011, there

U.S. PV INSTALLATIONS TO DATE

State	MWp	State	MWp
CA	1020	NC	41
NJ	260	MA	38
CO	121	TX	35
NV	105	CT	25
AZ	110	OR	24
FL	74	OH	21
NY	56	IL	16
PA	54	WI	9
HI	45	DE	6
NM	43	TN	5

All other states have < 5 MWp PV each

Figure 5-15. US utilities most active in the solar energy field (8).

are almost 12 gigawatts of announced solar projects estimated to come online by 2016 (8). More than half of the planned solar installations, however, are CSP based (some of which are already cancelled), and some will not materialize for a number of reasons. PV technologies are on the rise and moving to fill the gap left by CSP, so the battle is raging. No matter what happens, these are remarkable achievements, pointing to a very good start of the "solar revolution" in the US. Watch out, Germany!

PV System Types

PV technologies manufacturers, supply chain companies, engineering teams, owners, investors and all participants are looking forward to the opportunities presented by solar power. The technical aspects of PV technologies complicate matters and make it hard to navigate in this relatively new field, so understanding the structure, function and issues of available PV technologies is the first step to successful design, installation and operation of a PV installation.

The major types of PV power generating systems for the purposes of this writing are residential, commercial, and utility installations, with emphasis on the latter.

Residential PV Systems

Let's start with residential systems and projects, which although not a subject of this book, are an important factor in today's solar energy scenario. Utilities love the residential PV market for a number of reasons; but, to be fair, residential projects are not the solution to our energy problems. And we expect that with the decrease and disappearance of government incentives and subsidies we'll see less residential installations.

On the other hand, residential installations are mini-power plants which help us understand the bigger issues facing us in large field installations. They also help in the "green" movement, so we hope that their numbers increase for these two reasons, if nothing else.

Figure 5-16 is a simple electrical wiring diagram of a residential solar power installation—our virtual PV system. It's about 5 kW—good enough for running the electric meter backwards several kWh a day, while Mom and Pop are at work and the kids are at school. Putting the roof to work? Not a bad idea. But it took some doing to get the system on the roof, get it connected, tested and blessed. And now, when it is all ready to produce power and help pay the monthly power bills, there are some problems as we will see below.

Again, residential installations are not the object of this book, but they parallel large-scale installations closely, albeit at a different level of size, complexity and difficulty, so we just can't resist the opportunity to point out the basic function and issues of this simple installation and then transfer these to our much larger projects.

Residential PV systems are usually classified as off-grid, or more often as grid-tied systems. We will not dedicate much time to off-grid systems, and will only point out that they usually consist of a small PV array, either portable, or built-in, intended to power part of a house, a remote cabin or a small piece of equipment, such as a water pump somewhere in the remote fields of Kansas or Texas. These PV systems are usually roof- or ground-mounted and power appliances and equipment, feeding DC power directly, or converting it to 120 V AC.

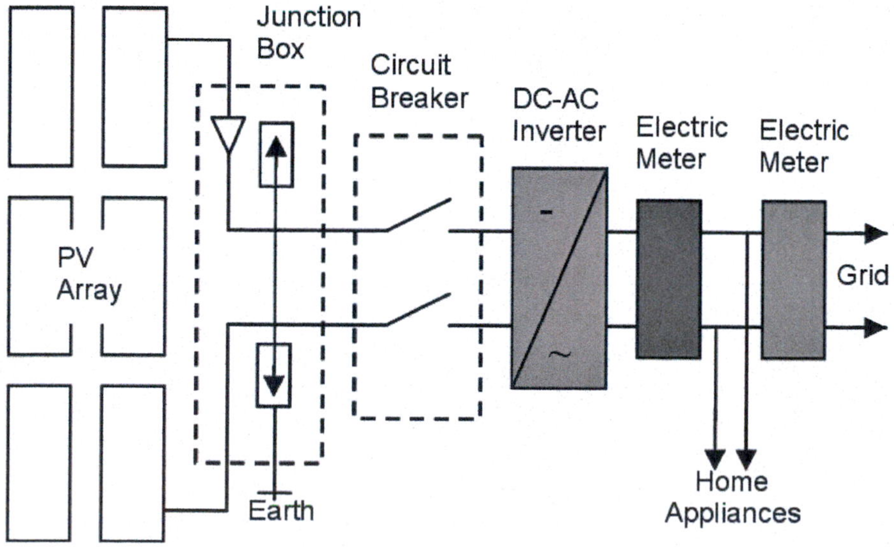

So, back to our residential installation. The PV array produces DC power and runs it through a PV disconnect for safety and maintenance purposes. The power then goes into a charge controller, which is a glorified battery charger. The DC power from the controller and/or the batteries goes to the house appliances that can run on DC power. Whatever is not used as DC in the house goes into the inverter, which converts it into AC power and sends it into the power grid. Simple, yes, but not trouble-free.

It can take months to agree on the design and file the forms to obtain the necessary permits and get

Figure 5-16. Residential PV installation

the installation completed. Once that is done, our new PV array is ready to go. It "wakes up" in the morning by generating a small amount of power because the sunlight is still too weak to activate the electrons asleep in their holes and get them energized for another day of running up and down the wires. But soon the sun rises and more juice flows into the circuitry and into the grid. Yes, the system is running and the higher the sun gets, the more power it generates.

All of a sudden, however, there is an alarm. The PV array is disconnected. A wet wire from last night's rain has shorted and the EGFP (equipment ground fault protection) device tripped, isolating the array from the DC loads and the inverter. After a couple of tries, the system is back on line and operational, but soon it is off again, this time because many such installations in the neighborhood have similar problems and the on-and-off mode of these has destabilized the grid. The inverter/controller is not capable of synchronizing, went into "islanding" mode, and our array is off the grid until the condition is remedied.

The next day the wind blows a bunch of leaves from the neighbors' trees onto the PV array and the power output is restricted by half. We are told that partially shaded cells in the modules are in danger of overheating and the module is threatened by a meltdown and even a fire, so somebody has to climb on the roof and sweep the leaves. Our small PV power plant is up and running again, when, a day or two later, the output seems to be very low for some reason. A quick test points to a module that has malfunctioned and is no longer producing full power. It could be a broken or shorted solar cell, but a sealed PV module cannot be repaired, so we call the maintenance crew to replace it. That effort is covered by contract or warranty. If not, the repair will cost us the last 6 months energy produced by the system.

Even with best of efforts and impeccable professional maintenance, we notice that the array produces ~ 5-6% less than normal after 4-5 years of operation. The modules manufacturer's warranty says that even after 20 years of operation, the modules are guaranteed to produce 80% of the electricity they produced when they were new. So a 5-6% drop in output in 4-5 years is within the warranty claims. Too bad, but there is not much we can do about that. It is a loss, which we should've estimated and allowed for at the very beginning.

Now, 20 years have gone by, the contract term is completed, and the array output is very low. The installation must be decommissioned and disposed of. Who is going to climb up on the roof to disconnect and disassemble it? What are we going to do with the modules and other hardware, which now are mostly junk and even considered hazardous waste (in the case of some types of thin film modules)? Can we throw them in the garbage? Nope. Many PV modules must be handled as hazardous waste, carefully disassembled, packed and shipped as such to the corresponding recycling center. What if the PV modules manufacturer is out of business now and the insurance or recycling fund no longer covers the disassembly and shipping charges? Impossible? Yes, possible! Look at the small print very carefully before buying the modules and/or signing the contract. There are many manufacturers here today and gone tomorrow, so utmost caution is needed.

Our virtual and somewhat troublesome PV system above is a good reminder that a good—or rather excellent—design and planning for everything and anything that the future might bring is indispensible in assuring good results from any PV system, be it residential or large-scale. And large-scale systems are, of course, many times larger so that any design error, production quality issue, or maintenance negligence will cost us multiples in delays, poor performance, failures and financial losses.

Now let's go to our larger, commercial type system.

Commercial PV Systems

A typical commercial PV system for a small business usually also consists of rooftop-mounted PV modules tilted at an angle of, or lower than, the latitude of the particular location. The biggest difference between residential and commercial PV systems would be the number of modules and the size of the inverters used.

Trackers are seldom used in such installations, and when they are used on the roof they are usually small structures, though larger trackers could be used in ground-mounted commercial applications.

As you see, the array in Figure 5-17 is very similar to the residential one we discussed previously. Both arrays would have similar design, installation and operation considerations, except that the commercial arrays are much bigger and usually more complex and expensive. The problems of the residential PV system described above are encountered in these systems too, so we must make sure that they are well designed and installed, and that appropriate O&M procedures have been implemented.

Utilities Scale PV Projects

Here we must provide a definition and scope of utilities-type PV projects. A "utility" is a company that generates and/or distributes electrical power. The pow-

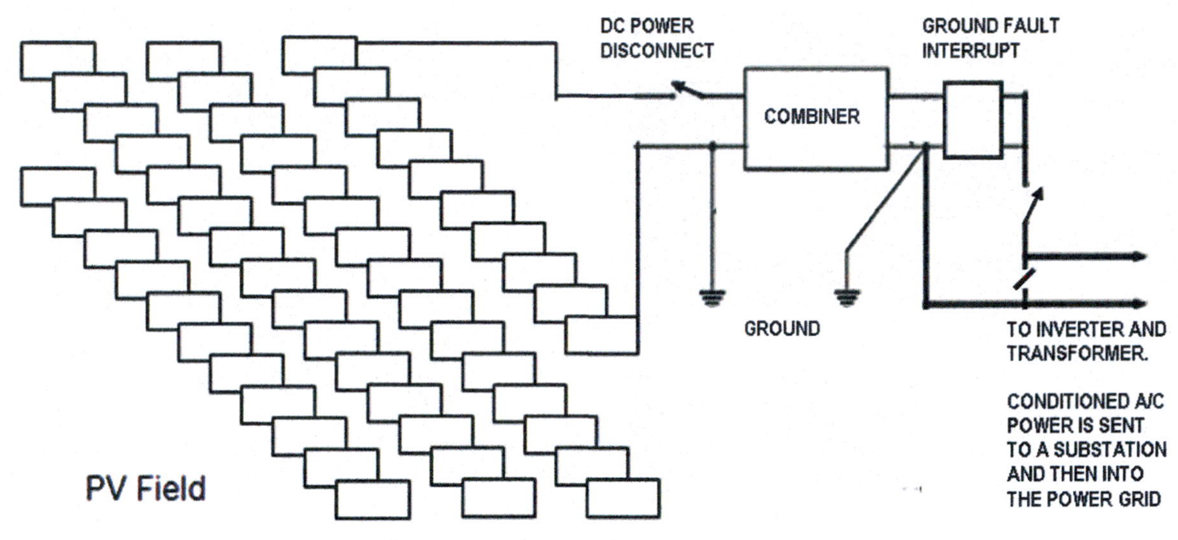

Figure 5-17. Commercial PV installation

er can be generated by the utility itself, or purchased from somewhere else, but in both cases the utility is responsible for the proper and safe distribution and delivery of the power to the customers interconnect points. Utilities-type PV installations usually deliver power exclusively to the utility's substation close to the solar field. The PV power fields could be as little as 1.0 MWp and up to a GWp or more.

For simplicity's sake we will refer to all commercial and utilities systems as "large-scale" installations in this text, keeping in mind that these could be of different sizes and designations which we will point out as needed. We will take a closer look at large-scale PV systems in this and the following chapters.

So what does it take to design and install utilities (large-scale) PV power plants?

Fixed Mounting

Fixed mounting is simply arranging and fastening the PV modules on a fixed frame, anchored to the ground or roof. The frame is tilted at a special angle, which is most efficient for the particular location, and sometimes for the season. The PV modules sit on these frames still—without any movement—and get whatever sunlight reaches them at each part of the day.

Ideally PV modules should be placed so that they are close to perpendicular to the noonday sun—due south and at an angle of inclination approximately equal to the angle of latitude, +/–15 degrees. They will function when mounted flat as well and, as a matter of fact, on overcast days or in areas with a lot of fog or clouds, this can be a better setup since the sunlight is diffused and reflected.

A steeper angle will enhance output during winter months when the sun is lower in the sky, at the expense of some reduced output in summer (a fixed mount would have an angle of latitude plus 15° from horizontal). A simple system, where the angle can be changed manually twice a year (mid spring & autumn) is best and does not involve much cost or effort. This option, however, also needs to be included in the initial design and the O&M procedures of the plans.

PV modules may be mounted on a pole, a ground support, a wall of a building, or a building roof. The main considerations are day-long access to unobstructed sunlight and wire lengths to grid-connect. Appearance and ease of access for cleaning, etc., should also be considered.

Mounting frames for fixed-mounted PV modules are usually made of aluminum or metals that are plated, to avoid electrolytic corrosion. Stainless steel nuts, bolts and galvanized or plated retaining hardware should be used,

Figure 5-18. Fixed mounted PV modules

including the mounts which should also be strong and capable of withstanding wind loads. (See Figure 5-18.)

A wide range of high quality standard frames and mounting kits are available for use at different locations (roof or ground) and setups (fixed, seasonably adjustable etc.). Some of these are pre-fabricated and pre-drilled to accept standard modules and usually include all-stainless steel fasteners for corrosion protection.

Spacing between the Rows

When installing PV modules in a fixed-mount position we take into consideration a number of factors, such as the type and size of modules, frame construction, latitude, seasonal and weather changes, tilt angle, hourly and seasonal utilities rates etc. Basically speaking, during the summer and winter seasons the sun takes a different path over the installation. It rises much faster and higher in the summer, while in the winter it takes its time and rises slowly and stays low on the horizon all through the day. This creates a problem in deciding what tilt to use, and how far the rows should be spaced.

Narrow spacing between the rows uses the maximum number of modules, which will produce the maximum amount of energy in summer. Narrow spacing, however, is not a good option for generating maximum power in winter, because of the heavy shade on each row—especially during early morning and late afternoon. Reducing spacing between the rows to cause 10-15% power decrease in summer will cause 20-30% power decrease in winter. Spring and fall would be somewhere in between.

Increasing the spacing between the rows will improve or eliminate the early AM-late PM shading, but will require significant reduction of the total number of modules installed. The number of modules installed in such case will be reduced according to the spacing increase, and the total daily power output decrease will be proportional to the number of modules removed.

Figure 5-19 gives a visual and mathematical approximation of the spacing between the rows.

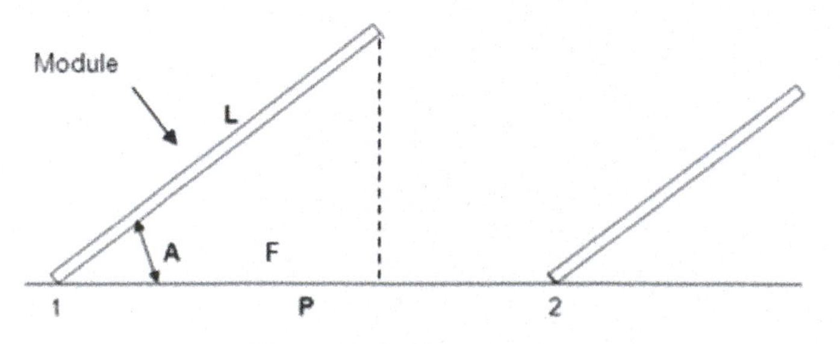

Figure 5-19. Row spacing

If we know the height of the PV modules (L) and the tilt angle (A) we can find the pitch (the distance between the front edge of the first (1) and second (2) rows), we use the formula:

$$P = L \div \cos(A)$$

Where:

P = Pitch (distance between the front of the first (1) and second (2) modules)
A = Latitude angle +/–15°
L = Module length

This will give normal incidence at the equinoxes with minimal or no shadowing between summer and winter equinox.

A practical (but maybe not the best) way of calculating the spacing between the adjacent rows is by looking for the lowest sun angle (solar elevation) at which all modules are 100% shade-free. This is a critical variable, which varies from place to place and with the seasons, playing a determining role in the tilt (or seasonal adjustment), and the spacing between the rows.

The formula for determining the spacing at a given sun angle (and time of day) at which all modules are 100% shade-free is:

$$P = \sin(Se + Tm) * L \div \sin(Se)$$

Where:

Se = the lowest solar elevation angle for shade-free operation
Tm = the module tilt angle
L = Module length

For example, if we decide to look at 20° as the lowest sun angle (or the earliest in the day possible) for shade-free operation of a 1.0 m long module (in portrait position), tilted at 30° in the Arizona desert, then P = 2.2 m. Or 2.2 m distance between the front of the first and second rows will be needed for shade-free operation at this sun angle and the associated time of day, which will be quite different for the summer and winter months.

$$P = \sin(20 + 30) * 1.0 \div \sin(20) = 2.2 \text{ m}$$

Subtracting 0.8 m the footprint (F) of the first row from the total distance, we determine that the actual spacing between the rows should be 1.4 m for 100% shade-free operation at this sun angle. If we decide on 30° as a lowest sun angle (later in the day) then P = 1.7 m and the actual spacing between the rows would drop to 0.9 m. This tilt angle and spacing might provide higher power output around the noon hour but will be under more shade during early morning and late evening.

These are critical calculations and should be carefully considered and meticulously executed, together with all other variables affecting the modules' performance, to obtain the best combination of performance and cost. In most cases the PV modules will be partially shaded shortly after sunup and before sundown, no matter how we space them, unless we are extremely liberal with land use, which could quickly become unreasonable.

The following logic is used to explain and justify the sun angle and spacing for the above modules mounted in Phoenix, AZ.

Summer Shading Estimate:

Shortly after sunrise on June 21, the sun starts moving up quickly. If the rows are east-west oriented, the shading will be minimal at first because the sun's rays will be almost parallel to the rows.

a. The shading caused by the minimum spacing between the rows, as calculated above for 20° solar elevation, will be ~10% around 6:30AM for some of the back rows (sun angle = 18° from the east).

b. Between 6:40AM (sun angle = 20° from the east) and 5:25PM (sun angle=20° from the west) the modules will be 100% illuminated and will generate maximum possible power at the location with this tilt.
NOTE: This is when the maximum power is needed by the utilities (summer peak hours are 1:00PM to 8:00PM) and we get maximum payback from generated power during this time.

c. After 5:25PM the modules will start getting partially shaded, with ~10% shading around 5:35PM (sun angle = 18° from the west) and slowly increasing after that.

Winter Shading Estimate:

Shortly after sunrise on December 23, the sun starts moving up slowly.

a. The shading caused by the minimum spacing between the rows, as calculated above for 20° solar elevation, will be approximately 10% around 8:50AM (sun angle=18° from the west).

b. Between 9:00AM and 2:50PM (sun angle = 20° from the west) the modules will be 100% illuminated and will produce maximum amount of power.
NOTE: This power, however, is not needed by the utilities (winter off-peak hours are between 9:00AM and 5:00PM) and we will get minimum payback for generated power during this time.

c. After 2:50PM the modules will start getting shaded progressively, with 10% shading around 3:05PM (sun angle = 18°), and increasing quickly after that.

Basically speaking, increasing the spacing between the rows will cause the power generation to start earlier in the morning and continue until later in the evening. This, however, will decrease proportionally the number of PV modules installed in the given land area, thus decreasing the total daily output. This simply means that, due to the great effects on the overall results, the spacing between the rows must be carefully considered and calculated.

NOTE: These are estimates used to clarify the row spacing concept only. The data contained herein should not be used in actual pitch or tilt angle estimates.

Keep in mind the fact that utilities have different summer and winter peak times and rate schedules. For example, summer peak hours in Arizona are from 1:00 to 8:00PM, while winter peak hours are 5:00-9:00AM and 5:00-9:00PM. Some utilities offer summer peak rate payback of $0.25/kW, while the winter peak rate pay is only $0.10/kW, and the off-peak is even less. So it is easy to see that our PV system will make more money (several times more) during the summer peak hours. In addition, the sun is brighter (at a higher angle), so the energy generation is maximized. On top of that, there are many more useful hours in a summer day, while the winter days are short.

Because of the importance of these variables, the design of our PV system is paramount. The PV technology and location we choose, the tilt (or tracking) of the modules, the spacing between the rows, and their variables must be thoroughly evaluated during the planning and design stages.

PV Trackers

On a standard fixed-mount PV tracker, the modules are always pointed at a fixed position all through the day—hopefully the best angle possible towards the sun—and collect whatever amount of sunlight falls on them. Trackers offer the advantage of "looking" directly at the sun all day, thus collecting maximum possible sun energy. The trackers consist of a steel or aluminum frame on which the modules are attached. Gear-drive mechanism(s) turn the tracker frame slowly, trying to keep the modules exactly perpendicular to the sun, or as closely as possible. This minimizes the angle of incidence of the incoming light falling on the PV modules, increasing the amount of sun energy gathered from the direct component of the incoming light, and maximizing electric energy generation.

Photovoltaic trackers can be grouped into classes by their number of axes—dual or single-axis (one or two-axis). A single-axis tracker could increase the annual output of a fixed mount array by 15-20%, or more, while a dual-axis tracker could add more efficiency, depending on its accuracy. This, however, also adds complexity and additional cost to the system, thus two-axis trackers are usually not used for these applications.

Two-axis PV tracking systems produce the most energy when pointed directly (90° angle) at the sun. So, if a tracker can follow the sun precisely at 90 degrees all day (which requires a two-axis tracker) with high accuracy, it will produce the most energy, equal to the accuracy of the positioning of the modules towards the sun. Trackers with ± 5° accuracy can deliver greater than 99.5% of the energy delivered by the direct beam and 100% of the diffuse light. Since such accuracy tracking is quite expensive, it is normally used only with special high efficiency PV systems called high concentration PV (HCPV). These systems are becoming more popular, but due to their complexity, and the presence of moving parts, it will be some time before we see them deployed in large numbers.

One- or two-axis trackers can be used with any type of PV technology, but are rarely used with thin film modules, due to their low efficiency which requires a much larger number of trackers, thus making the initial cost and O&M expenses prohibitive. Nevertheless, there are proposals to use TFPV modules on trackers, so we will watch this development closely.

Single-axis trackers have one degree of freedom that acts as an axis of rotation. The axis of rotation of single-axis trackers is typically aligned along a true north meridian. It is possible to align them in any cardinal direction with advanced tracking algorithms as well. There are several common implementations of single-axis trackers. These include horizontal single-axis trackers, vertical single-axis trackers, and tilted single-axis trackers, as follow.

Horizontal Single-axis Tracker

This tracker consists of a long metal tube, installed parallel to the ground, supported on bearings mounted on pylons which are cemented into the ground. PV modules are mounted on a frame supported by the tube. The tube with frame and modules mounted on it rotates on its axis to track the sun motion through the day. The axis of rotation for horizontal single-axis trackers is horizontal with respect to the ground. The support posts at either end of the axis of rotation (a row) can be shared between trackers to lower the installation cost. See Figure 5-20.

Figure 5-20. Horizontal single-axis tracker

The rows of trackers can be oriented S-N or E-W, with both options having their advantages and disadvantages. S-N orientation is the preferred orientation in many cases, but it is less efficient during the most productive hours of the day (around noon). E-W orientation, on the other hand, does not allow good elevation tracking, so some sunlight would be lost at the early and late hours of the day. The difference in orientations must be carefully considered, for optimum seasonal and daily utility load matching at the particular locations.

Horizontal trackers typically have the face of the module oriented parallel to the axis of rotation. As the modules track, they sweep a semi-cylinder that is rotationally symmetric around the axis of rotation. The distance between the tracker rows is important also, as discussed in the fixed-mount modules section above. The same considerations and calculations apply here as well, and should be thoroughly considered during the planning and design steps. Basically speaking, the allowable degree of shading of the modules during the early morning and late afternoon hours will determine the spacing between the rows. The closer the rows, the more densely packed the power field will be (more modules can be installed), but the shading on the modules will increase proportionally. This might result in decreased daily output, so finding the best balance between module shading and spacing between the rows is critical for obtaining the greatest power output at critical power production times.

Vertical Single-axis Tracker

The axis of rotation for vertical single-axis trackers is vertical (a pole cemented into the ground) with respect to the ground. The frame on these trackers is fixed at the local elevation angle, and the entire frame rotates from east to west over the course of the day, following the sun.

Figure 5-21. Vertical single axis tracker

Field layouts must consider shading by spacing the trackers in such a way as to avoid unnecessary energy losses and to optimize land use. Also optimization for dense packing is limited due to the nature of the shading over the course of a year. In some cases, densely packed trackers are used for optimizing the utility load requirements. This approach will produce maximum power around noon, while sacrificing some power output due to shading in the early morning and late afternoon. The distance between the tracker rows is similar to that discussed in the fixed-mount modules section above. The usual considerations and calculations apply here as well, and should be thoroughly considered during the planning and design steps.

Tilted Single-axis Tracker

This is a variation of the vertical single-axis tracker, with the axes of rotation fixed in tilted (not vertical) position in respect to the ground. Tracker tilt angles are often limited as needed to reduce the wind profile and decrease the elevated end's height from the ground.

Similar to the above examples, field layouts must consider the effects of modules' shading to avoid losses and to optimize land use. Optimized packing can be used here as well for optimizing noon hour power generation.

Figure 5-22. Tilted single-axis tracker

Dual-axis Trackers

Dual-axis trackers use x and y directional drives, which allows them to move up and down and left and right while following the sun. They have two degrees of freedom that act as axes of rotation. There are several common implementations of dual-axis trackers. They are classified by the orientation of their primary axes with respect to the ground. Two common implementations are tip-tilt trackers and azimuth-altitude trackers.

a. Tip-tilt Dual-axis Tracker

This tracker has its primary axis horizontal to the ground. The secondary axis is normal to the primary axis. The posts at either end of the primary axis of rotation of a tip-tilt dual-axis tracker can be shared between trackers to lower installation costs. Field layouts with tip-tilt dual-axis trackers are very flexible. The simple geometry means that keeping the axes of rotation parallel to one another is all that is required for appropriately positioning the trackers with respect to one another.

With backtracking, these trackers can be packed without shading at any density. The axes of rotation of tip-tilt dual-axis trackers are typically aligned either along a true north meridian or an eas-west line of latitude. It is possible to align them in any cardinal direction with advanced tracking algorithms.

b. Azimuth-altitude Dual-axis Tracker

These trackers have the primary axis vertical to the ground. The secondary axis is then typically normal to the primary axis. Field layouts must consider shading to avoid unnecessary energy losses and to optimize land utilization. Also optimization for dense packing is limited due to the nature of the shading over the course of a year.

This mount is used as a large telescope mount owing to its structure and dimensions. One axis is a vertical pivot shaft or horizontal ring mount that allows the device to be swung to a compass point. The second axis is a horizontal elevation pivot mounted upon the azimuth platform. By using combinations of the two axes, any location in the upward hemisphere may be pointed. Such systems may be operated under computer control according to the expected solar orientation, or may use a tracking sensor to control motor drives that orient the panels toward the sun. The orientation of the module with respect to the tracker axis is important when modeling performance. Dual-axis trackers typically have modules oriented parallel to the secondary axis of rotation.

For all practical purposes, azimuth-altitude (x-y) trackers are used, or considered for use, in larger installations, and especially when applied to high concentration photovoltaic (HCPV) equipment. They are very precise, thus allowing maximum exposure of the CPV modules to direct sunlight, which results in maximum power generation. They are, however, too expensive (initial and maintenance costs) to be considered for regular PV modules applications.

CPV optics require tracking accuracy increase as the system's concentration ratio increases. In typical high concentration systems tracking accuracy must be in the ± 0.1° range to deliver approximately 90% of the rated power output. In low concentration systems, tracking accuracy must be in the ± 2.0° range to deliver 90% of the rated power output. As a result, complex, expensive high-accuracy tracking systems are typically used.

Figure 5-23. x-y (dual) two-axis tracker

Backtracking

With backtracking*, the modules can be packed without shading perpendicular to their axis of rotation at any density. However, the packing parallel to their axis of rotation is limited by the tilt angle and the latitude.

At very low solar elevations (i.e. early in the morning and late in the evening) the backtracked azimuthal orientation of the modules is nearly perpendicular to the sun's azimuth. In other words, the modules are not "looking" directly at the sun, but are at an angle towards it, thus allowing sunlight to illuminate all trackers at the expense of increased incident angle and somewhat reduced power output.

Without backtracking, the rows in the back of the field will be seriously shaded, so backtracking opens channels between the trackers, and all modules get some sunlight at that time of day, and all through the day. With increasing solar height, the module azimuth is continuously readjusted; i.e., the difference between the module azimuth and the solar azimuth decreases continuously, but still ensures that the modules on all trackers are fully illuminated. When the solar height angle equals the shadow angle, this difference becomes zero; i.e., the module azimuth equals the solar azimuth, and backtracking is not needed around that time

As an example of fixed vs. tracking PV power generation, actual AC power produced by three 4 kW PV systems: a) fixed mount, b) seasonally adjusted (best fixed tilt), and c) one-axis tracked, was as follows: 4,900 kWh/y, 5,100 kWh/y and 6,470 kWh/y respectively. Obviously, if done right, tracking works best. "If done right" is the key, because if NOT done right, tracking can become a great hindrance. If, for example, a tracker stops, due to electro-mechanical problems, its power generation will drop according to the position in which it stopped. If it stops at an early morning position, the power for the rest of the day will be minimal. If it stops at the noon position, the power generated during the early morning and late afternoon hours will be minimal. And in some cases, it could shade the trackers around it, causing even greater decrease in power generation.

The literature cites tests, where dual-axis trackers are credited with only 6% increase of power output, when compared with single axis trackers for use with PV modules. Thus we do not recommend them for PV module tracking because the power increase is marginal while the maintenance cost is high.

Balance of System

Balance of system (BOS) is a generic term that describes all components needed to install, adjust and control a PV system, except the PV modules. This includes wiring, switches, support racks, inverters, and batteries in off-grid systems. Land, labor and O&M expenses and components are usually not included in the BOS calculations. See Table 5-8.

BOS components are key elements of any PV power plant model. Their particular design, overall efficien-

*NOTE: Backtracking is an important component of tracking systems, for it allows dense packing of the trackers with minimal loss of generated power. Backtracking is simply the difference between the actual solar azimuth and the particular azimuth orientation of the modules. Backtracking angle is also called shift angle, because it "shifts" the modules to a degree, as needed for better operation and packing density.

Table 5-8. BOS estimated cost of materials and labor

BOS Components Considerations and Prices		
Mounting Structure		
Mounting Structure Cost ($/M2)	G	$55.64
Savings unique to a-Si module	H	
TOTAL $/W	**I=G*B*(1-H)/1000**	**$0.40**
Combiner Box		
Inbound voltage (max)	J	600
No of modules/string	K=J/D	16
No of strings	L=F/K	27
# of poles	M	24
No of combiner boxes required	N=L/M	2
Combiner box cost per piece	O	$1,500
TOTAL $/W	**P=N*O/100kW**	**$0.03**
Cabling Cost		
Total cable required (meters)	Q	260.66
Cabling cost/m	R	$2.00
TOTAL COST $/W	**S=B*Q*R/1000**	**$0.01**
Inverter		
TOTAL COST $/W	**T**	**$0.49**
Installation Cost		
Installation hours per m2	U	0.50
Installation labor $/hour	V	$50.00
TOTAL COST $/W	**W=B*U*V/1000**	**$0.18**
Indirect Cost		
Engineering & Design	X	$25,000
TOTAL COST $/W	**Y=X/100kW**	**$0.25**
GRAND TOTAL $/W	Z=**I+P+S+T+W+Y** =	**$1.36/W***

*Note: Using one-axis trackers would add $0.35-0.45/Wp to the BOS estimate. The tracking PV modules would then produce 15-30% more power for the same active area, depending on weather conditions.

cy quality and actual field performance will determine the plant's success rate.

NOTE: We have not discussed "success rates" yet, because it is not a standard measurement unit, but in the framework of this writing success is determined by:

- Quality of location (sunshine availability and weather conditions)
- Quality of components; low vs. high quality manufacturers
- Quality of design, installation and O&M teams,
- Luck (lack of natural and man-made disasters)

Wiring

Wiring is part of the BOS, but because of its importance and effect on the installation's performance it deserves a closer look. There are miles of small- and large-diameter (gauge) wires in a large-scale power plant, so choosing the proper wire routing, and the type and size of the wires at the different legs of the installation is a major and very important undertaking.

Residential wiring is usually #10-, 12-, or 14-gauge standard AC wire. For the small inverters, 800 watts or less, #16 can be used, but the mechanical strength of small wire leaves much to be desired. Commercial- and

Table 5-9. Voltage drop vs. wire length and size

AWG/kcmil gauge	Conductor Diameter mm	Ohms per km (Copper conductor)	Voltage drop V at 1400ADC per1km	Voltage drop V at 1400ADC per100m
2000	35.92	0.02	47.6	4.8
1000	25.42	0.03	95.2	9.5
900	24.10	0.04	105.8	10.6
800	22.72	0.04	119.0	11.9
700	21.25	0.05	136.0	13.6
600	19.67	0.06	158.7	15.9
500	17.96	0.07	190.4	19.0
400	16.06	0.09	238.0	23.8
300	13.91	0.11	317.4	31.7
250	12.70	0.14	380.9	38.1
4/0	11.68	0.16	450.0	45.0
3/0	10.40	0.20	567.6	56.8
2/0	9.27	0.26	715.4	71.5
1/0	8.25	0.32	902.8	90.3
1	7.35	0.41	1137.9	113.8
2	6.54	0.51	1435.5	143.5

utilities-scale power plants require a number of different wire types and sizes, as specified by the power company and other authorities.

The voltage loss by extra cable length has a big impact on PV power generation, due to the increased resistivity, with the overall effect depending on the length of the wire and its specific resistivity. As a rule of thumb, if there is a 1% voltage loss from extra wire length, then the total power generation efficiency will also decrease by at least 1%. If the temperature of the cables is higher, as might happen in desert operations, then the loss of efficiency would be even greater.

For example, at 1 MW PV output with Vpm ~714V and current 1400 ADC, if cable length is about 100 m with cable size of 2000 kcmil, there would be less than 1% voltage loss overall. However, if the cable length is about 1 km, it is necessary to have more than 7 cables of 2000 kcmil in parallel to maintain the voltage loss within 1%. In reality, multiple cables could be used instead of a large one, since it is only necessary for the total (combined) cross section of the cable to increase proportionally in order to keep the voltage loss to a minimum.

For power generation efficiency and cable cost optimization, the DC cable length should be a maximum of 100 m from PV modules to inverter connect point. The use of cost efficient compact transformer stations that can be placed centrally in quadratic PV fields, results in minimization of cable loss on the DC side and on the AC low-voltage side as cable length between modules, inverter and transformer is minimized.

PV Power Plant Basics

Basically speaking, commercial and utilities PV power generating systems are fields covered with PV modules or other sunlight-to-electricity converting devices. Thus produced power is used by a consumer at the location, or is plugged into the power grid of the local utilities.

There are a number of different ways to produce PV electric power and Figure 5-24 shows several variations.

In this text we are focusing on commercial and large-scale installations which are mostly grid-connected operations. This is simply because it is more convenient to generate power at whichever time possible, send it into the grid for distribution and use wherever needed. The large-scale projects are most efficient and profitable in the desert regions with high solar insolation, which also present some challenges. We address those in this text. The utilities use the generated power, and it is their responsibility to regulate its fluctuations and to ensure constant supply any time of day and night. Win-win situation? Maybe in the long run, but now we have some problems to solve before the utilities

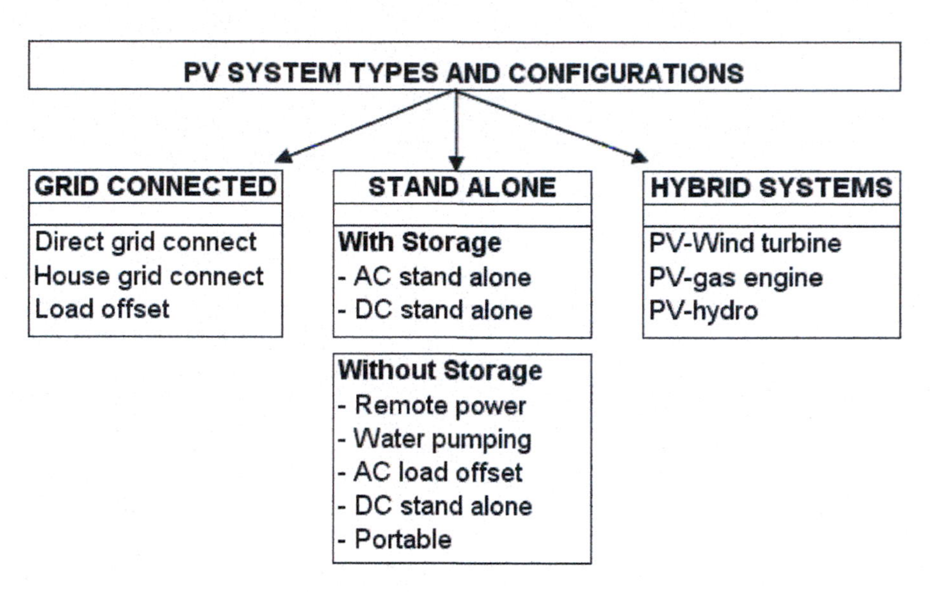

Figure 5-24. Types of PV systems

are totally happy with and willing to own and operate PV power systems.

Utilities are providing and servicing a somewhat variable electric load, where generation and use levels vary up and down, according to the power used at different locations, and which demand the power generators try to match. Adding another variable to this complex task (that of solar power generating plants which in themselves are uncontrollably variable power generators depending on sunshine and cloud covers) complicates the picture even more. Complex or not, PV power plants are here to stay and their number will only grow with time. We just have to make sure that we design, install, operate and use these to the best of our abilities. So let's see how we can do that.

PV Power Plant Planning

The objective of an efficient PV plant design is a) to generate power most efficiently at the lowest cost possible, and b) to obtain a high return on investment by generating maximum power output for the duration. First, this requires the use of PV modules, inverters and medium-voltage transformers with optimum efficiency, and limited wiring loss and losses due to shading and other obstructions, and with proper power generation and plant function monitoring and control. Second, materials and installation costs should be reduced as much as possible, to minimize capital expenditures and maximize ROI. It follows that proper selection of PV modules, inverters and other BOS is essential at the planning and plant design stages. Cheap but inefficient or troublesome equipment will defeat the objectives of efficient and cost effective power production. Complete understanding of the efficiency, durability, cost variables, and making the proper equipment selection will determine the proper and efficient plant operation and return of investment (ROI).

There are a number of design paramters and options and related combinations and permutations that must be evaluated during the PV plant planning and design stages. Understanding the technology and its use under different conditions is paramount to ensuring successful plant design. Key parameters must be considered such as management, location and actual field design. Let's take a look at some of these.

Management Team

The starting point and foundation of a properly and efficiently designed, installed and operated PV plant is its management team. Different PV plants have different management structure, but in all cases there is a chief executive officer (CEO), or general manager (GM), who is basically responsible for the selection of the different managers and other officials and employees, and for the overall coordination of the activities at hand, for the duration.

VPs of technology, finance and operations are part of the top management team, and report directly to the CEO. A number of different project managers, consultants, contractors and employees fill the ranks during planning, design, and installation, as well as during the operation and maintenance stages of the plant's lifetime.

The top management team is responsible for all aspects of the operation from cradle to grave, and its decisions have a major impact on the plant's proper and efficient implementation and operation. Thorough under-

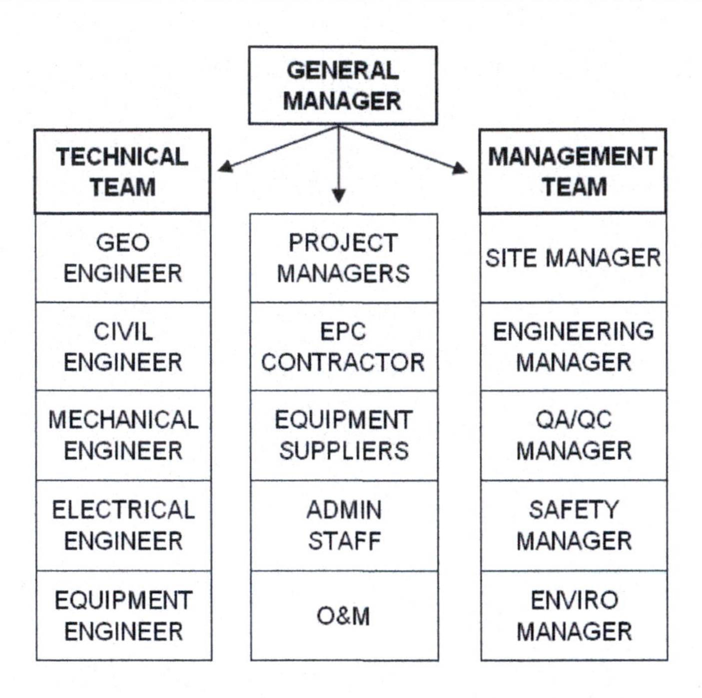

Figure 5-25. PV plant management structure

standing of the respective disciplines and related hands-on experience, as well as following established industry methods and procedures are essential in guiding the team through the different stages and successfully completing the tasks at hand. Figure 5-25 outlines a strong management structure.

Implementation and operation of a properly designed and efficiently operating PV plant requires highly capable and experienced team members functioning as efficiently and professionally as possible. The team is similar in structure and function to that of a hi-tech company working on a large project. Bringing the project to a successful end requires efficient daily coordination and communication among the managers, contractors, consultants and all other parties involved in the project during all stages.

With some variations, the team is usually led by a CEO who is in charge of all aspects of the operation from beginning to end. All VPs report directly to him and he reports to the owners and investors.

The day-to-day technical tasks during the design and implementation phases are managed and coordinated by a technical officer (VP of technology or CTO), who keeps track of the progress, meets with the different technical groups, consultants and contractors. In coordination with the CEO, the CTO makes the final decisions on all *technical* tasks and issues for the duration.

The day-to-day financial tasks during the design and implementation phases are managed and coordinated by a financial officer (VP of finance, or CFO), who keeps track of the daily expenses and other activities, and who in coordination with the CEO makes the final decisions on all *financial* tasks and issues for the duration.

The day-to-day operations (during the power plant's operation) are managed and coordinated by an operations manager (VP of operations, or COO), who keeps track of the daily operations and in coordination with the CEO (if CEO's mandate extends into the plant's operation phase), makes decisions and executes all daily *operations* and related tasks for the duration.

The PV power plant project is divided into sub-projects, which are different during each phase of the project. Each of these sub-projects has a lead, or a project manager (PM), who reports directly to one of the VPs, or to the CEO, and is fully responsible for the proper and efficient planning, design and execution of the sub-project and the tasks of which he is in charge. The PM, together with the CEO and the respective VPs and/or other responsible parties (consultants, contractors etc.), makes the decisions and then implements them accordingly on a day-by-day basis.

Location Selection

Designing, deploying and operating a large utility type system is not a textbook event. There are no standards or set-in-concrete procedures to plan, design and execute the numerous tasks at hand. Each PV project is different from technical, administrative, regulatory and financial points of view. Where do we start? Picking the right location is one of the first tasks on the list of responsibilities, because if it is not good enough, we have only ourselves to blame. If the PV plant is at a predetermined location, we will have to do our best to maximize the output by installing the most efficient PV system with highest output for the particular location, and ensure their proper operation for the duration.

In all cases, the planning, design and execution of the project is the responsibility of the designer/installer/investor teams. The responsible team members must be fully aware of the conditions at the site and evaluate all possible combinations and permutations of location, weather, technology, materials and labor, as well as properly use all management tools available to provide the best design and implementation possible. Every single detail should be taken into consideration, analyzed, calculated and entered into the overall formula, thus making proper decisions at every step.

What's in a location? There are a large number of conditions and variables to consider. Sunshine, or no sunshine? When, how long and how intense is the sunshine? Do clouds and fog visit the site frequently? Are there populated centers, commercial enterprises and roads nearby? Are there wild life in the area? Terrain issues? Environmental issues? And many, many more. Making sure that the proposed site has good sunlight during the day is a given. If it is located in a cloudy or foggy climate area, such as in mountains or near the ocean, the management team must calculate the energy input and decide how to proceed under the circumstances. A PV power plant in Portland, Maine, will be different in construction and performance than one in Portland, Oregon, and even more so than one in the Arizona desert. There are many parameters and issues that must be evaluated in a professional manner, and considered thoroughly in order to make sure that the proposed site is reasonably positioned for success, and that there will be no great obstacles in converting enough sunlight into AC power, as needed to maximize the returns. Of course, all other land-related conditions (permitting, regulatory and socio-economic) must be included in the planning and design effort.

In addition to sunlight and other technical characteristics, see the list below of administrative considerations that must be evaluated during the very early stages of the project development (mostly applicable for sites on public land in California) in order to make sure that the location (land) meets the requirements for a PV power generation site:

a. The renewable energy project is proposed to be located on land identified by REAT that is suitable for renewable energy development?
b. The project will not use fresh ground water or surface water for power plant cooling.
c. The appropriate biological resource surveys have been completed using the proper protocols during the appropriate season.
d. A draft biological assessment (BA), if required for the project, has been tentatively approved by FWS, DFG and the appropriate lead agencies.
e. The appropriate cultural resource surveys, assessments, and project impact mitigation measures have been completed following the proper protocols and standards.
f. Ensure that all BLM (if project site is on BLM land) requirements and Resource Management Plans (RMPs) have been addressed and incorporated in the project design, for projects located on BLM managed lands.
g. All the requirements of the local agency jurisdiction have been incorporated into the applications including but not limited to local zoning, general plan policies, land use, traffic, and height restrictions. The project will not be located on lands under a Williamson Act contract, require a zoning change, or General Plan amendment.
h. All of the requirements of the Department of Defense (DOD) and nearby military installations have been addressed and incorporated into a project's design.
i. The project site does not negatively impact ongoing transmission corridor planning.
j. Phase I site assessment ASTM E1527 or other equivalent assessment method deemed acceptable by the appropriate regulatory oversight agency for the project site and linear appurtenances is conducted in order to determine whether there are any environmental concerns.
k. The local fire protection district or if necessary, California Department of Forestry and Fire Protection (CALFIRE, Office of the State Fire Marshall) must be contacted to locate the proposed project site relative to fire hazard severity zones.
l. Soil surveys as needed to identify soil types and the typical silt content of soils in many locations are conducted.
m. Flood and fire zoning is conducted to determine whether the site is located within a Flood or Fire Hazard Zones and/or the development would result in flood or firefighting plain modifications.

This is not an easy or straightforward process, and that is why a knowledgeable and experienced project management team is essential to properly address and efficiently resolve all tasks and issues.

General Tasks

There is a long list of tasks to be completed, mountains of technical, regulatory and legal documents to go through, and many hours sitting in meetings in order to get a power installation up and running.

These tasks must be completed in order to deploy a PV installation:

1. General site evaluation, map, orientation, scale, photos,
2. 1st assessment of land; size, technology and annual production
3. Environmental impact study and assessment, regarding restrictions such as:
   a. Integral nature reserve
   b. Landscape conservation area
   c. Negative impact on natural scenery
   d. Priority area for agriculture, leisure, or flooding protection areas.

   NOTE: It is important to know the particularities of the regulation of application in the municipality and in the region.
4. Define and evaluate compensatory measures
5. Site visit:
   a. Evaluation of data received—visual evaluation
   b. Shading and site infrastructure evaluation
5. Field design
6. Bidding procedure for EPC (contractor/installer)
7. Final solar yield study (by external expert
   a. Approach and boundary conditions
   b. Evaluation of the technical concept and the components used
   c. Yield forecast (annual yield, performance ratio, etc.)
   d. Description of methods and calculation programs used
8. Formation of the required legal entity (corporation, JV etc.)
9. Identifying the equity capital
10. Getting the required permits
11. The inscription to the special power generation registry (REPE)
12. Grid connection point evaluation and negotiations
13. Final inscription to the REPE
14. Financing aspects
    a. Pre-construction finance
       i.Land purchase
       ii.Land permitting
       iii.PPA negotiations
    b. Construction finance negotiations
    c. Long-term finance options
15. Signing contracts with stakeholders
    a. Land rent/lease contract
    b. Equipment purchase contract
    c. Installation contract
    d. Performance assurance contracts
    e. Contract for monitoring the plant
    f. Contract for operation and maintenance of the plant
    g. Contract for tax consultant (annual Profit/loss declaration)
    h. Contract for project's technical and administration management
16. Construction phase
    a. Land leveling and preparation
    b. PV equipment installation
17. Finalizing the construction.
    a. Plant performance tests
    b. Plant acceptance procedures*
18. Inauguration of the plant
19. Power sale agreements take place
20. Plant O&M procedures take place

PV System Design Considerations

A number of technical factors and variables must be considered in the proper design of an efficient PV power generator. The performance and efficiency of the PV components, or of individual PV modules affecting the array performance must be well understood and used in the design effort. Some of these are:

Reflection

Not all the photons which hit the cell and module can actually pass through it, because some of them are

---

*NOTE: If the final inspection and acceptance tests are within spec, the power metering system is set to zero. From then onwards the produced electricity will be sold to the utility company as number of kW per day. This process is the same for small or large plants, but the length, magnitude and expense of the separate steps are proportional to the size.

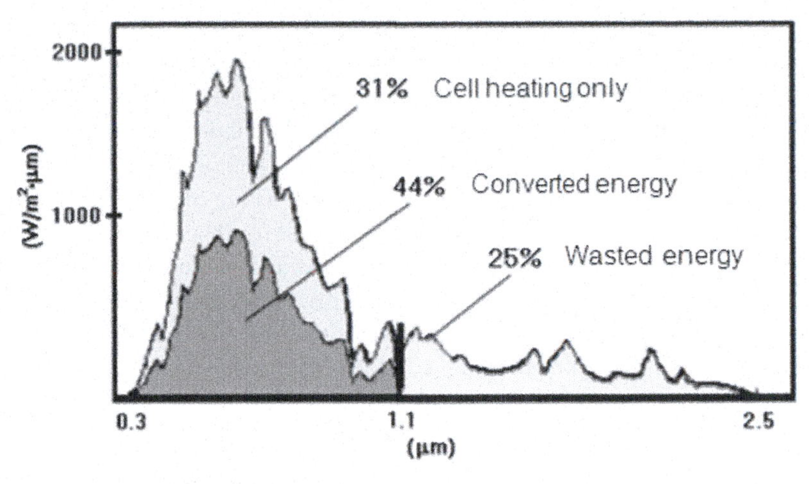

Figure 5-26. Solar conversion by PV cells

reflected by the cell AR coating and the module glass surface. Some of the sunlight hits the contact's metallic grid, and is also reflected. Reflection is different during different times depending on the location, technology used, time of day and the sun's intensity, angle and other climatic conditions.

Photons Energetic Levels

Not all incoming photons have enough energy to break the force holding the electrons into the electron-hole pair. Also, there are also some photons which bring too much energy, which could actually generate electron-hole couples, with the exceeding energy being converted into heat, which complicates the electron generation further. See Figure 5-26.

Recombination

Not all the electron-hole couples are collected by the junction electric field and then sent to the external load. Instead, they recombine with opposite charges encountered along their path from the generation point toward the junction and beyond. There are a number of reasons for this phenomena and these must be investigates in light of the technology choice, location and sunlight quality.

Parasitic Resistances:

Charges generated and collected into the depletion region must be sent into the outside load of the cell and module. The collection is performed by the metallic contacts, placed on both the front and rear sides of the cell. The manufacturing process creates an alloy between the silicon cell and the aluminum metal of the contacts and with it some interface resistance in the boundaries between the layers. This parasitic resistance causes heat dissipation, and further reduction of output, due to the resistance met by the electrons at the boundary. This condition is even more pronounced in amorphous silicon cells, because of the increased resistance, due to the random orientation of the crystals and atoms.

Temperature Coefficient

When the cell temperature gets higher than the standard test conditions (STC) of 25°C, its efficiency decreases around 0.5% per degree C temperature increase over STC. This phenomenon occurs because the conversion of the solar radiation into electrical energy by the PV cells is limited to only a part of the total radiation spectrum. The unused radiation is converted into heat which increases the cell temperature and reduces its output due to serial and shunt resistance effects.

Annual Degradation

A further problem of current PV technology is associated with the decrease of c-Si and thin film PV modules' performance over time. Generally speaking, the output of a PV module is reduced by ~0.5% to 1.5% per year. This phenomenon is mainly due to thermic variations among the different points of the cell, which could cause micro fractures in different regions, reducing the mobility of free electrons. Such micro fractures generate a resistance for charges and reduce the electrons' capability to move freely in the bulk and reach the electrode for extraction into the external load.

Another reason for such persistent degradation is normal changes in the silicon material and the diffusion layer properties. Also, changes in the encapsulating materials (EVA yellowing) could be expected, decreasing optical transmission of the material and interferring with incoming photons.

Permanent Failures

A major reason for permanent failure of solar cells and modules is the creation of micro-fractures in the bulk material as time goes by. The cracks are usually caused by torsions and flexions of the envelope materials (frame and encapsulants) due to mishandling, wind, and atmospheric agents. Wide temperature variations from extreme heat to below-zero freezing play a major role in the destructive processes as well. The worst irreversible effect, however, is the penetration of the elements into the module interior via defects in the encapsulation. The cells undergo chemical attacks in such cases and only time separates them and the module from an untimely and complete failure. Intimate contact and strong bonds between cells and the support materials, as well as reliable protection from the elements, are highly recommended, so careful design, manufacturing, installation and operating procedures must be followed to optimize performance and limit permanent failure rates.

PV System Performance Considerations

There are a number of factors to consider, estimate, calculate and implement in the design, installation and operation of a sound PV power plant. Some of them follow.

Latitude and Longitude

This is a critical factor, because it determines not only the geographic location of the plant, but everything that is related to it, such as climate variations, obstructions, and air pollution; i.e., a solar power plant was installed some time ago in a seemingly good location, just to discover after a while that heavy dust from a nearby cement factory settled onto the solar equipment, reducing and even shutting down its energy production.

Tilt Angle

The tilt angle is usually set at the latitude +/– 5 degrees, but variations of this rule should be considered for special effects, such as optimizing performance during certain parts of the day or season.

Seasonal angle adjustment makes sense under some conditions and for some types of modules but, after factoring labor into the equation, the difference might be negligible or even negative. One point in favor of seasonal adjustment is that it can be combined with the annual evaluation of field structure, PV equipment, inverters, wiring, etc., though this is seldom performed.

Tracking

To track or not to track? This is the question that many design engineers must answer at some point during the design of any PV power plant. Trackers, like seasonally adjustable tilt, come at a price of both initial capital investment and O&M later on. Is it worth spending additional money and effort to gain several additional MWh? The calculations are not straightforward, due to the climate changes, so in many cases it is a gamble.

As a rule of thumb; c-Si modules should be tracker mounted in desert regions only, while tracking with thin film modules cannot be justified under most conditions, due to the already high BOS cost.

Solar Radiation Levels

Radiation levels are directly proportional to the amount of energy generated by the power plant. A number of factors are to be considered, in order to optimize power output, but it starts with bright sunshine falling on the PV modules. Without it, the power plant is like a car running on empty. This is why the desert areas are the preferred large-scale power plant locations. They are where radiation levels are the highest, and where unobstructed direct beam radiation is most likely to be found.

Weather Conditions

Weather conditions at the particular location are the greatest unknown that faces power field designers. Climate variations can make the difference between a profitable and not-so-profitable PV power installation. Excess clouds, fog, smog, dust, strong winds, hail, heavy snow, etc. can bankrupt a PV plant, simply because the power output would be drastically reduced and, worse yet, the equipment itself might be damaged.

Historical weather data might help in the estimates, but no one can provide a precise long-term weather forecast, a science made even more difficult by the unpredictable world climate changes of late. Yet, there are regions that are more predictable than others, which bring us back to recommending desert areas as most suitable for large-scale power generation.

Temperature

Desert areas are good for power generation, but have their own serious problems. PV modules, installed in areas with normal temperature regimens will usually provide efficient and trouble-free operation for many years. This is not the case, however, for c-Si or thin film modules installed in desert areas. The extreme heat will add stress to the active structure and its laminating envelope, which will break first under the strong UV and IR radiation. This will allow water and gasses to enter the modules and disintegrate the active structure with time.

Strong sand storms and monsoon weather conditions are challenging to any type of PV design. These abnormalities must be taken into consideration when considering PV operation in open desert areas.

Humidity

Similarly so, c-Si and thin film modules installed in areas of high humidity (over 60%) of long duration, such as the several weeks of monsoon conditions in the Arizona desert, will suffer premature power degradation and failures, because the moisture could penetrate the laminate layer and attack the active PV structure, chemically disintegrating it with time.

Shading

The degree of shading is inversely proportional to the power output. Shades can come in the form of trees, bushes, or improperly installed rows of PV modules. Shading could cause the creation of "hot spots," which will reduce the output and might even damage the affected cells and modules.

As a rule of thumb, PV systems must be clear of any surface obstructions. Short of that, a thorough analysis of the shading type and level, including any future changes, must be performed and taken into consideration in the final design. In some cases, the manufacturer's warranty might be void if shading or surface obstructions cause any malfunctions or failures.

Load Mismatch

A common problem in poor performance or failure of PV systems is load mismatch due to poor design and/or installation, or equipment defects or failures. Cells, modules, wiring, inverters, stringing, combiner boxes etc. need to be precisely matched on paper and in the actual installation. Load mismatch can account for up to 20% power loss, and is often cited as a major cause for equipment failures.

Inverter Efficiency

Inverters have gotten very efficient and reliable lately—if properly installed and operated. Just like PV modules, however, they suffer from exposure to the elements (excess heat and moisture). Their efficiency will drop proportionally with temperature increases, and if not properly cooled most inverters will just simply stop working at exposure to extreme temperature. Equipment failures have been documented under such extreme conditions, as well.

Capacity Factor

A measure of performance of a power plant commonly used by utilities is its capacity factor. Another useful term is the annual kilowatt-hour produced per installed peak watt. These terms relate to each other as follows:

$$\text{Capacity Factor} = \frac{\text{Annual kWh/Wp}}{\text{\# of Hours/Year}}$$

The capacity factor for conventional systems is determined by utility system economic dispatch and unit availability on the following equation:

Generated Energy = (Insolation) x
(Conversion Efficiency) x (Time Interval)] – (Losses)

Where:

Insolation = the daily average kWh solar energy of the particular location, and

# of Hours/Year = the total of hours the system has operated under full insolation

Losses = anything that prevents the system from generating its nameplate power output

In-row Shadowing Losses

Shadowing of the panels in one row by those of the row in front, occurs primarily in the early morning and late afternoon hours and is most obvious during winter months, due to the sharp angle of the sun's rays. Its occurrence and extent are functions of the latitude, time of year, time of day, panel tilt, panel slant height, row-to-row spacing and east/west orientation. Shadowing could be completely avoided by careful module placement in relation to surrounding rows and objects, and by increasing the row-to-row spacing. Row spacing, however, increases land use, wire costs and losses, and overall capital cost.

Wire Losses

In calculating wire losses, we need to consider the current in each wire resulting from the insolation, as reduced by the shadowing losses. Both power and energy losses are calculated. To calculate the energy loss, it is necessary to know the time (in hours) that a given level of current flows. These data can be calculated and verified after the power plant is in operation. The power level is sensed and sorbed into power density bands from 0 to 11mW/cm in unit steps, during every 30-minute power output measurement, and stored in a file. When a full year's calculation is completed, the data stored for each power density band are multiplied by half an hour and divided by the level of the power band. Having derived these data, calculating the total wire energy losses becomes routine.

Misc. Losses

Other losses that need to be calculated as lost time include periodic maintenance and emergency repairs, wind, lightning and hail damages, etc. unforeseen events, which might force partial or complete plant shutdown.

Commercial PV Installations, Case Studies

Recently thousands, if not millions, of residential and small commercial PV installations have popped-up on roofs and empty lots around the country, and even many more in Europe and Asia. While most of them are well designed, and properly constructed and operated facilities, there are exceptions, which deserve a closer look. We don't have a complete set of data and information on the examples below, so we can only guess some of the details. Nevertheless, they all add to the wealth of our experience, so we consider it a worthwhile exercise.

Figure 5-27 shows an ambitious commercial installation of approximately 35 kW on an apartment house in California, consisting of both vertically and horizontally mounted panels. The vertical panels face southwest and do not receive direct sunlight until late each morning. Moreover, neither the vertical panels nor the horizontal panels at the right are tilted toward the south at the angle of latitude, thus their efficiency is cut in half or more. In addition, the shadows cast by the three palm trees and the eucalyptus tree (right) for the better part of the day almost certainly will have an attenuating effect on the energy output. How much effect would be a function of the internal series/parallel circuitry but could be determined with a simulated equivalent unshaded system.

There does seem to be a cleaning schedule in place judging from the crystalline appearance of the panels' surfaces, so this is a good thing.

The California Department of Transportation building in Los Angeles (Figure 5-28) has a system of panels sandwiched in a casing of bullet-proof glass on the south face, but each rank of panels shadows the one below. Moreover, there is no apparent cleaning schedule for the glass surfaces.

If one could depend on frequent inundations blowing from the south, then these panels would be periodically cleaned, but that kind of weather doesn't happen in southern California. We have lengthy periods without rain and when the storms do come they're more often in the form of vertical drizzles which will very definitely clean the uppermost rank of panels but do little good for the ones below, and may even make the surface worse by depositing more dust and spots on it.

A system consisting of 3,872 3080-watt panels (Schott ASE-300-DGF/50) yielding a rated power output of 1,162 kilowatts was recently installed on the campus of CSU Fresno over Parking Lot V (Figure 5-29). The general contractor for this installation was Chevron Energy Solutions. The owner of the panels is MMA Renewable Ventures with which the campus has entered into a

Figure 5-27. Excessive module shading

Figure 5-28. Self-shading of modules

Figure 5-29. Example of a good installation

20-year power purchase agreement at a starting rate of $0.16 per kilowatt hour and a 2% annual inflation adjustment. An examination of current rates paid by big users of electricity makes an average rate of $0.16/kWh appear to be a bit pricey.

The installation is near a busy road, so it would get dusty quickly. Since it doesn't rain often in this area, the modules need periodic cleaning which might be costly and create a mess at the location.

The Los Angeles Convention Center has a PV system which was installed by the L.A. Department of Water and Power (Figure 5-30). The panels were placed around the periphery of the building well below the roof line.

Figure 5-30. Modules shading by building walls

The panels which are mounted on the east and west sides receive no direct sunlight for about half of each day. The ones mounted on the west side and shown in the photograph are in the shade until early afternoon. We are willing to bet that their ROI is not that great. Not to mention that the whole thing looks kind of funny and out of place.

These spots appear on PV modules on another California installation (Figure 5-31). We are not sure what caused them, but whatever it was it very likely caused some damage to the cells too, and these modules will have reduced output and might fail with time too.

Figure 5-31. Module damage

PV POWER VARIABILITY

There are a number of considerations to take into account when designing and operating a PV system—be it a small commercial, or large-scale utilities power plant.

PV Plant Output Variability

Generally speaking, solar (PV) power plants are aninconsistent (variable) source of power, since they depend heavily on weather conditions (sun intensity and angle, cloud or fog cover, rain, snow), and electro-mechanical malfunctions. These variables determine the quantity and quality of the electric power these plants produce, and must be taken into consideration in all phases of the plants' design, installation and operation.

There is a defined relationship between sunlight intensity and frequency (clouds) vs. total or instantaneous PV output, but it is a complex variable of the following components: spatial-temporal effects (clouds, fog, smog, shadows etc.), inverter effects, incident angle effects and temperature effects.

Taking a closer look at the variables we see:

a. Special-temporal effects are short periods of non-linear sunlight intensity changes, most likely due to fast moving small clouds, casting small shadows onto the PV field. This affects the power output in a number of ways, all of which reduce it to an extent. Inverters are affected in a serious way by these phenomena.
b. Inverter effects are expressed in reduced and inconsistent output, due to the "clipping" of the ups and downs of the power output from the field, as driven by the cloud movement.
c. Incident angle effects are variables that depend on the tilt angle of the modules and time of day. Tracking systems harvest more energy, but are more sensitive to passing clouds.
d. Temperature increases negatively affect power output of PV arrays. Approximately 0.5% decrease of power from the PV modules is expected by each degree C increase of temperature. Similar decrease of efficiency is expected from inverters as well, when the air temperature exceeds 120°F and approaches 0 at temperatures over 150°F. Yes, the inverters overheat and even stop working at such high temperatures, unless special precautions, such as air-conditioning, are taken. This are expensive, but unavoidable precautions when operating in harsh desert climates.

PV Plant Variability as Function of Load and Climate Resources

Considerable differences exist in the technical characteristics from one form of solar technology to another and from one location to another. One important characteristic shared by all types of solar power is their diurnal and seasonal pattern (i.e. peak output usually occurs in the middle of the day and in the summer). This is an important characteristic as it is well correlated with the peak demand of many power systems.

It is obvious, from Figure 5-32a that a bright sunny day will consistently produce the amount of energy we expect from the power field; i.e., the power field will be at its peak performance. The contrary is true during a partially cloudy day, during which the generated power will vary according to the amount of sunlight the power field receives at any given moment (Figure 5-32b).

Another characteristic of solar energy systems is that their output may be complementary to the output of wind generators. Solar power is often produced during the peak load hours (mid-day) when wind energy production may not be available. Figure 5-33 illustrates this phenomenon and compares the average demand with the aggregate wind and solar plant output in California.

Variability around these average demand values for individual wind and solar resources can fluctuate significantly on a daily basis. However, as illustrated, the solar and wind plant profiles when considered in aggregate can be a good match to the load profile and hence improve the resulting composite capacity value for variable generation. There are other factors to consider too—wind speed, air temperature, modules surface contamination etc., but with everything else kept constant, the sunlight intensity is the final determining factor of the variability

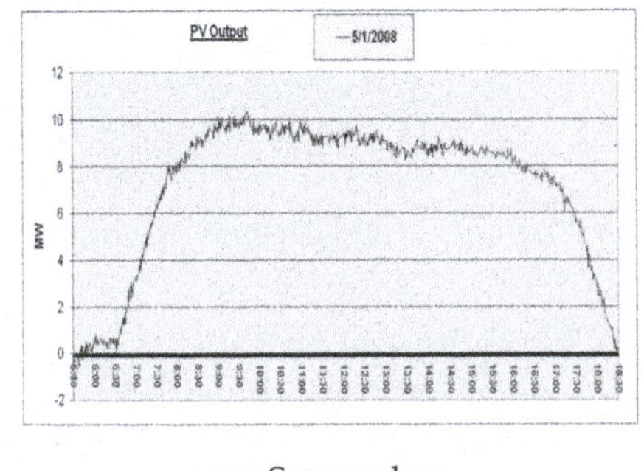

a. Sunny day

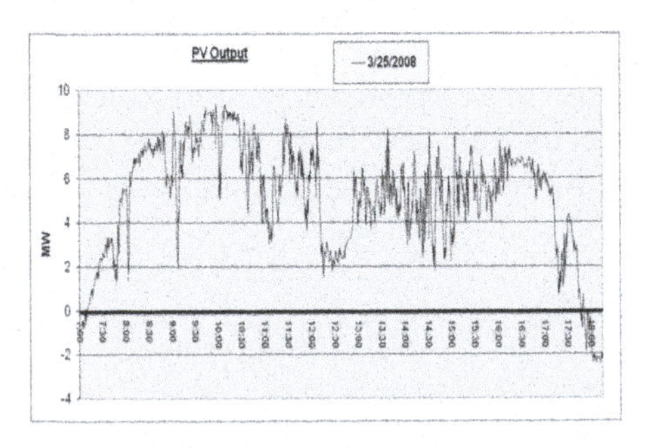

b. Partially cloudy day

Figure 5-32. PV plant output variability from day to day (17)

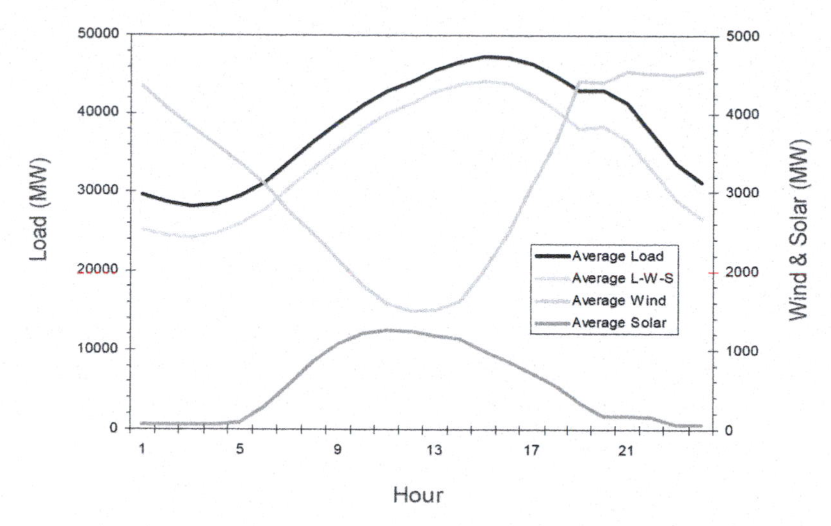

Figure 5-33. Variability and demand dependence (17)

and the amount of power generated.

Large photovoltaic (PV) plants, as have been proposed in the southwestern U.S. and Southern California, have the potential to place extremely fast ramping resources on the power distribution system. Under certain weather conditions, PV installations can change output by +/– 70% in a time frame of two to ten minutes, many times per day. Therefore, since this variable load is hard to manage, these power plants should consider incorporating the ability to manage ramp rates and/or curtail power output as needed by the power network. This is something that utility companies are looking into and which eventually should be implemented on a national level.

Conclusions

- Variability of PV power output is not a simple linear function of variability in plane-of-array point irradiance, especially on partly cloudy days.
- Preliminary results suggest that >10 min variability of multi-MW PV plants can be approximated by the variability of point irradiance averaged over a similar time window.
- Short- term (<10 min) variability is influenced by the size of the plant, with variability decreasing with increasing size.

Energy Storage

Energy storage is one way to deal with variability. Energy stored via electrical or mechanical means could be released in case of reduced solar radiation or no-wind conditions. We will take a closer look at energy storage options in the next chapter, but would like to overemphasize here the fact that the future of PV and wind energy generation depends on efficient energy storage for periods of time without sunshine, and after hours. There are a number of ways to store energy for later use, but all of them are expensive and under developed at this stage.

Presently there are more than 2,100 MW of energy storage sites in operation as follow:

1. Thermal energy storage capacity is over 1,140 MW
2. Battery energy storage is over 450 MW
3. Compressed air energy storage is over 440 MW
4. Flywheel energy storage is over 80 MW

We do expect that further development of these and many other energy storage technologies will take care of the variability of solar energy generation, thus making it more competitive with conventional energy sources.

POWER CONTROL

Controlling power is essential for the safe, efficient and profitable delivery and use of electricity by customers connected to the national power grid. The utilities

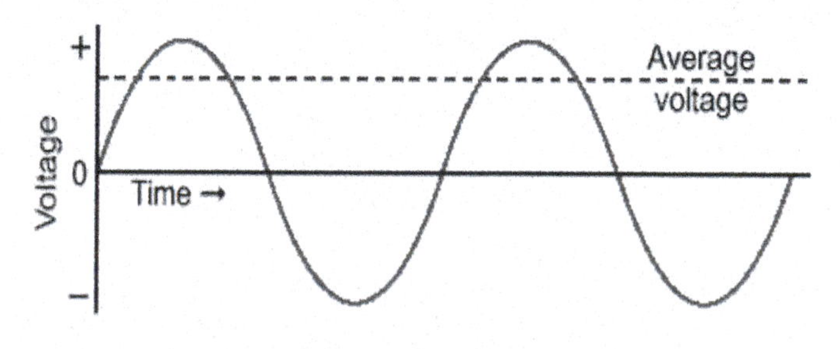

Figure 5-34. Alternating current

are responsible for ensuring the quality and safety of the grid components and the related power use, but the solar power generators also have their defined responsibilities and accountabilities in making this happen.

Let's take a brief look at the different aspects of power generation and control, keeping in mind that PV plants contribute (negatively or positively) to the quality of the electric power in local electric grids.

AC Power Quality

Power quality is a real concern for today's power grid and the loads it serves. Computer equipment, in particular, is sensitive to power quality problems, and the ubiquity of computers in today's manufacturing environment means high power quality is important to many commercial firms as well as the average homeowner and power user. Alternating current can be illustrated as a sinusoidal wave, as shown in Figure 5-34.

Over time, voltage oscillates between a positive and negative value that is slightly more than the average voltage. Larger deviations are unacceptable, although quite possible under different circumstances. The largest and most harmful deviations are in the form of voltage surge, and are caused by lightning strikes.

Other power quality problems include:

Voltage Sags and Swells

The amplitude of the wave gets momentarily smaller or larger, because of large electrical loads such as motors switching on and off.

Impulse Events

Also called glitches, spikes, or transients, these are events in which the voltage deviates from the curve for a millisecond or two (much shorter than the time for the wave to complete a cycle). Impulse events can be isolated or occur repeatedly and may or may not have a pattern.

Decaying Oscillatory Voltages

The voltage deviation gradually dampens, like a ringing bell. This is caused by banks of capacitors being switched on by the utility.

Commutation Notches

These appear as notches taken out of the voltage wave. They are caused by momentary short circuits in the circuitry that generates the wave.

Harmonic Voltage Waveform Distortions

These occur when voltage waves of a different frequency (some multiple of the standard frequency of 60 cycles per second) are present to such an extent that they distort the shape of the voltage waveform.

Harmonic Voltages

These can also be present at very high frequencies to the extent that they cause equipment to overheat and interfere with the performance of sensitive electronic equipment.

Other power quality problems may also be considered reliability problems because they occur when the transmission system is not capable of meeting the load on the system, such as:

1. Brownouts are a persistent lowering of system voltage caused by too many electric loads on the transmission line.
2. Blackouts are, of course, a complete loss of power. Unanticipated blackouts are caused by equipment failures, such as downed power lines, blown transformers, or a failed relay circuits.
3. "Rolling" blackouts are intentionally imposed upon a transmission grid when the loads exceed the generation capabilities. By blacking out a small sector of the grid for a short time, some of the load on the grid is removed, allowing the grid to continue serving the rest of the customers. To spread the burden among customers, the sector that is blacked out is changed every 15 minutes or so—and hence, the blackouts "roll" through the grid's service area.

PV Plant Power Controls

Controlling the operation of a photovoltaic plant basically means keeping the efficiency of its energy production as high as possible in the course of many years. Moreover, putting in of evidence the amount of produced energy gives the operator and customer the perception of investment worthiness.

Inverters also assist in monitoring and controlling the quality and safety of power going into the grid. This includes all measures required for the protection of people and devices in case of extraordinary occurrences and utility outages. Inverters are also programmed to immediately stop operating in case of such irregular grid conditions and to interrupt feeding the grid.

There are a series of possibilities for inverters to identify utility outage:

a. Voltage monitoring
b. Frequency monitoring
c. Measuring the line resistance (only with the IG fea-

turing ENS)

d. Intended feeding of a slightly modified frequency or voltage

Inverters are also able to carry out the monitoring procedures (applicable in each country) by themselves, without additional electronic measuring devices. This considerably reduces installation work and all related costs.

The best way to monitor the system's operation is through the addition of a remote data acquisition and controls system, such as in Figure 5-35.

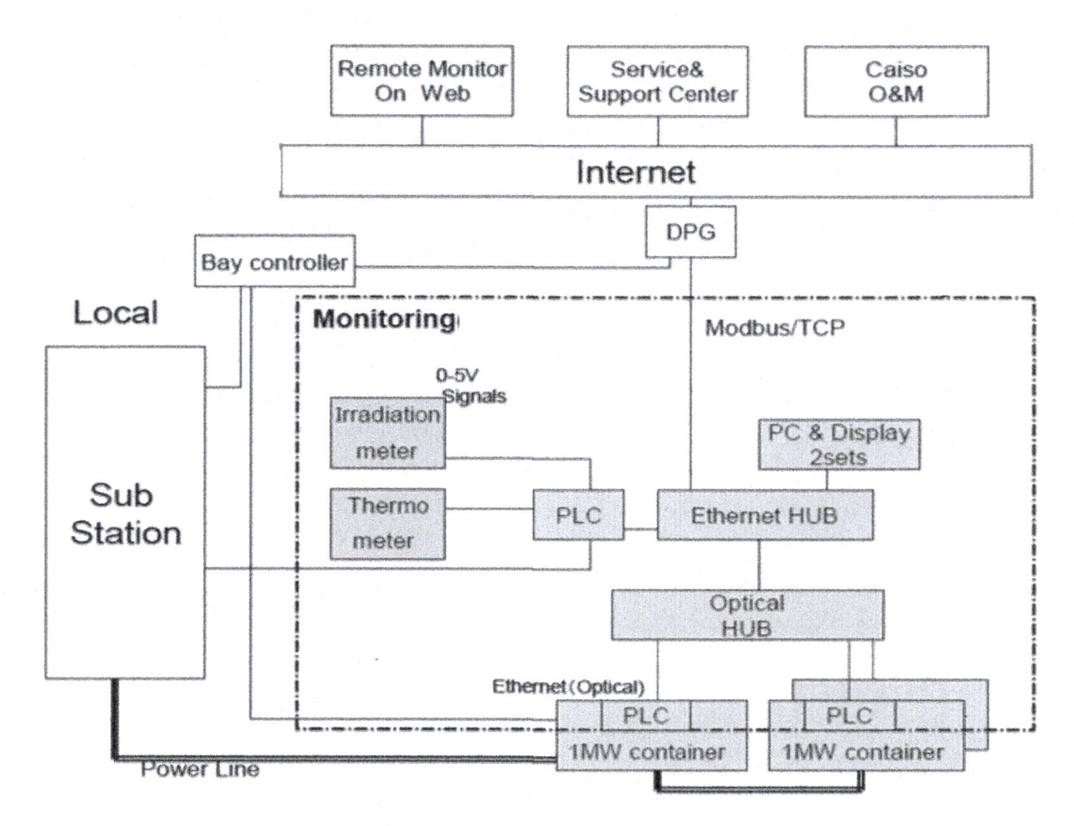

Figure 5-35. Monitoring station

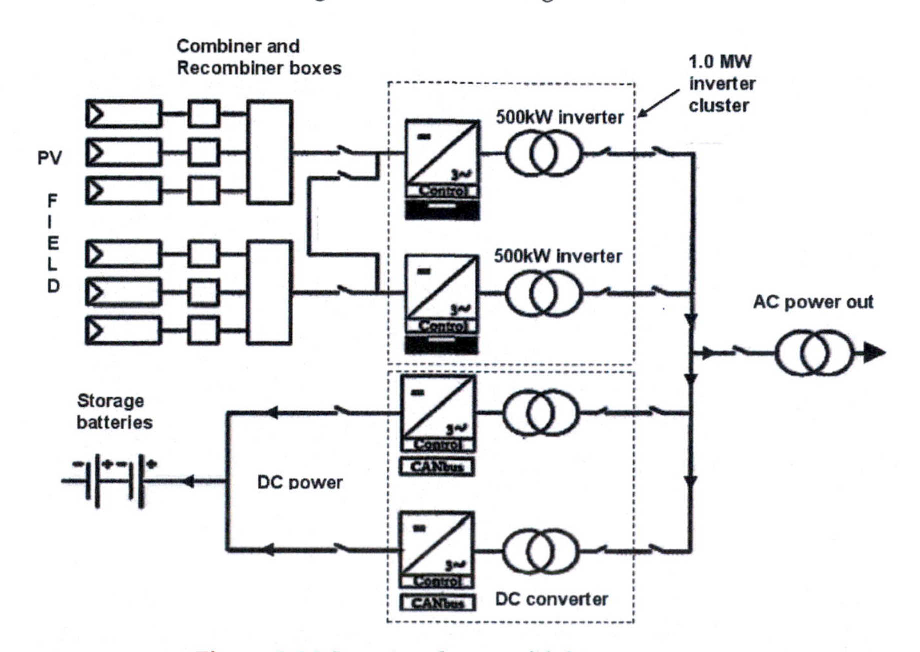

Figure 5-36. Inverter cluster with battery storage

The schematic represents a simplified power control system of a 2-MWp system (group of inverters totaling 1.0 MW per container). Signals from all monitors/sensors and power inputs are fed into the control system, and respective decisions for optimized operation are issued by the PLC to the different control systems. The information can be accessed remotely by internet, where it can be viewed. The automated control center can be bypassed, and optional remote commands can be issued to the system by off-side operators.

Figure 5-36 represents an inverter cluster for battery storage capability. Converters are needed to control the AC to DC power conditioning and battery pack's charge and discharge cycles 24/7. Batteries will provide the solution to badly needed PV energy storage when their prices fall enough.

PV Plant's Remote Controls Based on Satellite Technology

A number of different monitors and controllers are available on the market that allow for remote control of a photovoltaic plant by means of both real time monitoring of solar radiation through satellite and the acquisition of plant production data through a simple data logger installed at the plant.

These controllers are equipped with a modem that sends data about the hourly energy production to the central control station and/or website. Produced energy is sensed locally by a pulse energy counter (S0 standard). Basically, these controllers are capable of retrieving a plant's working status, without any need of interfacing to inverters or other sensors, and then presenting it on the web.

Besides this, the remote controllers are also capable of acquiring data from inverters and then providing PV plant performance analysis and diagnostics. For medium-size and large-size PV plants, the controllers include additional use of irradiance and meteorological sensors to be interfaced with the more sophisticated data loggers. Some of these are also equipped with a graphic display to allow for a check of PV plant data directly at the place. The controller evaluates the hourly producible energy and compares it to the counter's measure of energy produced. It does so by exploiting an accurate opto-electronic model of the PV plant, where plant parameters are those furnished by the installer.

Producible energy is derived from solar irradiance and other parameters like PV module temperature, ambient temperature, wind speed and direction, etc. The computers then present the plant efficiency level in terms of energy production and in terms of profitability, together with plant diagnostics (inverters efficiency, malfunction conditions, etc.). The operator can obtain the plant diagnostics and receive eventual malfunction alarms at anytime he wants, or at the instant these events occur.

These services can always be accessed on the web as well, which can even be customized for the particular use of the operator or customer. Remote controllers allow the detection of possible inefficiencies in energy production and hence the estimation of consequent economic losses. Thanks to satellite measurements, they also provide real time info on alarm conditions, reducing the risk of false alarms.

The main advantages of remote controllers are:

Affordability and Ease of Operation

The service is available also for small plants, so that even the single end user can verify at anytime the value of the investment made. The basic models are cheap and easy to install, requiring no interface with the plant, no connection to the internet, no router configuration. Connecting these to a counter will automatically transmit the data to central control or to a website, via GPRS link.

A Single Web Standard for Different PV Plants

The compatibility with the majority of inverter models ensured by some data loggers, allows management of PV plants of different sizes and features through a unique web portal shared by the various operators.

Reliability with No Maintenance

The monitoring of whole PV plant efficiency is done without the need to install sensors which would require periodic interventions for calibration and maintenance. The satellite data are accurate and their quality is continuously maintained.

Management of PV Plant Maintenance

Maintenance management by the company or the operator can also include the management of eventual plant malfunction alarms, thus suggesting specific interventions of maintenance.

Secure and Diversified Access to Data

Plant diagnostics data are presented on a user-friendly interface, directly accessible by the operator or by the end user through a personal password on the website of the distributor company.

Remote Modules Integration

By integrating remote control modules, the system follows the entire PV plant life cycle, from feasibility check to design, cost estimation, test, monitoring of working status, maintenance. Such approach aims at the highest satisfaction of the end user by means of global quality and high added value services.

Display and Data Communication

The high technical complexity of today's PV power plants requires a perfect display design, which is actually the "interface" with the user. It has to be uncompromisingly designed for easy operation and permanent availability of the system data. Inverters already feature a fundamental logging function for recording data on minimum and maximum values directly on the display—on a daily, yearly, and total basis.

As an optional feature, the plant power control display also allows displaying the following weather-related data:

1. Several different temperature values (e.g. temperature near or in the solar modules, and outside temperature in the shade)
2. Solar insolation
3. Wind velocity
4. Atmospheric pressure
5. Relative atmospheric humidity

In addition to the functions implemented by the inverters, a wide range of elements for data communication available now provide a large number of possibilities for data monitoring, recording and visualization.

Summary

PV technologies offer the most flexible and promising renewable, alternative energy power generation available today.

Large-scale PV power plants will continue to play a major role in bringing us to energy independence, while cleaning the environment at the same time.

The deserts present abundant sunlight radiation, most suitable for generating large, almost unlimited, amounts of electric power in the US and around the world.

PV technologies are still in their infancy, with low efficiency and unresolved reliability and longevity issues when faced with 30 years non-stop on-sun operation in the deserts.

Choosing the proper location and most suitable technologies for large-scale power generation are challenges which we must embrace and work on diligently, keeping in mind our long-term goals.

The solar industry (manufacturing, permitting, interconnections, installation, and O&M) are not standardized for ensuring best PV products and projects quality and efficiency.

Some PV technologies contain non-renewable and even toxic materials, which presents a dilemma as far as their use in large quantities in the deserts. More work is needed to make sure we don't get into irreversible materials shortages, or toxic contamination of large land areas.

Most PV modules are imported from foreign countries, which is not the way to true energy independence. Switching from importing one type of energy source (oil) to another (PV modules) is not the long-term solution we are looking for. Establishing independent, standardized, efficient and well supported (by the government, big business and the scientific community) US-based solar industry is!

Notes and References

1. A Study of Very Large Solar Desert Systems with the Requirements and Benefits to those Nations Having High Solar Irradiation Potential, SENI, 2006
2. New Network Topologies For Large-scale Photovoltaic Systems, G. Carcangi
3. Reprinted with permission, National Renewable Energy Laboratory, http://www.nrel.gov/rredc/pvwatts/changing_parameters.html#dc2ac (July, 2011)
4. 2009 global PV cell and module production analysis, Shyam Mehta, GTM Research
5. Photovoltaic Array Performance Model, D.L. King, Sandia
6. PVCDROM
7. Understanding PV Module Specifications, Justine Sanchez, Jan. 2009.
8. 2009 Utility Solar Rankings I May 2010, SEPA
9. Failure modes of crystalline Si Modules, BP Solar, J. Wohlgemuth, D. Cunningham, A. Nguyen, G. Kelly and D. Amin.
10. Commonly Observed Degradation In Field-Aged Photovoltaic Modules, M. A. Quintana and D. L. King, Sandia National Laboratories, T.J. McMahon and C. R. Osterwald
11. Long-Term Performance And Reliability Assessment Of 8 PV Arrays At Sandia National Laboratories, J.E. Granata, W.E. Boyson, J.A. Kratochvil and M.A. Quintana, Sandia National Laboratories, Albuquerque, NM Photovoltaic Systems Evaluation Laboratory
12. Study of delamination in acceleration tested modules., Dhere Neelkanth and Pendit Mandar.
13. Accelerated Stress Testing of Thin-Film Modules with SnO2:F Transparent Conductors, C.R. Osterwald, T.J. McMahon, J.A. del Cueto, J. Adelstein, and J. Pruett
14. IEC standards and recommendations: http://www.iec.ch/cgi-bin/procgi.pl/www/iecwww.p?wwwlang=e&wwwprog=sea00227.p&progdb=db1&ics=Item.
15. Weather Durability of PV Modules; Developing a

Common Language for Talking About PV Reliability, Kurt Scott, Atlas Material Testing Technology
16. Mora Associates, Internal research
17. Accommodating high levels of variable generation. NERC, April 2009
18. Comparison Of Degradation Rates Of Individual Modules Held At Maximum Power1 C.R. Osterwald, J. Adelstein, J.A. Del Cueto, B. Kroposki, D. Trudell, and T. Moriarty National Renewable Energy Laboratory (NREL), Golden, CO 80401
19. D. King, W. Boyson, and J. Kratochvil, "Analysis of Factors Influencing the Annual Energy Production of Photovoltaic Systems," 29th IEEE PV Specialists Conference, 2002.
20. D. King, J. Kratochvil, and W. Boyson, "Measuring Solar Spectral and Angle-of-Incidence Effects on PV Modules and Solar Irradiance Sensors," 26th IEEE PV Specialists Conference, 1997, pp. 1113-1116.
21. D. King, J. Kratochvil, and W. Boyson, "Temperature Coefficients for PV Modules and Arrays: Measurement Methods, Difficulties, and Results," 26th IEEE PV Specialists Conference, 1997, pp. 1183-1186.
22. D. King and P. Eckert, "Characterizing (Rating) Performance of Large PV Arrays for All Operating Conditions," 25th IEEE PV Specialists Conference, 1996, pp. 1385-1388.
23. D. King, J. Kratochvil, and W. Boyson, "Field Experience with a New Performance Characterization Procedure for Photovoltaic Arrays," 2nd World Conference on PV Solar Energy Conversion, Vienna, 1998, pp. 1947-1952.
24. C. Whitaker, T. Townsend, J. Newmiller, D. King, W. Boyson, J. Kratochvil, D. Collier, and D. Osborn, "Application and Validation of a New PV Performance Characterization Method," 26th IEEE PV Specialists Conference, 1997, pp. 1253-1256.
25. B. Kroposki, W. Marion, D. King, W. Boyson, and J. Kratochvil, "Comparison of Module Performance Characterization Methods," 28th IEEE PV Specialists Conference, 2000, pp. 1407-1411.
26. A.H. Fanney, et al., "Short-Term Characterization of Building Integrated Photovoltaic Modules," Proceedings of Solar Forum 2002, Reno, NV, June 15-19, 2002.
27. D. King, T. Hund, W. Boyson, and J. Kratochvil, "Experimental Optimization of the Performance and Reliability of Stand-Alone Photovoltaic Systems," 29th IEEE PV Specialists Conference, 2002.
28. R. Sullivan and F. Winkelmann, "Validation Studies of the DOE-2 Building Energy Simulation Program—Final Report," Lawrence Berkeley National Laboratory, LBL-42241, June 1998.
29. ASTM E 1036, "Testing Electrical Performance of Nonconcentrator Photovoltaic Modules and Arrays Using Reference Cells," American Society for Testing and Materials.

APPENDIX A

IEC Standards

The International Electrotechnical Commission (IEC) has issued a large data base of guidelines, standards and recommendation for use with modules and systems, and their design, tests, installation and operation. Some of these are:

IEC 60364-7-712 (2002-05)
Electrical installations of buildings—Part 7-712: Requirements for special installations or locations—Solar photovoltaic (PV) power supply systems

IEC 60891 (2009-12)
Photovoltaic devices—Procedures for temperature and irradiance corrections to measured I-V characteristics

IEC 60904-1 (2006-09)
Photovoltaic devices—Part 1: Measurement of photovoltaic current-voltage characteristics

IEC 60904-2 (2007-03)
Photovoltaic devices—Part 2: Requirements for reference solar devices

IEC 60904-3 (2008-04)
Photovoltaic devices—Part 3: Measurement principles for terrestrial photovoltaic (PV) solar devices with reference spectral irradiance data

IEC 60904-4 (2009-06)
Photovoltaic devices—Part 4: Reference solar devices—Procedures for establishing calibration traceability

IEC 60904-5 (1993-10)
Photovoltaic devices—Part 5: Determination of the equivalent cell temperature (ECT) of photovoltaic (PV) devices by the open-circuit voltage method

IEC 60904-7 (2008-11)
Photovoltaic devices—Part 7: Computation of the spectral mismatch correction for measurements of photovoltaic devices

IEC 60904-8 (1998-02)
Photovoltaic devices—Part 8: Measurement of spectral response of a photovoltaic (PV) device

IEC 60904-9 (2007-10)
Photovoltaic devices—Part 9: Solar simulator performance requirements

IEC 60904-10 (2009-12)
Photovoltaic devices—Part 10: Methods of linearity measurement

IEC 61194 (1992-12)
Characteristic parameters of stand-alone photovoltaic (PV) systems

IEC 61215 (2005-04)
Crystalline silicon terrestrial photovoltaic (PV) modules—Design qualification and type approval

IEC 61345 (1998-02)
UV test for photovoltaic (PV) modules

IEC 61427 (2005-05)

Secondary cells and batteries for photovoltaic energy systems (PVES)—General

IEC 61646 (2008-05)

Thin-film terrestrial photovoltaic (PV) modules—Design qualification and type approval

IEC 61683 (1999-11)

Photovoltaic systems—Power conditioners—Procedure for measuring efficiency

IEC 61701 (1995-03)

Salt mist corrosion testing of photovoltaic (PV) modules

IEC 61702 (1995-03)

Rating of direct coupled photovoltaic (PV) pumping systems

IEC 61725 (1997-05)

Analytical expression for daily solar profiles

IEC 61727 (2004-12)

Photovoltaic (PV) systems—Characteristics of the utility interface

IEC 61730-1 (2004-10)

Photovoltaic (PV) module safety qualification—Part 1: Requirements for construction

IEC 61730-2 (2004-10)

Photovoltaic (PV) module safety qualification—Part 2: Requirements for testing

IEC 61829 (1995-03)

Crystalline silicon photovoltaic (PV) array—On-site measurement of I-V characteristics

IEC/TS 61836 (2007-12)

Solar photovoltaic energy systems—Terms, definitions, symbols

IEC 62093 (2005-03)

Balance-of-system components for photovoltaic systems—Design qualification natural environments

IEC 62108 (2007-12)

Concentrator photovoltaic (CPV) modules and assemblies—Design qualification and type approval

IEC 62109-1 (2010-04)

Safety of power converters for use in photovoltaic power systems—Part 1: General requirements

IEC 62116 (2008-09)

Test procedure of islanding prevention measures for utility-interconnected photovoltaic inverters

IEC 62124 (2004-10)

Photovoltaic (PV) stand alone systems—Design verification

IEC/TS 62257-1 (2003-08)

Recommendations for small renewable energy and hybrid systems for rural electrification—Part 1: General introduction to rural electrification

IEC/TS 62257-2 (2004-05)

Recommendations for small renewable energy and hybrid systems for rural electrification—Part 2: From requirements to a range of electrification systems

IEC/TS 62257-3 (2004-11)

Recommendations for small renewable energy and hybrid systems for rural electrification—Part 3: Project development and management

IEC/TS 62257-4 (2005-07)

Recommendations for small renewable energy and hybrid systems for rural electrification—Part 4: System selection and design

IEC/TS 62257-5 (2005-07)

Recommendations for small renewable energy and hybrid systems for rural electrification—Part 5: Protection against electrical hazards

IEC/TS 62257-6 (2005-06)

Recommendations for small renewable energy and hybrid systems for rural electrification—Part 6: Acceptance, operation, maintenance and replacement

IEC/TS 62257-7 (2008-04)

Recommendations for small renewable energy and hybrid systems for rural electrification—Part 7: Generators

IEC/TS 62257-7-1 (2006-12)

Recommendations for small renewable energy and hybrid systems for rural electrification—Part 7-1: Generators—Photovoltaic arrays

IEC/TS 62257-7-3 (2008-04)

Recommendations for small renewable energy and hybrid systems for rural electrification—Part 7-3: Generator set—Selection of generator sets for rural electrification systems

IEC/TS 62257-8-1 (2007-06)

Recommendations for small renewable energy and hybrid systems for rural electrification—Part 8-1: Selection of batteries and battery management systems for stand-alone electrification systems—Specific case of automotive flooded lead-acid batteries available in developing countries

IEC/TS 62257-9-1 (2008-09)

Recommendations for small renewable energy and hybrid systems for rural electrification—Part 9-1: Micropower systems

IEC/TS 62257-9-2 (2006-10)

Recommendations for small renewable energy and hybrid systems for rural electrification—Part 9-2: Microgrids

IEC/TS 62257-9-3 (2006-10)

Recommendations for small renewable energy and hybrid systems for rural electrification—Part 9-3: Integrated system—User interface

IEC/TS 62257-9-4 (2006-10)

Recommendations for small renewable energy and hybrid systems for rural electrification—Part 9-4: Integrated system—User installation

IEC/TS 62257-9-5 (2007-06)

Recommendations for small renewable energy and hybrid systems for rural electrification—Part 9-5: Integrated system—Selection of portable PV lanterns for rural electrification projects

IEC/TS 62257-9-6 (2008-09)

Recommendations for small renewable energy and hybrid systems for rural electrification—Part 9-6: Integrated system—Selection of Photovoltaic Individual Electrification Systems (PV-IES)

IEC/TS 62257-12-1 (2007-06)

Recommendations for small renewable energy and hybrid systems for rural electrification—Part 12-1: Selection of self-ballasted lamps (CFL) for rural electrification systems and recommendations for household lighting equipment

IEC 62446 (2009-05)

Grid connected photovoltaic systems—Minimum requirements for system documentation, commissioning tests and inspection.

APPENDIX B

Inverter Specs
1.0MW inverter station for use in CA, consisting of 4 pcs. 250kWp inverters.

1. FUNCTIONAL SPECIFICATIONS	
Outdoor 1MW Substation Type (250kW Inverter x 4 pcs.)	
DC INPUT	
Maximum Voltage	600V
Maximum DC Power	1000KW
Rated Voltage PV Field	12MWp
MPPT Operating Range	450V – 850V
Number of Inputs	10 inputs per 250kW inverter
AC Output	
Nominal AC Voltage	480V
Nominal AC Frequency	50/60Hz
Power Factor	>0.99 (Rated Power)
Harmonic Distortion AC current	<3% (Rated Power)
Maximum AC Power	1000KW
Maximum Efficiency	98.4%
European Efficiency	98%
2. GRID INTEGRATION	
Grid Interconnection Requirement	Yes
Low Voltage Ride Through Requirement	Utility / CAISO specific
Slew Rate Control	Utility / CAISO specific
Real Power Control	Utility / CAISO specific
Dynamic Power Factor Adjustment	Utility / CAISO specific
Dynamic Power Factor Response Time	Utility / CAISO specific
Remote Start, Stop, Error Reset	Yes
Off Nominal Frequency Ride Through	Utility / CAISO specific
3. OTHER REQUIREMENTS	
Listing Requirement	UL
Noise Level Requirement	TBD
Operating Temperature Requirement (allowed derate)	NO de-rating to 190F air temperature (active cooling required)
Install location / Environment	Desert < 3000 feet above sea level
4. SUBSTATION REQUIREMENTS	
Maintenance Access requirement	Full
Lighting Requirement	TBD
Dimension Requirement (ease for transportation)	TBD
Weight Requirement	TBD
Foundation Requirement	TBD
5. COMMUNICATION	
INV Monitoring System Required?	YES
Communication Protocol and speed	Standard
Physical Communication Access	Yes
Required Commands	Standard
Required Production LOGs	Yes
Monitoring system coverage	Yes
6. Warranty	
Required Warranty Terms	Extended 5, 10 and 20 year options
Uptime guarantee Requirement	Yes, to be negotiated
PM and Service Work	EPC
Manufacturer Technician Response Time	1 day
Phone service for field Technicians	N/A
7. TERMS	
Delivery Terms	FOB Japan sea port
Lead Time / drop off scheduling	3-4 units per month after order
Payment	To be negotiated
Place of delivery	Port of Oakland, CA

Chapter 6

Large-scale PV Projects

"We can't solve problems by using the same kind of thinking we used when we created them."

Albert Einstein

"Large-scale commercial solar PV and central-station utility-scale PV will most likely become the dominant growth market" (EPRI, 2010).

Large-scale PV power generating (commercial and utilities type of installations) are the future of our energy independence, and one sure way to clean the environment. They consist of large land areas, usually in the world's deserts, covered with solar energy conversion equipment, such as fixed mount PV modules, trackers etc. With their large size and unlimited power generating capacities, they are on the extreme end of the PV power generation spectrum and represent the best and most efficient way to generate solar electricity.

When properly designed, installed, and operated, solar equipment can generate a large quantity of electricity via concentrated solar thermal power (CSP) or photovoltaic (PV) technologies. We will concentrate on the PV side of things in this text, because it is in the area of our expertise and long experience. PV technologies with their great versatility and flexibility will play an ever more important, if not determining, role in our energy independence and clean environment, so they do deserve a closer look.

In order to be cost effective, CSP technologies are limited to very large installations (over 50 MWp and larger) and also require a large quantity of water for cooling, which is simply not available in the deserts where this technology is most useful. They have also reached, or are very close to reaching, their efficiency peak so very little can be done to improve their performance, or reduce their cost. A number of cancelled CSP projects in the US, converted into PV installations lately, are a confirmation of the greater flexibility of PV power.

There are several different PV technologies to choose from, most of which can be installed anywhere and in any size. They need no water for cooling and have not yet reached their maximum possible output levels. With continuing research by private and government entities, we expect the efficiencies of some PV technologies to reach 60% before this decade is over, and 80% in the near future, which opens great new possibilities for energy generation. By increasing the size of the PV power plants, and adding many GWs of electric power production, the different PV technologies combined could easily surpass the generating capacity of the conventional energy sources in the long run.

PV technologies are our hope for mass electric power production without the negative effects of global warming and other environmental disasters we are seeing. The transition won't be easy. The battle has started and will rage on many fronts and for a long time. In the end, however, we will make the right choices and do the right things for our sake and for the sake of future generations.

PV TECHNOLOGY REVIEW

As previously mentioned, PV for large-scale installations vary in type, size and purpose. Some of the crystalline silicon (c-Si) and thin film PV (TFPV) technologies are mature enough to be used for large-scale power generation in commercial and utility type installations. Commercial power generating installations could be as small as 5 kW, and up to 1 MW and larger, while utilities require larger power generation; starting at 1 MW and going up to 1 GW or larger. At present the trend (set by California-based utilities) limits PV installations' size to between 1 and 20 MWp. We will refer to commercial and utility type power generators as "large-scale" installations in this writing.

Since we focus on large-scale installations in this text, we will take a closer look at the state of the art of PV technologies suitable for such use and the conditions for installing and operating them. We will attempt to provide a fair analysis of the technical, administrative, and financial aspects, focusing on the technical issues related to long-term operation of large-scale PV power plants in desert areas where they are most efficient and where most of them will be located in the future.

PV Technologies for Large-scale Installations

The major PV technologies suitable for commercial and utility-scale applications are c-Si (mono and poly crystalline silicon) modules, thin film PV (TFPV) modules, and high concentration PV (HCPV) tracking systems, as follow.

Poly-silicon PV Technology

This is a mature technology, which represents the largest segment of the PV industry today. The materials, equipment, and processes used today are almost the same as those used nearly 40 years ago, when the first 2- and 3-inch silicon solar cells were processed by hand. Some changes have occurred through the years, of course, where the quality of materials, the sophistication and efficiency of the production equipment and processes have increased significantly. The performance and efficiency of the finished product has increased some too, but is still below the theoretical limits of the respective technologies.

Commercial poly silicon solar cells operate on average at 16-18% efficiency, which means that 1 m^2 PV module packed with the maximum number of polysilicon solar cells will produce maximum of 160-180W of DC power at STC (or 1000 W/m^2 solar insolation at 25°C ambient temperature.).

Also, when operating under high desert temperatures in the 180-190°F range (85-95°C) as measured inside the modules, silicon solar cells lose efficiency, which causes significant drop of power output. Output losses due to operation under elevated temperature range from 0.4 to 0.6% drop per degree C, as measured in US deserts. This loss could amount to 20-30% power loss at a summer noon in the Arizona desert—not a small problem.

In addition, silicon solar cells work well in the beginning of their working life and slowly decrease in efficiency after that, losing ~1.0% per annum. This is widely accepted phenomena, so all PV modules come with a "20 years" performance warranty, which states that the modules will be at least 80% efficient after 20 years of operation.

The long-term loss, combined with the efficiency loss when operating in excess temperature regimes basically tells us that c-Si PV modules prefer cooler operating temperatures. Nevertheless, well made poly silicon cells from major and established manufacturers have proven worthy in long-term field tests, which means that the technology in its present form and at its best quality is capable of handling long-term exposure to the elements. This means that we must chose a reliable manufacturer if we want to use poly Si modules in a large-scale PV power plant.

Mono-silicon PV Technology

This is the most efficient and reliable PV technology today. With its 18-20% efficiency and going higher while prices are going down, it is the technology of choice in many PV projects. It is also the most mature technology, since it has been in use the longest. Its properties and behavior are best known, because mono-silicon (m-S) is widely used in the semiconductor industry too, where it is thoroughly researched and used in mass production.

And yet, TUV Rheinland test center in Tempe, AZ, issued a summary in one of their reports (1) as follows: "Out of 18 (9 mono-Si and 9 poly-Si) modules tested, 9 poly-Si and 7 mono-Si modules passed the hotspot tests of all three standards. Both failures (mono-Si with voltage limited and current limited cells) occurred in the UL method. One failure occurred within the first few minutes (current limited) and the other failure occurred about 70 hours of stress (voltage limited)."

It is undeniable fact that mono-silicon solar cells are the oldest, most stable and reliable PV technology available today, and we are most familiar with their behavior. Still, some of the m-Si modules (20%) failed within 70 hours, while others did not fail for the duration. Why did one of these modules (10%) fail shortly after initiating the test? The disturbing fact is that modules of this batch would not have lasted very long in the field—or even worse—would have lasted just beyond the product warranty period. Imagine 10-20% of the modules in our large-scale PV power plant failing a month or two after warranty expiration.

Was this anomaly caused by materials, quality or equipment issues, or was it due to manufacturing problems? We will never know the answers, but one thing is for sure: the quality of the most reliable technology coming from different manufacturers varies. This means that the failures are not caused by failing technology, but by failures of some manufacturers.

Conclusions: Mono silicon cells and modules are the oldest, best understood and most reliable solar energy conversion devices presently. Mono Si modules, made by reputable and well established manufacturers using proper techniques, have the quality and reliability needed for use in world energy markets. Some m-Si modules made by the new and unproven manufacturers, however, have quality issues which we should be aware of and watch for.

Thin Film PV (TFPV) Technology

This is the new kid on the block, with very bullish outlook for the solar industry and the energy markets in general. And this is good—very good! The energy industry which is dominated by conventional energy interests got a good kick in the pants via the quick and efficient deployment of TFPV production capacities, products and large-scale projects lately. TFPV manufacturers offer a new approach to solar modules manufacturing—that of elegant, quick, efficient and cheap mass production. Using over a half century of experience with related semiconductor device manufacturing processes, TFPV manufacturers are showing us how PV equipment of the 22nd century will be made.

Efficiency, however, is where TFPV technology stumbles for now. At 8-9% efficiency at the PV module and 5-6% (and we have seen reports of 2% efficiency) in the grid, TFPV modules have a long way to go in competing for the top efficiencies in energy markets. There are also problems with supplying raw materials for these technologies because some of them are rare, exotic, or toxic, or all of the above. This is exposing the TFPV industry to a serious risk of shortages, price variations, toxicity issues and regulations changes which jeopardize its very existence.

Hi-concentration PV (HCPV) Technology

The HCPV technology was developed in the 1980s and tested in small-scale projects during the 1990s. It was, however, put on the shelf soon thereafter, so it is still waiting to be introduced and proven in field operations. Several manufacturers have installed capacities, but they seem to be still mostly in the development stages.

HCPV systems are designed and built for operation under extreme desert conditions. They operate at 50-1000 times concentration of sunlight with over 42% efficiency at the cell. Efficiency in the grid is lower, of course, but still at least 2-3 times higher than any of the competing PV technologies. The CPV solar cells are made out of germanium, or GaAs, which are superior semiconductors not affected by temperature extremes. They are also made via sophisticated and precise semiconductor processes, which make them durable and reliable for long-term operation. The optics are made out of glass and silicone, which are also unaffected by sunlight and temperature extremes.

HCPV modules are mounted on trackers which have a number of moving parts, control mechanisms and electronics that need regular maintenance and tuning for optimum performance. This requires highly trained personnel and increases O&M expenses.

We can only hope that with its 40+% efficiency, which is increasing with time, HCPV technology will become an active participant in the energy markets of the US and the world in the near future. HCPV holds great promise of high efficiency (over 60% in the near future), combined with reliability of quality hi-tech components suitable for large-scale generation of electricity in the deserts.

Manufacturing Process and Product Quality

There has been a number of documented news recently of PV modules from different manufacturers failing soon after field installation. This is a good reason for concern, and deserves a closer look and in-depth analysis. As an example, according to AURIGA USA, LLC, "We are aware of a disruption in the manufacturing process that occurred in 2009 that resulted in some modules having significant degradation after being installed for just one month. Although the modules passed the initial flash inspection at the factory, they later degraded by as much 20% (from 75Wp to 60Wp) once installed in the field. We discussed the potential problem with several prominent installers, and we were satisfied that the problem is not an ongoing concern. We estimate that of the roughly 1.3GW production, only ~10MW has been identified as problematic."

Is this dramatic field failure due to low materials quality or manufacturing abnormalities, or both? It is hard to tell in this case, but we know that the manufacturer involved here is one of the most sophisticated and reputable US solar companies, using state-of-the-art equipment and processes, and superb testing and other quality control procedures, so whatever happened is not likely to happen again soon.

Just think of the consequences 10 MW defective modules would have on our brand new 100 MWp power field. This is 133,000 PV modules failing in different strings, different rows, and at different rates and times. The task of tracking, testing, and locating this large quantity of modules is overwhelming and will prove challenging to the project's resources and budgets.

The actual removal and re-installation of 133,000 modules alone will require serious outside help, which has to be managed and paid for. Who will pay for this work? The manufacturer, the insurance company, the investors? We don't know exactly how the above described case was handled, but we know that the manufacturer is a large and respected company, so they probably took good care of the failed modules. What would happen, on the other hand, if the modules were made by a new startup somewhere in Asia? Would they have the ability and will to take care of such a large problem? What are plant owners' and operators' options in such a case?

The possibility of this or a similar problem happening again with the above-mentioned manufacturer is slim, but cannot be disregarded totally. It will happen surely and more often to other, less sophisticated and less reputable companies—especially of the type we call "newcomers," many of which have sprouted around the world lately and whose product has not yet been proven reliable. We cannot expect full cooperation from them in case of a mass failure, which complicates things further. Even if they admit fault and send new modules, as the warranty is usually structured, the downtime and replacement of 133,000 modules might break the project's budget.

New and Unproven Manufacturers

According to GTM Research Report (7); "Breaking down 2009 production by region confirms what the industry has known for quite some time now: manufacturing solar cells and modules is now a game of low-cost production. This is especially true for standard crystalline silicon cells, which have become highly commoditized with little perceived differentiation across suppliers. Almost half (49%) of the cells made in 2009 came from China; when limited to crystalline silicon cells this number jumps to 56%. With regard to modules, these numbers are slightly lower (40% share for all technologies and 47% for c-Si alone)."

So there's no question about it—the race is on, with the Asian companies in the lead for now.

Take a close look at Figure 6-1. Asia is manufacturing more than half of the world's PV cells and modules. In the last 5 years they were able to surpass all developed countries as far as the quantity of produced c-Si cells and modules is concerned. They built new factories and converted a myriad of old ones into PV cells and modules manufacturing and assembly facilities. The quantity of produced PV cells and modules grew fast and is growing still faster. It's a money-making machine, yes; but, does this fast growth reflect high product quality too?

"However, the quality issues saw shipments of crystalline modules fall by nearly half to 26 MW, compared to 44 MW in the second quarter," says a report by one of the Asian solar cells and modules manufacturers. What does this mean? We will never know the answer, because the company won't release any further information. The more important question is what were the actual quality issues that caused the failures? Were they limited to the production floor? Were they caught in the field and, if so, how did the company deal with the customers to rectify the problem? We will never know the answers to these questions either, so we only have to hope that we never get into a similar situation ourselves.

Many PV cells and modules are made by new and unproven manufacturers and, since the worldwide PV manufacturing capacity has increased several fold in the last 4-5 years, the potential of lower quality product finding its way into large-scale PV fields increases proportionally. PV modules manufacturing is a huge business for many developing countries, where low labor cost and government support get many companies started in the solar business. Most solar companies in developing countries also benefit from lower SG&A, R&D, low peripheral costs and tax rates. Lately, there have been some hefty (with capital H) government benefits, incentives and subsidies, which allow them to maximize their production capacities which are geared mostly for export. Efficiency and quality of the manufacturing operation, however, is often measured differently than what we are accustomed to in the West—one of the main reasons for concern.

In addition, a brand new production equipment manufacturing industry popped up in Asia, which makes production equipment for a fraction of that imported from the West. A large part of this equipment has been copied from Western technologies and made with inexpensive local materials and labor. Government subsidies and incentives propped up this new industry as well, and it is prospering for now. The current generation of low-cost production equipment includes a variety of module laminators, wafer etch/clean baths, diffusion and coating equipment and crystalline silicon pullers.

The list of Asian companies providing supplies (MG and SG silicon, chemicals, gasses and other materials needed for manufacturing solar cells and modules) is growing daily. Their prices are drastically lower than the competitors'. Their quality, however, is not established. Of particular interest here is the availability and quality of the starting material of the solar cell manufacturing process—MG and SG silicon. This is the substrate from which silicon wafers are made and onto which solar cells are built. There were basically no quality suppliers of this material in Asia only 5-6 years ago. Just recently, large MG and SG silicon production capacities grew overnight, with dozens of new large silicon producers popping up during the last several years. They are now keeping pace with market demands, for *quantity*, thus the demand for mega tons of silicon is satisfied, but we know very little about the *quality* of these new enterprises.

Key factors influencing the quality of the final product (PV cells and modules) are the quality of raw materials, process chemicals and gasses, metal pastes, cover glass, metal frame, wiring, process equipment design, power availability and labor qualifications. Quality is important to everybody, but it means different things

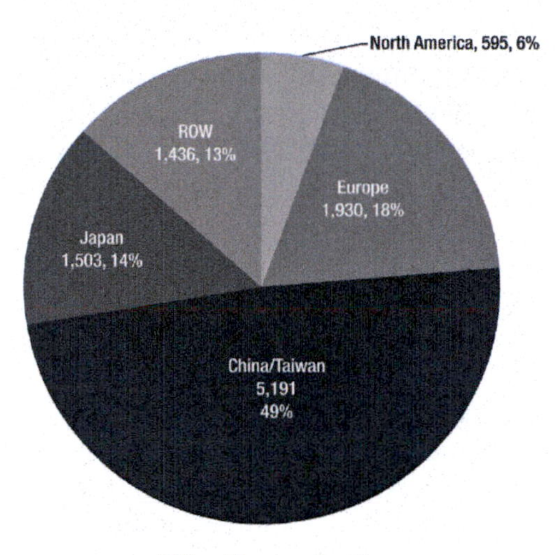

a. PV cells production

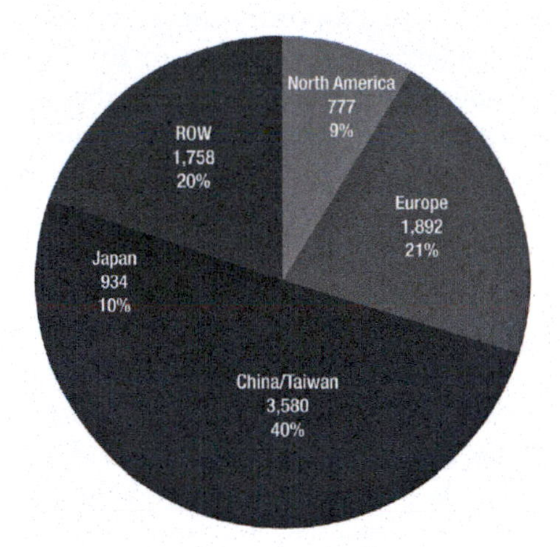

b. PV modules production

Figure 6-1. 2009 global production by region (7)

to different people and varies from country to country. No uniform, or standardized, quality control and quality assurance norms and procedures exists in most Asian PV manufacturing facilities, but individual enterprises adopt whatever technique is most suitable for their particular conditions.

We don't hear much talk about quality issues, and as a matter of fact, discussing quality is somewhat "taboo" among manufacturers and solar professionals. It is another "black box" which we try not to focus on for fear of being politically incorrect or bringing negative vibes. From our point of view, however, purchasing materials and components that are expected to last 30+ years in the fierce desert heat is not like buying disposable toys. It is a serious undertaking without precedent or guaranty of any sort, so yes, we must take a close look at the "black box," as any responsible professional would and should do, if his or her reputation is on the line.

So it is up to us, owners, engineers, contractors and investors involved in large PV installations to make sure we investigate and learn all we can about the quality of the products we plan to use. This is not easily done but there are specialists in the different areas of the manufacturing process who know what is required and how to find the discrepancies. These people must be consulted before any large purchases from new and unproven manufacturers are made. By locating and eliminating potential manufacturing problems, defects, and associated risks, we will protect the interests of all involved in the planned large-scale PV power fields.

To be fair, there are a number of PV cells and modules manufacturers in the US, Asia, and Europe, that are good, reliable, world class companies, who (save the few and far-between issues that plague any manufacturing operation) are exempt from the many negative points in this analysis. Because a large number of newcomers are blindly rushing into the PV market, producing large quantities of product of unproven quality, with the goal of making a quick buck, we need to learn to navigate this situation to ensure 25-30 years of efficient operation of the PV modules and other components in our 100 MWp PV power plant.

Using foreign made PV modules, there are only two ways to ensure success of a large-scale undertaking while considering unprecedented long-term operation in harsh climate areas:

1. Using only brand name PV modules made by well-established, world-class companies who stand 100% behind their products' quality and long-term performance, or

2. Convincing the manufacturer to become part owner and part investor in the project. That way, he'll work with the US technical consultants and contractors in ensuring that every process step is properly planned, supplied, executed, supervised, tested, QC'd and documented. Equally important, all parties will address any issues promptly, efficiently, and for the duration.

Author's note: Sharing long-term responsibilities as an integrated approach of large-scale PV installations is expected to dominate the PV energy markets in the future. It makes a lot of sense and the manufacturers who get into it first will benefit most on the long run.

One hundred percent manufacturer-owned installations are also considered by some, and even underway by others. We see this as a temporary, unsustainable trend driven by the urgency and confusion of the moment. Manufacturers have a significant role in the projects their products are used in, no doubt, but this role should be clearly defined and executed to benefit the particular project(s)—not the manufacturer or the shareholders.

Tracking vs. Fixed PV Systems

The future of large-scale power installations will be split between tracking PV and fixed-mounted versions of the different c-Si and TFPV technologies. The most widely used installations today are those with fixed module mounting, and while there are some good things to be said about them, they are a basically inefficient way to use land and site resources. Trackers offer much greater flexibility and efficiency, albeit at higher prices, so we will consider and compare the options as follow:

Fixed Mounting

Fixed mounting is simply arranging and fastening the PV modules on a fixed frame, anchored to the ground or roof. The frame is tilted at a special angle, which is most efficient for the particular location, and sometimes is adjusted differently for each season. The PV modules sit still on the frames—without any movement—and get whatever sunlight reaches them at each point of the day. Ideally, PV modules should be placed so they are as close as possible to perpendicular to the noonday sun, i.e. pointing due south, and at an angle of inclination approximately equal to the angle of latitude, +/-15 degrees. They will function when mounted flat as well, and as a matter of fact, on overcast days, or in areas with a lot of fog or clouds, this can produce more power, since the sunlight is diffused and reflected all through the day.

A steeper angle of inclination will enhance output during winter months when the sun is lower in the sky, at the expense of some reduced output in summer. A simple adjustable tilt angle system where the angle can be changed manually twice a year (in spring & autumn) is best and does not involve much cost or effort. This option and its cost needs to be included in the design configuration and plans.

Trackers for Large-scale PV Installations

PV trackers could be one-axis or two-axis structures onto which PV modules are mounted in order to follow the sun, thus getting more sunlight and increasing the generated power. Two-way trackers are large structures mounted on poles that are mounted on a pedestal cemented into the ground. They are complex and expensive to install and operate, and since we have seen only 6-8% increase of power over one-axis trackers under normal operating conditions, we will not consider them for PV modules tracking in this text.

The highest capacity factors (Cf) in a PV power plant are generated with trackers that follow the sun throughout the day to keep the modules optimally oriented sunward in order to receive maximum sunlight. Tracking has the benefit of generating more energy in peak electricity demand periods and also during early morning and late afternoon hours, when the sun is at a sharp angle in respect to the modules.

A tracker's benefit to a PV power plant's annual and summer capacity factors can be substantial. A tilted one-axis tracker can generate approximately 25-30% more energy than a fixed system on an annual basis. Additionally, during the summer peak season, capacity factors can exceed 38% with a horizontal one-axis unit, providing energy when the utility experiences maximum seasonal demand.

The LCOE model assigns an equal value to electricity generated throughout the year; however, electricity generated at peak periods is more valuable to the utility and has a higher rate of return. The use of solar trackers in summer can produce 40% more electricity in peak demand periods when energy is highly valued. During peak periods capacity factors from trackers can exceed 70%, directly offsetting the need for natural gas peaking plants and other alternative peaking power resources.

In addition to its location there are two fundamental drivers for the land consumed by a solar power plant: a) PV module efficiency, and b) ground-coverage ratio (GCR), which represents the ratio of land area covered by PV modules to the total land area.

Or generally speaking:

a. High-efficiency PV modules (around 18-20% overall efficiency) need 50-60% less land and infrastructure for a given capacity factor configuration than low-efficiency PV modules (10% efficiency or less).

b. High-efficiency PV modules mounted on trackers provide up to 20-30% higher capacity factors, while at the same time using a similar or lower amount of land than low- and medium-efficiency modules mounted on fixed-tilt systems. This means that lower LCOE and high-capacity factor configurations can be achieved without prohibitively increasing the amount of land required.

The main trackers' concerns have to do with cost of the hardware, complexity of installation, and additional maintenance, so trackers are not the solution in all cases. To make a sound decision, PV power field designers must have a good understanding of the structure and function of the different trackers and go through extensive and complex calculations, considering the particular site and available technologies.

Cost and Effectiveness of Tracking

Although the capacity factor benefit of tracking is clear, the decline in PV power plant prices, and using free (BLM) land, raise a question about the continued cost effectiveness of tracking systems. One could argue that low-cost PV modules mounted on fixed structures would yield superior economics to a high-capacity factor tracking system during a new era of low-cost and low-efficiency PV; i.e., mounting 8% efficient TFPV modules on trackers won't be economical in the long run, even if modules and land costs are low, so we cannot recommend it under most circumstances. And yes, we are well aware of the plans of some TFPV manufacturers and developers to use tracker-mounted TFPV modules in large-scale PV fields, but the numbers just don't match, so we just have to wait and see.

The question of tracking cost effectiveness can be answered with an application of the LCOE equation and its variables for each different case. As with any change that improves the capacity factor of the system, the increase in performance must be weighed against the incremental cost, if any. If the absolute change in capital cost is less than the absolute change in capacity factor, then economics suggest the implementation of the system that best improves the capacity factor. If this analysis is applied to a PV power plant located in the U.S. desert (southwest), even at a low system price of ~$4/Wp DC, the value of the tracker's 30% capacity factor improvement would be $1.20/Wp DC, far above the incremental capital cost of the tracker motor and control system and long-term O&M expenses. Higher efficiency PV technologies could benefit from trackers, and some cannot even operate without trackers, so the battle for higher efficiency and larger profit share is just now starting.

Another major consideration in this respect is the use of inverters. Studies show that using trackers can significantly reduce the inverter requirements of generating power in large-scale PV fields; i.e., 200 GWh of annual energy generated by a PV power plant would require ~100 inverters, each rated at 1.0 MWp, when using PV modules mounted on single-axis tilt tracker, versus using ~125 inverters with a fixed-tilt system for generating the same power at the same location. Using two-axis trackers would provide even more total generated power, but complicates the electro-mechanical structure and O&M procedures disproportionally, so these are not widely used now.

The use of a tracking system, albeit complicated, could provide versatility, offer more options, generate more power and significantly reduce capital cost of inverters and their maintenance expenses (the costliest portion of power plant's setup and operation). Again, tracking is not a fix-all solution, but a careful consideration must be given to this option in all cases.

HCPV Tracking Systems

As mentioned previously, HCPV systems use very accurate two-axis tracking, and this combined with other sophisticated components makes HCPV the most efficient, reliable technology to date. Its operation, however, is quite complex for use in today's energy market, as we know it, so it will take a while but HCPV technology will eventually find its place in energy markets.

HCPV is especially efficient, because in addition to over 42% efficient CPV cells, it tracks the sun extremely accurately—0.01% accuracy—which allows it to capture every single sun beam coming down from sunrise to sunset. The problem is that if there is any cloud cover or haziness in the sky, HCPV optics cannot handle the diffused light going through them, and the efficiency drops quickly and significantly. Basically, HCPV systems can be used only in bright, hot desert areas, where other technologies have a hard time surviving the harsh elements. But HCPV is a complex technology and some work is still left to be done before we see HCPV trackers installed in large quantities in the world's large-scale PV power plants. At that point the equipment cost can be reduced significantly by placing large orders for materials and components.

It is quite clear from Figure 6-2 that, everything considered, the power output is in favor of tracking systems.

The Long-term Field Effects

There are a number of potential issues that need to be identified and addressed when planning use of PV modules in large-scale installations in the deserts. Some of these are:

1. Antireflective (AR) coating (which is the most obvious thing we see on the cells) could degrade to a certain degree with time. This will cause several problems:
   a. Visual discoloration, which will show as faded or different color areas on the cells' surfaces.

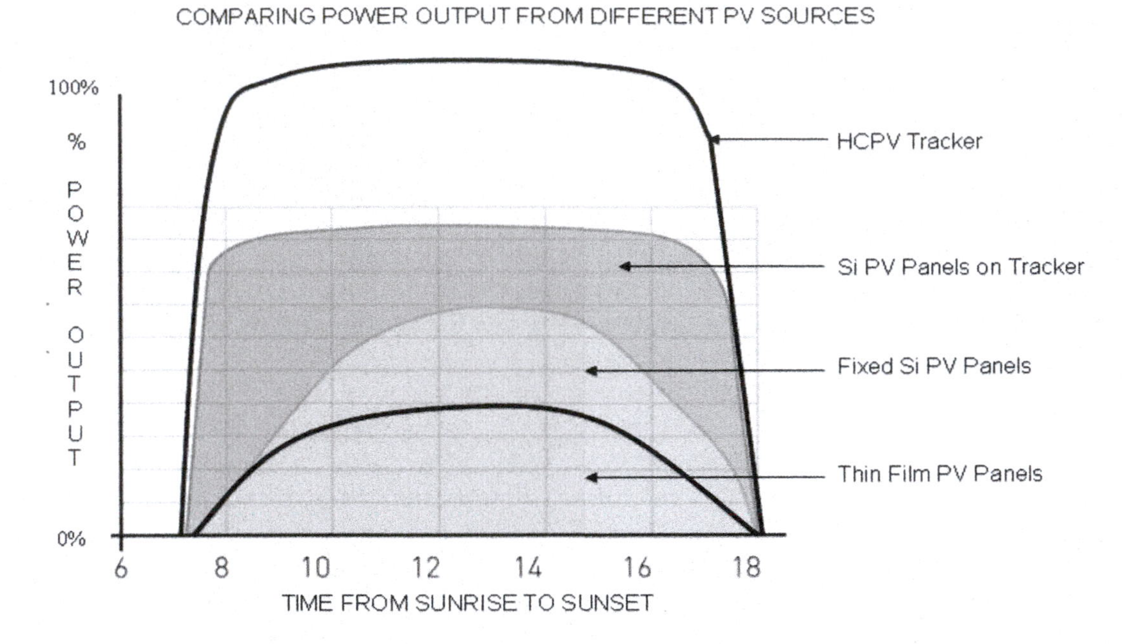

Figure 6-2. Efficiency of the different PV technologies

The discolored areas will decrease the esthetic appeal of the installation, and will bring a lot of questions by site workers, visitors, investors and owners alike.

b. The degraded AR layer will reflect more sunlight than before, thus reducing the cells' efficiency significantly.

c. Delaminated AR film will trap moisture and other harmful chemicals and gasses, which might cause even greater damage in the affected area.

2. The EVA layer (which is the most obvious thing we should *not* see in the module) also could degrade with time, causing:

   a. Visual discoloration, which will show as "yellowing" or different color areas on the modules' surface. These areas will decrease the esthetic appeal of the installation, and will reduce its power output.

   b. Delamination of the EVA from the cells' surface could create an optical gap, thus bringing the cell's efficiency down even more, and

   c. The degraded EVA might also allow harmful moisture and gasses into the module which would attack and destroy the cells and shorten the module's life.

3. c-Si PV modules lose power with increase of temperature. 0.5-0.6% power output is lost per degree C increase of temperature above STC (25°C). In the deserts this is equivalent to 20-30% loss of output several hours during, before, and after the noon hour. Some TFPV modules undergo slightly lower temperature degradation in the 0.35-0.45% range. Nevertheless, this is still a significant loss and must be considered in all design calculations.

4. Even the best made modules—those without any quality issues during manufacturing—undergo efficiency degradation of ~1.0% every year in the field. Some TFPV modules show significantly higher annual degradation rates (1.3-1.5%). This is not a fatal failure, but is going to cost us 10% decrease in output in 10 years. This translates into at least 10% decrease of profit in year #10. This amount doubles in 20 years, and triples in 30—not a small thing to consider!

   For a 100 MWp power plant, this 1.0% yearly loss is equivalent to losing ~50,000 PV modules and 10 MWp power capacity loss during year 10. This is equivalent to losing 18.0 GWh of energy production during year 10 alone, because the modules have lost that much of their ability to produce electric power. In other words, our 100 MWp large-scale PV power plant in the AZ or CA deserts, capable of producing ~200 GWh power during year one, will produce only 180 GWh during year 10, down to 160 GWh during year 20, and 140 GWh during year 30. These are substantial losses and must be taken into consideration in the plant design, financial calculations and contractual negotiations.

5. There is always a chance for the aging process to accelerate unexpectedly, as seen in field tests, where many modules fail without notice and we might lose 20-30% of the original power output within the first 10 years of operation. This is unthinkable, but not an impossible scenario. It is actually quite probable, if you take a look at the test failure rates and production quality incidents discussed in previous chapters and extrapolate those to 20-30 years of field operation. How can a plant survive financially with such great losses? And finally,

6. PV modules (and TFPV based especially) contain some toxic and otherwise harmful materials which must be fully accounted for during design, installation, and O&M stages of power plant construction. They must be also fully contained for the duration and recycled at end of life. The deployment of such modules in large-scale power fields is a serious undertaking, which must be evaluated from a Hazmat point of view and fully accounted for technologically, logistically, and financially. The risks for the duration of operation, decommissioning, and product recycling stages must be considered.

In summary, combinations of the above concerns and uncertainties are part of the reason we don't see many 500 MWp, or 200 MWp, or even 100 MWp PV power generating plants in US deserts as yet. And if we see any of these in planning or under construction, we need to ask the basic questions, "Were all above mentioned issues considered, calculated and justified? Were all team members in agreement that the key efficiency, longevity and safety concerns were properly addressed and answered?"

So the basic question remains: "Are present-day PV technologies suitable for long-term large-scale PV operation in the world's desert areas?" Addressing specific issues (which we reviewed in previous chapters), we ask, "Were the test and field problems caused by materials, manufacturing processes, process equipment, labor, management, or a combination of the above?" The answer would be different in each case, of course, but we know that PV products made by established and proven manufacturers, using quality equipment and proper techniques, will be predictably efficient and reliable, while those made by many newcomers and unproven manufacturers must be questioned when considered for long-term, large-scale PV use.

In any case, there are no examples or precedents of long-term operation of large-scale PV installations in US deserts, so we must be careful and thorough when dealing with the issues related to this type of application.

NOTE: PV installations financed by the manufacturers of PV modules used in large-scale installations is a new phenomenon which we have seen used in the US lately. In these cases, all issues and risks related to the design, installation and operation are the full responsibility of the manufacturer and the respective support entities in their specific areas of expertise and involvement.

Future Improvements

There have been a number of important improvements during the last decade in the quality of materials, consumables, processes and manufacturing techniques of solar cells and modules. These contributions to improved quality and efficiency have given us some confidence in their reliability and longevity. Nevertheless, the level of uncertainty remains high—much higher than what is needed to ensure their uninterrupted and efficient on-sun operation.

So we need to keep in mind that efficiency, price and quality are related, but are not one and the same. In fact, some of the new achievements in the efficiency increase of PV modules, on one hand, and price reductions on the other, have resulted in decreased quality and longevity. This is not a trend, but an indication that efficiency, price and quality are related in not such obvious ways. PV technologies are approaching their theoretical limits, so efficiency increases or price decreases are getting harder to obtain, but efforts are ongoing and the product is sent to market without proper testing at times, while long-term field testing is not even an option because the markets won't wait.

There are many efforts underway to achieve increased efficiency, higher levels of quality, reliability and longevity in PV cells and modules. See Table 6-1 for examples.

THE PLANNING PHASE

A number of issues must be considered before we start the design and installation of PV modules in the large-scale PV fields. Some of the critical topics and issues are discussed below.

The Planning Process

The PV power plant design process is a complex undertaking, which must be evaluated and executed by a team of professionals, specialized in land, environment,

Table 6-1. Proposed improvements to PV materials, components and processes

TASKS	SURFACE, MATERIALS AND PROCESS OPTIMIZATION AND R&D
Wafers	Thickness reduction, and surfaces cleaning, etching and texturing 100% initial material tests for bulk contaminants and mechanical problems New and/or improved solar materials for c-Si and TFPV cells and modules
Solar Cells	Front surface preparation and passivation for efficiency improvement Front and back metallization improvement via new deposition methods Integration of bypass diodes on the cell level in the module body
Compounds	Optimizing the use of copper, tin, steel, aluminum, etc. structural materials Soldering connections, metal welding operations, glue and bonding use Materials and processes for front and back contacts and integrated connections.
Front Cover	AR coated, self-cleaning and/or textured glass for improved optical performance Synthetic materials (polycarbonate, acrylic etc.) safe for use in some climates
Encapsulants	Optimization of EVA formulations and application methods Developing new autoclave free formulations (PVB, etc.) Optimization of silicones, resins, gels, lamination foils and casting resins Improved polymer systems for encapsulation per specific applications
Back Cover	Optimizing existing and developing new PET-PVF lamination systems Optimizing application of PET materials with vapor barriers New types and configurations of polymer, glass, steel and aluminum back covers
Junction Box	Developing new materials, such as polymers Improving the heat transfer and longevity of the materials and components Developing integrated electronics and automated placement process
Labeling	Introducing new labeling, embossing and embedded marking systems Developing new and more secure electronic marking and labeling methods

government, regulatory, and technical disciplines, as related to the broad spectrum of the tasks and sub-projects of the design process:

The preliminary work in PV power plant planning evaluation and considerations includes:

1. **PV field evaluation:**
   a. **Solar analysis**
      i. Local solar irradiation levels and issues
      ii. Local climate and weather conditions
   b. **Land evaluation**
      i. Ownership and lease/purchasing issues
      ii. Topography and development issues
      iii. Environmental assessment
      iv. Local, state and federal permits (if needed)
   c. **PV system planning and design considerations**
      i. Mechanical and construction services
      ii. Electrical service (PV module level)
      iii. Electrical service (field level)
      iv. Final electrical design considerations
      v. BOS components
      vi. Installation and O&M conditions
      vii. General system considerations
   d. **PV system verifications, calculations and final design**
      i.Mechanical and construction services
      ii.Electrical service (PV module level)
      iii.Electrical service (field level)
      iv.Final electrical design considerations
      v.BOS components
      vi.Installation and O&M conditions
      vii. Final design parameters

2. **Permits and PPA**
   a. Create minimum acceptable proposal for permitting authorities and utilities
   b. Set up and document as needed to proceed with permitting and negotiating a suitable PPA

Planning and pre-design efforts consist of understanding and considering major design elements, and those that influence the design. We'll take a look at some of them below.

Pre-design Proposal Writing

Planning, design, construction and operation of a commercial or utility type PV power generating installation is a daunting task, or rather a multiplicity of such tasks. It starts with an idea, an opportunity, and writing a report/proposal to a customer, investor, or utility company. Actually, utility companies have their own way to get the information they need (via RFPs), but some of the work needed for writing a proposal applies to the RFP completion as well.

There are no set rules and procedures for this effort, so we will take a shot at it based on our experience, by outlining the tasks that need to be addressed to come up with an accurate, thorough and convincing presentation.

Below is a list, in no particular order, of such tasks and activities which must be addressed in order to prepare a thorough report or presentation for a large-scale PV power generating installation:

a. Solar analysis

- Analyze NASA and other local weather data
- Optimize seasonal variations vs. output
- Perform remote or on-site assessment
- Determine available active area
- Evaluate field orientation options
- Determine magnetic declination
- Determine modules positioning and tilt angle
- Determine present and future shading
- Evaluate land/sky obstructions
- Enter all data in SAM, PVWATT or such and analyze the options
  + Analyze seasonal patterns
  + Determine optimum local conditions
  + Estimate power output variables
  + Evaluate performance vs. cost factors

b. Logistics

- Identify customer's needs and wants
- Determine project time frame
- Determine land type and outline performance related issues
- Identify jurisdictional issues
- Determine zoning and evaluate nearby business activities
- Determine the local road and RR system
- Determine transmission lines routes and kV sizing
- Determine substations location and capacity
- Determine fire marshal's requirements
- Determine tribal and reservation issues, if any
- Check city, county and utilities requirements
- Identify utilities conditions for interconnect and PPA

c. PV system pre-design

- Estimate array size based on max. kWp vs. available land area
- Evaluate max. power output considering different technologies
- Evaluate max. power output vs. cost considering different manufacturers
- Execute final predesign system analysis (location vs. technology vs. cost)
- Consider tracking, dual or single axis vs. output and cost analysis
- Perform temperature losses and annual degradation calculations and complete long-term power generation analysis
- Perform outside and inter-row shading and O&M analysis
- Evaluate inverters and BOS components for optimum performance
- Negotiate quality and warranty with module, inverter and BOS mfgrs.
- Negotiate warranty and guaranty conditions with all mfgrs.
- Negotiate insurance conditions and limitations
- Design and price array based on max. output and min. $/watt
- Evaluate and spec installation and O&M equipment and procedures
- Estimate life expectancy and spec end-of-life procedures
- Consider aesthetics and other visual/noise/safety values and issues
- Estimate potential price adders and incentives
- Develop price range and provide options (technology vs. methods, vs. cost)
- Finalize technology, methods and conditions estimates in final design

d. PV system design

i. Mechanical and construction services

- Identify major mechanical and construction component types
- Negotiate manufacturers' conditions and warranties

- Create components specs and installation and O&M procedures
- Determine soil composition for ground mounts or trackers
- Identify construction materials and methods
- Identify O&M requirements (PM intervals etc.)

ii. Electrical service (PV module level)

- Establish PV module handling safety parameters
- Establish PV module quality control specs
- Identify array strings location options
- Identify PV modules' electrical specs
- Calculate module temp. coefficient and voltage degradation over time
- Determine method of interconnection
- Determine power limits based on module specs and local elec. code
- Determine service rating current and voltage
- Identify buss bar and breakers location
- Identify line tap locations
- Specify grounding points and methods

iii. Electrical service (field level)

- Establish field safety parameters
- Establish field quality control specs
- Identify inverter type and manufacturer
- Negotiate terms with inverter manufacturer
- Identify inverter number and locations
- Identify AC/DC disconnects and fuse requirements
- Identify combiners and sub–combiners location
- Design and locate conduit runs (location and length)
- Design and locate lightning strike preventers
- Identify utilities connect/disconnect

iv. Final electrical design considerations

- Wiring systems layout and circuit design
- Conceptual substation design
- Transmission line conceptual and final design
- Feeder voltage drop analysis
- Feeder harmonic assessment
- Transfer trip system scoping
- Protective relay requirements definitions
- Determine communication type and path specification
- Power factor assessment
- Technical review of interconnection for feasibility and implementation
- Final conceptual designs for EHV generation tie-lines from the preliminary design
- Complete design solutions for the PV overhead and underground wiring systems
- Analyze and finalize substation requirements and parameters
- HV transmission line design
- Develop project scoping documents
- Develop project budget estimates vs. BOS components
- Establish safety parameters
- Establish quality control specs
- Spec frame tracking or tracking preparation and installation
- Identify fastening components and devices and vendors
- Identify type of monitoring system, broadband vs. wired etc.
- Identify all other passive components and devices and vendors
- Identify surface treatment and corrosion resistance options
- Provide a quote for BOS materials and services

vi. Installation and O&M

- Interview EPC companies or individual engineers (PE, EE, ME, AE)
- Negotiate with EPC and potential operators
- Establish a baseline of installation and operation procedures and issues

vii. General system considerations

- Estimate additional costs, including utilities hardware upgrades etc.
- Estimate additional costs for O&M purposes (see above)
- Identify Electrical P.E. to prepare electrical drawings per specs
- Identify Mechanical P.E. to prepare drawings per specs
- Identify contractor to provide installation estimate
- Identify contractor to negotiate O&M services

e. Proposal submission

i. Create minimum acceptable proposal, to include:

- Power generation estimate
- STC DC power rating, derating and estimated AC rating
- Include all conditions and estimates

- Total project cost estimate with rebates, tax incentives, and net cost
- Estimate of permit, interconnection fees, taxes and other costs
- List (and explanation) of variables and unknown costs
- Suggested payment (ROI) schedule
- Incentives paid over time (PBI, FIT, SRECs)
- Construction (installation) timeline and milestones
- List of major equipment
- List of special factors foreseeable issues and complications
  + Module temperature and degradation over time
  + Land issues (erosion, buildings and businesses nearby etc.)
  + System dependency on sun intensity and cloud cover
  + Expected output vs. system capacity
  + Instantaneous power vs. annual energy production
  + Potential cross-over of impact on overall warranty
  + Performance estimation and validation vs. actual output

If all of this seems overwhelming and somewhat chaotic, it is because it is so. The present-day system of "customers-suppliers-installers-investors-regulators-politicians-utilities and everyone else involved in a PV power installation" is work in progress, for lack of better words. The lack of proven engineering specifications and standardized procedures for manufacturing, design, permitting, financing, installation and operation of PV components and systems is quite obvious. There are a number of efforts in these areas, but they are fragmented at best, and it will take a long time to get them under control.

Location, Location, Location

OK, so we are going to build a large-scale PV plant. Smaller commercial systems are easier to design, install and operate. All we need in most cases is the owner's permission to hang some panels on the roof or in the back yard, get local permits and install the modules. Large-scale PV power plants are somewhat more complex to plan, design, permit, finance, install and operate. So where do we start?

The planning of the project is the responsibility of the whole owner/designer/installer/investor team. Team members must be fully aware of land, logistics, regulatory, and financial conditions (and limitations) at the site. Each member has specific responsibilities and accountabilities in their respective areas of expertise. Every detail should be taken into consideration, analyzed, calculated and entered into the overall formula—with the actual location in mind.

So what's in a location? There are a large number of conditions and variables to consider. Sunshine, or no sunshine, when and for how long? The presence and frequency of clouds, fog, dust, snow, storms, hail, etc. natural phenomena? Populated centers, industrial activities and roads close by? Wild life? Terrain issues? Environmental issues? Transmission lines? Substations? And this is just the beginning.

In most cases we don't have the luxury of choosing a location, because the land has been chosen by a customer or an investor, so we must do our best with what we have. In any case, there are a number of things to consider, and we will take a close look at the key parameters as we go along. For now, however, let's take a look at one of the most important parameters—the power grid issues at our location—which will determine and drive the power plant planning and design processes.

The Power Grid

Electric power is essential to modern society. Economic prosperity, national security, and public health and safety cannot be achieved without it. Communities that lack electric power, even for short periods, have trouble meeting basic needs for food, shelter, water, law and order. A prerequisite for an efficient and practical large-scale PV power plant is its ability to transfer the generated electricity from the source into the national grid, in order to be used as needed. So we have no choice, but to work with the grid, and we need to know all there is to know about it.

In 1940, 10% of energy consumption in America was used to produce electricity. In 1970, that fraction was 25%. Today it is 40%, showing electricity's growing importance as a source of energy supply. Electricity has the unique ability to convey both energy and information, thus yielding an increasing array of products, services, and applications in factories, offices, homes, campuses, complexes, and communities. The national power grid accomplishes that by its thousands of miles of power lines, substations and other equipment, which convert the power to voltages and frequencies that are adequate for transmission and use.

Looking at the map in Figure 6-3 we see that the national electric grid, just like the human body's circulatory

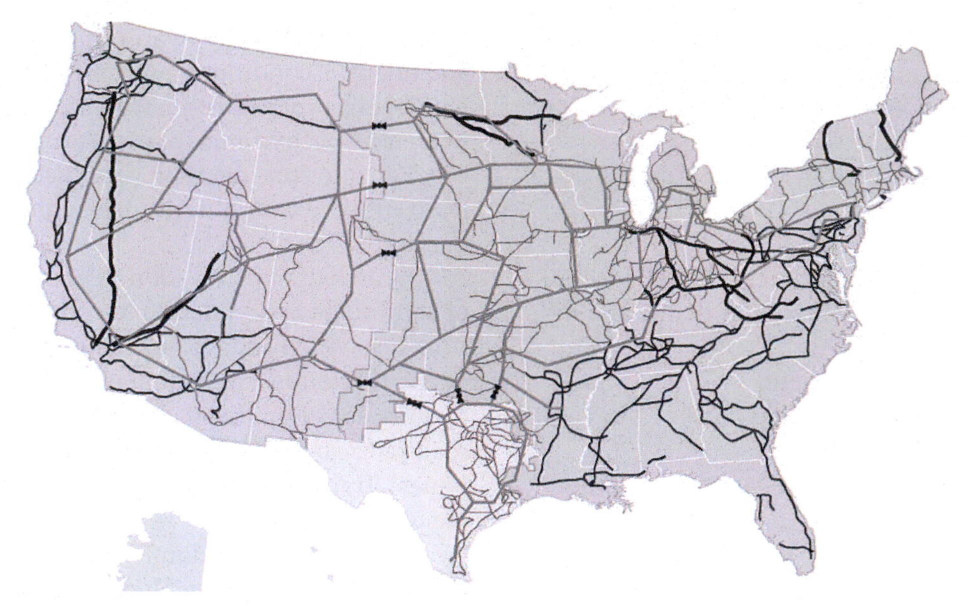

Figure 6-3. Map of the US Power Grid

system, carries life-sustaining energy to all the different parts of the country. Without it, most critical activities would simply stop. The economy would come to a hard and very expensive stop. Life as we know it would cease to exist. The consequences are impossible to even estimate, but nothing good can be expected to result from a drastic interruption in the national energy supply.

The Basics

The economic significance of generating, distributing and using electricity today is staggering. It is one of the largest and most capital-intensive sectors of the economy. Total asset value is estimated to exceed $800 billion, with approximately 60% invested in power plants, 30% in distribution facilities, and 10% in transmission facilities.

Annual electric revenues—the nation's electric bill—are about $247 billion, paid by America's 131 million electricity customers, which includes nearly every business and household. The average price paid is about 8-15 cents per kilowatt-hour, although prices vary from state to state depending on local regulations, generation costs, and customer mix.

There are more than 3,100 electric utilities in the US:

- 213 stockholder-owned utilities provide power to about 73% of the customers

- 2,000 public utilities run by state and local government agencies provide power to about 15% of the customers

- 930 electric cooperatives provide power to about 12% of the customers

Additionally, there are nearly 2,100 non-utility power producers, including both independent power companies and customer-owned distributed energy facilities.

The bulk power system consists of three independent networks: Eastern Interconnection, Western Interconnection, and the Texas Interconnection. These networks incorporate international connections with Canada and Mexico as well. Overall reliability planning and coordination is provided by the North American Electric Reliability Council, a nonprofit organization formed in 1968 in response to the Northeast blackout of 1965.

Power Generation

America operates a fleet of about 10,000 power plants, mostly thermal (coal and diesel) with average efficiency of around 33%. Efficiency has not changed much since 1960 because of slow turnover of the capital stock and the inherent inefficiency of central power generation that cannot recycle heat. Nuclear and hydro plants are more efficient, but much more expensive as well.

Power plants are generally long-lived investments, with the majority of existing US capacity 30 years old or older. They can be divided into:

a. Baseload power plants, which are run all the time to meet minimum power needs,

b. Peaking power plants, which are run only to meet power needs at maximum load (known as peak load), and

c. Intermediate power plants, which fall between the two and are used to meet intermediate and emergency power loads.

The roughly 5,600 distributed energy facilities typically combine heat and power generation and achieve efficiencies of 40% to 55%, accounting for about 6% of US power capacity in 2001 and several times more today.

A shift in ownership is occurring from regulated utilities to competitive suppliers. The share of installed capacity provided by competitive suppliers has increased from about 10% in 1997 to about 35% today. Recent data suggest, however, that this trend is slowing down. Also, cleaner and more fuel-efficient power generation technologies are becoming available. These include combined cycle combustion turbines, wind energy systems, advanced nuclear power plant designs, clean coal power systems, and distributed energy technologies such as photovoltaics and combined heat and power systems.

Because of the expected near-term retirement of many aging plants in the existing fleet, growth of the information economy, economic growth, and the forecasted growth in electricity demand, America faces a significant need for new electric power generation. In this transition, local market conditions will dictate fuel and technology choices for investment decisions, capital markets will provide the financing, and federal and state policies will affect siting and permitting. It is an enormous challenge that will require a large commitment of technological, financial and human resources in the years ahead.

PV power generating sources are constantly added to the already complex power generation and distribution system, but due to their location, size and volatile performance, it will take a long time for them to be fully integrated into it. Presently the utilities are doing their best to accommodate the additional unconventional load, but there are a number of problems related to interconnection, distribution, and control of variability that must be resolved before the new PV energy sources are truly part of the power generation and distribution system.

Power Quality and Reliability Issues

Power quality is a concern for today's power grid and the loads it serves. Computer equipment, in particular, is sensitive to power quality problems, and the ubiquity of computers in today's manufacturing environment means high power quality is very important to most commercial and industrial firms as well as the average homeowner.

Alternating current, which is the predominant way of transporting and delivering electric power, can be illustrated as a sinusoidal wave, as shown in the Figure 6-4. Over time, the voltage oscillates between a positive and negative value that is slightly more than average voltage.

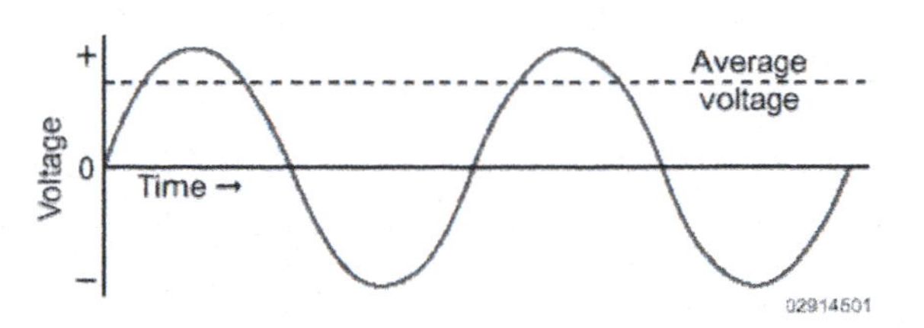

Figure 6-4 AC wave form.

Although alternating current is established as the world's standard, it still has unresolved problems and issues, some of which are important to understanding power quality control and the contribution of PV power generating plants. Below is a list of conditions and characteristic for AC power transmission and use:

1. Voltage sags and swells The amplitude of the wave gets momentarily smaller or larger because of large electrical loads such as motors switching on and off.

2. Impulse events. Also called glitches, spikes, surges, or transients, these are events in which the voltage deviates from the curve for a millisecond or two (much shorter than the time for the wave to complete a cycle). Impulse events can be isolated or can occur repeatedly and may or may not have a pattern. The largest voltage glitch, or surge, is caused by a lightning strike.

3. Decaying oscillatory voltages. The voltage deviation gradually dampens, like a ringing bell. This is caused by banks of capacitors being switched in by the utility.

4. Commutation notches. These appear as notches taken out of the voltage wave. They are caused by momentary short circuits in the circuitry that generates the wave.

5. Harmonic voltage waveform distortions. These occur when voltage waves of a different frequency (some multiple of the standard frequency of 60 cycles per second) are present to such an extent that they distort the shape of the voltage waveform.

6. Harmonic voltages. These can also be present at very high frequencies to the extent that they cause equipment to overheat and interfere with the performance of sensitive electronic equipment.

Other power quality problems may also be considered reliability problems because they occur when the transmission system is not capable of meeting the load on the system. These are:

1. Brownouts. These are persistent lowering of system voltage caused by too many electrical loads on the transmission line.

2. Blackouts. These are, of course, a complete loss of power. Unanticipated blackouts are caused by equipment failures, such as a downed power line, a blown transformer, or a failed relay circuit.

3. "Rolling" blackouts. These are intentionally imposed upon a transmission grid when the loads exceed the generation capabilities. By blacking out a small sector of the grid for a short time, some of the load on the grid is removed, allowing the grid to continue serving the rest of the customers. To spread the burden among customers, the sector that is blacked out is changed every 15 minutes or so—hence, the blackouts "roll" through the grid's service area.

How does PV power generation from millions of rooftops and larger PV installations affect all these factors, and what can be done to reduce the negative effects? This is a complex and pertinent question that can be answered by the experts, and which they are surely working on as we speak. One of the most discussed solutions is the implementation of the new and upcoming "smart grid."

The Smart Grid

"Smart grid" is a fairly new concept, with growing importance. It is a common denominator for a wide range of developments that make medium- and low-voltage grids more intelligent and flexible than they are, thus allowing them to be managed with ease. The main motive for smart grid initiatives is that such developments improve reliability of supply and/or support the trend towards a more sustainable energy supply. At this moment, most medium- and low-voltage networks can not be remotely observed and controlled, so when fully developed smart grid components will eventually solve that problem.

Various companies are developing technologies aiming at creating smarter networks. However, some of these developments tend to root in technological possibilities, rather than in a sound problem analysis and a structured approach toward its solution. In the recent past, a great variety of sensors, protocols, communication equipment and the like has been designed to support the move toward smart grids. However, many of them have not found wide application, which can be at least partly attributed to the fact that they simply did not provide significant solutions to the problems at hand. In other words, there was too much technology push and too little market pull. The fact that some manufacturers of unsuccessful technologies even blame network operators as conservative instead of improving the price performance ratio of their products further hampers a real take-off of smart grids.

In the longer term, smart grid technologies will play an important role in maintaining reliability of supply and improving sustainability. The complexity of electricity distribution increases, as new PV plants are connected into the grid, and the number of wind turbines and solar power plants increases as well. This also applies to small generators, such as roof top PV installations, where smart grids support these developments by continuously monitoring and controlling the grid and the generators.

The smart grid vision is becoming clearer over time, and ever greater efforts will be spent on developing appropriate and necessary smart grid technologies in cooperation with commercial energy companies, other grid operators, and suppliers. These efforts will also help to increasingly focus discussions between regulators and the government on the future energy supply and the role of smart grids in it. The energy future looks bright, and smart grids will play a significant part in it.

Power Transmission

Adequate electric generation in the US is hindered by bottlenecks in the transmission system, which interfere with reliable, efficient, and affordable delivery of electric power to the customers. America operates about 157,000 miles of high voltage (>230 kV) electric transmission lines.

Construction of transmission facilities has decreased about 30%, and annual investment in new transmission facilities has declined over the last 25 years. The result is grid congestion, which can mean higher electricity costs because customers cannot get access to lower-cost electricity supplies, and because of higher line losses. Transmission and distribution losses are related to how heavily the system is loaded. U.S.-wide transmission and distribution losses were about 5% in 1970, grew to 9.5% in 2001, and are even higher today, due to heavier utilization and more frequent congestion.

Congested transmission paths, or "bottlenecks," now

affect many parts of the grid across the country. In addition, it is estimated that power outages and power quality disturbances cost the economy from $25 to $180 billion annually. These costs could soar if outages or disturbances become more frequent or longer in duration. There are also operational problems in maintaining voltage levels.

America's electric transmission problems are also affected by the new structure of the increasingly competitive bulk power market. Based on a sample of the nation's transmission grid, the number of transactions have increased substantially recently. For example, annual transactions on the Tennessee Valley Authority's transmission system numbered less than 20,000 in 1996. They exceed 250,000 today, a volume the system was not originally designed to handle. Actions by transmission operators to curtail transactions for economic reasons and to maintain reliability (according to procedures developed by the North American Electric Reliability Council) grew from about 300 in 1998 to over 1,000 in 2000 and is much higher today.

Additionally, significant impediments interfere with solving the country's electric transmission problems. These include opposition and litigations by different groups against the construction of new facilities, uncertainty about cost recovery for investors, confusion over whose responsibility it is to build and maintain, and jurisdiction and government agency overlap for siting and permitting. Competing land uses, especially in urban areas, leads to opposition and litigation against new construction facilities.

In Figure 6-5, we get a glimpse into the complexity of the generation-transmission-distribution scheme of electric power transfer. The generator (coal or nuclear power plant) might be miles away from the point of use (POU)—residential or industrial customer. The power generated at the power plant is sent to a step-up transformer (substation), where it has to be transformed into higher voltage, as needed for long distance transfer. Some of this power is used by larger users who have their own sub-stations as needed for power use in their facilities. The rest of the power (most of it) is transported via the national power grid to step-down substations all over the country, where it is converted to lower voltage and is sent via the distribution power lines for use by residential and commercial customers, who also have their own sub-stations or transformers for converting the power to exact voltage they can use.

Power Distribution

The "handoff" from electric transmission to electric distribution usually occurs at the substation. America's fleet of substations takes power from transmission-level voltages and distributes it to hundreds of thousands of miles of lower voltage distribution lines. The distribution system is generally considered to begin at the substation and end at the customer's meter. Beyond the meter lies the customer's electric system, which consists of wires, equipment, and appliances—an increased number of which involve computerized controls and electronics operating on DC.

There are two types of distribution networks; radial or interconnected.

a. A radial network leaves the power generating station to its final destination with no connection to any other supply in the network. This is typical of long rural lines with isolated load areas.

b. An interconnected network is generally found in more urban areas and will have multiple connections to other points of supply. These points of connection are normally open but allow various configurations

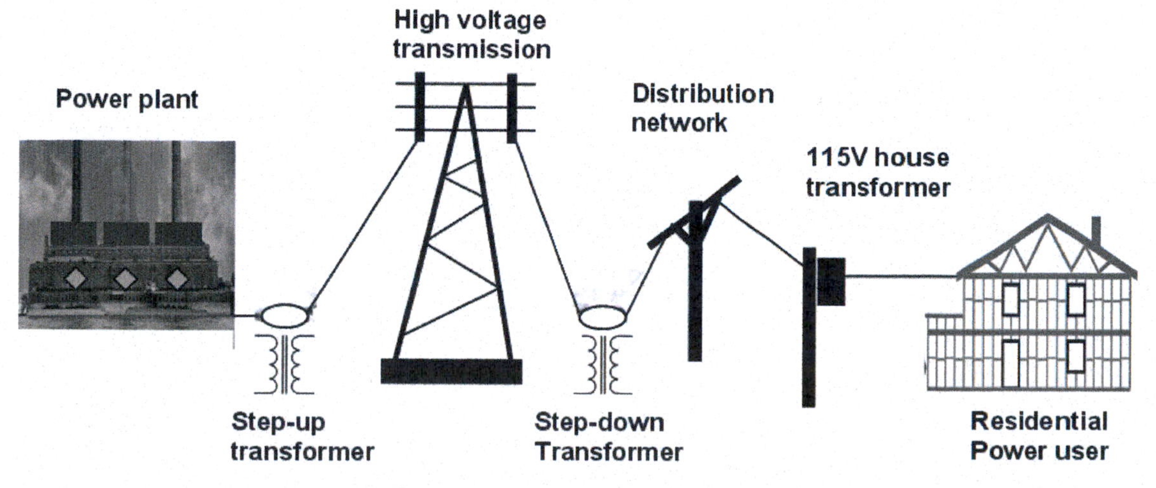

Figure 6-5. The electric power grid

by the operating utility by closing and opening switches.

The distribution system supports retail electricity markets. State or local government agencies are heavily involved in the electric distribution business, regulating prices and rates of return for shareholder-owned distribution utilities. Also, in 2,000 localities across the country, state and local government agencies operate their own distribution utilities, as do over 900 rural electric cooperative utilities. Virtually all of the distribution systems operate as franchise monopolies as established by state law.

The greatest challenge facing electric distribution is that of responding to rapidly changing customer needs for electricity; i.e., increased use of information technologies, computers, and consumer electronics has lowered the tolerance for outages, fluctuations in voltages and frequency levels, and other power quality disturbances. In addition, rising interest in distributed generation and electric storage devices is adding new requirements for interconnection and safe operation of electric distribution systems.

Finally, a wide array of information technology is entering the market that could revolutionize the electric distribution business. For example, having the ability to monitor and influence each customer's usage, in real time (part of the smart grid solution), could enable distribution operators to better match supply with demand, thus boosting asset utilization, improving service quality, and lowering costs. More complete integration of distributed energy and demand-side management resources into the distribution system could enable customers to implement their own tailored solutions, thus boosting profitability and quality of life.

The new PV power installations—millions of small residential and hundreds of large-scale plants—add a new dimension to the complexity of the problems that distribution lines are experiencing. They cause power variations and fluctuations that are hard to control, thus new power management schemes and controls must be developed to handle these newcomers. So the new smart grid issues are on the table and are discussed daily, with some work underway as we speak.

PV Power and the National Grid

As if the electric grid issues were not already too many and too complicated, we are now going to add large-scale PV power generating capacities to the mix (see Figure 6-6). This addition immediately brings a number of complications in the situation, the greatest of which (power availability and variability) are described below.

PV and wind power plants—small and large—are very different from the conventional power generating sources, such as coal-fired and nuclear power plants. The differences are rather great, and we will take a look at some of them in order to understand the PV integration in present-day power distribution and use.

Variability of PV Power Generation

PV plants produce most power when the sun shines brightly on an unobscured sky. In early morning and late

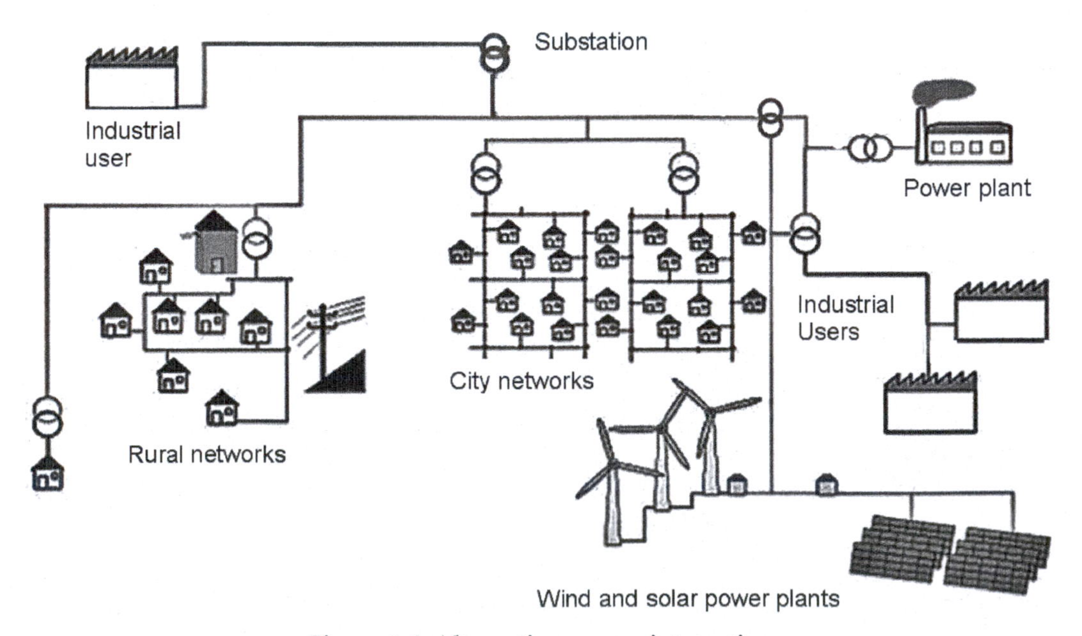

Figure 6-6. Alternative energy integration

afternoon, the sunlight falls on Earth at an angle, and its power is reduced several fold. When a cloud crosses the sunlight path, it reduces the sunlight power several fold as well. Fog, windy/dusty conditions and other climatic phenomena also contribute to reduction of the solar power coming to the PV modules, thus reducing their output dramatically. This makes them "variable" and unpredictable power generators.

Obviously, PV power plant output is not constant all the time because it depends on many factors. It actually varies from season to season, day to day, hour to hour and minute to minute. The variations during a bright sunny day (Figure 6-7a) are insignificant and will not pose any major concerns to grid operators who can predict and handle normal daily production. This includes the early-morning and late-evening low levels of power production, before and after full, bright sunshine days predominant in desert regions.

In Figure 6-7b however, the intermittent cloud cover forces the solar generation to fluctuate wildly. Thus greatly fluctuating power output is sent into the power grid which is supposed to absorb and distribute it efficiently and without variations. Grid power cannot fluctuate—period! This is not easily done, especially if we are considering the output from a large-scale solar power plant (over 100 MWp). Even worse, several such large plants in different areas of the country, where climate and weather conditions are drastically different, will introduce wild variations of power levels and quality sent into the grid from different locations at different times.

These up and down fluctuations could be very significant, with 100 MW power flowing on and off in the grid. This will stress the local grid and will challenge the power grid operators' ability to maintain steady power load conditions. This issue is of great concern today, and since we have no way of making the sun shine consistently, we must concentrate on providing level power output. The smart grid concept and other innovations, such as energy storage and backup will have to tackle and resolve the variability issues in the near future, if we are to consider PV a reliable power source. With many large solar power plants coming on line, it is becoming more important and even urgent to do that.

Combined Wind-solar Peak Load Considerations

Solar energy generation could be combined with other energy sources when constant output is the goal. Addition of solar power to conventional power sources (power plants) is one approach. Using energy storage devices (batteries or water storage) is another. A more natural and most efficient way, however, is combining PV power with wind at certain locations especially chosen for this purpose. At such locations PV power is complementary to the output of wind generation, since it is usually produced during the peak load hours when wind energy production may not be available. Variability around the average demand values for the individual characteristic wind and solar resources can fluctuate significantly on a daily basis. However, as illustrated by Figure 6-8, solar and wind plant profiles—when considered in aggregate—can be a good match to the load profile and hence improve the resulting composite capacity value for variable generation.

In this example, the average load (upper line) is closely followed during the day by the average output from the combined wind and solar generators (the second from top line) during the same time. This average is created regardless, and because, of the fluctuations of the individual wind and solar power generators. This is a marriage made in heaven, and this combined power generating combination will work quite well if wind and solar power outputs can be matched as closely as this one.

Although there are areas in the US and abroad that

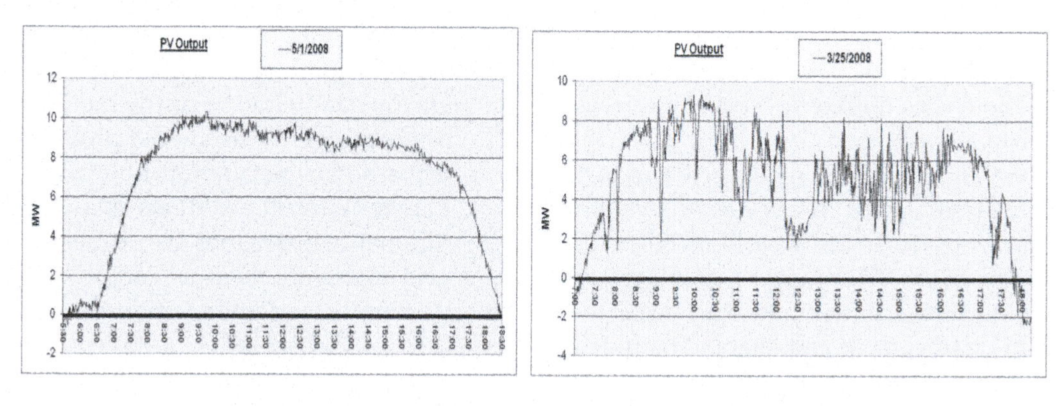

a. Clear day **b. Variable cloudiness day**

Figure 6-7. PV power plant output daily variations

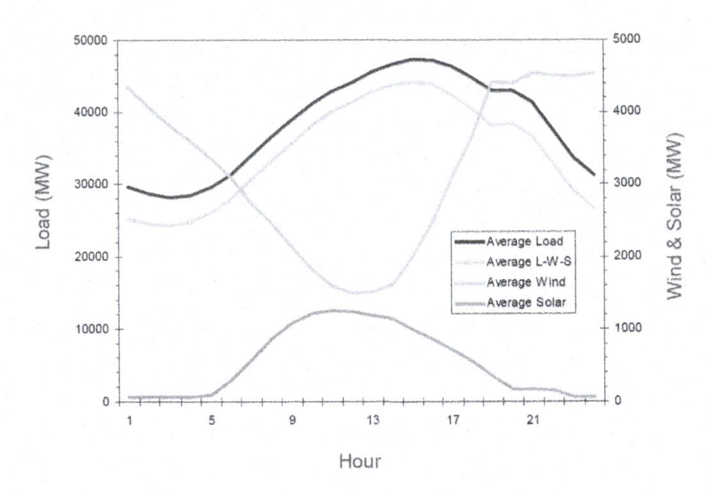

Figure 6-8. Simultaneous wind and solar generation (14)

match this wind and solar profile, the combined effort is usually hard to execute, because the best places for wind and solar are at different locations, often miles apart, and also because of lack of infrastructure at the most suitable locations. Because of that, it will require a great and very expensive effort to implement large-scale "wind-solar load matching" schemes anytime soon with existing technologies. Nevertheless, having as a goal matching wind and solar power outputs will force us to look for and find the most suitable locations and appropriate technologies for this match.

This is not going to happen overnight, but if we approach this solution seriously and intelligently, we will have a large-scale power output—nationally—that matches the grid power loads.

PV Industry's Large-scale Project Strategies

While on the subject of best approaches, we must also consider the way the PV industry itself operates today. Manufacturers, installers, operators and everyone else involved are part of the capitalist system, so they must consider the bottom line first. However, they seldom consider or care about the ultimate success of the projects in which their products are used.

For now we will just take a look at the PV industry, and especially its involvement in large-scale PV power generation and future trends in this area. Remember, these are the embryo years of the PV industry. It is still growing and shaping as we speak and will take a while before it steps on its feet in its final form and shape. Manufacturers are just now realizing that involvement in large-scale projects is the ticket to their success. This is where they can sell more products and make more money. The battle for more and more involvement in these large-scale projects, especially in the US, is just starting.

At this time PV products manufacturers of all types and sizes have one basic *modus operandi*: sell what you have. They are, however, becoming more aware of the fact that due to a number of technological, logistical, and financial considerations, they must get involved hands-on in the projects where their products are used.

Sell What You Have

This option is the simplest and is what manufacturers are best at—manufacture and sell their own products (modules, inverters, frames, etc.) to customers and developers. This is what most present-day suppliers are doing, in order to gain a share in the industry. They are constantly creating and introducing new products and services, tailored especially for large-scale projects (expecting large sale volumes). All these strategies, however, fall into one of two categories:

a. Products geared for large-scale installation (modules, inverters, etc.). At a minimum, most suppliers have introduced a larger series of modules and/or inverters especially for the utility market. The larger size or capacity of these special products is intended to drive down installation time and materials costs. Larger inverter products also often contain additional features that are attractive to utilities and grid operators such as control of reactive power, variability, and voltage ride-through.

b. Pre-designed (modular type) systems have been developed by a number of suppliers as integrated systems, intended for sale and ready to go as a single package. These systems are often shipped to the installation site pre-assembled in order to cut down installation time. At times, these are predesigned for the particular installation in order to drive down system design and engineering time and costs. The claim here is that these pre-engineered and well-made systems save time and money and improve the quality of the installation. The disadvantages range from extreme dependence on manufacturers' time tables and quality to higher initial price and maintenance costs.

Develop and Control the Entire Project

This is the second and still novel approach used by a few PV products manufacturers. It consists of the design and development of their own power plants from scratch, or in partial participation, in the projects at hand. These distinct approaches are under serious evaluation by a number of US, EU, and Asian manufacturers.

These approaches are new and are going through constant revisions and changes, but basically they could be divided into:

a. Full project integration. Here the manufacturer is playing the role of a designer, supplier, developer, installer, and operator. In the best of cases, products and services are owned and managed by the manufacturer/installer/operator entity. There are several examples of this type of *modus operandi*, and we feel that there will be many more. This is the best way for manufacturers to control their volume, price and quality.

b. Semi-integration is the most likely scenario simply because there are no (at least for now) manufacturers who make all products needed for a large-scale PV installation. Semi-integration means that the manufacturers maintain their own brands but have the option of sourcing third-party components. Parent entities' products have inherent advantage in all cases, of course. We have noticed that lately many manufacturers are looking into the second approach because it is the surest way of placing their excess inventory and ensuring increase in sales volumes.

Which approach will prove most successful remains to be seen, but it is the author's opinion, that manufacturers should get fully involved in project development without losing focus on their primary activity—providing quality product. Full involvement in PV projects, as in "a" above (of a single manufacturer-developer), or of several such as in "b" above, will provide the needed confidence in the efficiency and longevity of the products. This will eliminate a number of major technological, logistical, and financial barriers standing before the PV industry. More projects will see the light of day, and their long-term success will be fully ensured.

Even in the best of cases, however, a number of issues will still need to be resolved, the major of which are:

a.) The availability of land suitable for PV applications (including interconnect-ability), and

b.) Utilities willing to purchase the generated power.

There is a lot of land available, but it is not ready for PV, and the utilities have a limited appetite for new PV power, so they pick and choose from the flood of proposed PV projects. Clearly the ball is in the utilities' court, but most of them are not in a hurry, and many large-scale PV power projects will be delayed or cancelled because of that. The political and regulatory winds of late in the US and Europe are also introducing changes that add uncertainty to the solar energy future.

THE DESIGN PHASE

The preliminary investigative work on the PV power plant planning and design will follow established principals and methods, starting with looking into the basics, answering some preliminary diligence questions, and resolving issues identified during the pre-planning and actual planning processes. An outline of the tasks at hand and suggestions for their proper execution follows.

Technical, Administrative and Financial Considerations

The tasks ahead of us, during the power plant's design process (as applicable to California PV projects) can be summarized as follow:

General

a. Verify total useable acreage status and conditions
b. Verify location of solar irradiation and weather pattern characteristics
c. Annual MWh power output estimated for this location
d. PV technology to be used, its efficiency and cost
e. Tracking vs. fixed installation estimates
f. Total MWp of the chosen technology needed vs. MWh/year generated

Electrical

a. Local electrical transmission system condition
b. Ownership and condition of the nearest sub-station
c. Status of the available transmission and distribution lines
d. Electric transmission system interconnection study
e. CAISO/CPUC review and obtaining LGIA or SGIA status
f. Transmission system and interconnect estimate
g. PPA negotiations conditions and barriers
h. CEQA and or NEPA review and downstream transmission upgrades

Land Permitting

a. Permitting authority (CEQA or NEPA) involvement
b. RWQCB, air district, etc. permitting status
c. Biological, water and cultural resources assessment
d. Surveys for listed species and plants
e. Environmental review; ND, MND, EIR, EA/FONSI, EIS status
f. Environmental review and permitting status
g. Environmental impacts; issues; bio, cul, water, air, etc?
h. Mitigation efforts and costs
i. 2081 permit/CDFG or Section 10 consultation/ USFWS status

Financials

a. Power purchase agreement (PPA) status and conditions
b. Utility ownership and status (muni or investor owned)
c. PPA price point per MW/h
d. CPI escalator, term and investor's estimated IRR
e. Levelized cost of energy (LCOE) estimate
f. Key financing ratios; D/E, DSCR, interest/term on debt assumptions
g. Exit strategy if no PPA; market; assumed off-taker or other

The Design Process

The plant design, as used during the design stages of large-scale PV projects in California and Arizona, consists of the following steps:

Technology Selection and Preparations

Using some of the information and guidelines discussed above, we now need to get all necessary information on the technology we have chosen for this plant. This includes manufacturers' specs, information packages, and customer review information to date.

Our choices are actually limited to:

a. c-Si modules
   i. Monocrystalline silicon
      1. Fixed mount
      2. One axis tracker mounted
      3. Dual-axis tracking (not recommended in most cases)
   ii. Polycrystalline silicon
      1. Fixed mount
      2. One axis tracker mounted
      3. Dual-axis tracking (not recommended in most cases)
b. Thin film modules
   i. a-silicon fixed mounting
   ii. SIGS fixed mounting
   iii. CdTe fixed mounting
   iv. One axis tracking (not recommended in most cases)
c. HCPV modules on two axis trackers

PV Arrays Considerations

a. Location, orientation, mounting angle and height, tracking etc., will be carefully considered, and
b. a serious attempt will be made to provide an aesthetically pleasing layout by considering the modules' size and appearance, the shape of the field and the surrounding area characteristics.

Pre-engineered System Packages

a. Packages from different companies will be reviewed and those that contain the best options will be paid special attention, and
b. one will be chosen as a model, and then as a final package.

Product and System Warranties

a. Warranty conditions from each supplier will be carefully reviewed and discussed among the team members.
b. Follow-up discussions and negotiations will be conducted to obtain the best conditions possible for the products and services at hand.

Official Listings and Certifications

a. Listings and certifications as required by authorized officials and agencies (e.g. UL 1703, UL 1741, and any applicable evaluation reports from National Evaluation Services (NES) or International Conference of Building Officials (ICBO) Evaluation Services), and any other applicable such will be carefully reviewed and considered.

b. Only products meeting and exceeding these criteria should be considered, and
c. Additional checks and verification should be performed as well, to ensure the quality and integrity of the plant's building blocks.

System Options
a. All system options will be reviewed and considered, making sure the equipment meets the guidelines of local, state and federal programs.
b. State and federal programs will be reviewed and considered as well as needed to obtain the best conditions for project development.

Local Utility Companies
a. Local power companies will be interviewed and consulted, for the duration.
b. They will be kept posted on all developments and decisions as well.

Documentation
a. Documents will be generated as required by the QC program and reviewed periodically to ensure that the system meets local permitting, interconnection and other requirements.
b. Revisions and additions will be made periodically and/or as needed.

Equipment
a. The type of equipment to be used should be agreed upon, finalized and documented.
b. Equipment purchasing (type quantity and quality) will be done after the documentation package is approved by a consensus.

Buy-down

A power buy-down package should be completed and sent to the appropriate utilities and state authorities and regulators for review and decision.

Local, State and Federal Rules and Regulations

All local, state, and federal rules and regulations, including incentives and subsidies, should be reviewed and incorporated in the design.

Shading and Ground/Sky Obstructions
a. Impact of shading and other obstructions on the PV array layout will be reviewed and considered in the final design.
b. Any and all obstructions will be clearly shown on the final drawings.

Site Drawing
a. The distance between the estimated locations of all system components, including strings and grid interconnect points, will be measured and
b. A complete site drawing will be developed, including one-line diagram of PV system installation as needed for the permit package and for completion of the final field design.
c. The final design package will be certified by state certified local PEs (mechanical and electrical professionals).

Permits
a. A permit package for the local authority having jurisdiction over this project will be assembled and presented at the appropriate time.
b. It should include the following:
   i. Site drawing showing the location of the main system components; PV arrays, above ground and underground conduit runs, electrical boxes, inverter enclosure (housing), control room, critical load subpanels, utility disconnects, main service panel, utility service entrance, etc.
   ii. One-line diagram showing all significant electrical system components.
   iii. Cut sheets for all significant electrical system components (PV modules, inverter, combiner, DC-rated switches, fuses, etc.).
   iv. Copy of filled out utility contract (PPA agreement).
   v. Mechanical and structural calculations and drawings for the support system and structures.

Quality Control

The QC manager of the project is responsible for creation of training manuals, and actual training and retraining of all technical personnel, the installation and QC procedures, the documentation and records keeping, daily and periodic inspections of work sites, as well as the acceptance and performance evaluation of each step of the planning, design, and installation stages.

Plant Availability and Capacity Factors

Two pieces of information we need at the onset of the planning and design stages of a PV plant are estimates of its availability factor (Af) and its capacity factor (Cf), because these have a great influence on its final design and expected performance.

Plant Availability Factor (Af)

The availability factor of a power plant is basically

the amount of time it is able to produce electricity over a period of time, divided by the total amount of the time in the period. Partial capacity availability should be considered. The availability factor is not the same as the capacity factor.

So, a plant in maintenance mode is "unavailable" for the duration. Under this definition, most PV plants are over 95% available, although they idle during the night and in periods of no sunlight. Nevertheless, they are "available" only because the plant is "ready and willing" to go into action, weather permitting.

A more realistic definition that has to be considered in case of PV plants is considering the idle times of the plant during the producing hours of the day only. In this case, the above % Af would be much more meaningful, because it will tell us how available the plant is during the hours we expect it to operate at full capacity—i.e. from 6:00AM to 8:00PM in the summer months. Night hours are "dark" hours, so they should not be considered in any of the technical or financial calculations.

Plant Capacity Factor (Cf)

The capacity factor is a more useful and accurate measure of PV plant performance. It is the ratio of actual output of a power plant over a period of time, and its output if it had operated at full uninterrupted power the entire time (or nameplate capacity). So, as an example:

$$Cf = \frac{\text{Annual MW/h}}{\text{(Hours/year x nameplate)}} \text{ x } 100 = \frac{\text{1,800MW/h}}{\text{(8760h/y x 1MW)}} \text{ x } 100 = 20.5\%$$

In this case, our PV plant with 1.0 MWp nameplate capacity has operated at full capacity an average of 6 hrs/day during 300 days of the year. The plant produced a total of 1,800 MW/h during that time and when this is divided by the sum of the total hours of the year (8760 hrs.) and the nameplate capacity (1 MW) we get a capacity factor of 20.5, which is about average for the southwestern USA.

This can be also understood to mean that the plant was working at 100% capacity 20.5% of the time—or ~1/5 of the 24-hr cycle. PV plants in the Midwest have Cf of 12-15%, while similar plants in Arizona or California deserts would have CF of 19-22%. Wind farms average 40-45%, while nuclear or diesel fuel plants could have Cf of over 90%, simply because they can run unobstructed, non-stop and at full capacity all the time and for long periods of time.

PV plants cannot run non-stop and we must be very careful in estimating their Af and Cf, when designing the power field at a particular location with existing PV technologies. In estimating the Cf of PV plants we need to take a very good look at the Af and all other variables and abnormalities that are expected, or will probably occur, during the first few years and all the way to year 30.

Historical weather data of the location could be used to approximate the solar availability, and from it the total energy produced. A margin of error must be considered here, based on the variability of the historical data and different trends we observe in it. Historical data of equipment performance (from the manufacturers), along with utility company demand and supply records could provide information for estimating the expected power generation and losses. A margin of error should be considered here as well, based on the available data. Although the margin of error might be significant, these estimates are vitally important, and must be painstakingly worked out and dutifully incorporated in the design and finance process estimates and calculations. They could be also used for providing a baseline of the new plant's performance to be used in future O&M and financial analysis.

Solar Resource and Temperature

As part of the plant design's yield calculations and array sizing purposes, the solar irradiance data required by the performance model are of utmost importance. They determine the overall power production, and must be thoroughly researched. Some of the most reliable data are typically obtained from long-term meteorological models providing hourly averaged sunlight values, such as NASA's satellite weather data. These data could be complimented by more detailed weather data for the specific location obtained from local weather stations, which will allow more accurate calculations and predictions. Thus obtained solar irradiance and weather data can be manipulated using different methods to calculate the expected solar irradiance incident on the surface of a photovoltaic module positioned in an orientation that is determined by the power field design. The margin of error in all cases would be significant, because no one can predict the weather exactly—regardless of the accuracy of historical data—and there will always be surprises and errors. We can only hope to minimize these to an acceptable level, which at the end will determine the quality of our work.

The PV module design also has a lot to do with its performance, as far as temperature coefficient and overall efficiency and longevity are concerned, since these will vary from location to location. Does the manufacturer have such data? Is he willing to share them freely and assist in overall plant planning and design? In all cases, during long-term performance the solar irradiance in the plane of the module is often a measured value and should be used directly in the performance model. The level of irradiance is very important to the module's performance, especially with the quick expansion of PV plants in the deserts, so it must always be taken into consideration and dealt with carefully.

In Figure 6-9, note the 15-20% drop of Voc and Vmp at increasing the temperature from 25°C (room temperature) to 50°C (average—but not highest—desert day temperature). These results could be extrapolated for 85°C, which we measure in the modules during hot summer days in the deserts, and which will show a significant additional drop of output approaching 30% total drop of power.

Actual cell temperature and back-surface module temperature can be distinctly different for different PV modules, but particularly for concentrator type modules, since their temperatures are much higher—well over 150°C. The temperature of cells inside the module can be related to the module back surface temperature through a simple relationship, but is usually significantly higher. The relationship given in the equation below is based on an assumption of one-dimensional thermal heat conduction through the module materials behind the cell (encapsulant and polymer layers for flat-plate modules, ceramic dielectric and aluminum heat sink for concentrator modules).

Cell temperature inside the module is then calculated using a measured back-surface temperature and a predetermined temperature difference between the back surface and the cell.

$$T_c = T_m + E/E_o \bullet \Delta T$$

where:

T_c= Cell temperature inside module, (°C)

T_m = Measured back-surface module temperature, (°C).

E= Measured solar irradiance on module, (W/m^2)

E_o= Reference solar irradiance on module, (1000 W/m^2)

ΔT = Temperature difference between the cell and the module back surface at an irradiance level of 1000 W/m2. This temperature difference is typically 2 to 3°C for flat-plate modules in an open-rack mount. For flat-plate modules with a thermally insulated back surface, this temperature difference can be assumed to be zero. For concentrator modules, this temperature difference is typically determined between the cell and the heat sink on the back of the module.

Within module temperature variations and excesses are one of the key reasons for PV module efficiency decrease. They also contribute to annual degradation and excess failure rates of all types of PV technologies. Because of that, PV cells and modules manufacturers must account for these variations by issuing theoretical modeling as in the above example, and actual test data,

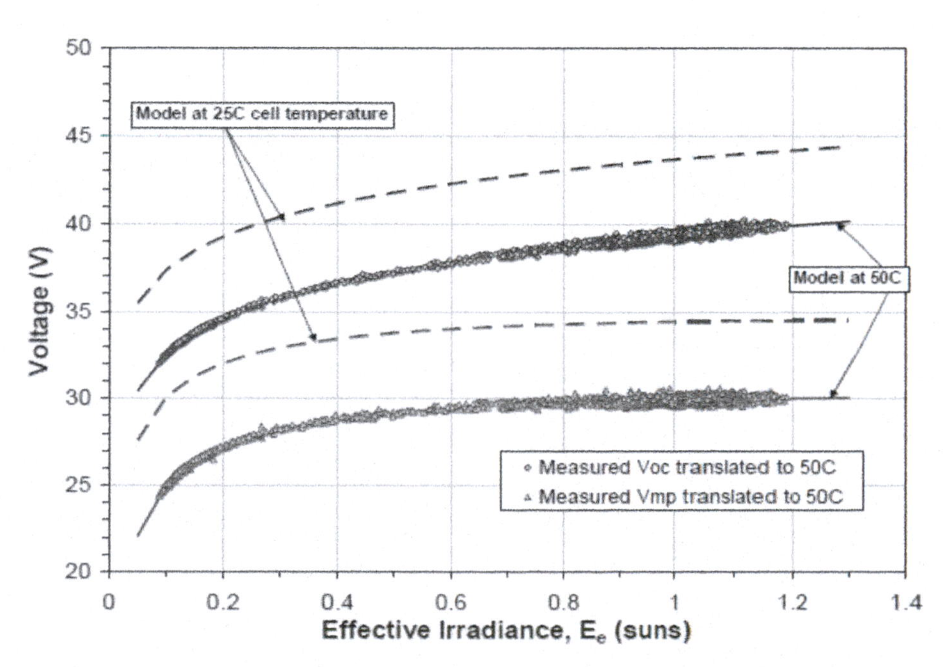

Figure 6-9. Over 3300 measurements recorded on five different days with both clear sky and cloudy operating conditions for 165-Wp mc-Si module (2).

to provide enough confidence in the product operating under different weather conditions—and especially under excess heat.

Short of that, the power plant design team must come up with the data by performing the above calculations and estimates as needed to provide a complete and clear picture of operating conditions in the specific PV field.

Power Field Performance

We discussed the QC/QA procedures and issues above, and have also seen some of these applied in some areas of PV power plants design, implementation and performance, albeit in a chaotic manner, or in desperate situations. The general consensus is that partial quality control is better than no quality control. Because there are no accepted, completely integrated, interlocking, all-encompassing quality control procedures or standards for PV components manufacturing, or PV power plants installation and operation at present, improvisations and on-the-fly modifications are the accepted *modus operandi*.

Planning and design efforts of a new PV power plant must start with an impartial and competent power field assessment—choice of location, weather conditions assessment, land improvement and permitting issues, PV technology selection, interconnect and PPA conditions etc. All this leads to estimates of the long-term performance of the field. There are several major factors to consider first in this process, but remember that these are just separate elements of a long and complex process that has to be standardized and globally accepted before it is fully functional.

Some of the key elements to be considered during a properly conducted, quality-oriented planning, design, installation and operation process are:

Final Power Yield

The final yield is a reflection of the bottom line of the power plant performance. It is derived by plugging many variables in a long estimate. It is the net AC energy output of the power field divided by the aggregate nameplate power of the installed PV array at STC (1000W/m2) solar irradiance and 25°C cell temperature.

$$\text{Final Yield} = \frac{\text{kWh AC}}{\text{kWDC}}, \text{ or}$$

It represents the number of hours that the PV array would need to operate at its rated power to provide the same energy. All UL-listed modules have a nameplate number on the back of the module that identifies the STC rated DC power. The aggregate array power can easily be determined by summing the nameplate power ratings for the array.

NOTE: While the power field nameplate is simply a sum of the modules' output tested at STC in a lab and is easy to calculate, the generated AC energy (kWh AC) in the equation is a multiple of many variables and factors. Thorough knowledge of the system operation is needed here, in order to come up with an accurate number during the planning and design stages. A team of experts is needed to consider and calculate each variable at this early stage, which might make the difference on the long run, because error in or overestimation of the plant's performance could be detrimental to its bottom line.

Reference Yield

The reference yield is the total in-plane solar insolation (kWh/m2) divided by the array reference irradiance. It represents an equivalent number of hours at the reference irradiance. The reference irradiance is typically equal to 1 kW/m2; therefore, the reference yield is the number of peak sun-hours.

$$\text{Reference Yield} = \frac{\text{kWh/m}^2 \text{ (Total insolation)}}{1000\text{W/m}^2}$$

NOTE: This variable is also critical, and here just like above, the lower number (STC conditions) is fixed, while the upper number (total insolation) depends totally on a combination of the climate/weather/seasonal conditions. Coming up with a precise number for it is impossible, but a thorough look at the historical data and the weather patterns, and with some luck, we could get a good picture of what to expect—give or take certain percentage.

Performance Ratio

The performance ratio is the final yield divided by the reference yield and is dimensionless. It represents the system's performance in terms of total losses due to equipment issues, restrictions or under-spec conditions. Typical system losses include DC wiring losses, cells/modules mismatch, bypass diodes issues, module temperature effects, annual degrading rate, inverter conversion efficiency degrading (due to temperature or age), transformers issues, etc..

$$\text{Performance Ratio} = \frac{\text{Final Yield}}{\text{Reference Yield}}$$

The proper and accurate estimate and representation of the above variables—major elements of the overall power filed yield assessment—are critical for proper design and efficient and profitable operation of our PV power field. They need to be looked at very carefully, especially when designing or evaluating a PV power plant or evaluating its operation. Again, impartiality and thorough understanding of the subjects and issues at hand is a must, in order to provide a reasonably accurate, quality yield assessment. Without that, serious problems will persist all through the PV plant's operation.

We will take a look below at some additional key variables and factors, determining the quality of the power field design and performance, starting with the solar irradiance and equipment requirements.

So, above we looked at one of the critical elements of our new PV power plant—the overall theoretical yield and performance assessment. To complete this effort, we also need to:

1. Verify that the location is appropriate for solar power generation,

2. Verify that the chosen PV technology works as planned by conducting actual field tests and evaluations, and

3. Establish parameters for performance and yield tests during the operation stage.

PV Modules and Strings Evaluation and Modeling

Executing all design and production steps perfectly means a lot of engineering time and usually results in a higher-cost final product. Nevertheless, some basic requirements must be kept in mind when looking into purchasing those 500,000 PV modules for our 100 MWp power field. In all cases, properly modeling the field performance of PV modules is an absolute requirement—during the design, manufacturing, QC, and final test stages. This should be verified during the design of the power field as well, for it will help us estimate the performance and longevity of the PV products (modules, inverters etc.).

The string design process must take into consideration that in large PV arrays, individual PV modules are connected in both series and parallel. A series-connected set of solar cells or modules is called a "string." The combination of series and parallel connections may lead to several problems in PV arrays. One potential problem arises from an open-circuit in one of the series strings. The current from the parallel connected string (often called a "block") will then have a lower current than the remaining blocks in the module. This is electrically identical to the case of one shaded solar cell in series with several good cells, and having the power from the entire block of solar cells lost. Figure 6-10 shows this effect.

One thing to keep in mind is that although all modules may be identical and the array does not experience any shading, mismatch and hot spot effects may still occur. Parallel connections in combination with mismatch effects may also lead to problems if the by-pass diodes are not rated to handle the current of the entire parallel connected array. For example, in parallel strings with series connected modules, the by-pass diodes of the series connected modules become connected in parallel, as shown in Figure 6-11. A mismatch in the series connected modules will cause current to flow in a by-pass diode, thereby heating this diode. The affected diode could overheat and eventually fail. In all cases it will also affect the module/string performance negatively.

The current may now flow through the by-pass diodes associated with each module, but must also pass through the one string of by-pass diodes. These by-pass diodes then become even hotter, further increasing the current flow. If the diodes are not rated to handle the current from the parallel combination of modules, they will burn out and allows damage to the PV modules.

In addition to the use of by-pass diodes to prevent mismatch losses, an additional diode, called a blocking diode, may be used to minimize mismatch losses. A blocking diode is typically used to prevent the module from loading the battery at night by preventing current flow from the power source through the PV array. With parallel connected modules, each string to be connected in parallel should have its own blocking diode. This not only reduces the required current-carrying capability of the blocking diode, but also prevents current flowing from one parallel string into a lower-current string and therefore helps to minimize mismatch losses arising in

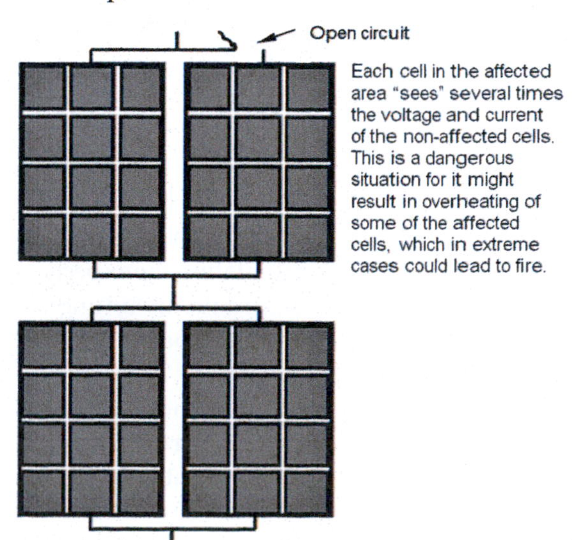

Figure 6-10. Open circuit effects in larger PV arrays.

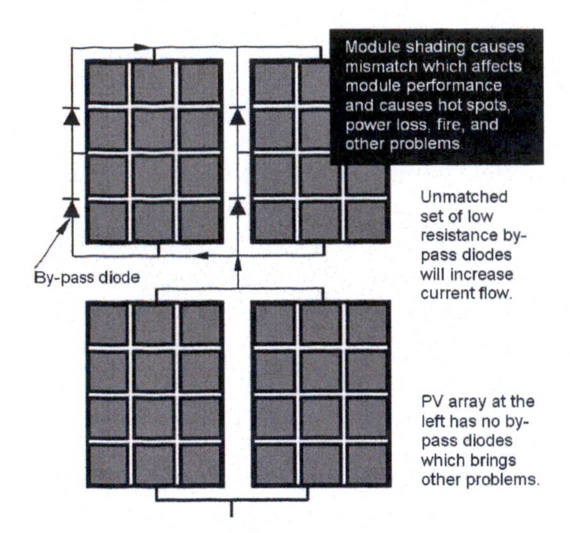

Figure 6-11. Bypass diodes in parallel connections

parallel connected arrays.

When modeling, we must start with theoretical suppositions and empirical formulations, to be followed by lab and field tests for final verification purposes. The theoretical estimates are intended to simulate the expected operating conditions and give us an idea of the expected performance. When performing final tests of the new product, we need to do over and over again, in order to verify the results and adjust our design and manufacturing operations. When we design a power field we need to look at the test data and compare it with our own data (from calculations and actual tests done with the modules to be used in the project). Appropriate modules-arrays system performance modeling is a must.

The following equations define a model used by Sandia for analyzing and modeling the performance of photovoltaic modules. The equations describe the electrical performance for individual photovoltaic modules, and can be scaled for any series or parallel combination of modules in any type or size array. The same equations apply equally well for individual cells, for individual modules, for large arrays of modules, and for both flat-plate and concentrator modules. (2)

The form of the model given by Equations 6-1 through 6-10 is used when calculating the expected power and energy produced by a module, assuming that pre-determined module performance coefficients and solar resource information are available. The solar resource and weather data required by the model can be obtained from tabulated databases or from direct measurements.

The three classic and key points of a PV module, current-voltage (I-V) curve, short-circuit current/open-circuit voltage, and the maximum-power point, are given by the following equations.

$$I_{sc} = I_{sco} \cdot f_1(AM_a) \cdot \{(E_b \cdot f_2(AOI) + f_d \cdot E_{diff})/E_o\} \cdot \{1 + \alpha_{Isc} \cdot (T_c - T_o)\} \quad (6\text{-}1)$$

$$I_{mp} = I_{mpo} \cdot \{C_0 \cdot E_e + C_1 \cdot E_e^2\} \cdot \{1 + \alpha_{Imp} \cdot (T_c - T_o)\} \quad (6\text{-}2)$$

$$V_{oc} = V_{oco} + N_s \cdot \alpha(T_c) \cdot \ln(E_e) + \alpha_{Voc}(E_e) \cdot (T_c - T_o) \quad (6\text{-}3)$$

$$V_{mp} = V_{mpo} + C_2 \cdot N_s \cdot \alpha(T_c) \cdot \ln(E_e) + C_3 \cdot N_s \cdot \{\alpha(T_c) \cdot \ln(E_e)\}^2 + \alpha_{Vmp}(E_e) \cdot (T_c - T_o) \quad (6\text{-}4)$$

$$P_{mp} = I_{mp} \cdot V_{mp} \quad (6\text{-}5)$$

$$FF = P_{mp}/(I_{sc} \cdot V_{oc}) \quad (6\text{-}6)$$

where:

$$E_e = I_{sc}/[I_{sco} \cdot \{1 + \alpha_{Isc} \cdot (T_c - T_o)\}] \quad (6\text{-}7)$$

$$\alpha(T_c) = n \cdot k \cdot (T_c + 273.15)/q \quad (6\text{-}8)$$

The two additional points on the I-V curve are defined by Equations 6-9 and 6-10. The fourth point (Ix) is defined at a voltage equal to one-half of the open-circuit voltage, and the fifth (Ixx) at a voltage midway between Vmp and Voc. The five points provided by the performance model provide the basic shape of the I-V curve and can be used for a close approximation of the entire I-V curve.

$$I_x = I_{xo} \cdot \{C_4 \cdot E_e + C_5 \cdot E_e^2\} \cdot \{1 + (\alpha_{Isc}) \cdot (T_c - T_o)\} \quad (6\text{-}9)$$

$$I_{xx} = I_{xxo} \cdot \{C_6 \cdot E_e + C_7 \cdot E_e^2\} \cdot \{1 + (\alpha_{Imp}) \cdot (T_c - T_o)\} \quad (6\text{-}10)$$

where:

- I_{sc} = Short-circuit current (A)
- I_{mp} = Current at the maximum-power point (A)
- I_x = Current at module $V = 0.5 \cdot V_{oc}$, defines 4th point on the I-V curve
- I_{xx} = Current at module $V = 0.5 \cdot (V_{oc} + V_{mp})$, defines 5th point on the I-V curve
- V_{oc} = Open-circuit voltage (V)
- V_{mp} = Voltage at maximum-power point (V)
- P_{mp} = Power at maximum-power point (W)
- FF = Fill Factor (dimensionless)
- fx = empirical functions quantifying the variation in the solar spectrum and optical losses due to solar angle of incidence variations
- N_s = Number of cells in series in a module's cell-string
- N_p = Number of cell-strings in parallel in module
- k = Boltzmann's constant, 1.38066E-23 (J/K).
- q = Elementary charge, 1.60218E-19 (coulomb)
- T_c = Cell temperature inside module (°C)
- T_o = Reference cell temperature, typically 25°C
- E_o = Reference solar irradiance, typically 1000 W/m2
- $\alpha(T_c)$ = 'Thermal voltage' per cell at temperature T_c. For diode factor of unity (n=1) and a cell temperature of 25°C, the thermal voltage is about 26 mV per cell.

Figure 6-12 illustrates the three key points (current-voltage (I-V) curve, short-circuit current/open-circuit voltage, and the maximum-power point), along with two additional points (Ix and Ixx) that better define the shape of the curve.

The formulas of the performance model can be re-written to 'translate' measurements at an arbitrary test condition to performance at a desired reporting (reference) condition. In addition, these equations are applicable to a single cell, a single module, a module-string with multiple modules connected in series, and to an array with multiple module-strings connected in parallel.

The equations use coefficients from the module database that are matched to the modules in the array being evaluated. The performance model was designed to make it unnecessary to account for the number of modules or module-strings connected in parallel. However, for the voltage translation equations to work correctly it is necessary to specify how many modules are connected in series in each module-string.

The "translation" from arbitrary to actual operating, or desirable condition is:

$$I_{sco} = I_{sc}/[E_e \cdot \{1+\alpha_{Isc} \cdot (T_c-T_o)\}] \quad (6\text{-}13)$$

$$I_{mpo} = I_{mp}/[\{1 + \alpha_{Imp} \cdot (T_c-T_o)\} \cdot \{C_0 \cdot E_e + C_1 \cdot E_e^2\}] \quad (6\text{-}14)$$

$$V_{oco} = V_{oc} - M_s \cdot N_s \cdot \alpha(T_c) \cdot \ln(E_e) - M_s \cdot \alpha_{Voc}(E_e) \cdot (T_c-T_o) \quad (6\text{-}15)$$

$$V_{mpo} = V_{mp} - C_2 \cdot M_s \cdot N_s \cdot \alpha(T_c) \cdot \ln(E_e) - C_3 \cdot M_s \cdot N_s \cdot \{\alpha(T_c) \cdot \ln(E_e)\}^2 - M_s \cdot \alpha_{Vmp}(E_e) \cdot (T_c-T_o) \quad (6\text{-}16)$$

$$P_{mpo} = I_{mpo} \cdot V_{mpo} \quad (6\text{-}17)$$

$$FF_o = P_{mpo}/(I_{sco} \cdot V_{oco}) \quad (6\text{-}18)$$

$$I_{xo} = I_x/[\{1 + (\alpha_{Isc}) \cdot (T_c-T_o)\} \cdot \{ C_4 \cdot E_e + C_5 \cdot E_e^2\}] \quad (6\text{-}19)$$

$$I_{xxo} = I_{xx}/[\{1 + (\alpha_{Imp}) \cdot (T_c-T_o)\} \cdot \{ C_6 \cdot E_e + C_7 \cdot E_e^2\}] \quad (6\text{-}20)$$

where:

M_s = Number of modules connected in series in each module-string

T_c = Cell temperature inside module, °C. This value can be refined using Eqn. (6-12) by starting with measurements of back-surface module temperature.

E_e = 'Effective' irradiance is the solar irradiance in the plane of the module to which the cells in the module actually respond, after the influences of solar spectral variation, optical losses due to solar angle-of-incidence, and module soiling are considered.

Note:

$$E_e = f_1(AM_a) \cdot \{(E_b \cdot f_2(AOI)+f_d \cdot E_{diff})/E_o\} \cdot SF$$

where:

E_b = $E_{dni} \cdot \cos(AOI)$

E_{dni} = Direct normal irradiance from pyroheliometer, (W/m²)

NOTE: Other parameters in the above formulas are the same as previously defined for individual modules.

So, plugging the corresponding numbers into the above equations, we could predict the performance of modules and arrays under different operating conditions, including excess temperature, voltage excursions caused by climate and other abnormal conditions. This is a useful tool when designing, or evaluating a PV installation, or

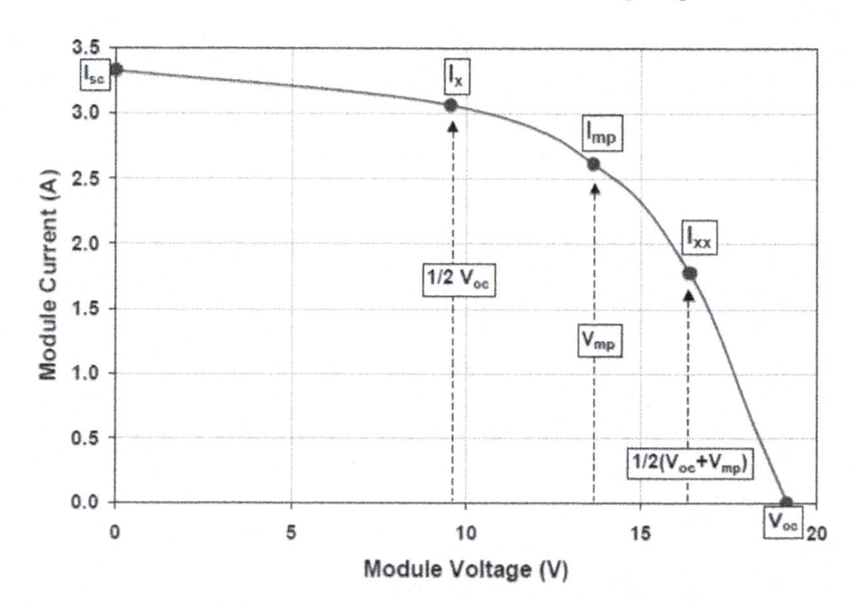

Figure 6-12. Module I-V curve showing the five points on the curve that are provided by the Sandia performance model (2).

for calculating the performance of individual modules for comparative purposes under different operating conditions and periods of time, provided that no extraneous conditions or manufacturing defects are allowed to affect the model.

Thirty years exposure to the harsh desert elements in a large field is quite extreme, to be sure, and very different from the short-term tests done on a lab bench during the certification or manufacturing verification tests. Theoretical calculations and estimates, using different variables, could help us estimate the performance and foresee some of the problems especially those related to high-temperature exposure. But can our calculations and modules withstand the test of time in the desert? Can the model help us predict the behavior and faith of the modules after 30 years of broiling heat? Can we calculate the longevity of the modules as a function of their performance under the particular conditions at the chosen location? These calculations are the responsibility of the power plant design team, because the entire project depends on the accurate answers to the above questions.

Has the manufacturer whose product we are planning to purchase, done these calculations and tests? Can we see them documented in a quality plan? Can we also see SPC charts of tests with actual product done in an attempt to confirm the initial performance/design considerations and theoretical estimates and calculations? If so, then we will have more confidence in the product. This is also the responsibility of the design team, and it is up to its members to figure out how to obtain the necessary information. This task is difficult, because the manufacturers' full cooperation is needed but is not easy to obtain in many cases.

Modeling is not always perfect, or as George Box said 20 or more years ago, "All models are wrong, but some are useful." Nevertheless, we must have a good idea of the performance characteristics of the particular type of modules we are buying for that special 100 MWp PV power plant we are planning and designing, or else we might be unpleasantly surprised on the long run.

NOTE: The trend today is to buy the cheapest available PV modules. We are so used to cheap, disposable products that this is accepted practice. While "cheap" might be justifiable in a small installation, it is absurd to even consider as a factor in a large-scale PV installation.

Just think about the difference between a disposable toy, which will end up in the closet after a week of play, and a PV module, which will be operating non-stop 30 years in the most inhospitable places on Earth. Just imagine the excess heat and UV radiation during the day, freezing at night, sand storms, hail, and whatever else Mother Nature throws its way. And boy, she knows how to throw things.

Optical and Visual Field Degradation

Tests done by the author and his associates with a number of different c-Si PV modules exposed to direct sunshine for long periods of time—over 10 years under direct desert sunlight—show significant visual signs of degradation. Some of the signs can be easily seen by the naked eye even during the first year or two. These range from discoloration (browning) of central region of cells, to stains (halos and rainbows) on the module's glass surface. The degradation, which causes the change in color of the EVA (yellowing) is demonstrated visually by producing the characteristic yellow or brown stains on its surface. The EVA discoloration process, according to some researchers, is most likely due to the formation of polyenes and carbonyl-polyenes compounds after long-term exposure to heat and the elements.

Visual inspection of the reflectance from the individual cells shows degradation also of the AR (anti-reflection) film as a discoloration and decrease of intensity of the blue AR film color. Degraded modules also show an overall increase of reflectance, which contributes to the measured electrical losses. The increase of reflectance was attributed at times to optical interference from EVA degradation which is most likely accompanied by production and emitting of chemical byproducts such as lactones, ketones and acetaldehydes. Once the module encapsulation materials start decomposing, in addition to generating harmful gasses, they will also allow environmental chemicals, gasses and moisture to penetrate into the module's interior. Inside the modules these foreign substances would have ample time and opportunity to attack the solar cells' structure thus causing gradual decrease of power and eventually total failure of the modules.

A disturbing trend lately is the sale and installation of millions of PV modules (TFPV mostly) which lack the much needed edge protection of side frames. Instead, these modules consist of two glass panes bonded together with organic thermoplastic materials. Although the quality of these materials has improved significantly lately, they are no match for the ferocious IR and UV radiation they will be exposed to during 30+ years of non-stop on-sun operation in extreme heat and moisture environments. All organic materials decompose with time under these conditions, which in the case of frameless TFPV modules opens the door for the harmful elements to penetrate into the module and damage the active PV structure.

So the question here is how to predict and prevent quick deterioration and premature failure of the protective encapsulation layers and the subsequent destruction of the active module components. What are the variables that we should keep in mind, and what should be done to ensure proper protection in preserving optimum performance of PV cells and modules in large-scale PV plants? What kind of quality control is needed to provide maximum quality of the final product? How can we attract the manufacturers' attention in a positive way, so as to obtain the relevant data and cooperation needed for the analysis of the quality of their products?

Our goal here is to address and answer these questions, or at least bring them in the open to initiate open and thorough discussions on these key issues.

QA/QC Verification and Enforcement

The performance and longevity of a PV power generating installation depends on the quality of its components, PV modules and BOS equipment, and the quality of the power field design, installation and operation. PV modules' quality is one of the first discussions taking place when considering a PV installation of any type or size. There are a number of PV module manufacturers with stellar reputations for quality and customer support. So how do we make sure that we *find* the best manufacturers, in order to make the right selections of materials, components and services for our PV project?

PV modules manufactured by world-class companies perform perfectly and for a long time without major degradation or failures. Why do these modules do so well, while others fail right out of the box? Installations with proper design and implementation are performing as specified and expected, while others don't fair as well. What is the difference? The simple answer is quality—quality of supply chain materials, equipment, processes and procedures; quality of the quality control; quality of the people and the management team. The quality of the entire cradle-to-grave process—from sand to decommissioning of the power field—is what makes the difference.

Asking the manufacturers to provide data on their quality control procedures usually results in silence, or garbled responses. So what is the solution? Of course, the old and proven, "Buy from reputable companies" is the best way to go. But since this is not always possible, what can we do if we must deal with a new company—especially for the purchase of large quantity PV modules?

Just imagine 500,000 PV modules are to be delivered for installation in our 100 MWp power field and we have no idea of their quality, except for what we see in the enclosed documentation provided by the manufacturer. What an overpowering feeling of helplessness this must be for any professional who knows the odds of success when using untested products with unverified quality from unproven manufacturers. How can anyone who is in charge of such a project allow installation of such a large quantity of products (which are supposed to last 30 years) without thoroughly verifying the product quality and testing its performance? Yet, it happens every day. It happens because we are so used to believing what we see on the product label. But the label is the last step, step #323, of a 6-months-long "sand-to-module" (semi-controlled) manufacturing process, done 6,000 miles away by people we don't know and who use equipment and procedures that are not thoroughly revealed to us.

So here are several options to be considered toward providing some assurance of the quality of the PV modules we plan to purchase from a particular manufacturer:

1. *Request a copy of the manufacturer's:*
   a. Engineering specifications and manufacturing procedures
   b. QA/QC system manual and documentation
   c. SPC program documentation
   d. Quality control inspections and tests documents
   e. Supply chain qualification and control
   f. Management of change documentation
   g. Equipment maintenance and calibration
   h. Technicians training/qualification program
   i. PM and safety programs
   j. Corrective action system
   k. Test and performance measurements documents
   l. Warranty policies workmanship and long-term performance warranty
   m. Customer feedback and suggestions log

2. *Request a site visit*

   A site visit with access to the production line; people, equipment, process specs, quality control logs, etc. If a visit is allowed, then take a close look at and request documentation of:
   a. Raw silicon supplier qualification, material testing and QC control
   b. Process chemicals (acids, bases etc.) supplier qualification and QC control
   c. Process gasses (Ar, N2, etc.) supplier qualification and control
   d. EVA, front glass and back cover supplier qualification and control
   e. Process specs list and compliance verification
   f. In-process quality control procedures and final QC tests data review
   g. Observe the manufacturing operations.

3. *Quality verification*

   If a site visit with a quality verification option is granted, then we need to:
   a. Set up a QC station at key steps of the process,

including the final test.

b. Verify proper execution of the steps and check the quality if possible.
c. Observe the final test procedure and verify final product quality

Far fetched, right? Yes, if we were buying a new toy, but not when buying 500,000 PV modules to be installed in the Arizona desert, where they are expected to operate properly and profitably for over 30 years. Not far fetched at all. As a matter of fact, it is our responsibility to find out as much as we can before making a final decision.

This would be the best way to ensure quality and save many long-term problems. Usually, however, it is difficult to even obtain permission for a site visit, let alone performing quality checks there.

Final Plant Considerations

We have discussed a number of subjects related to the manufacturing process and the PV products we plan to use, but control of the quality of the power field itself—the planning, design, installation and operation stages—is something we also need to look into. It's not an easy task, because while some quality control specs and procedures are used in some manufacturing operations, power field quality control is partial at best today. Its complete version is only a thought in the mind of some responsible professionals at this time. There are no established QC/QA standards for completely integrated control of the quality of all materials, procedures and services during the power field inception and duration. In most cases only certain areas are subject to some quality control and inspection procedures. It is fair to say that no large-scale PV project is under complete and undisputable quality control today, at least to our knowledge.

Quality control of the power field starts at the planning and design stages. It consists of a thorough knowledge and understanding of the major components, their interactions and the forces acting upon them that influence their operation.

Now we need to put the following on paper and use the results in the power field design:

a. Field design considerations and drawings
   i. Technology choice
   ii. Land location and preparation
   iii. Support structure choice
   iv. Mounting choice (fixed tilt, trackers)
   v. Packing density (row spaces)
   vi. Equipment positioning (inverters, combiners, etc.)
   vii. Wiring size, pathways and methods
b. Key equipment and control instrumentation selection
c. Installation, QC, and O&M procedures and manuals
d. Mechanical calculations, considerations and drawings
e. Electrical calculations, considerations and drawings
f. Performance charts and calculations
g. Aesthetics and socio-political considerations
h. Permitting points, issues and negotiations
i. NEC, OSHA and EPA considerations

After taking into consideration all technical, regulatory and logistical issues and having chosen the location and the technology to use, we can apply our knowledge to put down the final plant layout on paper. PV modules and arrays will be laid in their respective rows and mounted on the chosen structures. Proper size wires will be run from the rows or strings to the inverters which will then be hooked into the grid. A number of control and measurement systems will be interconnected as well—and the field is ready for operation—at least on paper.

Figure 6-13 is a sketch of our final design for a 1.0 MWp DC power modular unit for use in a large-scale PV power plant in California deserts. This particular design uses 4 ea. 250 MW inverters, but the same concept applies when using different number of inverters. Each of these units is basically a self-contained modular PV power generator. Several of these could be installed and interconnected in any field, using any type of modules, inverters and BOS. This package will work flawlessly, if there is sunshine to generate power and a nearby interconnection point to send the generated power to.

THE INSTALLATION PHASE

Installation of components in the large-scale power PV field will start upon agreement by all parties on the technical, administrative and financial conditions, and after thorough evaluation of the design plans, specs, bids, proposals and contracts.

The actual work on the components and systems installation will follow established principals and methods, some of which are outlined below. In the absence of one accepted standard, the text below is only a guideline—a glimpse into the complexity of the PV power plants' design and installation procedures.

The Pre-installation Process

The installation (construction) phase of the power plant starts immediately after approval of the design documentation, obtaining the necessary permits, PPA and financing. Before starting installation, however, the project

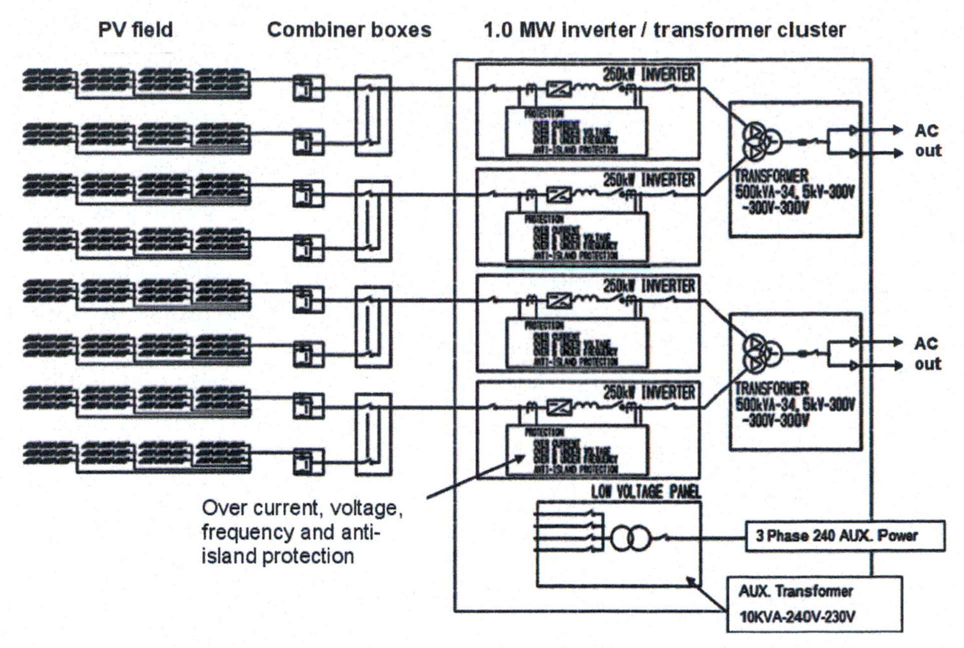

Figure 6-13. 1.0 MWp module of a large-scale PV power generating plant

manager and the coordinators of the respective areas of the installation process must make sure that all necessary steps of the planning and design processes have been completed successfully, and the relevant documentation has been properly executed.

The pre-installation process usually follows the steps outlined below:

1. Required permit documents and materials will be submitted to the authorities, and preliminary approvals will be issued prior to starting any construction activities.
2. Receiving schedules for plant equipment and materials, and the activities related to preparation for their installation will be finalized and followed through.
3. All equipment will go through a thorough initial inspection to verify quantity received, and to make sure that it has arrived without any modifications or shipping damage.
4. Actual field tests with some equipment (PV modules) should be performed to verify performance and compliance.
5. Installation instructions for each component will be finalized and reviewed, and the responsible personnel will be trained to become familiar with the equipment and the installation procedures.
6. Length of wire runs from PV modules to combiner and inverter will be verified and documented. Trial runs must be performed to verify the process steps.
7. Ampacity of PV array circuits will be verified to determine the minimum wire size for current flow.
8. Wire runs must be verified based on maximum short circuit current for each circuit and the length of the wire run, and as follow:
   a. The minimum wire ampacity for the wire run from modules to combiner is based on module maximum series fuse rating printed on the label.
   b. The minimum wire ampacity for the wire run from combiner to inverter is based on the number of module series strings times the maximum series fuse rating.
9. The size of the PV array wiring must be such that the maximum voltage drop at full power from the PV modules to the inverter is 3% or less, for
   a. Wire run from modules to combiner
   b. Wire run from combiner to inverter
10. Length of wire run from inverter to main service panel must be established and verified. The goal is 1% voltage drop for AC-side of system (3% maximum).
11. Main service electric panels must be checked to determine if they are adequately sized to receive the PV breakers or whether the panels must be upgraded.

Once these procedures have been completed, double-checked and signed-off by the responsible personnel—including the QC manager—the installation process can begin.

The Installation (Construction) Process

Installation of PV and BOS equipment should be done by a certified solar energy equipment installer/contractor. Several contractors, specialists in the different areas of the plant's construction could be employed as well.

The installation process usually follows the sample procedure below:

1. Review instruction manual and train technicians accordingly. The proper execution of this work effort will be supervised by the project manager and verified and documented by the QC manager and his crew.

2. The pre-installation check procedures include:
   a. Check all modules visually. Test the open circuit voltage and short circuit current of each module, before taking onto the field to verify proper operation. See checklist for detailed procedures.
   b. Check plug connectors and connector boxes.
   c. Check the attachment points and methods as indicated in the drawings.

3. The installation procedure includes:
   a. Mount modules on support structure and connect to conduit.
   b. Install PV combiner, inverter, and associated equipment to prepare for system wiring.
   c. Connect properly sized wire to each circuit of modules and
   d. Run properly sized wire (per drawings) for each circuit to the circuit combiners.
   e. Run properly sized wire (per drawings) from circuit combiner to inverter over-current/disconnect switch
   f. Run properly sized wire (per drawings) from inverter to utility disconnect switch
   g. Run properly sized wire (per drawings) from utility disconnect switch to main service panel and connect circuit to the main utility service.

4. Use the checklist and drawings to ensure proper installation throughout the system by visual inspection. Sign off each step of the installation and verification procedures.

5. Verify that all PV circuits are operating properly and the system is performing as expected. Double check the checklist and drawings prior to executing the system acceptance test.

6. Final inspections will be conducted by the QC manager, the local authorities, and the utilities. Once approval is received, the system is ready for operation

7. The buy-down request form, with all necessary attachments, will be sent to the appropriate authorities in order to receive buy-down payments per agreed conditions.

8. Quality Control. The QC manager is fully responsible for the proper training of the technical personnel in charge of the QC procedures at each step of the design and installation stages. The QC manager also coordinates the proper documentation and records keeping, daily and periodic inspections of work sites, as well as the actual acceptance and performance evaluation of each step of the installation phase. The QC manager is ultimately responsible for following proper preparation, installation, inspection, verification and test procedures.

Performance Evaluation

The performance evaluation phase will start immediately after the EPC contractor(s) signals completion of the installation and has performed start-up and initial performance and safety testing of each step of the process

A detailed step-by-step checklist of steps to be taken during the installation and performance evaluation procedures—all of which will be supervised and verified by QC manager and his QC inspectors—is to be developed ahead of time, as needed to ensure an efficient and complete quality installation.

Upon completion of each step of the installation procedures, the QC manager and his inspectors will conduct performance evaluation of the components and systems. Tests designed to verify and confirm compliance with the design specs and requirements, as well as to test the performance characteristics of the components and systems at each step of the installation process will be executed accordingly. The results of these tests will be verified, approved and signed off by the QC manager.

When all construction stages have been completed, and the related components and systems have been signed off, the final test of the entire power plant complex will begin. With all modules and strings activated and connected in the circuit, a number of tests will be performed with each string, to begin with, and recorded as a baseline for each particular sting under the existing weather conditions. The total power output—on the DC and AC sides—will be measured and the appropriate calculations of the plant's efficiency, yield, etc. will be made and documented as well. The resulting data will be used as a baseline of the power plant performance under the existing weather conditions and for comparative purposes in the future.

System Certification and Documentation

Thorough documentation and record keeping is the best way to ensure efficient implementation of all phases of this project. A number of documents and record-keeping media will be developed for the different segments of the project implementation, such as:

1. General system documentation for installation and operation phases
2. Technical drawings for the same
3. Training documentation
4. QC manuals and inspection documentation
5. Non-conforming materials and procedures documents

These will be needed for inspection and verification during each step of the project by different specialists and responsible parties. The complete package of documentation will be submitted to the inspecting authorities for final inspection, verification and certification of the facility.

At the end of the construction phase, and after all quality checks have been performed and signed off by the QC manager, the appropriate authorities are notified in order to execute certification checks and tests. These vary from state to state and from project to project, but basically consist of review of all construction and QC documents and verification of the final performance tests results.

Again, some operation and quality control policies and standards for the different stages of the PV products manufacturing, installation and operation do exist, but there is no single, all-encompassing and accepted by the worldwide PV industry standard for these. So, until such a standard is implemented, we are going to improvise in some cases, and compromise in others. A number of serious gaps still exist in the operating and quality control procedures of the sand-to-modules manufacturing procedures and the power plants' design-to-decommissioning processes.

Figure 6-14. Large-scale PV power plant

Since we don't see any such standard emerging, especially in the US, we will use this text to bring some of the key issues in the open for discussion and eventual resolution.

OPERATION AND MAINTENANCE PHASE

Upon completion of the construction and installation stages, the responsibility for the operation of the new facility is transferred immediately to a team of specialists, working for the plant owners, or an independent contractor who is paid a fee to operate the plant for a certain amount of time. The engagement conditions are carefully discussed and officially agreed upon in advance and an operation and maintenance (O&M) contract is signed prior to starting the actual O&M work.

The O&M team has a number of important responsibilities, which range from simple janitorial type work, to solving most complex issues and situations. The O&M management team is fully and irrevocably accountable to the project owners for all issues related to the proper and efficient operation of the plant.

System Check and Yield Monitoring

As part of our QC/QA plan discussed above, we tested and verified the integrity and longevity of the PV products purchased for our 100 MWp power plant. Once all equipment is installed, the power plant will be switched on, tested, and certified. When all the inauguration champagne has been drunk, we will start the final step of the project—transfer of the plan's long-term operation and management to the O&M team.

The results from the final QC and certification tests will be accepted as baseline of the plant's performance. Additional tests will be run under different weather conditions, which will tell us if the PV components conform to the manufacturers' specs, and if our planning, design and construction efforts were of the quality we claimed.

Initial O&M system tests consist of thoroughly checking and testing all components, identifying defects and malfunctions, and documenting all events and procedures. A final report of the initial O&M test must be issued by the engineering group conducting the tests. The report should include a thorough description of technical issues and the related financial impact—reflected in equipment malfunctions or yield loss.

System tests include visual examination of the

modules and infrared check of each module at peak performance. These tests, done by qualified personnel, will spot damaged modules (cracked glass, discoloration, etc.) and will identify thermal malfunctions like hot spots, which could reduce the output power and/or damage the modules. These tests are especially important, and an absolute must, if no pre-installation tests were conducted

I-V curves for the different strings and sub-strings must be generated at peak power and in different weather conditions as well, to establish a baseline for each of them to use as a reference in the future. I-V curve tests could reveal inefficient or faulty module performance, wiring and other abnormalities. The tests must be performed by well trained and qualified personnel, in order to avoid erroneous results and conclusions.

Yield assessment must be performed at the same time and by the same team as well, which will ensure more reliable results. The yield assessment is an even more complex variable, so it suffices to say that unless it is done with utmost care by qualified personnel, the results would be erroneous or unrepeatable at best.

Operational Considerations, Tasks and Issues

Once the initial tests are done and the power plant is given a clean bill of health, normal operation starts. The O&M procedures of a PV power plant are a complex, sometimes expensive, but very important task. Daily observations must be documented and analyzed periodically, with appropriate action taken if abnormal conditions or behaviors are noticed. Yield monitoring and assessment is also a daily task—especially in the early days of the plant's operation. The O&M team is in charge of these tasks, and is ultimately responsible for everything going on at the plant, including the prevention of natural or man-made incidents.

Yield monitoring has to be carried properly and precisely too, since yield variations are expected on a minute-by-minute, and hour-by-hour basis. Things can change in an instant or overnight. They surely change monthly and yearly. The dependencies are very complex, so their proper relation and documentation are very important. Even more important is their proper interpretation and the resulting corrective actions.

The PV power plant is a living entity that can be very dangerous if not treated properly, so only well trained and well qualified personnel can run it efficiently, safely and profitably. These three factors are also interrelated and require deep understanding of the behavior of each component and its function.

O&M team responsibilities include:

1. Materials and labor needed to operate efficiently
2. Solving mechanical and electrical problems
3. Voltage disturbances corrections
4. Frequency disturbances corrections
5. Islanding protection response
6. Power factor support
7. Reconnect after failure or P&M
8. DC injection into AC grid issues
9. Scheduled and unscheduled disconnects
10. Grounding and source circuit abnormalities
11. Ground fault protection failures
12. Overcurrent protection incidents
13. Containment of electromagnetic interferences of any sort
14. Personnel training and management of daily activities
15. Safety and accident prevention programs

PV plants are operated and maintained by an operator/contractor, who is trained and fully responsible for the proper and efficient system operation.

Some of the duties of the operator/contractor are:

1. Wash PV array at scheduled time intervals, or when build-up of soiling deposits is noticed.

2. Periodically inspect the system to make sure all wiring and supports stay intact.

3. Review the output of the system daily to see if performance is close to the specs and previous year's reading.

4. Maintain a log of these readings to identify system problems.

5. Daily monitoring of the system is a key component of its proper operation, since it provides feedback to the operator as to total power and actual energy production metering. Without proper metering the operator will never know whether the system is operating properly or not. A properly designed PV power plant will be equipped with the latest in remote monitoring equipment, which will register and record the power production of each string, as well as the total power pumped into the grid.

6. Weather watch in the form of a basic weather station is also needed to follow and record weather changes and alert the operators in case of drastic changes. Actual temperature and solar insolation at the site are also important and must be tracked constantly. The data will be analyzed periodically to calculate the overall system efficiency, to prevent system problems, and to take corrective actions.

PV Plant Output Variability

The amount of power produced by a PV power plant depends on several factors, with sunlight availability and intensity being the major variables. Sunlight can ramp up and down rapidly, which is important characteristic of large-scale PV power generation, but the utility system operators need to maintain a balance between the aggregate of all generators and loads.

General Principals

In all cases several general principles must be maintained, to ensure the integrity of the grid, following the basic rules of conduct:

a. National electric system reliability must be maintained, regardless of the generation methods and their variability.

b. Generating sources must contribute to system reliability and should not cause unnecessary complications and interruptions.

c. Power industry standards and operating criteria apply to all generators. They are transparent and based on proper and efficient performance.

Understanding the characteristics of variable PV output over large areas of PV installations and correlation to the load are critical to understanding the potential impacts of large quantities of PV injected into the national energy system.

PV variability can drive localized concerns, which typically manifest themselves as voltage or power quality problems. These issues are distinct from grid system level issues of balancing, and ought not to be confused. Management and remediation options for local power quality problems are generally different than options for maintaining a balance between load and supply at the system level.

The complexity of injecting variable load into a steady power transmitting system is not to be overlooked, or taken lightly. There are a number of things that the utilities could do to control the variability of injected power, but the responsibility to ensure trouble-free operation falls squarely on the PV power plant owners and operators.

The local utility company can help in most cases, but it will not step in to solve problems at the PV plant. Good understanding of the variability issues and their consequences must be a priority for the owners and managers of the plant.

Large-scale Power Plants Output Variability and Issues

Basically, photovoltaics fall under the broader category of variable generation. The experience with appropriate, unified approaches for managing variable generation will ease integration issues. The most important lesson is that the dialogue regarding PV variability requires, above all else, additional time-synchronized data from multiple PV plants and insolation meters over spatial scales ranging from a square kilometer to greater than 10,000 square kilometers.

In all cases, the data will need to cover at least a year of measurements and should be synchronized with comparable load data in order to understand the net impact on the variability that must be managed by the system operators. Certain questions, particularly those concerning power quality and regulation reserves, will require data with a time resolution as high as multiple seconds.

Analysis of data from multiple time-synchronized PV plants will allow detailed evaluation of the degree to which rapid ramps observed in point measurements will be smoothed by large PV plants and the aggregation of multiple PV plants. Such studies will help remove unwarranted barriers to interconnection and provide the basis for setting appropriate interconnection standards that will allow solar energy from PV plants to reach significant penetration levels.

The output variability observed by a single point of insolation measurement will not directly correspond to the total (daily or monthly) variability of a large PV plant, simply because one end of the plant might be sunny, while the other could be covered by clouds. A single point measurement ignores this and the sub-minute and sub-hour time scale smoothing that can occur within these multi-kW plants. This difference can be amplified when looking at sub-minute smoothing that can occur within very large-scale plants.

In summary, basically:

a. Extrapolation suggests that further smoothing is expected for short time-scale variability within PV plants that are hundreds of MW, but this needs to be confirmed with field data from large systems.

b. Diversity over longer time scales (10-min to hours) can occur over broad areas encompassed by a power system balancing area. Data from the Great Plains region of the U.S. indicate that the spatial separation between plants required for changes in output to be uncorrelated over time scales of 30-min is on the order of 50 km. The spatial separation required for output to be uncorrelated over time scales of 60-min is on the order of 150 km. The assumption that variability on a 15-min or shorter time scale is uncorrelated between plants separated by 20 km or more is supported by data from at least one region of the U.S. Additional data are required to examine this assumption in regions with different weather patterns.

c. Multiple methods will be used for forecasting solar resources at differing time scales. Weather variations (clouds, rain, snow, etc.) are the primary influence in the solar forecast. It is important to recognize that clouds (and their rate and direction of movement) are visible to satellites and ground-based sensors, so some successful forecasting can be expected. Over longer time scales clouds can change shape and grow or dissipate, so numerical weather modeling methods may prove necessary. As with wind forecasting, solar forecasting will benefit from further development of weather models and datasets. In all cases, complete and error-free forecasting is needed and is a thing of the not-so-distant future.

It is obvious from the above, however, that no matter what we do, power output variability is here to stay, and that accurate and much needed error-free forecasting and power control are still not available. We do expect that, with all the work underway in these areas, we will have good results soon, which will allow us to predict and manage PV plants' power output variability efficiently enough to assure their reliability as grid-connected energy sources, thus bringing us closer to our energy independence.

Future Plans and Developments

The future direction is clear: more effort and money will be poured into developing new renewable energy technologies, and a number of significant breakthroughs in the field have been documented during the last several years. If and when fully developed, some of these could contribute significantly to reducing the pollution and easing the pending energy crisis. The future looks bright for some of these technologies, if and when they get to market.

Some of the new developments are:

1. *IBM's solar cell created from "earth abundant" materials*

Researchers at IBM created an inexpensive solar cell from materials that are dirt cheap and easily available. The layer that absorbs sunlight and converts it into electricity is made with copper, tin, zinc, sulfur and selenium. The best part of the solar cell is that it still manages to hit an efficiency of 9.6 percent, which is much higher than earlier attempts to make solar panels using similar materials.

2. *MIT's Concentrated Solar Funnel*

A group of researchers at MIT devised a way to collect solar energy 100 times more concentrated than a traditional photovoltaic cell. The system could drastically alter how solar energy is collected in the near future as there will no longer be a need to build massive solar arrays to generate large amounts of power. The research work conducted has determined that carbon nanotubes will be the primary instrument used in capturing and focusing light energy, allowing for not just smaller, but more powerful solar arrays.

3. *Wake Forest University's Light Pipes*

Researchers at Wake Forest University in North Carolina made a breakthrough by developing organic solar cells with a layer of optical fiber bristles that doubles the performance of the cells in tests. The prototype solar cell has been developed by David Carroll, who is the chief scientist at a spinoff company called FiberCell.

The problem with standard flat panels is that some sunlight is lost through reflection. To reduce this effect, the research team took a dramatic approach by stamping optical fibers onto a polymer substrate that forms the foundation of the cell. These fibers, dubbed "light pipes," are surrounded by thin organic solar cells applied using a dip-coating process, and a light-absorbing dye or polymer is also sprayed onto the surface. Light can enter the tip of a fiber at any angle. Photons then bounce around inside the fiber until they are absorbed by the surrounding organic cell.

Louisiana Tech University's CNF-PZT Cantilever

Created by a research team at Louisiana Tech University, the CNF-PZT Cantilever is a breakthrough energy-harvesting device which utilizes waste heat energy from electronic gadgets to power them. The device features the use of a carbon nanotube on a cantilever base of piezoelectric materials. The carbon nanotube film absorbs heat and forces the piezoelectric cantilever to bend, generating an electric current in the material. The device is so small that thousands of CNF-PZT Cantilever devices can be designed into devices, allowing them to harvest their own wasted energy.

5. *New Energy Technologies' see-through glass SolarWindow*

New Energy Technologies developed a working

prototype of the world's first glass window capable of generating electricity. Until now, solar panels have remained opaque, with the prospect of creating a see-through glass window capable of generating electricity limited by the use of metals and other expensive processes, which block visibility and prevent light from passing through glass surfaces. The technology has been made possible by making use of the world's smallest working organic solar cells, developed by Dr. Xiaomei Jiang at the University of South Florida. Unlike conventional solar systems, New Energy's solar cells generate electricity from both natural and artificial light sources, outperforming today's commercial solar and thin-film technologies by as much as 10-fold.

6. *Innowattech's Piezoelectric IPEG PAD*

Innowattech recently created piezoelectric generators that can be used as normal rail pads, but generate renewable energy whenever trains pass on them. The company tested the technology by replacing 32 railway pads with new IPEG PADs, where the pads were able to generate enough renewable electricity to determine the number of wheels, weight of each wheel and its position. The speed of the train and wheel diameter could also be calculated. The company states that areas of railway track getting between 10 and 20 ten-car trains an hour can be used to produce up to 120 kWh of renewable electricity per hour, which can be used by the railways or transferred to the grid.

7. *CSIRO's Brayton Cycle Project*

Australia's national science agency, CSIRO, developed a technology that requires only sunlight and air to generate electricity. The system is ideal for areas that face acute water shortages. The solar Brayton Cycle project replaces use of concentrated sun rays to heat water into high-pressure steam to drive a turbine with solar energy to create a solar thermal field. The technology focuses the sun's rays projected onto a field of mirrors known as heliostats onto a 30-meter-high (98-ft) solar tower to heat compressed air, which subsequently expands through a 200 kW turbine to generate electricity.

8. *UW students develop low-cost solar-powered water purification system*

The idea of using solar energy for purifying water has been around for ages and is still practiced in different parts of the globe. Now a team of students from the University of Washington has come up with a bright idea, taking a little bit of technical help to devise a low-cost solution to a global problem. Using simple parts, the student team has developed a cheap set-up solution for drinking water in poor areas, which has won them a $40,000 prize.

The team reports that using parts from a keychain that blinks in response to light, they have created a device that monitors how light is passing through a water-filled bottle and how many particulates are obstructing the light. When enough particulates are removed, the sensor alerts that the water is now safe to drink. The student team believes that the device can be retailed at just $3.40, which makes it a viable solution for many parts of the world.

9. *New 'plant mimicking' machine generates fuels using solar energy*

A team of researchers in the US and Switzerland has created a machine that, like plants, uses solar energy to produce fuels, which can later be used in different ways. The machine makes use of the sun's rays and a metal oxide called ceria to break down carbon dioxide or water in fuels that can be stored and transported. Unlike solar panels, which work only during the day, this new machine is designed to store energy for later use.

Ceria or cerium oxide has a natural property to exhale oxygen as it heats up and inhale as it cools down. In the prototype, carbon dioxide and water are pumped into the vessel ceria rapidly stripping oxygen from them, creating hydrogen and/or carbon monoxide. Hydrogen produced by the machine can be used in fuel cells, whereas hydrogen and carbon monoxide can be combined to produce syngas.

While the prototype does show a possible use of solar energy for producing energy through the night as well, the machine suffers from an efficiency problem. The prototype harnesses only 0.8 percent of incident solar energy to produce more energy, while a major portion is lost through the reactor's wall or through the reradiation of sunlight back through the device's aperture. The research team is confident that the efficiency rates can be enhanced to 19 percent by making some improvements in insulation and by using smaller apertures.

In conclusion, the direction is clear; in addition to looking into improving the known energy sources—coal, nuclear, hydro, wind and solar—totally new approaches and technologies are being sought, and a lot of time and money is spent on their development. None of these, in our opinion, is a fix-all solution to today's energy and environmental problems, but drop by drop fills the bucket, as they say. So we should support all efforts to bring efficient and safe alternative energy sources into reality and to be able to use them in large-scale power generation. Our future depends on that.

ENERGY STORAGE

We've seen that variability of PV power plants' output is unavoidable. PV power might be combined with other energy sources (i.e., wind) to smooth the variations and even match the peak load, but even that is not a complete solution. During cloudy or rainy days the output will be limited and the only way to rectify that anomaly is to provide energy storage as a supplemental power generation. Stored energy can be used during periods of low energy generation or at night.

There are a number of potential energy storage solutions for use with solar power plants, some of which are applicable for PV power storage too. See below a complete list of presently available energy storage technologies, followed by a discussion of the most promising for use today:

1. Thermal storage
   Steam accumulator
   Molten salt
   Cryogenic liquid air or nitrogen
   Seasonal thermal store
   Solar pond
   Hot bricks
   Fireless locomotive
   Eutectic system
   Ice Storage

2. Electrochemical storage
   Batteries
   Flow batteries
   Fuel cells
   Electrical
   Capacitor
   Supercapacitor
   Superconducting magnetic energy storage (SMES)

3. Mechanical storage
   Compressed air energy storage (CAES)
   Flywheel energy storage
   Hydraulic accumulator
   Hydroelectric energy storage
   Spring
   Gravitational potential energy (device)

4. Chemical storage
   Hydrogen
   Biofuels
   Liquid nitrogen
   Oxyhydrogen and Hydrogen peroxide

5. Biological storage
   Starch
   Glycogen

6. Electric grid storage

We will review here only the most suitable for CSP and PV applications storage options:

Thermal Energy Storage

Presently, this is the most widely used method of energy storage in thermal solar (CSP) plants. Heat is transferred to a thermal storage medium in an insulated reservoir during the day, and withdrawn for power generation at night. Thermal storage media include pressurized steam, concrete, a variety of phase change materials, and molten salts such as sodium and potassium nitrate.

The most widely used heat transfer liquids today are:

1. Pressurized steam energy storage. Some thermal solar power plants store heat generated during the day in high-pressure tanks as pressurized steam at 50 bar and 285°C. The steam condenses and flashes back to steam, when pressure is lowered. Storage time is short—maximum one hour. Longer storage time is theoretically possible, but has not yet been proven.

2. Molten salt energy storage. A variety of fluids can be used as energy storage vehicles, including water, air, oil, and sodium, but molten salt is considered the best, mostly because it is liquid at atmospheric pressure, it provides an efficient, low-cost medium for thermal energy storage, and its operating temperatures are compatible with today's high-pressure and high-temperature steam turbines. It is also non-flammable and nontoxic, and since it is widely used in other industries, its behavior is well understood and the price is cheap.

   Molten salt is a mixture of 60% sodium nitrate and 40% potassium nitrate. The mixture melts at 220°C, and is kept liquid at 290°C (550°F) in insulated storage tanks for several hours. It is used in periods of cloudy weather or at night using the stored thermal energy in the molten salt tank to generate steam and turn a turbine, which in turn generates electricity. These turbines are well established technology and are relatively cheap to install and operate.

3. Pumped heat storage. Pumped heat storage systems are used in CSP power plants and consist of two tanks (hot and cold) connected by transfer pipes with a heat pump in between performing the cold-to-heat conversion and transfer cycles. Electrical energy generated by the PV power plant is used to drive the heat pump with the working gas flowing from the cold to hot tanks. The gas is heated and pumped into the hot tank (+50°C) for storage and use at a later time. The hot tank is filled with solids (heat absorbing materials), where the contained heat energy can be kept at high temperature for long periods of time.

   The heat stored in the hot tank can be converted back to electricity by pumping it through the heat pump and storing it back in the cold tank. The heat pump recovers the stored energy by reversing the process.

   Some power (20-30%) is wasted for driving the heat pump and during the transfer and conversion cycles, but the technology can be optimized for use in large-scale PV plants.

   In all cases, large heat energy losses accompany the energy storage processes. At least one third of the energy is lost during the conversion of stored heat energy into electricity. More is lost during the storage and following cooling cycles.

Battery Energy Storage

No doubt, this is the most direct and efficient way to store a large amount of electricity generated by PV power plants. The generated DC electric energy is stored as DC power in batteries for later use.

There are several types of batteries, the most commonly used as follow:

1. *Lead Acid batteries*

These are the most common type of rechargeable batteries in use today. Each battery consists of several electrolytic cells, where each cell contains electrodes of elemental lead (Pb) and lead oxide (PbO_2) in an electrolyte of approximately 33.5% sulturic cacid (H_2SO_4). In the discharged state both electrodes turn into lead sulfate ($PbSO_4$), while the electrolyte loses its dissolved sulturic acid and becomes primarily water. During the charging cycle, this process is reversed.

These batteries last a long time, and can go through many charge-discharge cycles, if properly used and maintained. They are affected, however, by high temperatures, when the electrolyte can boil off and destroy the battery. Since there is water in the cells, the electrolyte can freeze during winter weather, which could destroy the battery as well.

2. *Lithium batteries*

These are a mature technology, having been used widely for a long time in consumer electronics. They are actually a family of different batteries, containing many types of cathodes and electrolytes. The most common type of lithium cell used in consumer applications uses metallic lithium as anode and manganese dioxide as cathode, with a salt of lithium dissolved in an organic solvent. A large model of these can be used to store large amounts of electric power generated by a PV power plant, and due to their highest known power density they could be quite efficient—70-85%.

They are suitable for smaller PV installations, too, because scaling up to large PV plants would be a very expensive proposition.

3. *Sodium Sulphur batteries*

These are high temperature, molten metal, batteries constructed from sodium (Na) and sulfur (S). They have a high energy density, high efficiency of charge/discharge (89–92%) and long cycle life. They are also usually made of inexpensive materials, and due to the high operating temperatures of 300-350°C they are quite suitable for large-scale, grid energy storage.

During the discharge phase, molten elemental sodium at the core serves as the anode, and donates electrons to the external circuit. The sulfur is absorbed in a carbon sponge around the sodium core and Na+ ions migrate to the sulfur container. These electrons drive an electric current through the molten sodium to the contact, through the electric load and back to the sulfur container.

During the charging phase, the reverse process takes place. Once running, the heat produced by charging and discharging cycles is sufficient to maintain operating temperatures and usually no external source is required.

There are, however, a number of safety and corrosion problems, due to the sodium reactivity, which need to be resolved before full implementation of this technology takes place.

4. *Vanadium redox batteries*

These are liquid energy sources, where different chemicals are stored in two tanks and pumped through electrochemical cells. Depending on the voltage supplied, the energy carriers are electrochemically charged or discharged. Charge controllers and inverters are used to control the process and to interface with the electrical source of energy.

Unlike conventional batteries, the redox-flow cell stores energy in the solutions, so that the capacity of the system is determined by the size of the electrolyte tanks, while the system power is determined by the size of the cell stacks. The redox-flow cell is therefore more like a rechargeable fuel cell than a battery. This makes it suitable as an efficient energy storage for PV installations.

A number of additional types of batteries are under development, and some show great potential for use in larger PV installations in the near future. Most batteries have problems with moisture, high temperature, memory effect, and use of scarce and toxic exotic materials, all of which cause longevity problems and abnormally high prices. If and when all these problems are resolved, the energy storage problems of PV power plants will be resolved as well.

Compressed Air Energy Storage

Compressing air into large high-pressure tanks is one of the most discussed and most promising energy storage methods for use with PV power plants today. It is quite simple way of energy storage, using a compressor powered by the electricity produced by the PV plant compressing air into the storage tank. A lot of energy is lost by activating the compressor and, heat is wasted during the compression process, so there are several compressing methods that treat generated heat so as to optimize conversion efficiency.

Some of these are as follow:

1. *Adiabatic storage*

Adiabatic storage retains the heat produced by compression via special heat exchangers, and returns it to the compressed air when the air is expanded to generate power. Its overall efficiency is in the 70-80% range, with the heat stored in a fluid such as hot oil (300°C) or molten salt solutions (600°C).

2. *Diabatic storage*

Here extra heat is dissipated into the atmosphere as waste, thus losing a significant portion of the generated energy. Upon removal from storage, the air must be reheated prior to expansion in the turbines, which requires extra energy as well. The lost and added heat cycles lower the efficiency, but simplify the approach, so it is the only one implemented commercially these days. The overall efficiency of this method is in the 50-60% range.

3. *Isothermal compression and expansion*

This method attempts to maintain constant operating temperature by constant heat exchange to the environment. This is only practical for small power plants, which don't require very effective heat exchangers, and although this method is theoretically 100% efficient, this is impossible to achieve in practice, because losses are unavoidable.

There are a large number of other methods using compressed air, such as pumping air into large bags in the depths of lakes and oceans, where the water pressure is used instead of large pressure vessels. Pumping air into large underground caverns is another approach that is receiving a lot of attention lately.

Pumped Hydro Energy Storage

Pumped hydro energy is a variation of the old hydroelectric power generation method used worldwide, and is used quite successfully by some power plants. Energy is stored in the form of water, pumped from a lower elevation reservoir to one at a higher elevation. This way, low-cost off-peak electric power from the PV power plant can be used to run the pumps for elevating the water. Stored water is released through turbines and the generated electric power is sold during periods of high electrical demand. This way the energy losses during the pumping process are recovered by selling more electricity during peak hours at a higher price.

This method provides the largest capacity of grid energy storage—limited only by the available land and size of the storage ponds.

Flywheel Energy Storage

Flywheel energy storage works by using the electricity produced by the PV power plant to power an electric motor, which in turn rotates a flywheel to a high speed, thus converting the electric energy to, and maintaining the energy balance of the system as, rotational energy. Over time, energy is extracted from the system and the flywheel's rotational speed is reduced. In reverse, adding energy to the system results in a corresponding increase in the speed of the flywheel. Most FES systems use electricity to accelerate and decelerate the flywheel, but devices that use mechanical energy directly are being developed as well.

Advanced FES systems have rotors made of high-strength carbon filaments, suspended by magnetic bearings, and spinning at speeds from 20,000 to over 50,000 rpm in a vacuum enclosure. Such flywheels can come up to speed in a matter of minutes—much quicker than some other forms of energy storage.

Flywheels are not affected by temperature changes, nor do they suffer from memory effect. By a simple measurement of the rotation speed it is possible to know the exact amount of energy stored. One of the problems with flywheels is the tensile strength of the material used for the rotor. When the tensile strength of a flywheel is exceeded, the flywheel will shatter, which is a big safety problem. Energy storage time is another issue, since flywheels using mechanical bearings can lose 20-50% of their energy in 2 hours. Flywheels with magnetic bearings and high vacuum, however, can maintain 97% mechanical efficiency, but their price is correspondingly higher.

Electric Grid Storage

Grid energy storage is large-scale storage of electrical energy, using the resources of the national electric grid, which allows energy producers to send excess electricity over the electricity transmission grid to temporary electricity storage sites that become energy producers when electricity demand is greater. Grid energy storage is a very efficient storage method, playing an important role in leveling and matching electric power supply and demand over a 24-hour period.

There are several variations of this method, one of which is the proposed grid energy storage called vehicle-to-grid energy storage system, where modern electric vehicles that are plugged into the energy grid can release the stored electrical energy in their batteries back into the grid when needed. Far fetched, yes, but the future will demand many such ingenious approaches, if we are to be energy independent.

In conclusion, there are a number of other energy storage methods such as fuel cells, new types of batteries, superconducting devices, supercapacitors, hydrogen production, and many others under development, so the future looks bright in this area. In practice, however, there are a number of energy storage installations around the world, totaling over 2,100 MW, with the major technologies being:

1. Thermal energy storage is over 1,140 MW
2. Batteries energy storage is over 450 MW
3. Compressed air energy storage is over 440 MW
4. Flywheels energy storage is over 80 MW

Energy storage has many advantages, in addition to its unique potential to transform the electric utility industry by improving wind and solar power variability, availability and utilization. It can contribute to the overall energy independence and environmental clean-up by avoiding the building of new power plants and transmission and distribution networks. Experts consider energy storage to be the solution to the electric power industry's issues of variability and availability, opening new opportunities for wind and solar power use.

Complexity, safety, price and other restrains, however, will have to be worked out well before any of the energy storage methods become accepted reality for large-scale PV installations

SUMMARY

Transforming the sun's energy into electricity has evolved from providing power to satellites in space, to powering individual houses and small remote off-grid village applications, to very large-scale PV power generating plants in the world's deserts. Early schemes of large solar arrays showed the potential of using abundant, clean energy. Yet, system costs and low conversion efficiencies kept solar power at the kilowatt scale for the past three decades.

Now we are witnessing another solar revolution in the making. Would it be successful in changing our ways? Would it be successful in bringing us energy independence and a clean environment? These are questions, which will be answered soon, but we must be proactive and do our best in promoting solar and other clean energies.

The potential solar resource from desert regions is truly astounding. Several studies show that the entire global electricity demand could be provided from just 3% of the world's deserts, and several hundred square miles in the Nevada desert could power all 48 states. Many of the grand schemes place large arrays around the equatorial or high desert regions with high-voltage DC transmission lines delivering that energy to far away populated areas. On paper, it's a magnificent dream—that could become a reality.

Annual solar industry growth rates of 20% per year and entry by major energy companies has driven costs down and efficiencies up. Policymakers can assist the solar industry by cutting subsidies for fossil fuels and nuclear energy, providing tax incentives to solar purchasers, thus driving down the costs and creating a mass market for this clean fuel technology. For example, in the US, direct subsidies to nuclear energy amounted to $115 billion until 1999 with a further $145 billion in indirect subsidies. In contrast, subsidies to wind and solar combined during the same period totaled only $5.5 billion.

With energy demand projected to double in the next 30 years, mostly from developing nations, the business-as-usual energy scenarios predict dire consequences for the planet. Growing CO_2 concentrations in the atmosphere are driving climate change, and transitioning to carbon-free energy sources is a prime solution. Large solar arrays in the deserts (PV and CSP) provide the largest potential renewable energy resource to meet growing energy demands in a sustainable manner. High-voltage transmission grids can carry electricity over thousands of kilometers, from regions with abundant solar radiation to our cities and industries, for use by the population.

All things considered, the good old days of "fill-er-up" are gone. We have learned the lessons of the past and are now much smarter. We are able to take full responsibility for our future, and that of our children. Are we willing to do it? That's a different question, to which we have no answer yet, but which we try and try again to bring into the open for an honest discussion.

This book evaluates the technological potential of implementing large-scale power plants and the related socio-economic benefits to millions of people, which obtained through use of solar energy is a viable and sustainable way to solve the urgent energy, environmental, and socio-economic problems. We clearly see that large solar power systems could one day replace fossil fuels as the main energy source for our society. Only time, ingenuity, and hard work separate us from our goals and dreams for energy independence and clean environment.

We will let the reader make a conclusion on the suitability of the different technologies for large-scale PV use, based on the available data on each of the participating technologies. The future will tell which technology fits where, and we only hope that proper decisions, based on scientific facts and data will be made in the long run. We also hope that political and regulatory changes will be in favor of expanding the use of alternative energies, which are the answer to our energy independence and environmental cleanup. Large-scale PV power installations are the quickest and most flexible vehicle in this important effort.

Notes And References:

1. Hot Spot Evaluation of Photovoltaic Modules," by Govindasamy (Mani) TamizhMani and Samir Sharma
2. Photovoltaic array performance model, D.L. King, W.E. Boyson, J.A. Kratochvil Sandia National Laboratories, Albuquerque, New Mexico 87185-0752
3. Integrating Large-Scale Photovoltaic Power Plants into the Grid, Peter Mark Jansson, Richard A. Michelfelder, Victor E. Udo, Gary Sheehan, Sarah Hetznecker and Michael Freeman, Rowan University/Rutgers University/PHI/Suntechnics Energy Services/Exelon Energy
4. Executive Summary, 2009 Top Ten Utility Solar Rankings, SEPA, 2009
5. Modeling of GE Solar Photovoltaic Plants for Grid Studies, GE, 2009 Mesa, AZ, USA 85212
6. Mora Associates, internal
7. 2009 global PV cell and module production analysis, Shyam Mehta | GTM research
8. Large Solar Plant Woes: Will it Open the Door for Smaller Projects? Greentechsolar, January 2010
9. IREC Net Metering Rules http://www.irecusa.org/fileadmin/user_upload/ConnectDocs/NM_Model.pdf.
10. Decoupling Utility Profits from Sales, SEPA, Feb. 2009
12. Wikipedia
13. Photovoltaic Incentive Programs Survey, SEPA, Nov. 2009
14. Accommodating high levels of variable generation. NERC, April 2009.

Chapter 7

Utilities, Land, Regulatory, Finance, Legal, Oh My...

"Technological progress is like an axe in the hands of a pathological criminal."
Albert Einstein

In this chapter we will review a number of important subjects, which are the foundation of developing and operating PV power projects, focusing on commercial and utilities (herein referred to as "large-scale") power generation installations. The role of the utilities, land availability, federal and state politics and regulations, as well as the related financing and legal issues are some of the factors to consider when discussing, or working on, solar energy generating projects during their planning, design, development, and operation phases.

We will review the role of the utilities and the governing bodies who control the interconnect points (substations) and transmission lines, who make the rules and whose approval is needed in most (if not all) cases. We will take a look at land issues, and why land suitable for solar installations is scarce and not easy to convert for solar use. Of course, we will review the regulatory system and its complexity, which together with the financing issues have been an obstacle from the get-go. Last, but not least, we will take a look at the legal aspects which seem to be getting more important by the day. It's an interesting PV world we live in today—complex, changing and expensive.

Let's take a closer look at these aspects of solar energy in the 21st century in light of recent developments, keeping future developments in perspective as well.

US UTILITIES FUNCTION

Generally speaking, electric power utilities are companies who, in conjunction with state and federal governing bodies, own and control the electric power generation capacities and transmission and distribution networks. There are publically and investor-owned utilities in the US, as follow:

1. Publicly owned utilities (POU) include: a) municipal utilities who serve territories inside and outside the city limits and, b) cooperative utilities who are owned by the customers they serve, usually in rural areas. Public utilities are subject to public control and regulation, ranging from local community-based groups to state-wide government bodies.

2. Private utilities, also called investor-owned utilities (IOU) are owned by investors and operate under somewhat different rules than their public counterparts. As most other for-profit businesses, their primary goal and responsibility is the bottom line.

Utilities—regardless of type, function and location—have one great advantage in that they, one way or another, control (or at least have a major say-so in) electric power generation and distribution and related equipment, services and costs. Since everybody needs and uses electricity, this gives them some special powers and privileges most other businesses do not have. Like any other business, however, the utilities' goal is to sell the maximum amount of product (electricity and services) possible at the maximum possible price.

Some utility companies are small, some are large, some are regulated, and some are not. There are differences in their structures and *modus operandi* according to the state in which they are located. Generally speaking, however, a high volume of electricity sales means higher revenue, so utilities naturally have a strong incentive to increase electricity sales and limit activities (competition) that might reduce their sales volume. This is what we call the utility's "throughput incentive," which in some cases might prove contrary to any policies that support energy efficiency and distributed resources—such as the up-trend of residential and large-scale PV power installations today. PV installations have a great effect on the utilities' business, forcing them to put up with the complexity and hassle of unreliable and immature PV technologies and the variability of sunlight-dependent PV installations.

Throughput incentive is inherent in the way regulation works in many states, and the magnitude of its impact determines the way utilities operate and are capitalized. Reducing electric consumption via independent PV power generation reduces utility revenues. On top of that, customer PV generation requires additional management (typically load shifting and peak reduction) and increased maintenance all of which means additional expenses. The total result of these is quite unfavorable to the utilities' bottom line.

For a number of political and socio-economic reasons, the growing PV energy generation market is increasing in importance to the utilities, so they are trying to control these as much as possible. The battle, which the utilities are forced into by the newly developed "solar rush" is just beginning. Most utilities are showing good will and active participation in the programs, while others are going through the motions and secretly wishing all this solar hoopla and the hassles it brings would go away.

Utilities Cost Structure

Utilities have several key characteristics that make their profit margin especially sensitive to changes in any area of their operations. First, utilities typically have substantial operations and maintenance (O&M) and other financial costs, which are related to the extensive infrastructure of generation, transmission and distribution facilities operated by the utility. The infrastructure needs to be built, updated and maintained regardless of specific electricity quality of volume changes. Second, the capital structure of utilities tends to be significantly financially leveraged, which has the effect of making profits more sensitive to changes in revenues. Virtually all of a utility's revenue and profits are derived through the operation of its assets.

Adding a large amount of alternative energy generation to the mix has the negative initial effect of overburdening a utilities' assets and increasing related services. It also introduces variables in power quality, which cause technical difficulties to be dealt with on daily basis. All this adds many headaches to the already difficult task, and reduces further the bottom line of the utilities. This is not good for them, so it's no wonder many of them are not very excited about the solar developments and ambitions of late. We believe this will change when the solar technologies mature and become more efficient, reliable and less variable. Then the utilities might become more interested, and take a more active role in the development, implementation and operation of PV power generating capacities.

The PV Power Generation Effect

Utilities are obligated by law to add renewables to their energy-generating portfolios. Some of them have embraced that call and are making significant progress, while others are dragging their feet and looking the other way hoping that things will revert to the status quo they enjoyed for so long. There was a bill introduced in Arizona's legislation in the summer of 2010 to requalify nuclear power as "renewable" energy, thus allowing the utilities to bring additional nuclear capacities online, and which would've shelved all the solar development indefinitely. Sneaky, eh? And we hate to think that any of the utilities had anything to do with the introduction of that bill.

Like it or not, utilities' solar portfolios are on the verge of significant changes. Traditionally, solar electric markets have been distributed, consumer-focused, and solar-industry driven, but 2009 marked the beginning of significant change in energy market dynamics. A number of utilities across the U.S. are strengthening their solar portfolios according to the mandates. A survey of US utilities done by SEPA, came up with the top ten utilities in 2009 and found that their share of megawatts dropped from 88% in 2008 to 80% in 2009, indicating increasing solar activity by other US utilities, outside of the top ten.

PG&E led the pack with 85 MW installed in 2009, followed closely by SCE (both large California utilities) with 74 MW, which was an unprecedented 131% growth over the previous year's total. PG&E now has plans for 500-MWp PV projects in the 1-20 MWp size during the next 5 years.

Arizona's based utilities, APS and SRP also added PV capacity but in fairly small amounts (9.9 MW and 5.8 MW respectively), mostly residential and small commercial generators which is highly un-proportional to the availability of land and sun intensity in the Sunshine State. There are several new PV projects underway now, which will significantly increase Arizona's participation in the solar race and will bring it a notch up on the list of the ten. Even these developments, however, are far from what we expect from the sunniest state in the country. The long-term future of solar in the state is still veiled in uncertainties which could affect its direction, pushing it whichever way the political, regulatory and economic winds blow.

Why is it that some utilities enthusiastically embrace the new kid on the block, PV power, while others are staying on the sidelines? Well, we think there are several reasons that could explain and even justify that hesitation. The solar industry has a history of "disappearing" when things get rough, as has happened on a number of occasions in the not very distant past. There have been numerous opportunities to bring solar power into the energy mix, and every time (usually when oil prices fell) the political and business leaders chose to take the easy way out and stop solar development in its tracks. Will this happen again soon? We will see, but it is too early to disregard the possibility. We will have to wait at least until government subsidies dry up completely, as might happen during 2011-2012 in the US and even more likely in Europe. If solar power generators manage to operate on their own as reliable business without any outside help, the utilities will be much more open to include the alternative energy newcomers into the fold of their power generators, but this is not the case yet, and solar has a long way to go before becoming a major player in the energy market.

The PV industry is obviously still in its infancy, or better said in its embryonic state. Each large conventional energy generator alone has capacity several times larger than the entire PV installed capacity in the US, and this will not change any time soon, it seems. (See Figure 7-1.)

In addition to justifiable hesitation about solar being a reliable partner, utilities also worry about the variability solar brings into the mix, complicating their operations. Most importantly, solar seems to be challenging the utilities' very *modus operandi* by bringing uncontrollable competition into their power generation and distribution schemes. With everything said and done, utilities are in charge of the energy generation and distribution, and nobody can argue that point. So the PV infant needs to grow up before being allowed to share the playground with its older and more established cousins.

Grid Issues

Variability of solar power generation due to the sun's inconsistency (which is considerable even in desert areas) is a big problem for utilities. It introduces a number of negative effects into the already inadequate and overloaded power grid the utilities own and operate.

Some key issues with bringing large-scale solar power generators into the national grid are:

1. Voltage fluctuations in the grid, caused by varying input from large-scale PV output is a major issue. Cloud movement during the day could cause great power output fluctuations in PV plants. A 100-MWp plant could be at full power, then down to 50% after a minute and 75% the next minute. It happens that fast, and it is impossible to prevent or anticipate this with any degree of accuracy. Such instantaneous on-and-off variations of unpredictable level and duration could have serious effects on the grid and in extreme cases they might cause power disturbances and equipment failure, none of which would make grid operators happy and/or willing to bring more solar power on board. The specialists claim that larger fields are less variable, but this is

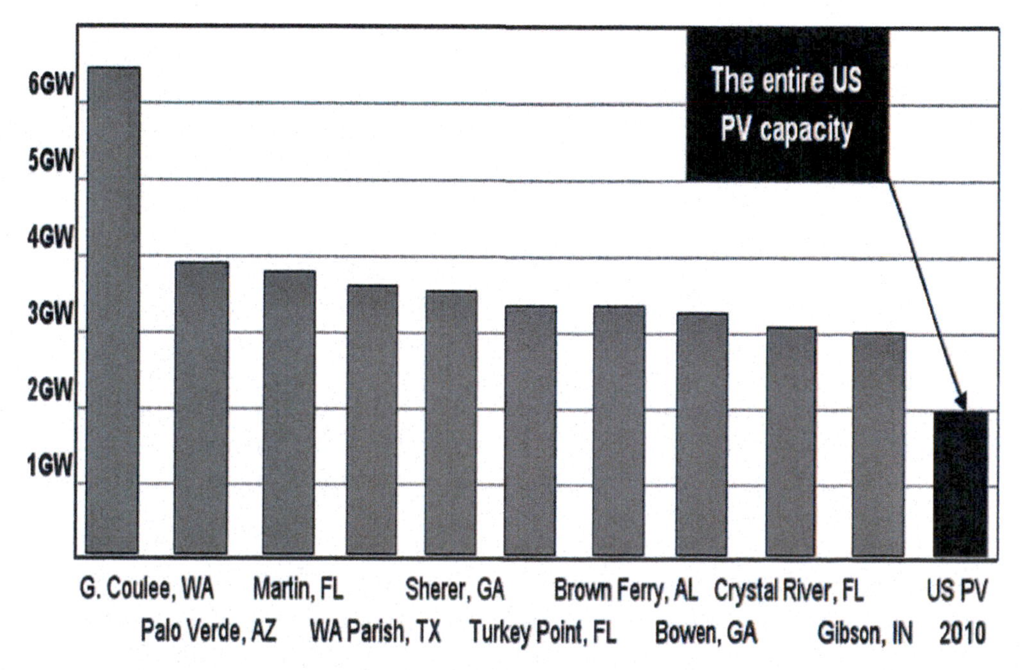

Figure 7-1. Conventional vs. PV power generation in the US 2009-2010

arguable because there is more chance of different weather phenomena in the different corners. In our experience, various-size PV power fields will behave differently in different geographic locations, with the variability depending mostly on the actual location and technology used and to a lesser extent on the size of the power field. The most stable power generation is expected by PV power fields operating in the deserts, but even then serious variability from season to season is expected and is unavoidable.

2. Serious power frequency variations are related to the varying output described above and are often caused by it. The supply/demand imbalance could contribute to serious harmonics and frequency variations which also could have damaging effects on grid power quality and performance and longevity of related equipment.

3. Electrical disturbances, caused by inverters' mechanical or electric malfunctions could cause power and frequency variations as well. An inverter or bank of inverters might shut down during low-level power management mode to compensate for a cloud system passing overhead, or to correct or avoid other disturbances. These transients could cause even greater damage than power variability (if their intensity is great enough) and are cause for great concern to grid operators.

4. Inverters are "smart" devices and know how to handle power effectively and safely. However, they don't have a common language with the grid as yet, and as their number and size increases so does their influence on the grid's proper operation and stable condition. Large numbers of inverters connected to the grid will influence it heavily, and the effect of harmonics will be more and more pronounced, forcing inverter manufacturers and utilities to expend a lot of effort and money on effective suppression and proper handling of these abnormalities. These headaches and expenses also contribute to the utilities' careful approach to the implementation of large-scale PV installations.

5. The board of directors of the Institute of Electrical and Electronics Engineers (IEEE)—USA approved a position statement on the formation of an electric reliability organization in the US late in 2002. In that statement they made some very clear points, as follow:

    a) reliability of the US electric power infrastructure has declined,

    b) the decline will affect the economy negatively,

    c) the decline is a consequence of under-investment in needed infrastructure to meet growing loads,

    d) there exists a serious misunderstanding in government and policy circles as to how the electric power system works, and

    e) there is significant misunderstanding regarding the design basis and capabilities of existing transmission in the US.

Perhaps the most telling quote from their position paper is the following: "The action of legislatures and regulators does not change the laws of physics: Transmission systems have only limited capabilities to perform beyond their design parameters."

In the absence of any near-term changes to government policies in this area, many utilities are responding by desiring to change the parameters by which they are measured. A fact of great importance here is that this very critical and timely conversation on reliability of electricity with potential causality linked to deregulation is becoming less public and is dominated instead by bureaucrats and utilities officials. It is widely believed among the experts in this field that major public policy action is required in order to avert a potential crisis, but this is just not happening.

Adding significant quantities of intermittent, dispersed renewable power to transmission lines over the next decades will only exacerbate current issues, so proper and decisive action is needed to avoid their escalation.

A number of solutions, in addition to the "smart grid" proposition are on the table presently:

1. Coupling PV power plants with diesel and gas-fired "peakers" (power plants that are fired only when needed), and/or

2. Collocating PV and wind power plants to level the peak load that could balance the overall load fluctuations and help level the critical peak load demand.

It's an expensive proposition, yes, but building new coal or nuclear power generating plants is equally expensive, while the continued use of large quantities of coal and oil will be even more expensive and damaging in the long run.

The Smart Grid Solution

There is a lot of talk lately about the new phenomena, called "smart grid," which is supposed to solve many of the issues of the old and tired national power grid. What is it all about? According to DOE's report, "The Smart Grid: An Introduction" (7), "A smarter grid applies technologies, tools and techniques available now to bring knowledge to power—knowledge capable of making the grid work far more efficiently.

1. Ensuring its reliability to degrees never before possible.
2. Maintaining its affordability.
3. Reinforcing our global competitiveness.
4. Fully accommodating renewable and traditional energy sources.
5. Potentially reducing our carbon footprint.
6. Introducing advancements and efficiencies yet to be envisioned.

Transforming our nation's grid has been compared in significance with building the interstate highway system or the development of the internet."

This basically means that the smart grid is a great thing to have, but it sounds complex and expensive so it is not going to happen any time soon. Like the interstate highway system, or the internet, it will take many years and billions of dollars to design and build. Actually, considering the complexities and magnitude of this undertaking, in terms of hardware, software, cost, regulations, permits etc., we suggest that it would be one of the most complex and expensive undertakings of the century.

Our century-old power grid, consisting of almost 10,000 electric generating units with more than 1,000,000 megawatts of generating capacity connected to more than 300,000 miles of transmission lines, decrepit substations and outdated, one-sided, regulations is the largest interconnected machine on Earth. Correspondingly, making it "smart" would be the greatest undertaking on Earth as well.

The US power grid is so huge, massively complex, and inextricably linked to human involvement and endeavor that it has been called an "ecosystem" by some. After we make it "smart" it will become the first "smart ecosystem" in the world too. Just imagine all power generators "talking" to each other 24/7, relating information on their load usage and needs, communicating with the substations and sending power where it is most needed—without any interruptions and errors. Imagine appliances in our homes and machines in the factories "talking" to the grid, telling it what they need and when, and the grid responding by effortlessly allocating needed resources on time. No grid jam, no inefficiencies, no worries—all taken care of by this giant thinking and self-serving machine. Wow! Yes, but we have to wait awhile for all this to happen, because judging from the slow development of this project, it will be a long time before we see significant progress in converting the old, dumb power grid into a new, smart and fully functional one. All of this requires a lot of money—money we just don't have right now—so the smart grid will be on the back burner for a long while.

RFPs and RFOs

Utilities are obligated by law to add renewable energies to their portfolios, and most are actively pursuing the options. Some are complying, and are much more actively involved in the process than others. Most have well defined plans to add renewables as soon as possible in order to meet the quota, but some are postponing major decisions in order to avoid risk.

RFPs and RFOs (request for proposal, and request for offers) for installation of PV power plants by interested third parties is one vehicle utilities use today to add PV power to the utilities' energy portfolios. This versus owning the generators seems to be working for there are many takers willing to jump at the chance. This way, these willing participants will take all the risks, while the utilities reap the benefits of their efforts without any risk exposure—capitalism in action. What appears to be a win-win situation on the surface might be misleading, so we need to take a much closer look before jumping into it.

Lots of progress was made in some states, and California in particular using this *modus operandi* where the major utilities PG&E and SCE and others issued a number of RFPs and RFOs during 2009 and 2010. The intention of this action was/is to install several hundred MW of PV power plants around the state using CPUC guidelines and with work done under the utilities' supervision. A number of these projects have been completed, and several are in planning or construction stages. This is a great achievement for the US solar industry, and other states are following the example. PG&E continues to lead the solar revolution by recently issuing an RFO inviting bidders to participate in the addition of 500-MW PV installations during the next 5 years.

As an example, the new (2011-2016) conditions for participation in the bidding process, according to a recent RFO by PG&E are (10):

To participate in the 2011 PV Program PPA RFO, Offers must meet the following eligibility criteria:

1. The generating facility must be a new photovoltaic electric generating facility and located within PG&E's service territory.

2. The nameplate capacity of the generating facility must be no less than 1 MW and no greater than 20 MW. Aggregation of facilities to meet the minimum 1 MW size requirement is allowed only if each individual facility is no less than 500 kW and the project comprised of the aggregated facilities interconnects within a single CAISO PNode.

3. The contract price must be no greater than $246/MWh (prior to adjustment for time of delivery).

4. The Participant and/or a member of Participant's project development team must have either completed or begun construction of a solar project that is at least 500 kW.

5. Participant must have site control and attest to site control as part of their Offer package.

6. Projects with executed PPAs must be online within 18 months following CPUC approval.

There are still some unanswered questions in the bidding process, and many additional glitches are expected before things start running smoothly, but things are slowly moving ahead. As a confirmation of the fragility of the situation, the first bidders' conference on the new RFO in February 2011 was cancelled in midstream due to a fire alarm. The building was evacuated and the conference had to be postponed. We just hope that the progress of solar energy programs doesn't follow this pattern of unexpected alarms and interruptions.

The Arizona utilities are following California's example, but from a distance, with much less enthusiasm, and with marginal results to date. There are a few commercial installations in operation and a few large-scale installations planned, while several large-scale solar projects were cancelled lately for a number of reasons. So the progress has been painfully slow and will most likely continue the same way for the foreseeable future.

One thing is for sure though; the utilities and regulators hold the key to the entry into the solar energy market. They set the rules; they are the judge, jury and executioner. Not much will happen without their blessing. But to play with them one needs a lot of patience, time and money, and must be willing to take risks. It would cost approximately $300,000 to get a 20 MWp PV project approved and permitted by the utility, and then we need much more money for land lease/purchase, assessments, permits and improvements just to get the project ready for construction. Another pile of money ($80-100 million total) would be needed for construction of the 20 MWp power installation. It's not an easy thing to do. We have not even discussed the legal barriers yet (caused by environmental and other concerns) which already stopped several large-scale projects in their tracks—even after all the effort and money was spent to get them ready for construction.

LAND AVAILABILITY

Land for solar power generation is available. Yes, there is a lot of land out there—federal, state, city and private—but you'd be surprised how difficult it is to obtain and convert it into a PV power generating installation. Money is not the only problem.

Land Basics

The consensus is that land for large-scale PV projects is available, and many large-scale PV projects are possible but are not probable for the near future, partially because of land issues. There are many reasons for this abnormality: technical, environmental, regulatory, political, financial, and combinations and permutations of these. Nevertheless, the results are the same: we are ready to go, but the door is only partially open, with many issues keeping it from being fully opened.

It all started several years back with the very quick growth of the solar energy market in the US and abroad. Solar "exploded" during 2007-2009; some of it is on paper only, because the actual large-scale PV installations have not come close to the planned, let alone to the maximum possible, size. Land prices went from $2,000 per acre in the Arizona and California deserts to up to $10,000 and even $20,000 per acre for land close to interconnect points. Acquisition of land adequate for solar and speedy permitting are the other serious problems, even under the present favorable political, regulatory and financial conditions.

The peak years of the solar boom were 2008-2010, with the US dragging its feet behind European countries. A number of soci-economic issues held back solar development in the U.S., with lack of financing being the major one. Confusion about what land to use and how to use it was, and still is, another major issue.

One of the peculiarities of the present situation is

Table 7-1. Total PV installations world wide

Country	INSTALLED CAPACITY (MWp)			
	2007	**2008**	**2009**	**2010**
Germany	1,100	1,500	3,800	7,200
Spain	560	2,500	75	850
Italy	75	420	550	2,200
USA	207	342	480	900
Japan	201	230	480	975
World	Balance	Balance	Balance	Balance
Annually	**~2,800**	**~5,600**	**~7,500**	**~16,250**
TOTAL	**~9,550**	**~15,150**	**~22,650**	**~38,900**

that the most suitable land for large-scale project development is desert land—isolated, remote, waste, desert land that is good for nothing else. It is abundant, cheap enough (or it used to be) and sees a lot of sunshine. The more sunlight, the more energy is generated, and we have a lot of it—much more than we need or can ever use.

Desert operation, however, creates problems:

1. Most PV technologies (PV modules, inverters etc.) cannot handle the extreme heat of the deserts and drop their efficiency, and/or simply fail with time. The more intense sunshine they are exposed to, the more quickly they deteriorate and fail.

2. Desert lands lack water sources, which creates problems for some solar technologies (i.e., CSP) which use steam heat to generate electricity, because these need water cooling for proper, efficient and cost-effective operation.
   Note: The dry cooling method used today for cooling CSP power plants (via large air conditioning units) is one way to avoid using cooling water, but it needs energy to run the coolers and requires maintenance, so it remains to be seen if this method will succeed.

3. Available desert lands are often far from large populated centers, thus they are far from interconnect points which require expensive modifications or additions of power transmission infrastructure. This causes delays and even cancellations of solar projects.

4. Desert land owners, seeing the potential of selling their land for solar use, increase prices to unrealistically high levels which makes private land prices too high for solar use, thus making solar too expensive and unaffordable on these lands.

5. Obtaining federal land requires a lot of time and money, which only some well positioned and rich companies can afford.

6. Bureau of Land Management (BLM), the governmental department responsible for allocating and managing the federal landmass, also charges solar installations as follow:
   a. First they charge rent for the land—up to $313.80/acre in Yuma, AZ, or $1.6 million per year rent on 100 MWp PV plant.
   b. Then, they charge a "megawatt capacity" fee—a charge based on the plant's total power output. The more power you generate, the more taxes you pay. Industry specialists claim that, "the fees are in many cases two times higher than market rates for private land." $5,256 to $7,884 per MW per annum, or $525k to $788k/year MW charge on 100 MWp power plant, depending on type of technology used.

7. Environmental groups delay and even stop development of projects for reasons that in some cases are difficult to justify. On one hand, solar is good for the environment, but on the other we won't allow it to proceed. Go figure... There are examples, including some at the end of this chapter which describe the delay and potential cancellation of the six largest projects in California, due to a lawsuit filed by local and environmental activists.

8. State and government agencies are often uncoordinated, and/or inefficient in their approaches, which causes delays or cancellations of projects. BLM, EPA and the other responsible agencies have been operating in uncharted territory lately, under

pressure from all sides, which has resulted in some hasty decisions. Since BLM is in charge of millions of acres of desert land, while EPA and other regulators control what can be put on it, they all collectively, and yet separately, dictate the conditions for land use and the solar installations on it. The results are inconclusive and even controversial at times.

9. 2008-2010 Lucrative legal proceedings have attracted an army of lawyers who look for gaps and cracks in the legal system and the local laws, including tribal and other groups' interests, and file lawsuits against solar projects. These actions seem to be growing and create serious delays and cancellations.

World Deserts

Desert lands, due to their high solar irradiation and favorable climate are best for large-scale solar power generation. Fortunately, there are many large deserts around the world that are perfect, or close enough to perfect, for this purpose. Only few parts of these deserts, however, are close to civilization and transmission lines, which renders the majority of desert land useless for practical solar power generation, at least for now. Nevertheless, significant amounts of suitable desert lands are still within reach, and efforts to install large-scale PV installations on them continue and will grow in the future.

World's deserts suitable for large-scale PV installations (infrastructure and other technical and socioeconomic barriers not withstanding):

1. Africa
   a. Arabian Desert, in Egypt and Iran
   b. Blue Desert, in Egypt
   c. Kalahari Desert, in Botswana, Namibia and South Africa
   d. Namib Desert, in Namibia
   e. Nubian Desert, in Sudan
   f. Sahara Desert, in central Africa
   g. Sinai Desert, in Egypt
   h. White Desert, in Egypt

2. Asia
   a. Betpak-Dala, in Kazakhstan
   b. Cholistan, in India and Pakistan
   c. Dasht-e Kavir, in Iran
   d. Dasht-e Lut, in Iran.
   f. Gobi Desert, in Mongolia and China
   g. Kara Kum, in Central Asia
   h. Kyzyl Kum, in Kazakhstan and Uzbekistan
   i. Lop Desert, in China
   j. Ordos, in China
   k. Taklamakan, in China
   l. Thar Desert, in Pakistan and India

3. Australia & New Zealand
   a. Gibson Desert, in Australia
   b. Great Sandy Desert, in Australia
   c. Great Victoria Desert, in Australia
   d. Little Sandy Desert, in Australia
   e. Rangipo Desert, in New Zealand
   f. Simpson Desert, in Australia
   g. Strzelecki Desert, in Australia
   h. Tanami Desert, in Australia
   i. Central Desert, in Australia

4. Europe
   a. Accona Desert, in Italy (semi-desert)
   b. Bardenas Reales, in Navarra, Spain (semi-desert)
   c. Błędowska Desert, in Lesser Poland Voivodeship, Poland
   d. Deliblatska Peščara, in Vojvodina, Serbia
   e. Oleshky Sands, in Ukraine
   f. Oltenian Sahara, in Oltenia, Romania
   g. Piscinas, South, in West Sardinia, Italy
   h. Tabernas Desert, in Almería, Spain
   i. Monegros Desert, in Aragón, Spain (semi-desert)

5. The Middle East
   a. Arabian Desert, on the Arabian Peninsula, composed of:
   b. Al-Dahna Desert, Empty Quarter, Nefud Desert and others
   c. Dasht-e Kavir, in Iran
   d. Dasht-e Lut, in Iran
   e. Judean Desert, in Israel and the West Bank
   f. Maranjab Desert, in Iran
   g. Negev, in Israel
   h. Wahiba Sands, in Oman

6. North America
   a. The Great Basin Desert, in Nevada, USA
   b. The Mojave Desert, in California, USA
   c. The Chihuahuan Desert, in Arizona, New Mexico and Texas
   d. The Sonoran Desert, in Arizona, California and New Mexico

7. South America
   a. Atacama, in Chile and Peru

b. La Guajira Desert, in northern Colombia
c. Monte Desert, in Argentina
d. Patagonian Desert, in Argentina and Chile
e. Peruvian Desert, in Peru and Chile
f. Sechura Desert, in Peru, South

US Deserts

The US deserts, especially those in south Arizona and southeast California, are unique in that elaborate, extensive and well maintained water irrigation systems were developed in their midst during the last 50-60 years, thus converting some parts into productive and very busy man-made agricultural and commercial areas. Many small and large populated centers are thriving there as well, with an excellent infrastructure (railroads, highways, electrical transmission lines, substations etc.) which is well established and functioning very efficiently.

This makes adding solar power plants in these areas an ideal solution to solar energy generation. All this should make adding solar capacity to the conventional energy mix an easy thing to do, or at least much easier than doing the same thing in the Sahara desert. The fact that the energy costs are increasing, while energy supplies are decreasing and the world is choking of pollution should add to the ease and desire of implementing solar energy in the US deserts.

We have been talking about this exact situation since the 1970s without significant practical results, so not much has changed during the last 40 years in the large-scale PV power generation area. Now, when energy independence and environmental cleanup are on the priority list and there's lots of talk about them, we still don't see many large-scale PV plants operating in the deserts. But the deserts are still there, reminding us about the abundant energy under our noses and that all we have to do is get it. Getting is not going to be easy, or quick, so we just have to persist in our goal of converting the US deserts into the energy source we so desperately need.

Here is a detailed list of US deserts suitable for large-scale PV installations:

a. Red Desert is in Wyoming
b. Alvord Desert is in eastern Oregon
c. Owyhee Desert is in northern Nevada, SW Idaho and SE Oregon
(Yp Desert is a portion of the Owyhee Desert in Idaho)
d. Black Rock Desert is a dry lake bed in northwestern Nevada
(Smoke Creek Desert is an extension of the Black Rock Desert)
e. Great Salt Lake Desert is in Utah
f. The Great Basin Desert is in Nevada
g. Tule Desert is in Nevada
h. Amargosa Desert is in western Nevada
i. Painted Desert is in Arizona
j. Mojave Desert is in California
(Death Valley is in California (also considered part of the Mojave Desert and part of the Great Basin.
k. Chihuahuan Desert is in Arizona, Texas, New Mexico, and Mexico
l. Trans-Pecos Desert is in west Texas
m. White Sands is in New Mexico
n. Sonoran Desert is in US and Mexico
o. Lower Colorado Desert is in California and Arizona
p. Low Desert of Southern California is in California, USA
q. Yuha Desert is in Imperial Valley, California
r. Lechuguilla Desert is in southwest Arizona
s. Tule Desert is in Arizona and Mexico
t. Yuma Desert is in southwest Arizona

The most suitable for large-scale PV power generation deserts are located mostly in the SW USA, and occupy a significant portion of the territory of the states of Arizona, California, New Mexico, Nevada and Texas. There are also some areas in Utah and Colorado with desert, or desert-like climate as well, which are suitable for solar installations.

The US deserts vary in climate, flora and fauna, and are mostly poor in natural resources, but are all blessed with one efficient, constant and unending energy resource—a lot of hot, bright sunshine—almost every day of the year, nearly all day long—precious sunlight, waiting to be harvested.

On the flip side of the coin, the deserts have a basically hostile, dangerous and even deadly climate with extreme temperatures (extreme heat during the day and extreme cold during the night), high humidity during monsoon seasons, lack of water and vegetation, and mostly poor infrastructure. The harsh climate is hard on the PV equipment intended for long-term operation forcing power degradation and failures. These problems are being worked on as we speak. It is only logical and practical to point out, therefore, that even small portions of the US deserts are capable of providing a major portion of the electric power to the US and neighboring countries in the future. The climate and behavior of the deserts have been studied for a long time, and is well understood, so with the necessary precautions and proper solar technology selection, design, installation and operation, deserts might—and should—become the hub of solar energy generation for the US, Mexico and Canada in the not-so-distant future.

Desert Power

The world produces and uses ~15 terawatts of electricity annually, which would require ~75 million acres of land to produce, assuming 5 acres per each MW generated at 15% efficiency, and provided that the PV modules will last long enough to make a difference. This means ~120,000 square miles of desert land are needed to power the world with this type of equipment.

The world's land surface area is ~55 million square miles. The Sahara Desert covers over 5 million square miles (almost 10% of the World's land area), so we would need only a small sliver of its (otherwise useless) land to power the entire world. This isn't possible at this time, for a number of technological, logistical and financial reasons, so we need to find ways to use the deserts close to the point of use (populated centers, enterprises etc.) where we could provide enough power to significantly reduce the use of conventional fuels consumed at these areas.

Per some estimates, the world's deserts take an area of nearly 18 million square miles (or close to 30% of the world's land area). And since this is where the best solar irradiation is found, we have no shortage of good energy sources to provide all the power we need. We only have to find a way to capture this abundant and constant source, and transport its energy to where we can use it.

The US generates ~1.1 terawatts of electricity, which would require ~8,000 square miles of 15% efficient PV modules installed in the desert to generate. This is an area of 80 miles by 100 miles in the Arizona or Nevada deserts. Imagine driving from Phoenix, AZ, to Las Vegas, NV, among a forest of solar equipment and installations feeding the national power grid. Far fetched, yes, but this is all the future generations will have left, after we've leached all the crude from the ground and polluted the atmosphere with coal to the point of no return, providing electric power as we know it today. Such a huge project, however, is not practical to undertake in its entirety today. Practical use of PV power today would be to locate smaller solar power generating plants near populated and industrial centers instead. This could provide enough electricity to reduce our dependence on foreign oil and partially eliminate the problems associated with its production, transportation, use, and the resulting pollution.

Now imagine large-scale PV fields where 40 or 50% efficient PV equipment is operating in the Arizona desert wastelands. This automatically reduces the land use by a factor of 3 or 4. Imagine that. Suddenly, things look more doable. Instead of 120,000 square miles, we are now talking about 30,000-40,000 square miles to power the entire world, and only 2,000-3,000 square miles to power the entire US. This is a huge difference but still far removed from today's reality, although some advances in the technology, design, manufacturing, and use might bring us close to the target soon. One ray of hope is the high concentration PV (HCPV) technology which is 42% efficient today. When all the bugs are worked out, HCPV just might take the lead in desert power generation.

Land and PV Projects Management

There are a number of large-scale PV projects in the US and Europe in the 20-50 MWp size and some

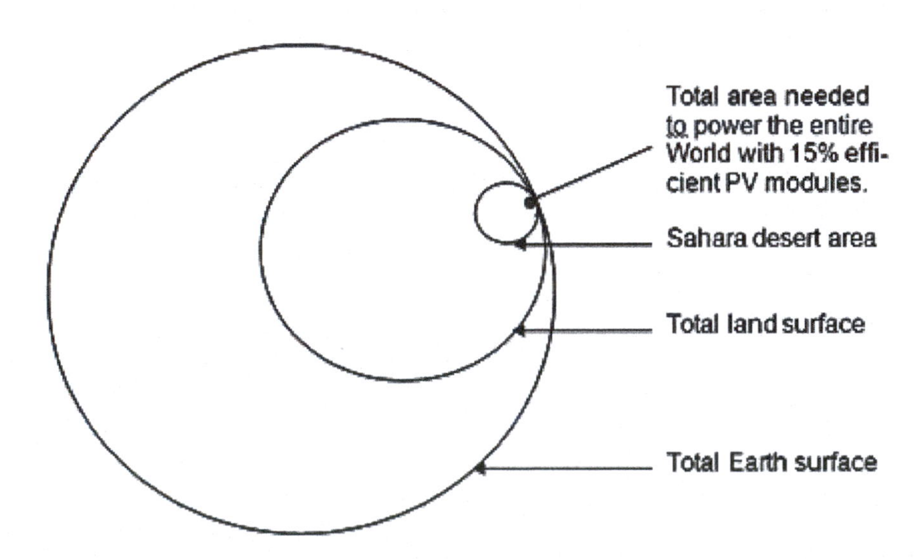

Figure 7-2. Land needed to power the world by PV (not to scale)

even larger are planned. The trend for 2011-2016, however, driven by CPUC in California is towards smaller PV projects in the 1-20 MWp size. The smaller projects require less land and investment, limited transmission upgrades, and have shorter construction lead times than the larger projects. They are usually sited closer to the load or interconnect point, so these projects typically have lower interconnection costs and lose less power in the transmission process. Smaller projects are easier to finance and control too, so we foresee that as the status-quo for the foreseeable future in the US.

Centralized large-scale solar projects are still in the plans, but they are much more difficult to develop and so their future is somewhat uncertain. BLM is presently fast-track reviewing 14 large-scale solar projects in several states (mostly in California). All these planned projects are over 100MWp each, and some got delayed and cancelled already. A larger number of smaller projects to be built on private land will be up and running while work on the larger plants is under consideration. The solar run continues... with California still in the lead by far.

The permitting process speed may favor the smaller plants too, since smaller projects are easier to permit and handle in general. For example, in California, the three major IOUs (investor owned utilities) are still short of the 2015 goals, so we may see a large number of smaller PV plants pop up soon.

Another trend in the making in California is that of converting large CSP (solar thermal) projects into smaller PV installations. So when several large CSP projects were cancelled (and several are following) in 2010-2011, the large-scale projects might be broken into smaller pieces and permitted for PV power generation. So, a 200 MW CSP plant might be converted into 10 x 20 MWp PV power plants. Negotiations on several such transactions are ongoing.

US REGULATORY AND POLITICAL SYSTEMS

Regulation is in the foundation of the US government bureaucracy, while politics and politicians determine the direction and speed the regulators move. Regulators regulate almost everything that has to do with generating revenue: material goods, services, and most anything in between. Sometimes their work is needed, useful and appreciated. Often, however, the results from their efforts are mediocre.

Solar power is a prime example of this mediocrity. Since the 1970s the different US government and regulating agencies have been wandering in whatever direction the political winds blow, trying to decide how to handle the extent and complexity of solar power deployment in the US. We knew even back in the 1970s that solar and wind power are the ultimate solution to our energy crisis. There was no political will, however, and until recently responsible agencies did not take the initiative to force the development of alternative energy sources. They are doing so now, only as a desperate last resort response to serious energy and environmental crises. We just have to wait and see how today's solar boom plays out, hoping it is different this time around.

In 2008 the Obama Administration forced a move towards speedy implementation of alternative energy projects in the US, and the resulting mish-mash of half-baked rules, hastily made decisions and poorly executed tasks by the responsible agencies is mind boggling. The combination of these mishaps has caused the delay, or cancellation, of a number of major projects and who knows how many smaller ones. Nevertheless, the completion of many alternative energy projects in California and the overall progress of solar energy deployment in the US and around the world are undeniable and may be unstoppable.

Political and Regulatory Aspects of Solar Energy

US Department of Energy Secretary Steven Chu says the U.S. lags behind other countries in the race for clean technology even though it has the greatest "innovation machine" in the world. He doesn't say why this is so, because he doesn't want to point a finger at the culprits, but it is not hard to see, if one reads carefully between the lines. The situation reminds us of the old fairytale of an eagle, a shark and a crab all pulling a cart in different directions. One is pulling up, the other to the left, and the third to the right. The cart is still sitting where it was when they started. The development of PV energy in the US has not moved very far either, as compared to the great need for clean, US made power sources.

So large-scale solar projects which are the future of our energy independence face a number of challenges; i.e., Sen. Dianne Feinstein recently introduced legislation seeking to block development of large-scale solar and wind projects on 1 million acres of federal land in California. In January, environmental groups challenged BrightSource Energy's 440 MW Ivanpah Solar Power Complex, because the development interferes with the habitat of two dozen rare tortoises. BrightSource cancelled another large-scale CSP project in the Mojave Desert in September, 2010 citing a desire to avoid wrangling with Sen. Feinstein over her proposed legislation. Two large solar projects in Arizona were cancelled in 2010 for a number of logistical, technical and financial reasons as well. These and dozens of similar examples demonstrate the complexity and difficulty of starting a large-scale solar project in the US even now when we need cheap, clean power and jobs more than ever, and when government subsidies and incentives are the greatest ever.

The situation is different, but equally difficult in other countries as well. Germany is, or was, the exception. Solar energy in Germany is an unlikely phenomena, considering that Germany has an average of just 4.8 hours sun exposure each day, mostly under cloudy or foggy skies. Despite its being solar-challenged, Germany produces over 12 percent of the electricity it uses annually from solar installations, which is many times more than any other country in the world. Things are slowing in Germany and many other EU countries now, due to the severe economic downturn in Europe, so we will just have to wait for the final tally.

In contrast, Israel has twice the sunlight of Germany (9.5 hrs. daily), and most of it is clear bright sunlight. Nevertheless, in a recent poll by Global Green USA, Israel got a D– for its solar electricity production efforts compared to Germany's A– and California's B. A careful look reveals that politics is what drive the solar energy discrepancy, with German politicians and business in total support of solar, while Israel has other priorities, with solar far back on the back burner regardless of the acute need and the favorable solar climate.

The US is somewhere in between. Many of the stakeholders are lukewarm and partially—almost unwillingly—interested in solar energy, the development of which is hanging in the balance and could go either way. Solar power plant development is under the administrative and financial control (and strict supervision) of the US regulatory bodies with several different responsibilities.

Below are several key agencies responsible for making and enforcing the solar power rules.

DOE

The United States Department of Energy (DOE) is a cabinet-level department of the United States government in charge of energy and safety of nuclear materials. Its responsibilities include nuclear reactor production for the United States Navy, energy conservation, energy-related research (including wind, CSP and PV), radioactive waste disposal, and domestic energy production. DOE also finances and oversees basic and applied scientific research funded through the US DOE National Laboratories, such as NREL in Albuquerque, NM, Sandia in Colorado Springs, CO, LLNL in Livermore, CA, and others.

The agency is administered by the United States Secretary of Energy, and its headquarters are located in southwest Washington, DC, and Germantown, MD. DOE has been heavily involved with solar energy materials, devices and procedures in the past, and is still the leading funding instrument for solar energy development in the US. A number of energy programs have been initiated and financed by DOE and some have brought real progress to the US energy industries. Some major decisions of late, however, lack the scientific and financial savvy required from such an important high-profile organization.

BLM

The Bureau of Land Management (BLM) is the US government agency that controls the federal land utilization, including permitting the installation of solar power plants. Since 2008 BLM has been under the gun (by the Obama administration) to issue as many permits as possible. This has caused some discrepancies, mostly due to the fact that BLM bureaucrats are not that familiar with the solar industry and are not accustomed to moving quickly. This abnormal situation has also shown the inadequacy of this body to adapt quickly to change and handle complex undertakings.

BLM has made some changes to facilitate the permit process, but it is still a very difficult, extremely expensive, lengthy and basically uncertain process. This has made raising capital for the construction of large-scale projects quite difficult, and has contributed to delay in the development of this segment of the U.S. solar market over the last two years. The challenges are significant, and BLM together with other governing bodies has been rethinking and redesigning its efforts to develop renewable energy capacity in the US. There are a large number of success stories, and an equal amount of failures, so it remains to be seen how efficient and successful BLM will be in this new situation.

EPA

EPA has a major role in developing solar energy in the US, but thus far it has been staying in the background, acting as an observer, and stepping in only in rare cases; i.e., an EPA letter sent to the California Energy Commission (CEC) on the Ridgecrest solar project states that the CEC assessment/draft environmental-impact statement does not provide sufficient information regarding the viability and effectiveness of proposed mitigation measures that are intended to reduce the impacts to below the level of significance. "We believe approval of a right of way for this project on such an ecologically valuable site, and with the potential for such significant environmental degradation, would set an unwise precedent for the many renewable energy right-of-way applications under consideration by the Bureau of Land Management which, collectively, could result in severe and immitigable impacts to desert ecosystems." For these reasons, EPA rated the preferred alternative as Environmental Objections, and classified it as "Insufficient Information," which only shows EPA's importance in deciding our energy future. It also shows how easily EPA could put a project out of business. As a matter of fact, the certification application was withdrawn in January 2011, in response to concerns over the Mohave ground squirrel habitat (not EPA decision, but following EPA' guidelines).

EPA was officially recognized by the US Superior court as "best suited to regulate GHGs, so EPA is in the driver's seat...again. In a press release on Nov. '10, EPA official said, "In January 2011, industries that are large emitters of green-house gases (GHGs), and are planning to build new facilities or make major modifications to existing ones, will work with permitting authorities to identify and implement BACT [best available control technology] to minimize their GHGs." These provisions would only affect large GHG producers, not farms or restaurants, according to the release, but will benefit solar power development in the near future.

EPA is in the unenviable position of having its fingers in too many pots, with energy being just one of them. Since the complexity and speed of the energy developments of late are unprecedented, EPA is having a hard time making the rules and being in full control of the situation—trying as hard as it might. This has made EPA the scapegoat of a number of energy project delays and failures, and we do expect that there will be many more added to this list before it's over.

CPUC

California Public Utilities Commission (CPUC) regulates the activities of the utilities in the state, and is adapting to changes quite efficiently. It came up with a proposal lately for a market-based feed-in tariff (FIT) for California energy projects, where CPUC would mandate IOUs in California to purchase the revenue requirement equivalent of 1,000 MW over a four-year period through a reverse auction, in which the utility would set a ceiling price and accept project bids up to that price and capacity allocation. All projects between 1.5 MW and 10 MW would be given a priority, while those between 10 MW and 20 MW would be left at the utility's discretion.

This is a serious initiative and significant challenge for the solar industry, which sets an example for other states and countries to follow. So far, so good. CPUC seems to be trying hard and is in charge of the situation up to the extent of its authority. But —adding 1,000 or 2,000 MW during the next 4-5 years is just a drop in the mandate of adding 30% alternative energy by 2020 and 80% by 2035. Is this the best we can do, or is it just another power (no pun intended) game?

Multi-agency Planning and Permitting Processes

Planning and permitting renewable energy power plants in California often involves one or more federal, state, or local government entities. For example:

- The California Energy Commission issues licenses (and imposes conditions) for constructing and operating thermal power plants, including solar thermal, geothermal, natural gas, and biomass projects for more than 50 MW in peak generating capacity.
- Local governments issue permits for renewable energy projects that fall below the Energy Commission's 50 MW siting jurisdiction for thermal power plants, and for wind projects and photovoltaic solar projects on privately owned land and some public lands.
- The U.S. Bureau of Land Management (BLM) issues right-of-way permits for energy projects on BLM lands.
- The California Department of Fish and Game (CDFG) or the US Fish and Wildlife Service, or both, issue permits setting conditions for projects that harm or may cause harm to endangered species and their sensitive habitats.
- The U.S. Army Corps of Engineers (USACE) issues permits for impacts related to federally designated waters, and may conduct environmental review, as does CDFG for other waterways and wetlands not covered by USACE.

To facilitate the permitting process, the California Energy Commission is proposing a new program that would provide local governments with planning and permitting assistance to help them evaluate and expedite renewable energy development in their jurisdictions. The proposed Renewable Planning and Permitting Program (RP3) is expected to provide local governments assistance from state agencies with planning and permitting experience and expertise.

AZCC

The Arizona Corporation Commission (AZCC) is involved in, and/or in charge of, a number of business activities in the state, in addition to regulating the local electric utilities and acting as a final decision maker in permitting solar energy projects statewide.

AZCC had a "solar team" for awhile which was working towards promoting solar in Arizona and making it the "solar capital" of the world. There were lots of meetings, discussions, interviews, etc. during 2008-2010, and some results to show, albeit somewhat humble for the solar capital of the world.

AZCC has done a good job in attracting out-of-state solar business, but Arizona is still way behind a number of states, some of which have marginal sunshine. The activities of late indicate that solar in Arizona is headed into a difficult period of changes and hesitation, driven by economic uncertainties and utilities issues.

One thing is for sure; Arizona is not going to become the solar capital of the world any time soon, and we can only hope that at least it retains its mid-field position in the U.S. solar rankings.

NCCP

It is impossible to separate the political from the regulatory aspects of alternative energy generation and their effects on its development. There are a number of obstacles to completing regulatory procedures, as needed for financing, permitting and finishing PV projects. The political and regulatory aspects play a great role from the initial approach to the final step of the process. So, potential candidates should be fully aware of these steps and related issues affecting the installation of PV plants on private or government lands.

For example, if you intend to apply for a permit to install an alternative energy generating plant under the jurisdiction of the Natural Community Conservation Planning (NCCP) Phases program, which is, or will be, implemented in most US states, you must complete a number of steps first. Some of these steps are:

Planning Agreement

Planning agreements are developed with interested jurisdictions, landowners and other parties. The purpose of the planning agreement is to specify the roles and responsibilities of the participants in developing the NCCP plan, identify the scope, natural communities, focus species, processes for scientific and public input, and an interim process for project review.

Committees

The NCCP process is facilitated by the formation of a variety of committees to accomplish all the tasks. There is usually a steering committee made up of the primary negotiators representing all the interests groups. Often there is a biological technical committee which includes biologists from all the agencies and project consultants. In addition, there may be other committees to deal with specific issues such as funding, mitigation strategies, data management (GIS), land management, etc. The department usually participates in all committees. Often toward the latter stages of planning, a smaller negotiating team is formed with just the applicants, their consultant, and the wildlife agencies.

Independent Scientific Input

NCCPs must be based on the best science available. NCCPs are developed based on a set of conservation guidelines developed by independent scientific input. Independent scientists use the principles of conservation biology and species conservation to develop the foundation for a habitat conservation system. The scientific advice also helps to identify any data gaps in current knowledge.

Biological Data Collection

Species specific biological field data are collected, natural community types are mapped and field verified, and habitat evaluation models are developed if needed. The data are input into a geographic information system for further analysis. The data collected in this phase of plan development help guide all future decisions for development of the plan.

Preserve Design

Preserve design involves the use of conservation biology principles, land ownership patterns, and species and habitat distribution information. Preserve design can be contentious, as private landowners come to the realization that a portion of their lands may be needed for establishing an adequate preserve. Public involvement in this phase is critical but also can result in significant delays or reanalysis of the preserve design. Guidance from independent scientists is recommended.

Development of Draft Plan

The draft plan will contain all the conditions and mechanisms to assure the conservation of the species and make the plan work on a daily basis. It will include a preserve map, a preserve implementation strategy (project mitigation requirements, public land set asides, monitoring and adaptive management programs, etc.), funding assurances, a time table for implementation, and a draft implementing agreement (the contract).

Public Review of Draft Plan

Although the draft plan is usually developed with significant public input, the draft plan is the first time the public and elected officials see a comprehensive document. Public hearings are conducted by the local plan participants. The draft plan is also accompanied by a draft environmental impact report and draft environmental impact statement (DEIR/S). The public review draft also has to be reviewed by the local elected officials.

Final Plan Development

Following public review of the plan, the draft plan is revised based on the direction from local elected officials and public comments received during public review. A final implementation agreement is developed, and the EIR/S is finalized.

Jurisdiction Approval of the Final Plan

The local jurisdiction holds an additional public hearing on the final plan and makes a decision to either adopt it or do further revisions. If the plan is adopted, it is then submitted to the Department of Fish and Game and the U.S. Fish and Wildlife Service for review and approval.

State and Federal Permit Issuance

The Department of Fish and Game and the U.S. Fish and Wildlife Service make final reviews of the plan and determine if it meets the standards for issuance of an NCCP Permit and Federal Endangered Species Act 10a Incidental Take permit respectively.

Implementation

Following the issuance of the state and federal permits, the local jurisdictions may approve individual projects consistent with the plan without further requirements for individual projects to receive state and federal approval (although compliance with other laws such as Streambed Alteration Agreements or Clean Water Act permits may require further consultations).

Permit holders must track habitat loss and protection, and monitor for permit compliance and biological effectiveness. Lands put into the preserve system must be protected in perpetuity and managed to benefit the covered species. Biological information collected through monitoring and research is used to modify management activities to ensure conservation success (adaptive management).

Finally, the permitee must provide annual reports to that effect to the wildlife agencies and the public.

Fun, eh? We almost went through the letters of the entire alphabet above, and it is much easier and faster on paper. It is actually quite painful in real life. Really! And we did not even fully explore the environmental impact work, related reviews and other procedures, which add another dimension to the already complex picture.

Permitting of Public Lands

Just in case things are not clear enough, here is a partial list of steps and tasks that must be completed for permitting of public lands in California. The process is similar to that of other states with some specific regulatory differences of the agencies involved in the process.

The procedure is as follow:

a. Applicant's plan of development (POD) is to be submitted to the Bureau of Land Management (BLM).
b. California Energy Commission (CEC), who actually has jurisdiction only over large solar thermal projects over 50 MWp, reviews the application for certification (AFC). If OK, the process moves on; otherwise, additional or more relevant information will have to be supplied. Local governments issue permits from smaller CSP projects, as well as for wind and PV.
c. BLM publishes notice of intent in Federal Register
d. CEC staff issues identification report (IIR)
e. CEC/BLM staff file data requests as needed
f. Informational hearing and site visits and BLM scoping meeting
g. CEC/BLM staff data request workshop
h. Applicant provides data responses to any remaining questions
i. Data response and issue resolution workshop
j. Initiate consultation with State Historical Preservation Offices (SHPOs)
k. Administrative Staff Assessment (SA) and Environmental Impact Statement (EIS) drafts circulated for agency staff review
l. Selected draft SA sections posted on CEC and BLM websites
m. Notice of availability (NOA) of SA/Draft EIS (DEIS) entered in Federal Register
n. SA/DEIS filed (90-day comment period begins)
o. BLM submits biological assessment (BA) to US Fish and Wildlife Service (FWS). Start of 135-day Section 7 consultation
p. BA determined adequate by FWS
q. Prehearing/evidentiary hearings start (evidentiary record remains open until key items such as FWS' biological opinion are received)
r. Close BLM comment period
s. BLM and FWS consultation

t. Prepare responses to comments and staff assessment errata (SAE) and final EIS (FEIS)
u. Administrative SAE/FEIS circulated for agency staff review
v. FWS issues biological opinion
w. NOA of SAE/FEIS is entered in Federal Register
x. SAE/FEIS distributed
y. Expedited governor's review period ends
z. BLM plan amendment protest period ends
a1. CEC committee files presiding member's proposed decisions (PMPD)
b1. CEC comment hearing on PMPD
c1. Close of public comments on PMPD
d1. Addendum/revised PMPD
e1. BLM record of decision/right-of-way issued
f1. CEC decision issued

If CEC decision is OK:
g1. Notice to proceed issued
h1. Plant construction starts
i1. Plant operation starts upon construction completion

If CEC decision is not OK, then the process is repeated.

This time we DID run out of alphabet letters. Just imagine the amount of paperwork, effort, money and time needed to go through all these steps—even if only half of these agencies must be contacted, and only half of the tasks completed. You do need an army of engineers, clerks and lawyers to do this right and finish on time.

Note: The above steps pertain to permitting solar thermal projects on federal lands under the jurisdiction of CEC. Similar processes, albeit with different agencies, must be followed when permitting PV installations.

Commercial PV Installations Permits

Figure 7-3 depicts the major steps of the permitting process of a smaller (residential or commercial) PV system in oversimplified form, with each of the steps containing many sub-steps, forms, documents, signatures, reports, visits, meetings, payments, etc., depending on the location, type and size of installation. This process seems more doable, but even then it takes months of paper filing, meetings, etc. to get all of this done.

Large-scale PV Projects Considerations

Provided that we have available land, financing opportunity, a good chance to get a local interconnection in the grid, favorable rent/lease and PPA rates, the questions we (and our partners and associates) need to answer in preparation for the design, permitting and implementation effort of any large-scale PV project in California, or anywhere in the US, are:

A. Land and Environmental
   1. Acreage size, shape and its status and history?
   2. Environmental permitting work done on the land, if any?
      a. The permitting authority detail: CEQA, NEPA, or...
      b. Other permitting agencies: RWQCB, air district, etc. and process status
      c. Permit documents: ND, MND, EIR, EA/FONSI, EIS and process status
      d. Approximate project approval date (all permits)?
      f. Environmental impact issues: bio, cul, water, air, to be addressed
      g. Total estimated mitigation costs and lead times?
         i. Biological resources (flora and fauna)
         i. Protocol surveys for listed species and rare plants
         ii. Listed species affected and scope of impact studies
         iii. 2081 permit, or Sec. 10 USFWS consultation

B. Technical
1. Estimated rated capacity in MW and annual MW/h output?
   a. Location sunlight intensity and availability
   b. Historical seasonal weather data
2. PV technology description, efficiency and cost in $/MWp?
   a. c-Si, thin films, HCPV etc.
   b. Fixed mounting or tracking
3. Electrical transmission system description?
   a. What entity owns the point of interconnection (POI)?
   b. Is the POI a transmission line or a distribution line?
   c. Status of Phase I electric transmission system interconnection study?
   d. Estimated schedule for completing CAISO/CPUC review and obtaining LGIA or SGIA?

The PV system Permit application submittal package includes:

1. Completed general permit application form as required by State and Federal laws.
2. Site plan of all equipment clearly identified on the drawings, as follow:
   a. PV array detail, including combiner and junction boxes and wiring
   b. Disconnects, inverters, meter and service panel and tie-in locations
   c. General footprint of structure and array as related to property lines
   d. Roof plan with location and physical size of PV modules and attic vents
3. PV equipment OEM spec sheets (PV modules, inverter, panel, j-boxes, wire etc.)
4. Complete 3 line electrical diagram, including:
   a. Module manufacturer name and module model number
   b. Module specs @ STC including VOC, VMP, ISC, and IMP
   c. System size, DC @ STC "nameplate" rating taken from OEM spec sheets
   d. Array specs (@ STC; max VOC (w/ temp corrections), VMP, ISC, and IMP
   e. Number of module strings
   f. String fuse rating (if available and appropriate)
   g. Current carrying conductors, indicating size and type
   h. Grounding point and conductor size and type
   i. Over current protection/disconnects, V and A rating and inverter "integration"
   j. Inverter manufacturer and model number, rated V AC volts and A max.
   k. Grid interconnection location, showing AC load panel with back-feed breaker rating (voltage and amperage),
   l. Panel rating (busbar and main breaker rating), and line / load side tap.
5. Mounting / racking system details shall include
   a. Manufacturer's spec sheets with uplift capacity for wind loads of 130 mph, 3 second gust and snow loads of 55 psf and better
   b. Attachment details (type, size, and spacing of fasteners etc.)
   c. Stamped engineer's verification letter.
   NOTE: For custom racking, a site specific engineered design is required.
6. Structural information:
   a. Size, type and spacing of roof framing members,
   b. Type and thickness of roof sheathing,
   c. Type and number of layers of roofing materials (flush or tilt mounted array.)
   d. Detailed method of locating framing members.
   NOTE: Additional information or engineering is required in some cases
7. Special Considerations:
   a. The installer shall be licensed as a Class C subcontractor.
   b. Electrical wiring shall be performed by a licensed electrician.
   c. Permit will be issued only upon licensed electrician's signature.
   d. Permit fees are based on the value of materials and labor before rebates.
   e. The valuation consists of racking/mounting and electrical wiring sections.
8. The permitting process starts by submitting a Permit Application form to the Program office, where it is reviewed and approved. The Applicant is then issued a notice and submits the above documents package (drawings, calculations, etc. info). These are reviewed and a building permit is issued. Upon completion of the PV installation a final inspection is conducted, after which the PV system can be used.

Figure 7-3. The permitting process simplified

e. How much has been paid towards deposits applicable to upgrades?

f. Total estimated transmission system upgrade costs? Of the total how much is assumed to be reimbursable by other users?

g. Have the downstream transmission system improvements been included in the Projects CEQA and or NEPA review?

h. Single environmental document being relied on by CPUC or is CPUC producing a separate doc?

i. Were all properties along the t-line route surveyed for biological and cultural resources; and waters of state and or US?

C. Financing

1. Estimated Revenue Stream?

a. Power purchase agreement status and probability?

b. If so, is it with public or investor owned utility (IOU)?

c. PPA price point per MW/h, CPI escalator, term; and investor's estimated IRR?

d. Estimated levelized cost of energy (LCOE)?

e. Assumed key financing ratios (D/E, DSCR, interest/term on debt)?

f. If no PPA available, then what is the exit strategy; market; assumed off-taker, etc.?

2. How much money have we invested to date and plan to invest in the future?

3. What exactly are we looking for in this project?

4. What are we offering (land, finance, equipment, expertise, etc.?

5. Do we have any upcoming funding deadlines that affect viability of the project?

6. Any other significant considerations (e.g. extraordinary site costs, etc.)?

Once the questions are answered we could evaluate our readiness and start planning the rest of the permitting, financing and implementation steps.

Interconnection Standards

Each state regulates the process by which an electricity generator can connect to a distribution grid by establishing interconnection standards and policies. These seek to maintain grid stability and safety, and usually require all generators to comply with the standard interconnection rules established to provide clear and uniform processes and technical requirements that apply to utilities within a state. These rules reduce uncertainty and prevent time delays that clean distributed generation (DG) systems can encounter when obtaining approval for electric grid connection.

EPA assessed the DG friendliness of existing interconnection rules through the following criteria:

1. Favorable:

Indicates that there is a well-defined interconnection policy in place that has at least one or more beneficial attributes:

a. Standard forms
b. A reasonable timeline for application approval
c. Low or no additional insurance requirements
d. Allows for >10 kW residential and >100 kW commercial units to interconnect
e. May have additional positive attributes

States with favorable interconnection standards: CA, CT, DE, IN, MA, MI, NH, NJ, NY, NV, OH, OR, PA, VT, WA

2. Neutral:

Indicates that there is an interconnection policy, but overall the policy cannot be considered either beneficial or detrimental to DG.

States with neutral interconnection standards: AZ, CO, HI, MN, MO, NC, SC, TX, UT, VA, WI, WY

3. Unfavorable:

Indicates that the interconnection policy may be available, but has unfavorable requirements such as:

a. Only allowing small units to interconnect (<10 kW residential and <100 kW commercial)
b. Having high liability insurance requirements
c. Requiring owners/operators to pay large interconnect study fees
d. May have other burdensome requirements

States with unfavorable interconnection standards: AR, FL, GA, KY, LA

States with no functioning policy in place: AL, AK, DC, ID, IA, IL, KS, ME, MD, MS, MT, NE, NM, ND, OK, RI, SD, TN, WV

The level of implementation of the interconnection standards reflects amazingly well the willingness of the respective utilities and regulating bodies (in most cases) to participate in the new wind-solar dance. Many of them do think it is important enough, some don't care much, while others consider wind and solar too much hassle.

PV Industry Standardization

"Standards in photovoltaics are essential for the industry in order to lower trade barriers and to reduce the cost of ownership for cell and module manufacturers of their production facilities. Both elements are key to reach competitiveness of the photovoltaic industry in a global energy market."—Dr. Ossenbrink, Unit Head Renewable Energies of the European Commission's Joint Research Center.

Presently the PV industry in the US and the world follows a number of standards, and a number of rag-tag rules and regulations, but the industry has no universal, unified standardization system. So, different countries, local entities, users and manufacturers make their own rules and standards, which creates ever changing and even chaotic situations.

PV modules now come with long-term warranties, but the terms often raise questions about the quality of the PV components, systems, installations, and after-sales maintenance and support. The only way to solve this dilemma is by establishing worldwide standards for the solar industry. A number of attempts to standardize the industry are underway, but their completion, implementation and enforcement are still a bit far.

Some of these programs are.

IEC International Standards for PV

The most successful effort in this area is that of the International Electrotechnical Commission (IEC) and its International Standards for PV, and its Worldwide System for Conformity Testing and Certification of Electrotechnical Equipment and Components (IECEE) which provide both benchmarks and proofs of quality for industry and government. IEC's TC (Technical Committee) 82 prepares international standards for all elements of those systems—everything from the light inputs to a PV cell to the interface with the systems to which the electrical energy is supplied.

Comprised of leading industrial and governmental experts from 40 countries, IEC's TC 82 has prepared international standards for terms and symbols, PV module testing, design qualification and type approval of crystalline silicon and thin-film modules, and characteristic parameters of stand-alone systems, among other elements.

IEC TC 82's current work includes:

— System commissioning, maintenance and disposal.
— Characterization and measurement of such as new thin film photovoltaic module technologies, i.e., CdTe, CIS, CuInSe2, etc.
— New technology storage systems.
— Applications with special site conditions, such as tropical zone, northern latitudes and marine areas.
— Recommendations for small renewable energy and hybrid systems for rural electrification, including PV systems.

In addition, IEC TC 82 is addressing the safety of grid-connected systems on buildings and utility-connected inverters, as well as the protection of people and the environment from such things as radiofrequency and electromagnetic pollution and the toxic materials that need to be disposed of during the PV manufacturing processes.

PVECI (PV-2) PV Equipment Communication Interface Standard

A standard that is urgently needed today is the so-called PV-2, intended to define the communication and interface standard for PV manufacturing equipment. PV-2 Equipment Communication Interface Standard for PV Production Systems is based on GEM/SECS (SEMI E30), and will establish the operating parameters and detail standards for critical PV manufacturing operations: diffusion, etch, screen printing, cell testing and sorting, lamination, and final tests. Thus standardized protocols and procedures will be applicable to all equipment sets used for manufacturing PV cells and modules. Upon completion, they will be similar to those used in the semiconductor industry, and are expected to bring some normalcy, reliability and professionalism into

Key standards to be designed, developed and implemented.

Equipment Standards	Process Standards	Product Standards	Power Plants Standards
Communications	Process Specs	Module Design	Planning
Operation	Steps Execution	Materials Quality	Field Design
Control	QC Inspections	In-house Tests	Installation
Qualification	Final test	Field Tests	O&M
Maintenance	SPC & Document	Reliability	Safety
Training	Training	Training	Training

the present-day mix of improvised equipment and recipes used by scores of Asian manufacturers in their effort to flood the market with all sorts of PV contraptions before the doors slam shut.

PV-2 will provide uniform and enforceable standards for: a) optimization and verification of manufacturing specs for key process steps, and b) uniform equipment communications needed to coordinate the processes, and improve productivity and quality.

Challenges for the manufacturers in complying with a PV-2 standard are:

- Providing communication interfaces and languages for PV manufacturing equipment and systems
- Providing data configuration for external interfaces
- Effective logging capabilities
- Time-to-market issues, and
- Cost issues

Actually we see many other challenges and resistance accompanying PV-2 implementation, similar to those we witnessed in the semiconductor industry in the 1980s and 90s. A lot of water will run under the bridge before we see complete, comprehensive, universal PV industry standards in black on white, fully embraced by the industry and properly enforced around the world. Until then we simply have to understand the basics and do our best when manufacturing, purchasing, installing or operating PV products.

The CPUC Guideline

It is hard to navigate in today's solar energy climate, with all the changes and the daily step-ahead/step-behind developments. So every time we get a glimpse from the top, as in this case, we get a clearer picture of where PV stands and where they are going. See below a number of points taken from a CPUC (California' Public Utilities Commission) document, named "Decision Adopting a Solar PV Program for PG&E Company," which clarifies the situation and points the utilities and the PV industry in the right direction. (8)

This document is a guideline for the California utilities, PG&E in this case, and the PV industry, to follow in their drive to install up to 500 MW PV capacity by 2015 (250 MW UOG and 250 MW PPA type power plants), ranging from 1 to 20 MW in size, at an estimated cost of $4,312/kW DC installed, including a 10% contingency (all in 2009 dollars). This is a significant step towards overhauling the US energy markets, and California, as always, leads the pack.

Excerpts from the document:

- New renewable projects and transmission additions face a variety of risk factors, including permitting and financing challenges that may result in contract failure or delays.
- Smaller projects may avoid many of the risk factors that impede the timely development of larger scale renewable projects to the extent these smaller projects do not have the same land impacts, do not require the same level of project financing and permitting, and do not depend on large capacity transmission additions.
- The output profile of solar PV largely coincides with periods of peak demand.
- The final RETI Phase 1B Report identifies PV technology specifically as having significant potential for capital cost reductions in the future.
- The Commission has expressed its interest in utility proposals for utility-owned renewable projects.
- Few renewable UOG proposals have come forward to date.
- Under the RPS program as currently implemented, smaller scale projects, while likely to offer greater viability and speed of deployment relative to large-scale projects, are unlikely to be selected owing to their higher price, which may limit the extent to which smaller scale projects participate in the RPS program.
- A fixed PPA price may result in overpayment by ratepayers.
- The Commission has not yet implemented SB 32 and it is not known at this time how the price the Commission develops will impact the deployment of projects between 1 and 3 MW.
- The price of solar PV is anticipated to decline in the years ahead.
- A price cost cap for PPA projects is a reasonable way to ensure that the costs of the solar PV Program are not excessive to ratepayers.
- Land deposits prior to purchase of the land do not qualify as acquired or owned property for recovery in PHFU account.

The above points give us a general idea where the solar game is headed in the state of California and basically tell us that, according to CPUC:

a. There are risks associated with PV energy installation and transmission.

b. Smaller projects may be less risky.

c. PV is expected to contribute to peak-load leveling.

d. PV has potential for capital cost reductions in the future.

e. CPUC is interested in utility-owned solar projects.

f. There are only a few UOG (utility-owned generation) projects to date.

g. Smaller solar projects (1-3 MW is our guess) are less likely to be selected due to their higher price.

h. A fixed PPA will not be considered.

i. It is still unclear how the new prices will affect projects of 1-3 MW in size.

j. PV prices will likely decline in the future.

k. PPA-based projects will have a price cost cap.

l. Land deposits do not qualify as owned property until land is actually purchased.

All this sounds quite reasonable, albeit complicated, so we just have to hope that CPUC and PG&E will continue moving steadily in the direction they have chosen. We also hope that other states will follow in their wake—especially those of the southwestern US with lots of sunshine in their backyards.

SOLAR PROGRAMS, INCENTIVES AND SUBSIDIES

A large number of programs aimed to encourage and promote the development of solar energy in the US and abroad were undertaken during the critical 2008-2010 time period. These programs have been the driving force behind the quick addition of many megawatts of PV power generating installations around the world.

Some of these programs are:

Government and Private Incentives and Subsidies

There are a number of government-, state- and utilities-sponsored programs, incentives and subsidies today, intended to assist and support the development of alternative energy (including PV) during these hard economic times and in the near future. Some of these in force during 2010 are:

Business Solar Investment Tax Credit (IR Code §48).

The bill extends the 30% ITC for solar energy property for 8 years through Dec. 31, 2016, and it allows the ITC to be used to offset both regular and alternative minimum tax (AMT) and waives the public utility exception of current law (i.e., permits utilities to directly invest in solar facilities and claim the ITC). ITCs are critical for the further development of solar projects in the US.

The new 2010-2012 extension will provide 30% cash back to qualifying PV projects.

NOTE: The five-year accelerated depreciation allowance for solar property is permanent and unaffected by passage of the eight-year extension of the solar ITC.

SRECs

Solar renewable energy credits (SRECs) are a vehicle that allows for a market mechanism to set the price of the solar-generated electricity subsidy. In this mechanism, a renewable energy production or consumption target is set, and the utility is obliged to purchase renewable energy or face an alternative compliance payment (ACP) fine.

The power generator is credited an SREC for every 1,000 kWh of electricity produced. If the utility buys this SREC and retires it, they avoid paying the ACP. In principle, this system delivers the cheapest renewable energy, since all solar facilities are eligible. Uncertainties about the future value of SRECs have led long-term SREC contract markets to give clarity to their prices and allow solar developers to pre-sell/hedge their SRECs.

Utilities Programs

The California Public Utilities Commission (CPUC) has approved a number of solar programs to promote wholesale distributed generation using what they call a reverse auction as the method of procurement. To date, the CPUC has approved programs to procure 700 megawatts over five years using this mechanism. In addition, they have issued a draft decision to approve a program to procure an additional 1,000 megawatts over two years through a new reverse auction mechanism program. These new programs are among the first in the world that deal with the majority of solar subsidy concerns in an elegant and efficient manner. These programs are similar to the feed-in-tariff programs used in Europe and Canada, and are designed to attract investment capital via 20-year financeable cash flow.

New Clean Renewable Energy Bonds ("CREBs")

Clean renewable energy bonds (CREBs) may be used by certain entities—primarily in the public sector—to finance renewable energy projects. The list of qualifying technologies is generally the same as that used for the federal renewable energy production tax credit (PTC). CREBs may be issued by electric cooperatives, government entities (states, cities, counties, territories, Indian tribal governments or any political subdivision thereof), and by certain lenders. CREBs are issued—theoretically—with a 0% interest rate. The borrower pays back only the principal of the bond, and the bondholder receives federal tax credits in lieu of the traditional bond interest.

$800 million of new clean renewable energy bonds will finance facilities that generate electricity from renewable resources, including solar, wind, closed-loop biomass, open-loop biomass, geothermal, small irrigation, qualified hydropower, landfill gas, marine renewables and trash combustion facilities.

This $800 million authorization is allocated as follows: 1/3 will be used for qualifying projects of state/local/tribal governments; 1/3 for qualifying projects of public power providers; and 1/3 for qualifying projects of electric cooperatives. In March 2010 Congress enacted H.R. 2847 (Sec. 301) permitting new CREB issuers to make an irrevocable election to receive a direct payment—a refundable tax credit—from the Department of the Treasury equivalent to and in lieu of the amount of the non-refundable tax credit which would otherwise be provided to the bondholder. This option only applies to new CREBs issued after the March 18, 2010 enactment of the law.

Extension of Energy-Efficient Buildings Deduction

Current law allows taxpayers to deduct the cost of energy-efficient property installed in commercial buildings. The amount deductible is up to $1.80 per square foot of building floor area for property installed in commercial buildings as part of: (i) interior lighting systems, (ii) heating, cooling, ventilation, and hot water systems, or (iii) the building envelope. Expenditures must be certified as being installed as part of a plan designed to reduce the total annual energy and power costs with respect to the interior lighting systems, heating, cooling, ventilation, and hot water systems of the building by 50 percent or more in comparison to certain established standards. The bill extends the energy efficient commercial buildings deduction for five years, through December 31, 2013.

Qualified Energy Conservation Bonds

The bill created a new category of tax credit bonds, "qualified energy conservation bonds" (QECBs) to finance state and local government initiatives designed to reduce greenhouse emissions. QECBs can be issued to finance capital expenditures incurred for: (1) reducing energy consumption by at least 20%; (2) implementing green community programs; and (3) rural development involving the production of electricity from renewable resources. The bonds can also be used to finance research facilities and provide research grants for, among other things, technologies to reduce peak use of electricity. There is a national limitation of $800 million, allocated to states, municipalities, and tribal governments.

Research and Development Tax Credit

The bill would extend the research and development tax credit equal to 20 percent of the amount by which a taxpayer's qualified research expenditures for a taxable year exceed its base amount for that year. In addition, the proposal would increase the alternative simplified credit from 12% to 14%, and repeal the alternative incremental research credit for the 2009 tax year. The proposal is effective for amounts paid or incurred after December 31, 2007. Thus, research expenditures incurred by the solar energy industry would qualify for the credit.

Mandatory Renewable Portfolio Standards

Since 2008, the engine of solar energy development in the US has been the Obama Administration's mandatory renewable Portfolio Standards, which basically forces the utilities—like it or not—to include a certain amount of alternative energy in their power generating and distribution portfolios.

So here you have it—the wind and solar energy future of the US (see Table 7-2). No doubt that improvements are on the way—maybe not as soon as scheduled, but soon enough. Unless a catastrophic event pushes wind and solar energy development programs into the background... as happened several times already. Maybe there will be a follow-up mandatory portfolio standards to elevate mandatory percentages to 50-60% for each state. Only then will we be able to talk seriously about energy independence... but solar has to show some good results before we see anything major happen.

Net Metering Rules

Net metering is often cited as one of the most significant policies to advance solar PV use. Net metering is essentially electricity billing arrangement that provides an important incentive for consumers to install solar energy systems. Net metering enables customers

Table 7-2. Mandatory Portfolio Standards, 2008

State	Requirement
Arizona	15% by 2025
California	20% by 2010 (30% by 2020)
Colorado	20% by 2020 (IOUs)
	10% by 2020 (Co-ops & Municipals)
Connecticut	23% by 2020
Delaware	20% by 2019
District of Columbia	11% by 2022
Hawaii	20% by 2020
Illinois	25% by 2025
Iowa	105 MW by 1999
Maine	40$ by 2017
Maryland	20% by 2022
Massachusetts	9% by 2014
Minnesota	25% by 2025
	[Xcel 30% by 2020]
Missouri	11% by 2020
Montana	15% by 2015
Nevada	20% by 2015
New Hampshire	23.8% by 2025
New Jersey	22.5% by 2021
New Mexico	20% by 2020 (IOUs)
	10% by 2020 (Co-ops)
New York	24% by 2013
North Carolina	12.5% by 2021 (IOUs)
	10% by 2018 (Co-ops & Municipals)
Ohio	12.5% by 2024
Oregon	25% by 2025 large utilities,
	5-10% by 2025 small utilities
Rhode Island	16% by 2019
Vermont	20% by 2017
Washington	15% by 2020
Wisconsin	10% by 2015

Source: Berkeley Lab [Wiser and Bolinger 2008]

to use their own roof-top solar generation to offset their electricity consumption and send excess on-site solar generation back to the grid and receive a payment for that generation from the utility. One kWh generated by the electric utility customer has the exact same value as one kWh consumed by the customer.

Net metering is an incentive that allows the owner of a solar power generating system—residential or commercial type—to receive credit from an electric utility.

Net metering customers receive credit for energy their system sends to the utility grid, minus the energy they pull down from the grid. Balance of credits is reconciled periodically and a customer receives a final balance.

All public utilities in the United States are required to provide customers with net-metering service. However, the exact rules of how the net-metering is calculated and regulated vary significantly from state to state, and even from utility to utility. Too, there have been a number of modifications to the rules lately, delaying or simply stopping development of PV energy projects in some states.

Feed-In Tariff

Feed-in tariffs (FIT) encourage alternative energy businesses and investors to produce electricity through renewable sources by guaranteeing that the produced electricity will be bought at a guaranteed rate and for a fixed period of time by the established utility companies. The most important elements of every feed-in tariff are the level of the tariff and the length of time for which it is guaranteed. Obviously the higher the price, the more likely it is that business is going to be interested in developing renewable technologies.

The time factor is also important. Developers need the security of knowing that they will be able to achieve a guaranteed rate over a long period of time. This ensures proper compensation for the risks and costs of research, development and start-up that accompany renewable technologies.

FITs were first developed in America in the 1970s and are now used in different forms as an integral part of the alternative energy generating process. FITs are widely used in Europe and Canada, where they are the foundation of the solar industry development. The amount changes annually, and some slowdown has been noticed lately, but the progress to date is undeniable.

Power Purchase Agreement

Power purchase agreements (PPAs) are financial vehicles that enable residential and commercial customers to take advantage of the benefits of a solar energy source, often with small, or no, capital outlay. It is a legal contract between a residential or commercial electricity generator (PV system owner) and a power purchaser (local utility). A third party, usually a financial institution, will own, operate, and maintain the solar power system for terms of 20-25 years and sell the electricity to the utility at a price agreed on in the contract.

The customer provides a place (roof or land) for the PV system to be installed and buys power at agreed-upon, long-term, below market rates, and/or gets some financial benefits. Solar companies work with the leading PPA providers on behalf of the customers, to ensure most competitive options for their particular solar project and conditions, and facilitate the installation and operation of the PV plants. This arrangement has been implemented very successfully by several solar companies in the US and abroad and thousands of solar installations have been completed as a result.

Nevertheless, the majority of PPAs offer low payments, which often render the planned projects unprofitable for implementation and operation—even with the incentives and subsidies of today.

THE FINANCE DANCE

Financing a PV power project is a complex undertaking with a large number of hoops and loops that one must jump over and squeeze through in order to get financing to design, build and operate a PV system. Nowadays (and since 2008) the driving force behind the financing of solar and wind projects in the US and around the world, have been the federal, state and local incentives and subsidies provided by the respective entities and the utilities. It remains to be seen how the finance game will be played when most of these incentives and subsidies are reduced or gone.

For now financing PV projects in the US is a somewhat confusing undertaking with many people involved. The solar energy business grew quickly during the last 2-3 years and many businesses got into it, which is good in the short term because it provides jobs and brings some economic stability to the shaky local economies. In the long run, however, some of the materials and components, as well as the procedures used to install millions of PV modules, have not been standardized or regulated, so some of these lack the quality needed for hi-tech installations intended for long-term operation.

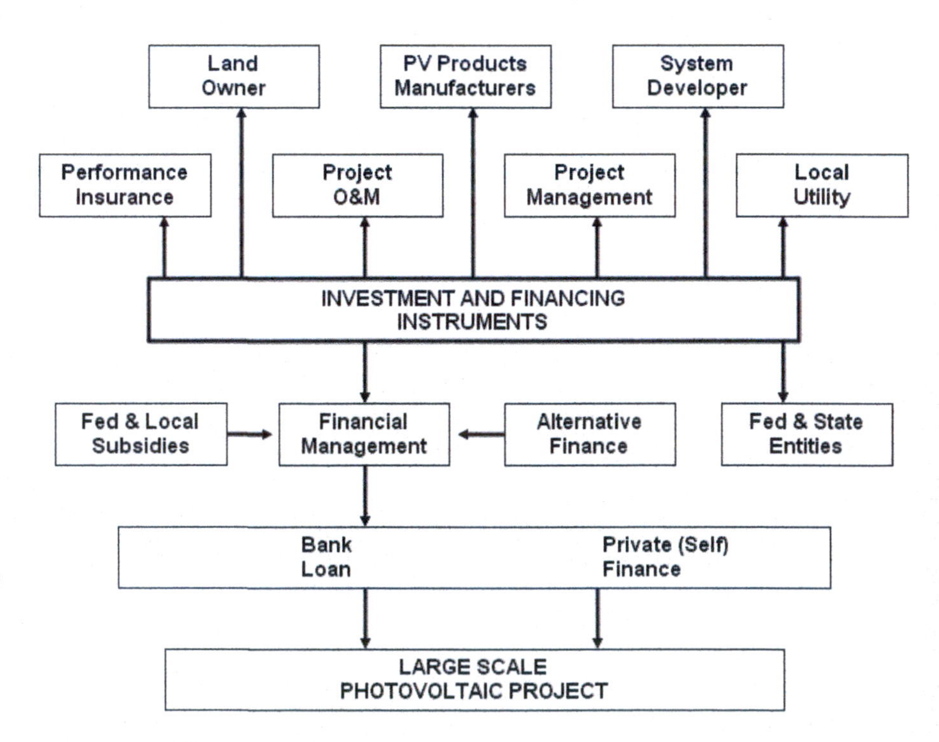

Figure 7-4. Structure of a PV system finance scheme

The new manufacturers (mostly Asian), together with the new solar specialists (installers and operators) don't have the necessary experience, and are improvising as they go. The combination of materials reliability and quality of services might have some unintended effects over time, when poorly designed and executed projects start to degrade quickly and fail prematurely.

Nevertheless, the number of PV installations (residential and large-scale) is growing regardless of the technical, administrative and financial issues. Prices have been slowly going down as well—hopefully not at the expense of quality. Financing some of these projects is a risky proposition, to say the least, yet many of them get financed by private and government sources.

Finance Sources

There are a number of ways to obtain financing for large-scale projects in the US, with private and government finance being the primary vehicles presently.

Private Finance Sources

Money, or lack thereof, is the biggest problem in building more PV power plants today, but the world is well aware that our energy independence and wellbeing is directly related to using solar energy. So, one way or another, financing is made available. Private financing comes from individuals (with deep pockets) to institutional investors, banks and such.

Some of the world's largest pension funds recently unveiled their plans to CREATE public-private climate funds as a contribution to ONGOING efforts to raise hundreds of billions of dollars for climate change financing. At a meeting in Seoul, South Korea, the P8—a grouping of major public pension funds convened by the UK's Prince of Wales—launched a listed fund, seeded with investments from P8 members and governments, that will provide financing to help mitigate climate change in developing countries. "We're looking at climate change as a big opportunity for investment," Aled Jones, of the University of Cambridge Programme for Sustainability Leadership (CPSL) and the P8 facilitator, told Environmental Finance on the sidelines of a Responsible Investor conference in London. So, air pollution is now officially a big business. We just have to wait and see how this business venture develops; hoping that it won't turn the solar business into a race of greed.

A large investment potential is the Desertec project in Europe, whereby wind and solar energy generated in the North African deserts will be supplying electric power to European countries. We estimate that most of the investment for this huge project will come from private investors. These plans will take time to materialize, but the pendulum is swinging in that direction, and we do believe these programs will be adequately funded and implemented in the near future.

Private financing has been affected also by the world's economic slowdown, but has kept pace with solar developments thus far, though somewhat lagging behind the expectations. Significant changes, mostly for the best, in this area are expected in the near future. According to a study by Allianz Global Investors, 78% of investors think green technology could be the "next great American industry," and 97% of investors believe the development of alternative energy sources will remain important even if oil prices remain relatively low. And the statistics bear that out.

The 2010 figures show that venture capital and private equity investment rose 28 percent from the 2009 total to reach $8.8 billion. This was still lower than 2008's record figure of $11.8 billion. Public market investment rose 18 percent from the previous slow years to $17.4 billion in 2010.

The most notable feature of the 2010 energy boom was 49% increase of solar power investments to $89.3 billion, due mostly to distributed generation projects in Europe, where investment grew by 91% to $59.6 billion. Over 86% percent of investment in small-scale solar took place in European countries where favorable FIT prices were introduced. The largest investment category in 2010 was, as usual, asset finance of utility-scale projects such as wind farms, solar parks and biofuel plants. These investments grew 19% to $127.8 billion.

R&D spending on clean energy technologies grew to a record level in 2010. Corporate R&D was higher than the $12.8 billion invested in 2009, to reach $14.4 billion. With that, the total for global clean energy R&D was approximately $35.5 billion.

Solar investments in the US have been spotty for the last several years. Investors have a lot of excuses and quote a number of risk factors. It seems that lately some foreign investors are more willing to invest in the US than the traditional US investors, which is surprising. Nevertheless, over $800 million was invested in several projects during the first part of 2010 in the US in the form of loans and loan guarantees. Some of these were: Solyndra, a California-based thin film company raised $175 million from existing investors instead of following through with its planned IPO. BrightSource Energy, a California-based developer of utility-scale solar thermal power plants, raised $150 million in Series D funding from new investors Alstom and the California

State Teachers Retirement System (CalSTRS) as well as existing investors. The deal followed a conditional commitment from the U.S. Department of Energy for $1.37 billion in loan guarantees that was made in February. Amonix, a California-based developer of concentrated photovoltaic (CPV) solar power systems, raised $129.4 million in a Series B round led by Kleiner, Perkins, Caufield & Byers. Lately the Spanish solar giant Abengoa secured a $1.45 billion loan guaranty from the US government for building a 280 MWp CSP power plant in Arizona.

Private investors have been blamed for not taking a more active role in the solar developments. Actually we believe they are taking a big part—bigger than they realize. Some of them are staying away which sends a clear signal that something is wrong with solar and forces us to look around in order to figure out the problem. Some investors are jumping into it semi-aware of the risks, thus taking a chance for whatever reasons investors take chances, blindly believing in the success of the venture as it was presented to them. Only few of the investors enter the solar market game fully aware of the technological and other issues predominating in the solar business. They are aware that some PV technologies are just not efficient and reliable enough, and that there are risks with some manufacturers who are too new and inexperienced. They are also aware of the environmental risks posed by some PV technologies containing toxic materials that have not been proven safe. And they understand the management and logistics risks—the long-term uncertainties with the outcome of PV projects due to management inadequacies and the lack of successful long-term PV power installation precedents.

As we see it, successful investors must be aware of the fact that some of the risks are foreseeable, quantifiable and manageable, while some are not. This is a complex picture that requires a thorough understanding of the technologies and their use, as well as the markets these are supposed to service for a very long time. Because of that, investors must be armed with substantial knowledge and overall understanding of the function and proper use of solar technologies for large-scale operation, with the operation of these in desert areas as a central focus, for the reasons which we emphasize in this text.

Government Finance

The Obama Administration has offered a number of incentives and subsidies which were extended and in full force for 2011. 2012 might bring some surprises in this area, if the Republicans' plans are implemented.

DOE and its associates EERE, NREL, etc. are financing a large number of R&D and production expansion projects in the solar energy area and other undertakings important to energy industries. A lot of this money, however, goes to finance foreign companies (Chinese, German, Japanese, Spanish) who are paid US tax-payers' dollars to import their wind and solar technology into the US.

A coalition paper issued in 2010 calculates that energy stimulus spending will peak at $400 billion, shrink to about $120 billion in 2011 and be virtually gone by 2015. Today, private investment is not strong enough to fund a trillion-dollar overhaul of the energy sector by itself. Too many households are repaying debt, too many building owners are strapped for capital, and the recession has taken such a deep cut out of energy demand that many existing power companies won't scrap carbon-intensive generation for clean energy on their own.

World-wide R&D spending on clean energy technologies by companies and governments grew to a record level in 2010. Government financed R&D reached $21 billion, up from $15.8 billion in 2009, to a global total of clean energy R&D investment of approximately $35.5 billion.

In 2010 the US administration awarded $1.5 billion loan guaranty to the Spanish solar company Abengoa

FUNDING AREAS	$ BILLION
Energy Efficiency and Renewable Energy	16.80
Loan Guarantees for Renewable Energy	4.00
SmartGrid and Efficient Electrical Transmission	4.50
Carbon Capture and Large Scale Sorage	3.40
Scientific Research in All Energy Areas	1.60
Advanced Research Energy Projects(ARPA)	0.40

Figure 7-5. DOE funding of energy projects

for the construction of the 280 MW Solana CSP power plant in Gila Bend, Arizona. And $400 million were pledged to another cadmium telluride thin film manufacturing company to expand the manufacturing of 8% efficient, toxic cadmium-containing PV modules. These actions bring a number of questions to mind: Why are we financing foreign companies for projects that can be done by US companies? Why are we investing in the further propagation of unproven toxic PV technologies? Is this a wise exposure to risk of taxpayers' money (money we just don't have today)? That is a question many are pondering.

If the Senate fails to get the 60 votes necessary to pass climate and energy legislation, clean energy proponents are looking at possibilities of tying key measures to tax or jobs legislation that has better prospects for passage. Backers of proposals for a new federal "green energy" financing bank say they have the answer. This new federal financing administration might change the equation. The backers offer an example of two higher-cost wind energy projects, one paid for solely with private financing, the other backed by federal loan guarantees. It remains to be seen how far this effort will go... if it goes anywhere any time soon. Solar is also waiting for some financial help, which is needed to put it on the right path. In any case, government financing is not a permanent solution to our energy problems, so the solar industry must find a way to support itself in the long run. New, more efficient and cleaner, PV technologies is one answer.

The US Administration's ambitious plans of achieving 80% clean energy by 2035 means that a lot of investment is expected to flow into these clean projects, solar being one of them. The majority of this finance avalanche will come from private investors, so we expect another "gold rush" to start soon in the US. Already, a number of large private financing programs are in the planning stages for large-scale solar and wind power fields in the US.

Prices for complete installation (including permitting etc., minus the land lease or purchase) were in the $5-$7/Watt range. 2010 brought significant decrease in module prices (mostly from low-cost Chinese manufacturers), while the prices of established, high-quality providers declined slightly. Due to fierce competition, however, many PV installers did drop their prices, as it can be seen below.

The following are several examples of PV installations, circa 2010, with a detailed cost breakdown and discussion of the conditions and issues at hand:

Example 1. 300kw Commercial PV Power Plant

A 300-kW PV power generating plant in central California was estimated to produce 522 MWh electric power for its owners annually, based on 300 sunny days and 5.8 hrs. daily average sunlight at the location. At a total cost of $1.485 million, and considering all federal and state subsidies and RECs, it was estimated to generate profit within 8.5 years, under ideal conditions and with no major disasters. (See Table 7-3.)

Total final cost for this PV system was $4.95/Wp installed. This is down from $6.25/Wp just a year ago for a similar PV system in California. This is also a top quality installation, using highest quality Sharp made PV modules with excellent manufacturer warranty and long-term guaranty. BOS is also of quality suppliers with equally excellent long-term guaranty.

As an option B, the price was lowered by 0.65/Wp by using low-cost Chinese made modules and wholesale BOS. This way the customer had a choice between prices and qualities. Lower-cost PV modules and BOS components bring the initial cost down substantially, but might increase O&M expenses and related headaches in the long run.

Example 2. 80.5kwp PV Power Generating Installation

An 80.5-kWp roof-mounted PV power generating plant in Arizona (Table 7-4), based on c-Si PV modules was estimated to generate the equivalent of 68.2 kW AC at a location of heavy electric power use in central Arizona.

The PV array was estimated to generate ~140 MWh electric power each year, assuming 325 sunny days and 6.4 sunny hours every day. Total cost of the installation was $460k, or $6.75/Wp before subsidies, rebates, etc. Under ideal conditions and with everything considered, the installation will be paid for and generate actual profit within 9.6 years.

PV Installation Examples, 2010

An estimate of a PV power plant installation (2009 prices)

• Modules	~$1.80-2.25/Watt
• Mounting frame	~$0.35-0.45/Watt
• Inverter	~$0.30-0.40/Watt
• Electrical	~$0.40-0.50/Watt
• Site preparation	~$0.75-1.25/Watt
• Engineering etc.	~$1.50-2.50/Watt
Totals	~$5.10-7.35/Watt

Table 7-3. 300 kW commercial PV installations in 2010

JOB DETAILS:
INSTALLATION OF 300 kW SOLAR SYSTEM (ROOF-MOUNTED SYSTEM)

FINANCIAL BREAKDOWN OF PROJECT

ITEM QUANTITY	DESCRIPTION	UNIT PRICE	P.O.P.	TOTAL AMOUNT
(1)	Hardware costs:			
	solar modules	2.20	45%	$660,000.00
	racking, mounting, splice rails, oateys	.46	9%	$138,000.00
	ballasted system	.15	3%	$45,000.00
	conduit, combiners, AC main gear	.15	3%	$45,000.00
	balance of system supplies	.30	6%	$90,000.00
	sales tax 8.75%	.28	6%	$85,500.00
	TOTAL LABOR	.80	16%	$240,000.00
(2)	CAPITALIZED costs:			
	Engineering consulting	.20	4%	$60,000.00
	Marketing, accounting, legal	.25	5%	$75,000.00
	Project management & admin cost	.10	2%	$30,000.00
	Sub-total amount	.55	11%	$165,000.00
	G & A Expense	.06	1%	$17,000.00
"All in" TOTALS (4.95 p/W):		4.95	100.00	$1,485.000.00

Example 3. 150kW Commercial Installation

Table 7-5 contains the data of a medium-sized commercial installation in Arizona, using fixed, ground-mounted PV modules. The initial investment is approximately $650,000 which includes the land, PV modules and BOS materials, and all permitting processes.

DC to AC derating is estimated to bring the power output from 150 kW DC to 121 kW AC at peak hours. The estimate is made on the basis of 7.5 peak hours per day and 309 days full sunshine (120,778 kW AC x 7.5 hrs/day x 309 days/yr). Both estimates, however, are exaggerated, because there were about 250 full-sunshine days in Arizona in 2910. Since we don't know how that affected the peak hour distribution, and since it is impossible to predict the sunshine in the future, we'd suggest more conservative numbers for both: the number of daily peak hours and full-sun days in the year. An acceptable compromise would be 6.0 peak hours/day and 300 days/yr, which brings the AC power generated by this 150 kWp commercial installation to 217,400 kWh/year (vs. the estimated the utility company's estimated 279,614 kWh).

At PPA of $1.147 kWh this PV installation is projected to bring in $41,103,26 gross income annually, or $31,957 according to our compromise estimate. It should be paid in 15.8 years (or in 20.3 years according to the new estimate), if no major problems are encountered. However, considering minimum, 1% annual degradation of the modules and a percentage of premature and total failures, in addition to regular O&M expenses, our estimate is that the net income would be half of that projected.

Some Arizona utilities will not allow commercial PV power generators to be connected to the grid. New commercial PV power sources can operate as stand-alone or load offset mode only. We do believe that this is a temporary situation, given what we know about the downside of burning fossil fuels and rising fuel cost.

Example 4. Large-scale Utilities Projects

PV projects in California are moving in a new and positive direction under the guidance of CPUC, and with the active participation of PG&E and SCE utilities. A number of RFOs (request for offers) were issued in 2010 and others follow in 2011. A recently issued long-term RFO, is looking for bidders for 250 MWp large-scale PV power plants (in 1-20 MWp range size) to be installed in California during the next 5 years.

A critical factor here is that some utilities will pay ~$240 per MWh generated in summer peak hours, but only $105 per MWh generated in winter. This, combined with the fact that there are almost twice as many sun-

Table 7-4. 80.5 kW commercial PV installation. Summer of 2010

Year	Average Utility Rate ¢/kWh	Utility Savings $/year	Net Utility Savings, sRECs, PBIs, etc., after tax &maintenance	Tax & Rebate Incentives	Cost/Payback Schedule (Cumulative Cash Position)
0				$75,000	($383,045)
1	15.0	$21,502	$20,357	$115,914	($246,774)
2	16.0	$23,223	$22,042	–$280	($225,012)
3	17.1	$25,080	$23,863		($201,149)
4	18.3	$26,686	$25,431		($175,718)
5	19.6	$28,394	$27,100		($148,618)
6	20.9	$30,211	$28,877		($119,742)
7	22.4	$32,144	$30,769		($88,973)
8	23.9	$34,201	$32,783		($56,189)
9	25.6	$36,390	$34,928		($21,261)
10	27.3	$38,719	$37,212		$15,951
11	29.2	$41,197	$39,643		$55,595
12	31.2	$43,834	$42,232		$97,826
13	33.4	$46,639	$44,988		$142,814
14	35.7	$49,624	$47,921		$190,735
15	38.1	$52,800	$3,275		$194,011
16	40.6	$56,179	$54,369		$248,380
17	43.6	$59,775	$57,909		$306,288
18	46.6	$63,600	$61,676		$367,965
19	49.8	$67,671	$65,687		$433,652
20	53.3	$72,002	$69,957		$503,608
21	56.9	$76,610	$74,501		$578,109
22	60.9	$81,513	$79,339		$657,448
23	65.1	$86,730	$84,488		$741,937
24	69.5	$92,281	$89,970		$831,906
25	74.3	$98,187	$95,804		$927,710

Table 7-5. Estimate of a small PV installation

EQUIPMENT INFORMATION					
PV Modules					
Array	**Manufacturer**	**Model Number**	**Model Rating**	**Quantity**	**Array Rating**
1	SolarWorld	SW245 mono	245 W-DC	630	154.35 kW-DC
Inverters					
Array	**Manufacturer**	**Model Number**	**Model Rating**	**Quantity**	**Efficiency**
1	SMA America	SC250U (480V)	250,000 W-AC	1	97.0%
System Orientation					
☑ Fixed ☐ Single-Axis Tracking ☐ Dual-Axis Tracking					

PROJECT INCENTIVE CALCULATION		
System Rating (kW DC-STC)	**154.350 kW DC-STC**	
Module Efficiency	**89.633%**	
Inverter Efficiency	**97.000%**	Average CEC Inverter Efficiency Rating
Other Losses Derating	**90.000%**	
Total DC to AC Derate Factor	**78.249%**	
System kW AC Rating	**120.778 kW-AC**	kW AC for entire system
Design Factor	**100.000%**	Based on Orientation, Location, and Shading
Estimated Annual Energy Production	**279,614 kWh/yr**	
Incentive Level	**$0.147/kWh, 20 years**	
Requested Incentive	**$822,065.16**	Incentive Amount: Incentive Rate x Estimated Production x Incentive Term **$0.147 x 279,614 kWh/yr x 20 years = $822,065.16**
Estimated Annual Payment	**$41,103.26**	

light hours during the summer months, makes a huge difference in the bottom line. Because of the importance of this variable, it has to be taken into account in the initial plant design, including selection of most appropriate technology, and its installation and operation.

As an example, a 15 MWp PV plant in Central California operating 8 hrs. a day at full capacity will generate ~3.720 MWh in the month of July. The utility will pay $818,400 for the purchase of this power. In winter, however, the same PV capacity would be lucky to operate 4 hrs. a day at full power, generating 1,860 MWh at best in January, and the utility will pay only $195,300 for the purchase of this power. The difference between the summer and winter monthly income is huge: $818,400 vs. 195,300—over 4 times the difference.

This, of course is just a theoretical calculation, which does not take into account many of the factors and variables we have been talking about in this text. But, comparing apples with apples we must emphasize the huge difference the climate and seasons make in the bottom line, which should force us to take a serious look at proper PV technology choice and efficient use. Variables among the types of PV modules, tilt angle, tracking, maintenance, etc. are the determining factors in obtaining the highest efficiency from the system and the largest possible profit ratios.

PV System Cost Estimate

Table 7-6 is a sample calculation of components, including BOS that could be used for a cost estimate of a commercial or large-scale PV installation:

BOS components are key elements of any PV power plant model. Their particular design, overall efficiency quality and actual field performance will determine the plant's success rate.

NOTE: We have not discussed "success rates" yet, because it is not a standard measurement unit, but in the framework of this writing success rate is determined by:

1. Quality of location (sunshine availability and weather conditions),
2. Quality of components—low vs. high quality manufacturers,
3. Quality of the design, installation and O&M teams, and
4. Luck (lack of natural and man-made disasters).

One key rule to remember, when working on PV power plant finances is that price per Watt installed capacity is usually inversely proportional to the size of the project, the modules efficiency, and if using tracking or fixed-mount modules.

In other words, and all things considered,

1. Choosing desert sunshine locations,
2. Building larger size plants,
3. Using more efficient PV modules, and
4. Using single-axis trackers are the most efficient and cheapest ways to maximize use of solar energy and generate profit from a PV installation with c-Si modules.

This is partially due to the fact that lower prices can be negotiated when ordering large quantities of materials and components. Also, in larger projects, labor can be properly trained and utilized more efficiently, which would bring the overall cost down.

Note: Using thin films PV modules on trackers is not a recommended procedure as discussed in detail previously.

Levelized Cost of Energy

Any PV plant has to deal with financial issues from the planning to the decommissioning stages. There are a number of parameters to be considered in the financing model, with the levelized cost of energy (LCOE) calculations being one of the key elements of financial calculations. They are commonly used to represent the relative cost of electricity generated by a PV plant, which ultimately represents the rate of ROI.

Table 7-7 clearly shows the discrepancies, with solar and wind being several times more expensive (LCOE) than conventional fuels. As the new kids on the block, they are just now trying to carve a niche in the energy markets. This makes alternative energy solutions comparably clumsy to fit in the mix and expensive to build and use, so they cannot be considered overly accessible or profitable energy sources, at present. In the long run, however, they will emerge as the technologies that will power this world—and there is no way around it, simply because none of the conventional sources can compete with the clean and renewable nature of these. Also, the conventional technologies' prices will continue to increase (some sharply) in the near future, which would bring them close to the LCOE level of their wind and solar cousins.

Note: One thing that must be considered here is the fact that all conventional energy generators (coal, oil, nuclear) get subsidies and incentives from government, state, local and other sources. We could not obtain reliable information on the amounts of these perks, or how they are determined and distributed, but we know that these numbers are substantial—in the billions of dollars range every year. Over 5 billion US taxpayer dollars are given to oil companies yearly as subsidies and tax breaks, while during 2005-2006 Australia spent ~$10 billion on energy subsidies with 96% going to conventional

Table 7-6. Sample calculation components for cost estimate
100kWp PV system cost estimate, including BOS

Module Specs		**c-Si**	
Module Efficiency	A	14.05%	
Density (M2/kWp)	B=1/A	7.12	
Watts Peak	C	230	
Open circuit Voltage (Voc)	D	37	
Maximum Voltage (Vmp)	E	30	
No. of modules used	F=100kW/C	435	
TOTAL $/W (quoted price of $2.00/Wp)			**$2.00**
Mounting Structure			
Mounting Structure Cost ($/M2)	G	$55.64	
Savings unique to a-Si module	H		
TOTAL $/W		**I=G*B*(1-H)/1000**	**$0.40**
Combiner Box			
Inbound voltage (max)	J	600	
No of modules/string	K=J/D	16	
No of strings	L=F/K	27	
# of poles	M	24	
No of combiner boxes required	N=L/M	2	
Combiner box cost per piece	O	$1,500	
TOTAL $/W		**P=N*O/100kW**	**$0.03**
Cabling Cost			
Total cable required (meters)	Q	260.66	
Cabling cost/m	R	$2.00	
TOTAL COST $/W		**S=B*Q*R/1000**	**$0.01**
Inverter			
TOTAL COST $/W	T	$0.49	
Installation Cost			
Installation hours per m2	U	0.50	
Installation labor $/hour	V	$50.00	
TOTAL COST $/W		**W=B*U*V/1000**	**$0.18**
Indirect Cost			
Engineering & Design	X	$25,000	
TOTAL COST $/W		**Y=X/100kW**	**$0.25**
GRAND TOTAL $/W		**Z=I+P+S+T+W+Y =**	**$3.36/W***

Notes:

1. Balance of system (BOS) includes all components needed to install, operate and control a PV system, except the PV modules. This includes wiring, switches, support racks, inverters, and batteries in off-grid systems. Land, labor and O&M expenses and components are usually separate items.
2. Using one-axis trackers would add $0.35-0.45/Wp to the BOS estimate. The tracking PV modules would then produce 15-30% more power for the same active area, depending on weather conditions.

energy generators. We have to conclude that the situation is similar in most developed countries.

This does not include the tremendous amount of money spent on conventional energy R&D through the years, a trend that continues and even escalates in the US, where the search for "clean coal" and "safe nuclear" is taking priority. These are additional $$$ billions going to the multi-billion dollar international conglomerates whoh control the energy markets. This makes the data look like comparing apples and oranges. The blatant discrepancies in it should force the scientific community to take the skeletons out of the closet for an honest discussion, because our children's future is at stake; we should not leave it in the hands of agenda-driven politicians and multi-national fossil burning conglomerates.

LCOE is the key determining factor in the financability of a solar power plant, and achieving it is a major contributor to utility-scale solar development. LCOE is used to analyze different types and sizes of PV technologies and plants, operating under different

Table 7-7. Estimate of LCOE for different technologies by 2016 (DOE/EIA-0383(2009)

Plant Type	Capacity Factor (%)	U.S. Average Levelized Costs (2008 $/megawatthour) for Plants Entering Service in 2016				
		Levelized Capital Cost	Fixed O&M	Variable O&M (including fuel)	Transmission Investment	Total System Levelized Cost
Conventional Coal	85	69.2	3.8	23.9	3.6	100.4
Advanced Coal	85	81.2	5.3	20.4	3.6	110.5
Advanced Coal with CCS	85	92.6	6.3	26.4	3.9	129.3
Natural Gas-fired						
Conventional Combined Cycle	87	22.9	1.7	54.9	3.6	83.1
Advanced Combined Cycle	87	22.4	1.6	51.7	3.6	79.3
Advanced CC with CCS	87	43.8	2.7	63.0	3.8	113.3
Conventional Combustion Turbine	30	41.1	4.7	82.9	10.8	139.5
Advanced Combustion Turbine	30	38.5	4.1	70.0	10.8	123.5
Advanced Nuclear	90	94.9	11.7	9.4	3.0	119.0
Wind	34.4	130.5	10.4	0.0	8.4	149.3
Wind – Offshore	39.3	159.9	23.8	0.0	7.4	191.1
Solar PV	21.7	376.8	6.4	0.0	13.0	396.1
Solar Thermal	31.2	224.4	21.8	0.0	10.4	256.6
Geothermal	90	88.0	22.9	0.0	4.8	115.7
Biomass	83	73.3	9.1	24.9	3.8	111.0
Hydro	51.4	103.7	3.5	7.1	5.7	119.9

Note: "Apples to apples" comparison between conventional and solar power generating technologies is simply not possible today. The conventional technologies (coal, oil, hydro and nuclear) have benefitted mightily from billions of dollars (taxpayers' money) spent on their R&D and deployment during the last 100 years. Solar has been (and still is) the industry's "step child" who is looking for its place under the sun (no pun intended) and needs a lot of help before becoming an equal members of the family. Because of this and many other reasons, LCOE calculations favor heavily the conventional energy sources.

investment, management, and other key parameters. It could also be used for comparing the cost of energy generated by a PV power plant with that of wind or fossil fuel systems.

The LCOE calculation is the net present value of total life cycle costs of the project divided by the quantity of energy produced over the system life, as shown in the following equation:

$$\text{LCOE} = \frac{\text{Total Life Cycle Cost}}{\text{Total Lifetime Energy Production}}$$

This oversimplified equation contains many components, some of which are easy to calculate, while others must be estimated. These components can make the difference between an efficient and an inefficient power plant. They determine the difference between a profitable operation, and one that is operating at a loss.

The different parameters of the below equation can be explained as follow (3):

1. Initial investment

The initial investment in a PV system combines the total cost of the project plus the cost of construction financing. The capital cost is driven by:

- Area-related costs that scale with the physical size of the system, namely the panel, mounting system, land, site preparation, field wiring and system protection.

- Grid interconnection costs that scale with the

$$\text{LCOE} = \frac{\text{Initial Investment} \sum_{n=1}^{N} - \frac{\text{Depreciation}^n}{(1+\text{Discount Rate})^n} \times (\text{Tax Rate}) + \sum_{n=1}^{N} \frac{\text{Annual Costs}^n}{(1+\text{Discount Rate})^n} \times (1-\text{Tax Rate}) - \frac{\text{Residual Value}}{(1+\text{Discount Rate})^n}}{\sum_{n=1}^{N} \frac{\text{Initial kWh/kWp} \times (1-\text{System Degradation Rate})^n}{(1+\text{Discount Rate})^n}}$$

peak power capacity of the system, including electrical infrastructure such as inverters, switching gear, transformers, interconnection relays and transmission upgrades.

- Project-related costs, such as general overhead, sales and marketing, and site design, which are generally fixed for similarly sized projects.

2. Depreciation tax benefit

The depreciation tax benefit is the present value of that benefit over the financed life of the project asset. Public policy, which enables accelerated depreciation, directly benefits the system LCOE since faster depreciation translates to faster recognition of the depreciation benefit.

$$\sum_{n=1}^{N} \frac{\text{Depreciation}^n}{(1+\text{Discount Rate})^n} \times (\text{Tax Rate})$$

3. Annual costs

In the LCOE calculation, the present value of the annual system operating and maintenance costs is added to the total life-cycle cost. These costs include inverter maintenance, module cleaning, site monitoring, insurance, land leases, financial reporting, general overheads, periodic maintenance and field repairs.

$$\sum_{n=1}^{N} \frac{\text{Annual Costs}^n}{(1+\text{Discount Rate})^n} \times (1-\text{Tax Rate})$$

4. System residual value

The present value of the end-of-life asset value is deducted from the total life cycle cost in the LCOE calculation. Silicon PV modules carry performance warranties for 25 years and have a significantly longer useful life. Therefore, if a project is financed for a 10- or 15-year term, the project residual value can be significant.

$$\frac{\text{Residual Value}}{(1+\text{Discount Rate})^n}$$

System Energy Production

The system lifetime energy production is calculated by determining the first-year energy generation as expressed in kWh (AC)/kWp (AC), then degrading output over the system life based on an annual performance degradation rate. System degradation (largely a function of PV panel type and manufacturing quality) and its predictability are important factors in life cycle costs since they determine the probable level of future case flows. This stream of energy produced is then discounted to derive a present value of the energy generated to make a levelized cost calculated.

The First Year kWh/kWp is a Function of:

a. The amount of sunshine the project site receives in a year.

b. The mounting and orientation of the system (i.e., flat, fixed-tilt, tracking, etc.).

c. The spacing between PV modules as expressed in terms of system ground coverage ratio (GCR).

d. The energy harvest of the PV module (i.e., performance sensitivity to high temperatures, sensitivity to low or diffuse light, etc.).

e. System losses from soiling, transformers, inverters, and wiring inefficiencies.

f. System availability largely driven by inverter downtime.

System's Degradation Rate

This is the inevitable and accepted annual performance degradation, expressed in % of power output loss. The degradation rate varies from 0.5% to 1.5% for modules from different manufacturers. There is additional degradation attributed to performance deterioration of other components in the PV power field, such as inverters, transformers, etc.

Finally, the system's financing term (n) will determine the duration of cash flows and affect the assessment of the system residual value.

Financing Term

The system's financing term determines the duration of cash flows and affects the assessment of the system residual value. The price of energy established under power purchase agreements (PPAs) or by feed-in-tariffs (FITs) may differ substantially from the LCOE of a given PV technology, since PPAs and FITs may represent different contract or incentive durations, and the inclusion of incentives such as tax benefits or accelerated depreciation, financing structures, and in some cases, the value of time-of-day production tariffs.

Ultimately, the lowest LCOE of PV power plants will be defined by:

1. Proper plant design, including best location and technology selection
2. Use of high-efficiency modules and BOS to minimize power plant capital costs through the reduction in the number of modules, and support structure, as well as land required.
3. Innovative higher efficiency manufacturing and use, based on more efficient use of silicon and larger scale manufacturing operations and power plants, as needed to drive continued higher-efficiencies and cost reductions.
4. Life cycle O&M costs that can be lowered by use of high-efficiency tracking PV, which can result in much higher energy production per active area in properly selected locations.
5. Higher system residual value for a silicon PV plant, which will reduce total life cycle cost.

Financial Formulas

There are a number of financial factors and formulas that can be used to get a good idea of the financial status and viability of any PV power plant.

Some of the most useful financial formulas:

1. Present worth.

The formula for the single present worth (P) of a future sum of money (F) in a given year (N) at a given discount rate (I) is

$$P = F/(1 + I)^N$$

2. Uniform present worth.

The formula for the uniform present worth (P) of an annual sum (A) received over a period of years (N) at a given discount rate (I) is

$$P = A[1 - (1 + I)^{-N}]/I$$

3. Modified present worth.

The formula for the modified uniform present worth of an annual sum (A) that escalates at a rate (E) over a period of years (N) at a given discount rate (I) is

$$P = A\{(1+E)/(I-E) *[1 - [(1+E)/(1+I)]^N]\}$$

4. Annual payment on a loan.

The formula for the annual payment (A) on a loan whose principal is (P) at an interest rate (I) for a given period of years (N) is

$$A = P\{I/[1 - (1 + I)^{-N}]\}$$

Cost of Energy Generation

Generating electric power is an expensive proposition, regardless of location, destination, or source. Initial cost of equipment and installation expenses to start with and the cost of operation and maintenance (O&M) in the long run are staggering, with many risks and headaches. Established power sources, like hydro, nuclear and coal are expensive to set up too, but are the cheapest to generate and deliver power to the customer, because they use cheap fuels, have been in existence for decades, and have established operation and distribution channels. They also have a lot of supporters in high places, who don't hesitate to provide a helping hand anytime one is needed. The risks associated with ownership and operation of conventional power generators are slim to none for the above and many other reasons.

On the other hand, the initial cost, O&M, insurance, labor, decommissioning, recycling and everything in between makes PV power plants complicated, expensive and risky undertakings which require careful design and implementation, as well as consistent and careful follow-through of proper O&M procedures during 25-30 years of non-stop operation. Mistakes in the initial design, installation, and/or O&M operations would lower energy output, distort financial calculations and cause a serious profit loss or worse. Remember also that the weather or a natural disaster (sand storm, hail, flood etc.) could put an end to an otherwise well-thought-out and executed venture.

Considering that large-scale PV installations are the new kid on the block, with many lose ends to take care of and with not so many supporters in high places (yet), PV will be going through hard times for the foreseeable future. As solar energy technologies develop and mature, they will become much more reliable and cheaper to build and operate, which will add to their attractiveness as a power generator. We need to mention here that new disruptive PV technologies are on their way to making a splash in the energy markets and adding a new dimension to the above estimates. An example is the so-called high concentration photovoltaic (HCPV) trackers. These are 42% efficient at the cell and

~30% efficient in the grid. They also do not suffer from excess temperature and longevity issues plaguing the c-Si and TFPV technologies today. These are significant advantages which will contribute to bringing high efficiency and reliability to energy markets in the near future.

OPERATING CONDITIONS AFFECTING FINANCES

A number of complex technical and financial parameters need to be taken into consideration when calculating the performance and profitability of a PV power plant. Some of these are listed below:

Operational Factors

Several factors characteristic of the setup and operation of PV power plants determine their efficient operation and reflect directly onto the bottom line.

1. Availability factor

This is the amount of time that a PV installation is able to produce electricity over a certain period of time, divided by the total amount of time in the period. A solar power plant is considered "available" when it is capable of producing power (it is not damaged or down for maintenance). If there is no sunlight during that time, however, it could be counted as "available," or "unavailable," depending on how we have structured our calculations and what we are trying to get.

If it is considered "available" all the time—even during the night and non-sunshine hours—then its availability factor would be 100%. Usually, however, certain portions of the plant, or the entire facility, are shut down for maintenance or repair, during which time the plant would be less than 100% available for production even if there were full sunshine—i.e., the availability factor (Af) of a PV plant shut down for repairs 6 hours during the day would be:

$$Af = \frac{(24\text{-}6)}{24 \text{ hrs./day}} = 75\%$$

This simply means that the plant was fully capable, ready and willing to operate all but 6 hours during the 24-hour time period. However, if we consider only the 12 hours of daylight as the total amount of time in the time period in question, then the plant's Af would be only 50%, because it was down for repairs half of the day. This, we believe is a more meaningful expression of the "availability factor" for PV power plants.

2. Ground cover ratio

This is a fixed driver for the land consumed by a solar power plant, which usually impacts directly the initial investment in land and materials, and to a lesser extend the O&M expenses for the duration.

Ground-coverage ratio (GCR) basically represents the ratio of the area covered by PV modules to the total land area of the plant. PV modules, for example are lined up in rows with some separation between them. In many cases the separation (width of the rows between the modules) is as much as the area of the modules. In such a case, and assuming that the total plant area is 10 acres, the GCR would be:

$$GCR= \frac{\text{5 acres (active area*)}}{\text{10 acres (total plant land area)}} = 50\%$$

Note: Active area is the actual land area covered with PV modules. This is the area that produces power, while the rest of the land is used for spacing between the rows of modules, installation of auxiliary equipment, etc.

3. Capacity factor

The capacity factor is used by the utility industry to measure the productivity of energy-generating assets and is a key driver of a PV power plant's economics. Since the greatest expense of a PV power plant is fixed capital cost, LCOE is strongly affected by the power plant's output, which is related to the plant's performance, efficiency and availability.

The net capacity factor (Cf) for a PV power plant over a 12-month period can be calculated as:

$$\%Cf = \frac{MWh/y}{Hours/y * Nameplate} \times 100$$

So, the capacity factor of a power plant is the ratio of the actual output over a period of time vs. its output if it had operated at full nameplate capacity the entire time. PV power plant capacity factors vary greatly depending on location, PV technology type, and the design of the plant.

NOTE: The capacity factor should not be confused with the availability factor or with the overall efficiency of the system.

Example: the capacity factor (Cf) of a 1.0 MWp PV power plant in Arizona could be calculated as follow:

$$Cf = \frac{\text{1,900 MWh (total annual power generation)}}{\text{365 days x 24 hrs/day x 1.0 MW}} = 21.69\%$$

The average Cf for a PV plant is in the 15-25% range, but is higher in certain locations (usually in the deserts) and by using more efficient technologies such as trackers. The capacity factor of a PV system operating under the cloudy skies of Berlin, Germany, or Portland, Maine, would be 10-12%, while in Arizona it could reach 25% by using trackers.

Here again, we'd like to explore the idea of using the daylight hours only as the basis of our calculations. This is important, because unlike conventional power plants, PV operates only during the daylight hours (the production period), so we really don't care what happens at night. So, if we use a 12-hour daylight time period in the above calculation, then the Cf will double to 43.4%. This is a more meaningful number and reflects more closely the operating conditions, showing us clearly that the plant was operating at the average 43% of its full capacity during the year.

This method is even more important if we measure Cf for the different seasons; i.e., summer seasons have almost twice the daylight hours of the winter seasons, so the Cf will be quite different in each case. Summers in the Arizona desert have 14 hours of useful sunlight, while winters have less than 8. This is a big difference, which has great impact on the Cf, and we must take it into consideration.

The capacity factor is directly affected by a number of variables, as follow:

1. The intensity of solar insolation at the location,
2. The weather conditions during the time period in question,
3. The mounting and orientation of the module, or tracking method,
4. The system's overall electrical efficiencies (including BOS components),
5. The availability of the power plant (during sun hours) to produce power,
6. Quality of O&M procedures.

The capacity factor's economic impact can be substantial, and each of the above mentioned variables have great influence on it.

Life Cycle Cost

The life-cycle cost (LCC) of a project can be calculated using the formula:

$$LCC = C + M_{pw} + E_{pw} + R_{pw} + D_{pw} - S_{pw}$$

Where:

pw subscript indicates the present worth of each factor.

C is the capital cost of a project which includes the initial capital expense for equipment, the system design, engineering, and installation. This cost is always considered as a single payment occurring in the initial year of the project, regardless of how the project is financed.

M is the sum of all yearly scheduled operation and maintenance (O&M) costs. Fuel or equipment replacement costs are not included here. O&M costs include such items as an operator's salary, inspections, insurance, property tax, misc. penalties, and all scheduled maintenance.

E is the energy cost of a system, or the sum of the yearly fuel cost and related energy expenses or uses. Energy cost is calculated separately from operation and maintenance costs, so that differential fuel inflation rates may be used.

R is the replacement cost, or the sum of all repair and equipment replacement cost anticipated over the life of the system. The replacement of an inverter is a good example of such a cost that may occur once or twice during the life of a PV system. Normally, these costs occur in specific years and the entire cost is included in those years.

D is the decommissioning, transport, recycling* and disposal of the old equipment all of which must be considered in the LCC and in the initial component purchase and operational contracts. The proper execution of these tasks is a must because if not done right we might end up with junkyards in the deserts.

*NOTE: Some PV modules contain hazardous materials (Cd, In, AS etc.), which require special treatment and which will double and triple the decommissioning expenses. This must be entered in the equation and measures taken for its proper execution.

S is the salvage value, of the system or its net worth in the final year of the life-cycle period. It is common practice to assign a salvage value of 20 percent of original cost for mechanical equipment that can be moved. This rate can be modified depending on other factors such as obsolescence and condition of equipment.

Note: All future costs must be discounted because of the time value of money. One dollar received today is worth more than the promise of $1 next year, because the $1 today can be invested and earn interest for the duration. Future sums of money must also be discounted because of the inherent risk of future events not occurring as planned.

Several additional factors should be considered when choosing the period for an LCC analysis:

1. Life span of the equipment

This is a variable which assumes that PV modules should operate for 20-25 years or more without failure (including no excessive efficiency degradation). This estimate, however, could change rather quickly. A 3- to 5-year test period is the minimum time required to give us a good idea of the durability, reliability and longevity of a PV system. Thirty years of non-stop operation is expected, so many of the components will change and degrade, and some won't last that long, so replacement costs must be factored into the initial calculations.

There is, however, simply no way to estimate the reliability and longevity of most present-day PV products, because the manufacturers have not been around long enough to have done long-term on-sun tests. Some have not even been around for 3-5 years, let alone having done 15 or 20 years of successful testing. So we are guesstimating and hoping for the best. Keep your fingers crossed that the components will work that long and work well.

2. Discount rate

The rate selected for an LCC analysis has a large effect on the final results. It should reflect the potential earnings rate of the system owner. Whether the owner is a national government, small village, or an individual, money spent on a project could have been invested elsewhere and earned a certain rate of return. The nominal investment rate, however, is not an investor's real rate of return on money invested. Inflation, the tendency of prices to rise over time, and other factors will make future earnings worth less. Thus, inflation must be subtracted from an investor's nominal rate of return to get the net discount rate (or real opportunity cost of capital). For example, if the nominal investment rate was 7 percent, and general inflation was assumed to be 2 percent over the LCC period, the net discount rate that should be used would be 5 percent.

Different discount rates can be used for different commodities as well. For instance, fuel prices may be expected to rise faster than general inflation. In this case, a lower discount rate would be used when dealing with future fuel costs. In the example above the net discount rate was assumed to be 5 percent. If the cost of diesel fuel was expected to rise 1 percent faster than the general inflation rate, then a discount rate of 4 percent would be used for calculating the present worth of future fuel costs.

3. Future rates estimate

We must consider the future rates variable, realizing that an error in our guess can have a large effect on the LCC analysis results. If you use a discount rate that is too low, the future costs will be exaggerated; using a high discount rate does just the opposite, emphasizing initial costs over future costs. You may want to perform an LCC analysis with high, low, and medium estimates on future rates to put low-to-high limits on the life-cycle costs.

Weather-related Conditions

Local weather conditions—such as the frequency, quality and amount of sunshine that a site receives—are the major drivers of the capacity factor and final profit levels. Desert sites can achieve capacity factors several times higher than those seen in less sunny northern states and countries. This will certainly maximize the energy output and ROI, since sunshine is what generates power. PV installations, even those in the desert, are variable power generators. Their output depends directly on the amount and quality of sunshine to which the active area is exposed for the duration.

The most obvious impacts on power output include:

1. Ambient temperature (excess highs and lows),
2. Wind (frequency, speed and direction),
3. Frequency and severity of storms, hail etc., and
4. Frequency and degree of soiling of module surfaces.

PV modules operate on the optical transmission principle, which requires clear front glass on top of the module to transmit the incoming sunlight fully. If that surface is contaminated by dust, bird droppings, water spots and such, the sunlight will be reflected or absorbed by the surface contamination. Some locations are more prone to such incidents than others, and this must be taken into consideration in all cases. In case of extreme soiling, much less sunlight will reach the solar cells underneath the glass, so the power output and our profits will take a dive. Dust and water spots are unavoidable problems in all situations, and all we can do to obtain maximum output is to schedule periodic cleaning of the modules' front glass.

With that said, however, we can take a number of

steps during the plant's design, installation and operation stages to ensure the integrity of the front glass surface. Locating dust-generating sources in the proximity of the installation and investigating the wind speed and direction through the year, as well as the weather pattern (rain and storms) is a good start. Moving the site to a less dust- and rain-prone location might be recommended. If this is not possible, then a thorough estimate of dust and water spot contamination to be expected at the site must be made and included in the plant's performance, O&M, and financial estimates.

Water spots are especially hard to remove from a glass surface, so special procedures must be implemented, if this is a serious and consistent problem. Some claim that PV modules do not need ANY cleaning for years at a time. This cannot be further from the truth. One single monsoon storm in the Arizona desert deposits enough dust, mud, and water stains on the modules' surface to reduce their output to 10-20% or more. So, not cleaning the modules is an option, only IF you can afford significant power losses for long periods of time.

Air temperature is a major factor on module efficiency, and has a great effect on the final power output and profit; i.e., typical PV modules experience 0.5-0.6% power output drop per each degree Celsius above the standard test condition of 25°C. On a hot desert day, modules' temperature can reach 95°C, resulting in a loss of power output of over 30% lower than the modules' standard test condition power rating, (95%-25%)*0.5=35%. Because of that, c-Si and thin film PV modules are much more suited for use in cooler and less sunny climates, such as those in the Northeast and in central Europe.

Some solar technologies perform better in desert-like high operating temperature conditions (i.e., solar thermal and HCPV trackers). As a matter of fact, these particular technologies can operate efficiently only under those conditions. Germanium-based multi-junction solar cells used in HCPV (high concentration photovoltaic) trackers are quite stable under high operating temperatures. Their performance degrades only slightly under the extreme temperature elevations to which they are exposed. This and their very high efficiency (over 40%) make them most suitable for long-term use in deserts.

On the other hand, some thin film PV modules are more efficient than c-Si when operated under cloudy and foggy conditions. This also has to be considered if the location generally doesn't have much direct sunlight radiation.

Lifetime Energy Balance

Lifetime energy balance (LEB) is another factor, which needs to be kept in mind when calculating the financial benefits from installing and operating a PV power plant.

$$LEB = \frac{E_{prod} + E_{trans} + E_{inst} + E_{use} + E_{decom}}{E_{gen}}$$

Where:

- E_{prod} Energy used during production of materials, wafers, cells and modules
- E_{trans} energy used to transport materials, modules and BoS to PV site
- E_{inst} energy used to assemble and install the PV power plant
- E_{use} energy used to operate the PV power plant
- E_{decom} energy used to decommission, transport and recycle the PV field
- E_{gen} energy generated during the life of the PV power plant

In all cases E_{gen} must be much higher than the sum of the other sources of energy used, for the system to be an effective energy source.

LEB may not be overly important to an investor or a customer because most of the variables are included in the initial energy valuation and price tag of the products and are no longer performance or financial factors, so the emphasis is on the remaining, and hopefully controllable variables, E_{use} and most importantly E_{gen}. This changes the formula to present and future values only, thus discounting major energy use and expenses in the past. Nevertheless, the LEB formula has real value when considered in its entirety and would serve us well in comparing the different PV technologies and the resulting total energy balance of PV systems built with it. These data can also help in estimating the carbon credits and other variables of the plant.

Lifetime CO_2 Balance

Similarly, a system's lifetime CO_2 balance (LCB) is a factor that takes into consideration the CO_2 used during the manufacturing of the PV components and their use in the field vs. the amount of CO_2 which is saved by using PV instead of coal- or oil-fired power generation.

$$LCB = C_{gen} - (C_{prod} + C_{trans} + C_{inst} + C_{use} + C_{decom})$$

Where:

- C_{gen} CO_2 saved by using PV instead of coal or oil (this is the difference between the CO_2 produced by a conventional coal or oil power plant and that produced by the PV plant).
- C_{prod} CO_2 produced during production of materials, wafers, cells and modules.
- C_{trans} CO_2 produced during transport of materials to the PV site.
- C_{inst} CO_2 produced during assembly and installation of the PV power plant.
- C_{use} CO_2 produced during operation of the PV power plant.
- C_{decom} CO_2 produced during decommissioning, transport and recycling the PV field.

Usually the quantity of CO_2 saved by C_{gen} is much higher than the sum of the CO_2 generation sources (C_{prod}, C_{trans}, C_{inst}, C_{use} and C_{decom}), which is a prerequisite for the system to be a viable clean energy source and effective CO_2 reducer. In all cases, the PV plant will receive carbon credits for generating power without emitting CO_2.

Carbon credits represent some of the benefits of, and profits from, generating clean PV power today. "Carbon credits" is a generic term for any tradable certificate or permit representing the right to emit carbon dioxide (CO_2) or carbon dioxide equivalent (CO_2-^{e}). Carbon credits, and the newly developed carbon markets are attempts to mitigate the increase of greenhouse gases. One carbon credit is equal to one ton of carbon dioxide, or in some markets, carbon dioxide-equivalent gases.

Carbon trading is an application of an emissions trading approach. Greenhouse gas emissions are capped and then markets are used to allocate the emissions among the group of regulated sources. The goal is to allow market mechanisms to drive industrial and commercial processes in the direction of low emissions or less carbon intensive approaches than those used when there is no cost to emitting carbon dioxide and other greenhouse gasses (GHG) into the atmosphere. Since GHG mitigation projects generate credits, this approach can be used to finance carbon reduction schemes between trading partners and around the world.

This mechanism can be quite successfully used by trading (or selling) credits generated by our PV power plant with companies that need them, thus adding value to our plant and ultimately maximizing the ROI, since the PV plant will receive carbon credits for the CO_2-free power generated during its lifetime.

PRE-CONSTRUCTION FINANCE

"Shovel ready" projects (or projects that have completed all steps required for their construction) are what everybody is talking about today, and what investors are looking for. The pre-construction phase is the input of time, effort and money that has to be invested in a project to get it ready for construction (or "shovel ready"). But who has a "shovel ready" project, and how do we get to that stage? This is a catch 22 proposition, because we need to put a lot of time and money into a number of elaborate activities: planning, engineering design, land acquisition, environmental assessments, permitting process, interconnect and PPA negotiations, etc. to get a project ready for construction. These activities require an enormous amount of time, resources and money which most developers and customers just don't have. Provided that we have the time and resources, we need money to complete all the tasks needed to get the project ready for construction, and this is where pre-construction financing comes into the picture.

Presently there are only a hand full of players who are successful with large-scale power PV plant development in California and Arizona. One of the reasons for this is that financing of the pre-construction activities is hard to do, risky, and one of the greatest barriers to solar development in the US today. This is a complex effort consisting of putting technical and management teams together, purchasing or leasing land, designing the plant, going through the initial land assessment and permitting stages, and preparing for negotiating interconnection and PPA agreements.

As an example, a 20-MW PV power plant in California requires $4.0 million pre-construction financing and 12-18 months effort to bring it to construction readiness. Upon completion of this work, the project would need additional financing for the construction phase (~$80 million), and then we need to find long-term finance. The pre-construction finance effort, however, is where everything starts ...or stops.

Investors will listen to our pitch with a critical ear, and we must convince them that we know all there is to know about the project at hand (who, what, where, when and how). They need to feel comfortable with us and believe that our team members are specialists in all areas of the project design, installation and operation. We must know everything about the products and procedures we plan to use and everything in between. No gaps, no hesitation, no miscalculations, no mistakes of any sort are allowed.

Below are some of the most important subjects and

tasks that we need to be familiar with and which must be addressed as thoroughly and professionally as possible with the investors.

The Principal Stakeholders

We'll be doing business with the principals behind our solar project's development for the project duration, and especially during the pre-construction stage. Who are these principals? Let's meet some of them and see how they do business, so we can best present to our investors the issues at hand.

Note: Most of the issues below are discussed in more detail in the previous chapters, so to avoid repetition, we will only bring the related questions to your attention in order to be prepared with appropriate answers if and when needed.

The principals of any PV project planned for implementation in the US are:

PV Products Manufacturers

The principals in this category are the companies manufacturing silicon, thin film and all other raw materials, chemicals, gasses and consumables used in the PV manufacturing process; PV cells and modules manufacturers; and inverters and BOS components manufacturers. These companies hold the key to the quality of the PV products we will use, so we must know what to look for when working on a PV project. We must convince our investors that the entire process is under control.

The questions we need to ask the manufacturers, to get a complete understanding of their operation and possibly the quality of their products, are:

a. How long have they been in the solar business?
b. What is their management structure—technical and administrative departments and personnel?
c. What is their annual production quantity per product type?
d. What manufacturing technology do they use—production equipment type, make, quality and function?
e. What type of manufacturing processes and procedures do they use?
f. What is the type and origin of their raw materials and consumables?
g. What type of process and quality control and documentation do they have?
h. Are their PV products UL or CE certified? By whom and when?
i. Have they done any long-term on-sun tests and what are the results from these?
j. Are they planning to modify or improve the production equipment, process sequence, or final product designs? What product lines would be affected by that, and would they follow management of change procedures?

These are important questions, and although few manufacturers will answer all of them thoroughly, just by the completeness of their answer to each question we can draw a number of conclusions about their manufacturing process. If we ask the right questions of the right people, we can get a good picture of the quality and durability of their final products as well.

The risks of buying a product with no history or precedent are enormous. Investors will not put money into a product that we are not totally familiar with nor one for which we have not presented a complete and convincing management plan.

Engineering Procurement and Construction Contractors

The success of any solar energy project depends on engineering, procurement and construction (EPC) contractors' ability to design and develop it properly. As the solar plant nears construction, the owners hire an EPC contractor with some idea in mind as to what exactly is needed or expected. There are no set rules or standards for the work the EPC is supposed to do, so each case has its own peculiarities. Very often this effort is governed by a trial-and-error project development methodology. How well or poorly this works depends on how experienced the EPC company, its managers and engineers are.

So how are EPC contractors planning to make our project a success? Here are some key questions to be answered by EPC contractors we interview for our large-scale PV project:

a. What general experience do they have with PV installations?
b. What is their level of understanding, know-how, and experience with and confidence in the technologies we plan to use?
c. What is the priority of EPC tasks and issues that need to be addressed during the planning, design, implementation, and operation stages of our solar plant?
d. What are the best construction methods for solar plants, and what kinds of assurances and warranties can the project owners and investors have that everything will work as specified for the duration?
e. Are the technical and financial risks in developing a solar project well understood?
f. Do they understand that they will have to assume

some of the risks and provide a number of warranties and guarantees?

The EPC contractor must answer these questions clearly and thoroughly. There should be no unanswered questions or unclear answers. We must be able to convey these answers to the investors, in order to gain their confidence in us and the project.

The Utilities

Utilities are a decisive factor in the development of any PV project. Utilities' perspectives on, and attitude towards, large-scale solar projects is still evolving as we speak. Significant changes have occurred, over the past several years in the approaches utilities take when they assess their options in the quickly changing energy environment. Their decisions, however, are driven by the fact that they are obligated by law to consider adding solar and wind power generation to their capacity.

Our understanding of the role and function of the utilities is critical for presenting a proper picture to the pre-finance investors. Following are some of the issues we must be aware of and questions to ask the utilities:

a. Where are the closest and most convenient interconnect points (substations)?
b. Is right-of-way needed, and is it possible?
c. What other electric network issues must we be aware of?
d. What is the local utilities' attitude towards, and role in, our project?
e. What do they need or want to achieve by working with us?
f. How do they determine PPA pricing? Would it apply to our case and how?
g. What type of assistance could they offer to help at the pre-construction phase, if any?
h. How can they assist with the proper and timely implementation of our project and ensure its long-term success? How would that relate to reducing risks?

Working with the utilities is not that simple, because they have limited personnel willing and able to answer all these questions. Some persistence is needed to get to the right person(s) and be able to get a good picture of the local situation and its relation to our project. Showing thorough knowledge of the utilities' operations and presenting a good plan of how to approach them for interconnect and PPA negotiations is a must in our pre-construction finance effort. The investors must be convinced that there will be no problem with the utilities and that we can secure good working relations with them.

Utility-scale PV Issues

The list of utility-scale PV power plants continues to grow, with more than 6,500 MW of planned and announced projects all over the US. Several new small and medium size utility-scale solar power projects were introduced and many are planned for implementation in the US during the next several years. Others, much larger, are coming up—250 MW and 550 MWp PV projects are planned for installation in the US deserts in the near future.

So, is utility-scale PV on the verge of unprecedented growth? What pieces of the puzzle need to be in place to ensure the success of the large-scale PV installations?

Investors need answers to these general questions which are before the PV industry:

1. What is the near future direction of the PV industry in the local area?
2. What are the key factors driving the growth of utility-scale PV projects?
3. What is the mark of a successful PV project, and what is the track record?
4. What land areas are suitable for use for PV projects?
5. What PV technologies are used today, and are they efficient, reliable, and safe enough for utility-scale applications?
6. How are we going to ensure the quality of materials and final product (PV modules, inverters, etc.).
7. What do we know about the product's quality, performance, reliability and longevity, and what requirements do we make of these issues?
8. Why is there no adequate standardization of materials and manufacturing and quality control procedures?
9. What effects will the toxic materials (Cd, In, As, etc.) in some thin film modules have on the development and use of these technologies?
10. What roadblocks must be overcome to get the planned PV projects completed?
11. What business models are best for large-scale PV projects development, and what are the pros and cons of these models?
12. What are the financial challenges PV project devel-

opers are experiencing in the current markets, and what strategies are they using to overcome these?

13. What are the pros and cons of using federal vs. private lands?
14. How are we going to assure the investors that the risks are manageable?
15. How are present-day incentives and subsidies used, and what will happen when they are reduced or no longer available?

The answers to these questions are not straightforward, but many are addressed in this text and can be found from other sources. These items must be discussed confidently with the investors, who might not be specialists in this area and who would benefit from an honest and professional discussion. This will also give them a level of confidence that we have the understanding and know-how to lead the project to a successful end.

Technical Issues

This is the trickiest part of any presentation to potential customers and investors. The most challenging issues confronting the PV industry today are related to the efficiency, long-term performance, and longevity of PV modules and BOS components in harsh areas like deserts. Most manufacturers have no experience dealing with new, unproven PV technologies for large-scale (desert) applications that are supposed to operate 30+ years under extreme environmental conditions. This is a new area for many of them, and although there are discussions on the risks associated with different PV technologies—their performance under excess heat, annual efficiency degradation, and total failures—there is no uniform, standardized approach to handling these. But the way we handle these technical issues will determine the volume and success rate of large-scale PV installations in the future. So how do we present all this in a way that shows we are in control and that the risks will be minimized or eliminated?

Most PV products are attractively cheap, and perform well during the qualification test sequence. But, extrapolating the short-term test results to 30+ years of non-stop IR and UV radiation bombardment, related mechanical stress, and chemical attacks in US deserts is almost impossible. Accelerated certification tests represent 60 to 90 days of actual on-sun performance, and most new manufacturers have not had enough time for on-sun tests, so we must guess as to the performance and longevity of their modules.

Also, lately there have been many changes and modifications to the manufacturing processes in the race for more efficient and cheaper products. Most of these products, however, are too new, untested and not officially proven reliable for long-term use. So how do we know what will happen on the 91st day? Or how about day 3,091, 6,091, or 9,091? Who will be responsible for the drop in efficiency for the duration? Who will pay for the loss of output when a failed string of modules is shut down? Who will pay for and replace the failed modules or inverters? While there are some answers, such as limited manufacturers and long-term warranties, they don't cover the entire spectrum of technical and financial responsibilities and the associated risks we take when commissioning a 100-MWp power plant. The number of unknown, or known but uncontrollable, variables is large, and so are the risks.

Everyone involved in a solar project's design, implementation and operation (owners, designers, EPCs, O&M staff, product suppliers, and investors) has their own ideas and expectations on the different issues. None has the entire set of issues sorted, figured out and resolved. Instead, all are struggling (usually individually) to find a solution that might help our respective area of responsibility in the project at hand. A collective solution for the entire solar industry as a whole is on the table, but solutions are few and far between. So we need to present a clear and thorough picture of the technical issues, and how we are going to resolve or control them.

A sample list of these follows:

a. Which manufacturer and product do we propose to use and why?
b. Do we have a second choice, and what would be the difference in initial price, quality and long-term ROI?
c. Do manufacturers' long-term warranties, and service agreements adequately cover the long-term risks?
d. How long will the chosen PV modules last and how do we guaranty that?
e. How long will the chosen BOS last and how do we guaranty that?
f. How do we minimize the risk of accelerated degradation and power loss, especially in the deserts?
g. How do we avoid catastrophic field failure risks?
h. What are the gaps in the expectations of customers, designers, equipment suppliers, EPCs, and investors?
i. What can be learned from the experience of the wind and CSP industries?

j. What solutions are available in segmenting product and service risks, and applying warranty and insurance solutions for their management?

k. Why would a wise investor put his/her money into a PV project, if and when there are a number of uncertainties, and unproven technical issues?

These are some of the critical issues when discussing a large-scale project with a potential investor. Thorough knowledge of these issues and clear explanations of risk-management approaches are paramount. We cannot and should not hide the problems, so it will be up to the investors to figure out if we are knowledgeable and capable enough to handle them.

Foreign Participants

We cannot possibly review the developments of the PV markets of late without noticing the great impact of foreign companies on the growth of PV installations in the US and abroad. Asian and European manufacturers and engineering companies are very interested in the US energy market. They are trying to get involved on all levels in the work at hand, or planned here.

There are a number of questions that we need to answer when discussing the possibility of working with foreign products suppliers on our PV project. Some of these are:

a. What is driving foreign companies into the US PV markets?

b. What do foreign companies bring to the US energy markets?

c. What types of projects are they looking for?

d. How should U.S. companies structure the relations with them, in order to control the efforts and retain our energy independence?

e. Are the foreign manufacturers willing to participate in other aspects of PV projects (with technical or financial assistance), or are they just interested in selling us their products without any long-term participation?

Again, there are more questions than answers, but this is what happens when a new field of opportunities opens. We must agree that large-scale PV, in its present form, is a brand new field, one that grew very quickly, and seems as if it will be growing even faster. This is a new phenomenon, a business opportunity that is larger and with longer-lasting repercussions than anything we have seen, or could've imagined, before. Foreign companies see the opportunity, too, and are ambitiously pursuing market penetration options. In light of the new global economy, we cannot avoid that, and must adjust to the situation in order to find ways to manage it.

Many foreign suppliers, however, have different methods of operation. Some of them don't know or care about Western-type quality control, or anything Western for that matter. How, then, do we know that their quality is comparable to our requirements? There are many instances of foreign-made toys causing injuries and deaths. PV products are not disposable toys, but are serious equipment destined for long-term use under the world's most extreme climate conditions. We need to make an effort to know as much about them as possible.

Here again, investors must be convinced that we understand the situation, and the issues at hand, and that we can make educated choices. We will need a team of professionals to do it right, but it can be done. How well we do it will usually determine if we get pre-construction finance, or not.

THE LEGAL ASPECTS

Like any other business operation, the solar power plant transactions (PV included) and the related negotiations, as well as the subsequent equipment purchases, construction, installation and operation and maintenance (O&M) phases are governed by the local, state and federal laws of the US, or similar authorities in the respective countries.

A number of filings, agreements and contracts are needed to properly define and establish relationships between various stakeholders and authorities of any solar project. While these are absolutely needed, they are complex legal documents which are executed in the near vacuum of the non-standard procedures of the PV energy markets, and they may or may not help protect the owner's or investor's cause. Their form and size vary considerably, depending on project type and size and the legal entity drafting the contracts.

Below is a very rough idea of the legal aspects and challenges facing anyone involved or interested in a solar power plant project in the US:

Real Estate Contracts

The planning of any solar project starts with identifying a piece of property, or land, where the solar equipment will be installed and operated. The process then goes through the normal property or land lease or purchase steps, including signing a lease or purchase contract. Signing a real estate contract, and reviewing the related issues are the first and one of the most important initial considerations when planning and designing a solar energy project. Other tasks and issues are project siting, land surveys, acquiring interests in land (purchase, lease, and/or easement), land title and title insurance, green leasing, permitting and approvals, zoning/land use, mineral/water rights, and securing access to essential resources.

These are key components of a real estate property purchase, which requires specific legal documents, or special consideration at the very least. The subject of resource acquisition is one of the most disputed and might encompass water rights for hydroelectric power generation or irrigation, geothermal heat source, or access to the wind or sun, or to the roads.

In all cases, a real estate purchase contract is negotiated and signed. It contains a number of clauses that detail the buyer's and seller's rights and obligations. It usually includes the following:

a. Participants: The full name of the parties must be on the contract. In a sales contract, the parties are the seller(s) and buyer(s) of the real estate, who are often called the principals to distinguish them from real estate agents, who are effectively their intermediaries and representatives in negotiation of the price. If there are any real estate agents brokering the sale, they are typically listed also as the real estate brokers/agents who would earn the commission from the sale.

b. Real estate (property): At least the address, but preferably the legal description must be on the contract.

c. Purchase price: The amount of the sale price or a reasonably ascertainable figure (an appraisal to be completed at a future date) must be on the contract.

d. Official signatures: A real estate contract must be entered into voluntarily (not by force), and must be signed by the parties, to be enforceable.

e. Legal purpose: The contract is void if it calls for illegal action.

f. Competent parties: Mentally or substance-impaired persons cannot enter into a contract. Contracts in which at least one of the parties is a minor are voidable by the minor.

g. Meeting of the minds: Each side must have a clear understanding of and agree to the essential details, rights, and obligations of the contract.

h. Special consideration: Consideration is something of value bargained for in exchange of the real estate. Money is the most common form of consideration, but other consideration of value, such as other property in exchange, or a promise to perform (i.e. a promise to pay) is also satisfactory.

i. Notarization. This is not required for a real estate contract, but many recording offices require that a seller's or conveyor's signature on a deed be notarized to record the deed.

j. Recording: The real estate contract is typically not recorded with the government, although statements or declarations of the price paid are commonly required to be submitted to the recorder's office.

k. Lawyer review period: Sometimes real estate contracts will provide for a lawyer review period of several days after the signing by the parties to check the provisions of the contract and counter-propose any that are unsuitable.

l. Agents and brokers: If there are any real estate brokers/agents brokering the sale, the buyer's agent will often fill in the blanks on a standard contract form for the buyer(s) and seller(s) to sign. The broker commonly gets such contract forms from the real estate association to which he/she belongs. When both buyer and seller have agreed to the contract by signing it, the broker provides copies of the signed contract to the buyer and seller.

If the real estate contract is properly designed and executed, the property changes hands and the new owner takes possession. At this point, the new owner is legally in control of the land destined for solar power plant development, and can start the permitting and other procedures needed to get the land to a "shovel ready" state.

In some pre-construction cases, the solar project manager gets a conditional authorization (temporary lease), in order to reserve the land for this project while going through the preliminaries, such as information gathering, permitting, team creation and construction finance. At the end of the temporary lease, the project manager must lease or purchase the land, or else lose possession. In some special cases, they might be given a chance to extend the temporary lease as needed to complete the pre-construction process.

Land Lease Contracts

There are a number of ways to lease a piece of land for solar project development. These are:

Long-term Lease

A long-term (20-30 years) lease is a common method

of acquiring land for large-scale solar and wind plants. Leasing is the safest and cheapest way to start a solar project; i.e., a 100 M PV project requires ~600 acres land, which at $10k/acre will cost over $6 million. This might not be a lot, considering the $450M tag of the entire project, but it is significant, especially in the pre-construction stage, when money is not readily available. As a matter of fact, many projects never see the light of day due to the lack of capital in the pre-construction stage, so the lease might go through several negotiation stages and extensions, but never be signed.

There are, however, a number of issues to beware of when signing a lease contract. For example, many commercial lease contracts executed in the last decade contain a "rent escalation" clause. The intent is to protect the landlord's investment from the ravages of inflation. Another trend is for landlords to draft leases where the tenant pays lease/rent plus reimbursement for operating costs, such as energy, water, janitorial, insurance and taxes. Some of these costs and expenses might be hidden at the time of lease negotiations and signing, so a very careful at the location is absolutely necessary. Real estate professionals call this a "net lease." In effect, it transfers operating risks from the landlord to the tenant. Doing so reduces the landlord's incentive to apply efficient management or sustainable principles to cut operating costs, since these costs are now excluded from calculating a property's net operating income (NOI); this is the basis for calculating property's economic value. For years, the operating expense and tax clause, or common area maintenance (CAM) clause, that requires tenants to pay for any increase in the property's operating expenses, has been a secret profit center for landlords.

Following is a list of the most common overcharges:

a. Capital costs incorrectly allocated as repairs.
b. Costs specifically excluded in the lease are lumped into "miscellaneous" expenses.
c. Tax abatements, rebates, or refunds not passed through or credited to tenants.
d. Double billings for utility service paid by an individual tenant but also included in a general utilities category.
e. Accountants' fees for keeping track of and preparing the landlord's tax returns.
f. Failure to pass through expense reductions as credits.

Green Lease Contracts

Green lease contracts are a new development that is designed to encourage, or support, the introduction and/or use of green technologies in buildings and projects. These are basically incentives for companies who are involved in energy savings, environmental protection and other such "green" acts. There is no simple way to design and enforce "green" lease" contracts, so solar projects developers negotiating a "green lease" should be extremely cautious.

There are a number of horror stories lately that show how confusing the issue is and how it can lead to a number of misunderstandings, so we can only advise extra caution when seeking a trouble-free building or land lease or purchase under a "green lease" contract. "Buyer beware" warnings should be printed in large red letters across each page of the pre-signed lease or purchase contract.

The U.S. Green Building Council's Leadership in Energy and Environmental Design (LEED) standards apply to an extent to green leases, but the market is too immature and is still sorting out the details of the "green lease" contracts, with people intentionally or unintentionally misusing the contracts. Currently, all participants in a solar project must beware of, understand, address, and agree fully and thoroughly on the key issues when drafting a green lease contract, as follows:

a. Lease/purchase of structure or land, and the related expenses,
b. Energy generated or used by tenant/new owner,
c. Operational performance of energy generating or conserving systems,
d. Hazardous materials used or generated during installation or for the duration,
e. Green cleaning and maintenance, and
f. Recycling at end of life.

Finally, regardless of what type of contract we are signing we must discuss it thoroughly with a legal expert before agreeing and signing. A lease, or a purchase agreement, is a complex legal document. It is also a tool that can be used to define parameters for sustainable operations, or can be used against us and the project. This is critical for project managers interested in occupying and running a solar power plant in a trouble-free manner under all possible scenarios—including laws and ownership changes over the course of the life of the project.

EPC Contracts

Engineering, procurement and construction (EPC) companies specialize in the development of construction projects, including solar and wind. A contract between the EPC company and the solar project owners and investors is the preferred instrument of establishing the relationship between the parties, with corresponding responsibilities and accountability.

In an EPC contract, the EPC contractor (EPCC) agrees to deliver the keys of a commissioned plant to the owner for an agreed amount, just as a builder hands over the keys to a building to the purchaser. The EPC way of executing a project is gaining importance worldwide, but it is also a way that needs good understanding, by the EPCC, for a profitable contract execution.

The owner, or investor, decides on an EPC contractor for several reasons:

a. The EPC company is basically in charge of, and fully responsible, for everything during the design, products acquisition and installation stages.
b. The owner puts in minimum effort and/or has no responsibilities in the project.
c. EPC gives the owner one main point contact, in order to easily monitor and coordinate the work effort.
d. It is easy for the owner to get post-commissioning services.
e. Experienced EPCs ensure quality and resolve problems under their control.
f. Owner is not affected by the market fluctuations.
g. Investment figure is known at the start of the project.

The project owner defines (or agrees) on the following in the EPC contract:

a. Scope and specifications of the plant and operational capacity and capability
b. Quality of materials, components and labor
c. Project efficiency, duration and cost
d. The amount to be paid to the EPC
e. Exemptions and non-compliances

EPC contracts must be negotiated to consider and include the following:

Coordination and Communication

In an EPC contract, coordination with various agencies is a demanding task, especially in item rate contracts. Often agencies try to put the responsibility for delays, etc. on each other. The coordination is more difficult during commissioning and post-commissioning times, because each agency tries to correlate its own performance with the others'. It is very likely that, in an EPC contract, the communication channels get overlapped, short-circuited and confused. The EPCC is the prime functionary and is fully responsible for the clarity of communications with all parties.

Global Arena

An EPC contract is a complex instrument. It involves various agencies and characteristics of products and services coming from different countries. So the EPC contractor needs thorough understanding of the global aspects of the operation. The EPC officials must know about the various factors that will affect the success or failure of the contract, in the global arena. The EPC must have data and expertise in all the required fields. A thorough knowledge of many aspects is required, some of which are the subject of this text. Some important areas are:

a. Local (where the plant will be located) market conditions for the materials supply and labor availability and performance.
b. Local code, statutory etc., requirements
c. Availability of local supervisory personnel
d. Availability of local and global engineering services
e. Local and global sub-contractors, their experience, performance and product quality.

Cost Variation

Another important factor that can affect the EPC's performance is cost variation. An EPC contract has no price escalation clause. The cost variation to the EPCC can occur on two main counts:

a. Foreign exchange fluctuation
b. Variation in the prices of materials and labor and services

These are the inherent risks, and the EPC must be careful. The owner has committed a fixed price and is free from the variation of market prices. At the time of the commitment to the owner, the EPC, too, must have a similar agreement with various agencies, but the scope and quantities of the services must be known. Again, thorough understanding and experience with the global PV suppliers and their products is critical.

Monitoring by Owner and EPC

The following points will be helpful to the owner for monitoring the project:

a. Define performance and guarantees well

b. Define scope and quality very carefully
c. Define milestones meticulously
d. Have the LD/penalty clauses well-defined

Handling of an EPC Contract

This is a complicated and complex function for the EPCC management. Some important points are:

a. Have very specific payment terms
b. Have similar terms and conditions regarding quality, guarantee etc., with the product vendors
c. Do not keep terms open-ended
d. Coordinate vigilantly to reduce chances of errors at the site

Project Owners

The project owner (or project manager) must clearly define the project, as any changes after the EPC contract has been signed will be costly or simply impossible. In order to ensure quality, the owner must select a reputable and experienced EPC contractor. The owner should also utilize a third party as a project manager, or at the very least an in-house consultant—specialist in PV installations—to keep track of and verify the design of major structures and to inspect the main plant equipment as work progresses. A full-scale total quality control system must be implemented when building large-scale power plants.

PV Products Warranty Contracts

All PV products—PV modules, inverters, electronic components, BOS, etc.—come with some sort of materials, labor and long-term performance warranty. PV module warranties are the trickiest, while the warranties of the BOS equipment are more straightforward. PV modules usually come with several warranties, and the conditions vary from manufacturer to manufacturer, but the example below is typical for foreign-made PV modules:

NOTE: The underlined portions of the text below denote what customers and EPC engineers must watch for, verify and clarify with the manufacturer, before accepting the terms and conditions of the warranty:

Limited Product Warranty

This is usually a 1-, 2-, 3- or 5-year repair, replacement or refund/remedy policy. Manufacturer usually warrants its PV modules, including factory-assembled DC connectors and cables, to be free from defect in materials and workmanship under normal application, installation, use and service conditions. For a period ending the agreed number of years from the date of sale, as shown in the invoice to the direct customer, the defective or failing modules will be taken care of by the manufacturer or his representative. Manufacturer, at its option, will either repair or replace the product or refund the purchase price as paid by the direct customer. The repair, replacement or refund remedy shall be the sole and exclusive remedies provided under the "Limited Product Warranty" and shall not extend beyond the agreed time period from the date of sale. The "Limited Product Warranty" usually does not warrant a specific power output, which shall be exclusively covered under the "Limited Peak Power Warranty" clause hereinafter.

NOTE: "Normal" application above needs thorough clarification, for there are no "normal" conditions in the Arizona deserts. "At its option" needs to be examined, negotiated, and agreed upon in writing.

Limited Peak Power Warranty—Limited Remedy

This is usually a 10-, 20- or 25-year warranty.

a. 10-years warranty. If, within a period of ten (10) years from the date shown on the invoice to the direct customer any module(s) exhibit a power output less than 90% of the minimum "Peak Power at STC" as specified on the date of invoice in manufacturer' information sheet, provided that such loss in power is determined by manufacturer (at its sole and final discretion) to be due to defects in material or workmanship, manufacturer will replace such a loss in power by either providing additional modules to the direct customer to make up for such loss in power or by replacing the defective module(s), at the option of manufacturer.

b. 20-year warranty. If, within a period of twenty (20) years from the date shown on the invoice to the direct customer any module(s) exhibit a power output less than 83% of the minimum "Peak Power at STC" as specified on the date of invoice in manufacturer's product information sheet, provided that such loss in power is determined by manufacturer (at its sole and final discretion) to be due to defects in material or workmanship, manufacturer will replace such a loss in power by either providing additional modules to the direct customer to make up for such loss in power or by replacing the defective module(s), at the option of manufacturer.

c. 25-year warranty. If, within a period of twenty-five (25) years from the date shown in the invoice to

the direct customer any module(s) exhibit a power output less than 75% of the minimum "Peak Power at STC" as specified on the date of invoice in manufacturer's product information sheet, provided that such loss in power is determined by manufacturer (at its sole and final discretion) to be due to defects in material or workmanship, manufacturer will replace such a loss in power by either providing additional modules to the direct customer to make up for such loss in power or by replacing the defective module(s) at the option of manufacturer.
NOTE: "At the option of the manufacturer" must be clearly defined, negotiated, and agreed upon in writing.

Exclusions and Limitations

Read carefully the small print. It might make a big difference in how non-conformities are handled in the future. Some of these are:

a. In any event, all warranty claims must be filed within the applicable warranty period, and

b. The "Limited Product Warranties" and the "Limited Peak Power Warranties" do not apply to any modules which have been subjected to:
   1. Misuse, abuse, neglect or application
   2. Alteration, improper installation or application
   3. Non-observance of manufacturer's installation and maintenance instructions
   4. Repair or modifications by someone other than an approved service technician of manufacturer
   5. Power failure or surges, lightning, flood, fire, accidental breakage or other events outside manufacturer's control

c. Both the "Limited Product Warranties" and "Limited Peak Power Warranties" do not cover any transportation charge, customs clearance or any other cost for return of the modules, or for return shipment of any repaired or replaced modules, or costs associated with installation, removal or reinstallation of the PV-modules.
NOTE: These conditions could prove exceedingly expensive to the project owner, if the terms are not clarified and agreed upon.

d. Warranty claims will not be honored if the type or serial number of the modules has been altered, removed or made illegible.

Limitation of Warranty Scope

Manufacturer shall have no responsibility or liability whatsoever for damage or injury to persons, property, or for other loss or injury resulting from any cause whatsoever arising out of or related to the modules, including any defects in the modules or from use or installation. Under no circumstances shall manufacturer be liable for incidental, consequential or special damages, however caused. Loss of use, loss of profits, loss of production, and loss of revenues are therefore specifically and without limitation excluded. Manufacturer's aggregate liability, if any, in damages or otherwise shall not exceed the invoice value as paid by the direct customer, for the module unit(s).

NOTE: This overall refusal of responsibility is questionable, and must be well clarified, negotiated, and agreed upon in writing.

Filing Warranty Claim

If you feel you have a claim covered by warranty, you must promptly notify the dealer who sold you the module with the associated claim. The dealer will instruct and advise you how to handle the claim. If additional assistance is needed, you must contact manufacturer for instructions. You may find contact information for manufacturer listed in the documentation. The customer shall submit a written claim, including documentation of module purchase, serial number, and product failure. Manufacturer will determine in its sole judgment the adequacy of such claim. Manufacturer may require that the product subject to a claim be returned to the factory at the customer's expense. If the product is determined to be defective and is replaced but is not returned to manufacturer, then the customer shall submit sufficient evidence that the product has been recycled or destroyed.

Note: "Sole judgment" needs clarification. Returning a truck load of PV modules to Asia might be as expensive as buying new ones, so the above terms must be well understood and agreed upon before signing a purchase contract.

Severability

If a part, provision or clause of this "Limited Warranty for PV Modules," or the application thereof to any person or circumstance, is held invalid, void or unenforceable, such holding shall not affect other parts, provisions, clauses or applications, thus this "Limited Warranty for PV Modules" shall be treated as severable.

Disputes

In case of any discrepancy in a warranty-claim, a first-class international test institute such as Fraunhofer

ISE in Freiburg, Germany, TÜV Rheinland in Cologne, Germany, or ASU Arizona State University shall be involved to judge the claim finally. All fees and expenses shall be borne by the losing party, unless otherwise awarded. The final explanation right shall be borne by manufacturer.

Note: The above mentioned entities can judge only the technical aspects of a claim. A local legal authority must be cited for resolving logistics and any other disputes. Also, a local test institute must be chosen and agreed upon before signing the final purchase contract. Fees and expenses clause must be also revised according to the specific conditions.

Various

The repair or replacement of the modules or the supply of additional modules does not cause the beginning of a new warranty term, nor shall the original term of this "Limited Warranty for PV Modules" be extended. Any replaced modules shall become the property of manufacturer and made available for their disposal. Manufacturer has the right to deliver another type (different in size, color, shape, and/or power) in case manufacturer discontinued offering the replaced modules at the time of the claim.

Note: We'd argue that if a replacement batch of modules is installed today, its warranty period should also start today. The reason for this exception, as well as the type of replacement modules, must be clarified and agreed upon ahead of time.

Warranty Transfer

The warranty is usually transferable when the product remains installed in its original location at the warranty registration.

Force Majeure

Manufacturer shall not be responsible or liable in any way to the customer or any third party arising from any non-performance or delay in performance of any terms and conditions of sale, including this "Limited Warranty for PV Modules," due to acts of God, war, riots, strikes, warlike conditions, plague or other epidemics, fire, flood or any other similar cause or circumstance beyond the reasonable control of such manufacturer. In such cases, performance by manufacturer of this Limited Warranty shall be suspended without liability for the period of delay reasonably attributable to such causes.

Validity

This "Limited Power Warranty for PV Modules" is valid for all modules dispatched within a certain period of time.

Notice: "Peak Power at STC" is the power in watts that a PV-module generates in its maximum power point. "STC" are as follows:

a. Light spectrum of AM 1.5
b. Irradiation of 1000 per m2
c. Cell temperature of 25°C

The measurements are carried out in accordance with IEC 61215 as tested at the connectors or junction box terminals, as applicable, per calibration and testing standards of manufacturer valid at the date of manufacture of the PV modules.

Note: This is tricky. Buyer beware! Some manufacturers may claim that PV modules in the AZ and CA deserts are "operating above STC and out of warranty conditions," since they see temperatures well above 80°C. It is basically impossible to run under STC conditions in the desert, so EPC officials, owners and investors must understand the difference well and negotiate the terms carefully.

Utilities Agreements and Contracts

Every large-scale project starts and ends with discussions with the local utilities, who own the nearby transmission lines and substations. Nothing will happen if there is no agreement with the local utility. You cannot interconnect to their network, nor would you get paid for the produced power, if the utility has not blessed your project with all its details and peculiarities. And don't forget that the utilities are the judge and executioner. If they don't like your technology, then you better find a better one. If you don't have enough money to comply with all their demands for payment and liability assurance, then just forget it—you won't get help from them. The rules are simple: a) show us that you can do this project, and b) show us the money to pay for it. There are no ifs or buts on these two conditions.

So, in order to get approved for your solar project you need to discuss in detail the conditions and obligations and sign a number of agreements and contracts with the utilities who set the rules—then make and execute the decisions.

Some of the subjects to discuss and hurdles to overcome are:

Interconnection Agreement

These are one-way agreements offered by most US utilities. In all cases the project owner and investor are fully responsible for the interconnection—whatever it means, and whatever it takes—and for the duration. In most cases the utility won't make a great effort to assist any more than absolutely necessary.

These are some of the requirements, set forth by the utility company for any takers to evaluate:

a. Construction. The seller (that's the project owner) shall have the obligation to cause (and pay for) the construction of the electrical interconnection facilities, including metering and sub-metering facilities, and transmission upgrades, and cause them to become operational. This usually means millions of dollars and many months spent on ensuring that the new PV plant can get online.

b. Maintenance of electrical interconnection facilities.

To the extent required to achieve the initial delivery date and at all times during the services term, seller shall maintain and/or cause to be maintained, at its expense, the electrical interconnection facilities such that the electrical interconnection facilities are capable of delivering the products that can be generated or produced using the maximum contract capacity in accordance with the terms of this agreement to and at the electrical delivery point during each month as applicable (in addition to such other output of the facility as the electrical interconnection facilities are required to transmit) in accordance with the terms of this agreement.

How many lawyers, and how long, do you think it took to come up with the above statements, which are not designed to discourage any wannabe project owners. No, they are only intended to let them know whose necks are on the line at all times. No exceptions!

PPA Contracts

A power purchase agreement, also known as a PPA, is a legal contract between an electricity generator (the solar plant operator) and a power purchaser (the utility company). The power purchaser purchases energy, and sometimes also capacity and/or ancillary services, from the electricity generator. PPAs allow purchasers of solar energy to take advantage of its benefits, without some of the risks.

Under a typical PPA, the owner or investor (also known as an independent power producer) will finance, install and own the power plant. The utilities usually pay nothing, for the installation or the operation. Then, as the modules generate clean, solar electricity, the provider sells the power to the utility company and bills it for the electrical output of the PV module system at a previously agreed upon rate.

A PPA is a contract between the generator and the utility for use of the solar electricity. This generally includes detailed provisions and important considerations concerning cost, insurance, taxes and more.

Key provisions typically included (in residential, but applicable to large-scale installations) pertain to:

a. Term: How long is the contract? Often, the term of a PPA is 15-20 years.
b. Assignability: What if you sell your business, or your property? What if the provider changes corporate structure? Will the PPA continue? If so, under what terms?
c. Cost: A very important consideration. How much do you get for power generated by the PV module system? Are you paid monthly? Quarterly? How often does the cost increase? Some contracts include escalations on an annual basis, or every 3-5 years.
d. Payment: Who makes the payments? Is there a grace period? As with most contracts, if payment is not made on time, does this represent breach of contract?
e. Performance: The solar project must produce energy as agreed and signed in the PPA contract. Any and all exemptions must be in the contract, or else penalties will follow. The utility requires a certain amount of power, and it must be delivered, or else. Very careful analyses during the design stages are needed to do this right.
f. Taxes and Insurance: Providers usually bear the cost of taxes and insurance for the system for the duration. Make sure that these charges are included in the initial plant design estimates. Also verify that the insurance is properly structured and the limits are adequate to cover the cost of replacement under different conditions, if and as needed.
g. Destruction and Damage: If a storm hits, or if vandals destroy the PV system, the PPA contract should specify under what circumstances the provider will replace and restore the modules. It also should spell out all other ramifications, such as the plant's failure to generate power.
h. Termination: When does the contract end? Do you have a right to extend the term? Who has the responsibility to remove the modules and restore the property?

Again, you need a good lawyer to evaluate the contractual conditions from all aspects of the laws in force at the time. To complicate things, the laws change, so moving carefully is the key here.

O&M Contracts

Upon completion of the PV power plant construction, an operations and maintenance (O&M) team takes over the plant's day-to-day operation and maintenance procedures. The team, as well as its responsibilities and accountability has been discussed and negotiated ahead of time, and an O&M contract is signed with the O&M entity. The contract must contain a number of conditions, some of which are listed below. The O&M team's management responsibilities include:

a. Plan and manage all processes required to get a functional O&M team ready to accept responsibility for the project immediately upon completion of construction.

b. Establish and communicate clearly the management's mission and vision statements.

c. Prepare an action plan and justify a reasonable budget for all expenses related to O&M team hiring, training and deployment, and manage the execution of the program.

d. Recruit, interview and hire complete technical and administrative staff.

e. Lead, train and manage the staff into technical competence, individual responsibility, personal commitment and efficient teamwork.

f. Provide an experienced plant manager, who is responsible for all aspects of the day-to-day operation for the duration.

g. Set up and enforce a complete set of technical and administrative policies, procedures and programs.

h. Set up a complete operations, maintenance, materials and labor management program.

i. Select, requisition and manage the spare parts and components inventory.

j. Create a comprehensive set of technical and administrative procedures manuals.

k. Provide a detailed management plan for efficient and profitable commercial operation, with realistic management goals and plans to accomplish the goals.

l. Provide a set of programs, geared to issue comprehensive and meaningful monthly budget variance reports and production information.

m. Provide intensive training to the plant management on the policies, procedures and philosophy of management, which will ensure efficient and profitable operation of the facility for the duration.

Basically, the expectations of the project owners and investors, together with the methods they are planning to use to supervise and keep track of the O&M team's performance, must be clearly understood, agreed upon, and officially incorporated into the contract. Our friendly, and hopefully knowledgeable and capable, lawyer is indispensable here too, because this is an important long-term commitment to an effort that will determine the success and profitability of the project.

Litigations and Pending Legal Actions

2009-2010 solar developments were marked by an increased number of planned and active litigations, filed, or to be filed, by different interests and local citizens' groups.

Winning a permit was once the end goal for each PV project developer. This, however, is no longer the case. At some point—even after permitting is completed and a PPA agreement has been signed—environmental or local community groups could, and surely have, strongly objected to the mega projects, particularly if they require a lot of water, or if they are proposed for installation on virgin lands.

Enter the US courts. They will play an increasing role in providing the "last word" on many large projects. The courts will decide which projects will be cleared for construction. There is a significant political push to get lots of PV projects done quickly, but it cannot prevent litigation, which comes from independent public interest groups and individuals. So far, most of the planned projects go through state and federal government approval, mostly because the developers want to take advantage of federal land and subsidies, which can be avoided if building on private land with private financing.

If a project is constructed with PV modules, instead with any of the solar thermal technologies, the owners could avoid some of the state permitting procedures as well, because in such cases the county government has the last word. Negotiating with county governments isn't that easy either, because another set of concerns and regulations is in play. County officials are usually much more sensitive to the wishes of their local constituents because they are directly elected by them, and this introduces other variables into the permitting process.

Different counties also have different approaches and put different barriers in the solar developers' path; i.e., Kern and San Bernardino counties are relatively easier to negotiate a solar project, while San Luis Obispo County is more restrictive and particular in their approach to new solar plants on their territory. In all cases, the final outcome will most likely be that of a) using more private lands for solar plant development, and b) shrinking the large power plants from 1,000 or 800 MW down to 100-200 MWp, which is more palatable for the local populations. The trend in California is for even smaller PV plants—in the 1-20 MWp range.

One particularly serious large case worth mentioning here, as an example of the power of the different local activist groups, is a federal court's injunction against a large Imperial Valley solar project, which was then taken up by several local groups, who filed a complaint in federal court against the largest CSP projects in California (8). Generally speaking, the lawsuit's intent is to stop development of six large solar thermal projects taking place on federal lands, as follow:

Ivanpah Solar Electric Generating System Project and Associated Amendment to the California Desert Conservation Area Plant ("Ivanpah Project"), approximately 3,472 acres in size; *Genesis Solar Energy Project and Amendment to the California Desert Conservation Area Plan* ("Genesis Project"), approximately 1,950 acres in size; *Imperial Valley Solar Project and Amendments to the California Desert Conservation Area Land Use Management Plan* ("Imperial Project"), approximately 6,360 acres in size; *Chevron Energy Solutions Lucerne Valley Solar Project and Amendment to the California Desert Conservation Area Plan* ("Chevron Project"), approximately 4,613 acres in size; and *Blythe Solar Power Project and Amendment to the California Desert Conservation Area Plan* ("Blythe Project"), approximately 7,025 acres in size.

This situation reflects the growing complexity of the immature and underdeveloped solar industry in the US. It presents a serious problem to the already approved projects (future solar power plants), where a lot of time, money and resources were spent to bring them to the "shovel-ready" point. Getting to the shovel-ready state today—in addition to the hundreds of complex steps and hurdles one has to overcome—requires a court decision. This is another complex and expensive step, which is totally unrelated to the initial designation and goals of the affected projects.

So what will happen to the above projects, if the court stops their development? There are not many options to choose from, because large CSP projects cannot be easily shrunk to a smaller size. Converting them to PV power generation might be one option, so most likely, they will sell the pipeline of projects to somebody who is more equipped to handle smaller, PV-based projects. Or not...

So the solar game continues, and is getting more complex and more interesting by the day.

Notes and References

1. Integrating Large-Scale Photovoltaic Power Plants into the Grid, Peter Mark Jans son, Richard A. Michelfelder, Victor E. Udo, Gary Sheehan, Sarah Hetznecker and Michael Freeman, Rowan University/Rutgers University/PHI/Suntechnics Energy Services/Exelon Energy
2. A Study of Very Large Solar Desert Systems with the Requirements and Benefits to those Nations Having High Solar Irradiation Potential, SENI, 2006
3. Minimizing utility-scale PV power plant levelized cost of energy using high capacity factor configurations, Matt Campbell, Sun Power Corp.
4. Integrating Large-Scale Photovoltaic Power Plants into the Grid, Peter Mark Jansson.
5. CEC website. http://www.dfg.ca.gov/habcon/nccp/publications.html
6. CPUC website. http://www.cpuc.ca.gov/puc/
7. The Smart Grid: An Introduction. http://www.oe.energy.gov/DocumentsandMedia/DOE_SG_Book_Single_Pages.pdf
8. http://dockets.justia.com/docket/california/casdce/3:2010cv02664/340990/
9. Pacific Gas and Electric Company 2011 PV Program PPA Update. http://www.pge.com/b2b/energysupply/wholesaleelectricsuppliersolicitation/Pvrfo/
10. Decision Adopting A Solar Photovoltaic Program For Pacific Gas And Electric Company http://www.pge.com/includes/docs/pdfs/b2b/wholesaleelectricsuppliersolicitation/117115.pdf

Chapter 8

Environment and Safety

"Any intelligent fool can make things bigger, more complex, and more violent.
It takes a touch of genius—and a lot of courage—to move in the opposite direction."
A. Einstein

THE ENVIRONMENT STATUS

It is only proper, once we are so deep into the subject of PV power generation, to take a close look at the environmental effects from the perspective of solar energy generation and use. It is also important to understand and analyze the problems caused, or that might be caused, by PV systems manufacturing and operation.

To start with, we will focus on the basics of the present day environmental puzzle. A number of serious environmental disasters opened the eyes of the world's scientific community to the consequences of acting recklessly and violating the laws of Mother Nature. This string of disastrous events established the foundation of the modern environmental movement and put the alternative energy industry on a respected and solid foundation. There is grave news about global warming. The ozone layer depletion, the ozone hole, ice cap melting, unusual floods and draughts, ocean level changes, and other such cataclysmic events are a daily confirmation of the fact that something big and not very good is happening. They have confirmed the fact that we are on the wrong path and need to consider making some changes.

Solar energy is shaping up as one of the energy sources that could help us achieve energy independence and at the same time clean the environment—a win-win situation, if everything is done right. But if not, we might just add to the environmental debacle we live in. So, we will take a look at solar technology's cradle-to-grave contribution to the environmental issues of today.

Brief History

Below is a chronological list of key environmental events that deserve mention due to their effect on, and importance in dealing with, the environmental problems:

1969—Blow-out at the Union Oil Platform A at the Santa Barbara, CA, coast line spews an estimated 80,000-100,000 barrels of crude oil into the channel and onto the beaches, fouling the coastline from Goleta to Ventura, as well as the northern shores of the four northern channel Islands. (Oil and tar deposits still linger on the local beaches.)

1970—National Environmental Policy Act signed, thus creating the Council on Environmental Quality (CEQ), which gives the President advice on environmental issues.

- General Motors president Edward Cole promises "pollution free" cars by 1980 and urges the elimination of lead additives from gasoline in order to allow the use of catalytic converters. (Catalytic converters are in, but pollution free cars are still in the queue and still a long way from reality.)
- Earth Day celebration in San Francisco organized by John McConnell.
- First nationwide Earth Day organized by Sen. Gaylord Nelson and Dennis Hayes.
- Clean Air Act passed.
- Natural Resources Defense Council created.
- Friends of the Everglades founded by Marjory Stoneman Douglas.
- Lake Michigan Federation founded.
- Environmental Protection Agency signed into law.
- Occupational Health and Safety Administration (OSHA) bill signed into law.

1971—Chamber of Commerce warns of dangers arising from enforcing pollution regulations.

- Passage of Animal Welfare Act and Wild and

Free Ranging Horse and Burro Protection Act.
— President's CEQ acknowledges racial discrimination negatively affects urban environment.
— Greenpeace founded in Victoria, B.C., to oppose atomic testing in Alaska.

1972—W. Eugene Smith completes his essay on the crippling effects of mercury pollution.
— First regional treaty to regulate dumping of radioactive wastes in Europe.
— EPA announces all gasoline stations required to carry nonleaded gasoline.
— Buffalo Creek disaster occurs in West Virginia, where strip mining kills 125 people.
— Congress passes: Federal Water Pollution Control Act, Coastal Zone Management Act, Ocean Dumping Act, and the Marine Mammal Protection Act.
— Toxic Substances Control Act (TSCA) law passed.
— First bottle recycling bill passed in Oregon.
— Supreme Court supports Sierra Club over Disney Inc. in battle over development.
— United Nations Conference on the Human Environment convenes in Stockholm, Sweden.
— UN Environment Program (UNEP) acts on the recommendations of Stockholm meeting.

1973—Eighty nations sign the Convention on International Trade in Endangered Species (CITES).
— Arab oil embargo panics US and European consumers; prices quadruple.
— Congress approves Alaska Oil pipeline.
— A group of Himalayan villagers stop loggers from cutting down a stand of hormbeam trees.
— Endangered Species Act passed by Congress.
— Tellico Dam controversy; Endangered Species Act blamed for stopping project.

1974—F.S. Rowland and M.J. Molina blame CFCs for breaking up ozone in a catalytic cycle.
— Congress passes Safe Drinking Water Act to be administered by EPA.
— K. Silkwood dies in a suspicious accident, involving Kerr-McGee nuclear weapons facility.
— Worldwatch Institute founded.

1975—Atlantic salmon return to Connecticut River after 100-year absence.
— Congress passes Hazardous Waste Transportation Act.
— Greenpeace leads the Great Whale Conspiracy battle.
— Standoff over logging in Brazil's Amazon region by local rubber tappers.
— Federal court says EPA has authority to regulate leaded gasoline.
— Catastrophic failure of Grand Teton Dam in Idaho causes 14 deaths and lots of damage.
— Chemical explosion in a Milan, Italy, spreads dioxin, causing chloracne in 300 school children.
— National Academy of Science report on CFCs gasses warns of damage to ozone layer.
— Congress passes: Resource Conservation and Recovery Act (RCRA), Federal Land Policy Management Act, and the Whale Conservation and Protective Study Act.
— *Urquiola* oil spill, La Coruna, Spain.
— Liberian tanker *Argo Merchant* crashes by Nantucket Island, leaks 9 million gallons of oil.
— The Land Institute founded in Salinas, Kansas.
— The International Primate Protection League formed in Thailand.
— American Museum of Natural History forced to halt cat experiments.

1977—U.S. Department of Energy is created by President Jimmy Carter.
— Congress passes Soil and Water Conservation Act; and the Surface Mining Control and Land Reclamation Act.
— Ecofisk oil well blowout occurs in the North Sea.
— U.S. Supreme Court upholds the 1973 Endangered Species Act and stops construction of Tellico Dam.
— Allied Chemical Company and state of Virginia settle lawsuit over extensive contamination of James River.
— Federal Clean Air Act amendments require review of all National Ambient Air Quality Standards by 1980.
— Congress adds additional protection for Class I National Park and Wilderness air quality.

1978—Propylene gas explosion occurs in Tarragona, Spain.
— The *Amoco Cadiz* wrecks off the coast of France and loses 68 million gallons crude oil.
— Energy Tax Act creates federal ethanol tax incentive of 5 cents per gallon.
— Lois Gibbs and her neighbors form the Love Canal Homeowners Association.

- Robert Bullard begins investigating Triana, AL, where DDT had contaminated a stream. Environmental justice movement is born as a result.
- US Congress passes: National Energy Act, Endangered American Wilderness Act, and the Antarctic Conservation Act.

1979—Three Mile Island nuclear power plant loses coolant and partially melts down.
- IXTOC I oil well blowout occurs in Bay of Campeche, Mexico; large area contaminated.
- Earth First! organized by Dave Foreman, Howie Wolke, and Mike Roselle.
- Bean v. SWM lawsuit filed, challenging the siting of a waste facility.
- EPA suspends and later bans domestic use of 2,4,5 T, Agent Orange component.
- Appropriate Community Technology demonstration—one of the first alternative energy exhibitions on the national and international level—is held on Washington, DC, mall.
- Greenpeace vessel rams the Portuguese pirate whaler *Sierra* on the high seas.
- J.J. LaFalce and D.P. Moynihan propose "superfund" legislation.

1986—Chernobyl nuclear plant reactor number four shows unusual and (as it turns out) unsafe systems test at low power. A sudden, rapid growth in power output takes place, and when an attempt is made for an emergency shutdown, an unexpected and more extreme spike in power output occurs, leading to a reactor vessel rupture and a series of explosions. The disaster is an unfolding one.

1989—The *Exxon Valdez* ran aground, resulting in the second largest oil spill in US history, estimated at 500,000-750,000 barrels and listed as the 54th largest spill in history.

2003—Summer heat wave in Europe takes 35,000 lives.
- Earthquake in Iran kills 40,000.

2004—Hurricane Jeannine kills 3,037 people.
- Asian tsunami kills 250,000 people.

2005—Hurricane Katrina devastates New Orleans and kills 1,836 people.
- Earthquake in Pakistan kills 75,000 people.

2008—Myanmar cyclone kills 146,000 people.
- China earthquake kills 70,000 people.

2009—Global swine flu creates world-wide chaos and kills 11,800 people.

2010—BP oil spill from Deepwater Horizon oil rig disaster spills millions of gallons of crude oil into the Gulf of Mexico. (The disaster is still unfolding, and the oil will be damaging the environment and threatening life forms and humans in the Gulf for many years to come.)
- Haiti earthquake devastates several cities and kills over 300,000 people.
- Iceland volcanic eruption paralyzes European air traffic for several days.
- Mining accident in Chile buried 30 miners and riveted the world's attention to the sage of their survival and eventual rescue.

2011—Japan earthquake and tsunami in March 2011 devastates several hundred miles of populated coastal area. Thousands of people were killed and many are still missing. Four nuclear reactors were damaged and still leak radiation. Japan is confronted with long-term uncertainty and large recovery expense.

This list is much longer if we include all reported environmental events during the last several decades. It suffices to say, however, that we are surrounded by uncontrollable (and at times unexplainable) environmental events and disasters, some of which are caused by human activity. How much of this activity could be avoided by the production and use of solar energy technologies is a multi-lateral question, which take scientists a long time to address and even longer to answer.

Alternative energy manufacturing and use also has an impact on the environment, but since this is a fairly new industry (especially that of large-scale installations), the impact has not been fully qualified or quantified yet. Invested parties (manufacturers, users and politicians) can tend to dismiss environmental issues and related dangers, so we are trying to at least bring the issues in the open and start honest discussion on the most pressing subjects.

Present-day Environmental Status Report

"There are only two ways to live your life. One is as though nothing is a miracle. The other is as though everything is a miracle. —*A. Einstein.*

There are many miracles around us, but the biggest is the way Earth is positioned at the right place in the universe, with the perfect equilibrium between the sun's energy coming down to Earth and the amount of energy used and radiated back into space. It is an amazing phenomenon that has kept life on Earth going for many millennia. If Earth or its atmosphere were slightly out of place, or if the atmosphere were not composed of the proper amount and type of gasses and particles, life on Earth would be quite different, or non-existent. This is a marvelous and yet fragile combination, which needs to remain this way in order to preserve human life and provide us with the comforts we enjoy daily.

The radiation reflected back into space is regulated by the amount and types of gases and particles in the Earth's atmosphere. In the absence of atmosphere, the temperature on Earth would be about –16°C, and life would cease. Carbon dioxide (CO_2) in the atmosphere absorbs the reflected radiation and keeps some of its energy in the atmosphere, thus warming the Earth enough to keep life at its present levels. This equilibrium maintains the Earth's temperature at ~15-16°C on the average.

CO_2 is absorbed mostly in the 13-19 μm wavelength band, while water vapor (also abundant in the atmosphere) is absorbed in the 4-7 μm wavelength band. Therefore, most outgoing radiation (70%) that escapes into space is in the "window" between 7-13 μm. If that window were filled with other gasses, then the escaping energy would trapped. And if that window were packed with these harmful gasses, temperatures would rise and life on Earth would change drastically, as we have heard time after time lately.

Since our "comfort zone" lies in a very narrow temperature window, and our maximum and minimum temperature tolerance is only slightly wider, we depend on the atmosphere to keep us alive and comfortable. We also live in a borrowed place, on borrowed time, and are still learning how to live in it safely and yet productively. We are learning that many "natural" events (and others out of our control) could modify and even drastically change life as we know it. Just imagine a large meteor hitting Earth and pushing it slightly out of orbit. What would that do to gravity (if we survive the impact), the already narrow energy exchange window, the ocean tides, or the sunlight falling on Earth and its effect on living things?

During the last century, human activities have contributed significantly to changing the environment. Increase of "anthropogenic" gases into the atmosphere is one of the results. These gasses are the most dangerous for the environment because they absorb in the 7-13 μm wavelength range. They are the particularly harmful gasses—carbon dioxide, methane, ozone, nitrous oxides, and chlorofluorocarbons (CFCs). These gases disturb and even prevent the normal exchange and escape of energy and eventually lead to an increase in the temperature of the Earth and its rivers and oceans, thus changing the climate and weather patterns. There is now scientific evidence that CO_2 levels will double by 2030, which will most likely cause global warming, increasing temperatures 1~4°C on average. And this increase might accelerate with time, taking the global warming to unpredictable and unimaginable levels.

This accelerated warming trend will change wind patterns and rainfalls, and as a result may cause the interior of continents to dry out, while the Earth's oceans increase their levels. These harmful effects will only increase with time unless the trend of an increasing release of anthropogenic gases is interrupted.

Clearly, human activities have now reached a scale where they are impacting the planet's environment and its attractiveness to humans. The side-effects could be devastating, and technologies with low environmental impact and no greenhouse gas emissions are becoming increasingly important over the coming decades. Since the energy sector is the major producer of greenhouse gases via the combustion of fossil fuels, technologies such as wind and solar that can substitute for fossil fuels must be considered seriously and used broadly.

Rapid increase of CO_2 levels have been blamed for recent environmental changes and particularly for global warming. We don't know how much of this is true, and how much is just media hoopla, but the chart in Figure 8-1 cannot be disputed; it accurately presents the events of the last century and a half. On the other hand, there is evidence that these events are transitional and a part of the natural evolution of the Earth. Just like there was an Ice Age, there might be time for a "Boiling Age." During such a time, the daily temperatures would get so high that life on Earth might cease to exist all together. Just like our ancestors froze to death (and we don't know why exactly) our future generations might boil to death (and never know exactly why as well). Then new deposits of vegetation could be laid into the Earth's crust, to be used by the new generations millions of years from today. Far fetched? Maybe, maybe not. There is no absolutely solid evidence either way, just theories for now.

There are, however, numerous arguments against the man-made global warming theory, some of which make sense and deserve a second look. (NOTE: The author supports neither theory, nor is he involved in any related debates or activities. As a matter of fact, we believe that the glass is neither half full nor half empty. Instead, it is observed that all work is in progress and that things can go either way, depending on human activities and Moth-

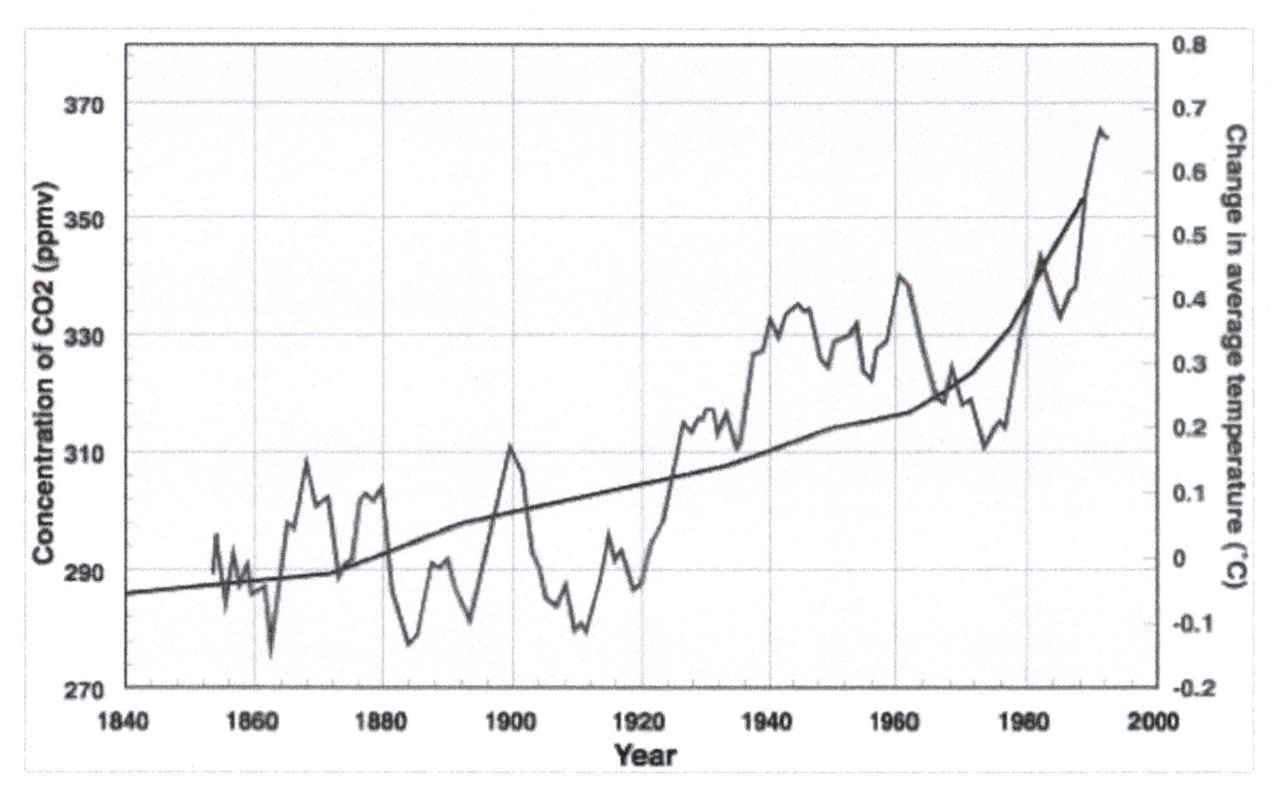

Figure 8-1. CO_2 and global temperature increase

er Nature's plans. There is a natural balance that man has not been able to figure out, nor maintain properly as yet. Nevertheless, environmental issues are serious matters, with lots of serious people involved on both sides of the debate who have some plausible pros and cons, so they all deserve equal time and respect.)

The views and facts supporting environmental decay are overwhelming, while the counter-views are not as well-advertised and are sometimes even laughed at. These counter-views are supported by 17,000 respected scientists who have signed a petition circulated by the Oregon Institute of Science and Medicine, claiming that there is no convincing scientific evidence that human release of carbon dioxide, methane, or other greenhouse gases is causing catastrophic heating of the Earth's atmosphere (1).

For fairness sake, here is a partial list of the global warming counter-claims:

1. Satellite readings in the troposphere (the most affected area) show no warming signs during the last quarter century. Land-based temperature measurement stations have been known to reflect the increase of temperature in their immediate locale, which could be affected by human presence (paved roads, houses etc.).

2. Global warming calculations and predictions are based on a computer model, using dubious "flux adjustments" methods, which can be significantly off at times.

3. The UN IPCC, whose reports are most often used in defense of global warming, says, "The Earth's atmosphere-ocean dynamics is chaotic: its evolution is sensitive to small perturbations in initial conditions. This sensitivity limits our ability to predict the detailed evolution of weather; inevitable errors and uncertainties in the starting conditions of a weather forecast amplify through the forecast. As well as uncertainty in initial conditions, such predictions are also degraded by errors and uncertainties in our ability to represent accurately the significant climate processes." Clear enough?

4. There was a period in human evolution (~ 5000-3000 BC) known as the "climatic optimum," when temperatures were much higher than today's. Civilization, nevertheless, progressed. People just wore less clothing in those days. Maybe it will benefit us; if the temperatures rises some, we will just have to wear less clothing too.

5. Efforts to quickly reduce the greenhouse effect, without having a solid understanding of the situation, might have side effects that we cannot even imagine, let alone foresee, right now. The different governments are approaching and handling these efforts in their different and, at times, clumsy and partial ways; the final cumulative effect of such efforts is even more complex to understand and control than the present situation.

6. Lately, new scientific data (2010 tests and observations) show that the ozone hole is shrinking, even though human activities and related damage are on the increase, so we cannot draw a parallel and don't know why exactly this is happening and/or if this process will continue in this direction.

We must agree that, as responsible citizens of Mother Earth, we must do our best to contribute to its well-being and long-term survival. The short-term solution is to live healthy, natural lives by considering the environment in everything we do in/with it first. Crushing the gas guzzlers, installing efficient light bulbs, and turning off the AC and heaters when not at home (sorry Spot and Kitty), might be a good first step in the right direction. Stopping coal and oil burning are the next steps, but big business won't agree to that, so a long battle awaits us. And, of course, using solar energy efficiently and safely is the best way to provide energy while reducing the pollution. This is also going to be a long and difficult road.

In conclusion, we all agree that the climate is changing somewhat and that this is a serious problem if it continues changing uncontrollably. We just don't agree entirely as to how fast it is changing and why. We all look for ways to find out what's going on and offer solutions to fix the problems. We have not reached an agreement on the methods; thus, the debate continues. This is the first step towards making progress, so with the skeleton now out of the closet, we can agree that we have problems to solve, bring them out into the open, and look for their proper solutions.

Energy Related Facts

Different fuels, and materials used as fuels, produce different amounts of energy and accordingly emit different amounts of pollutants. Some examples follow:

1. One MT (metric ton) of coal emits 745 kg. carbon as CO_2 and similar gases. One ton is 1,000 kg, so the coal to pollutant ration is 1.0: 0.75. Or, basically speaking, only ¼ of the coal is converted into energy, while the rest goes up in the atmosphere as CO_2 or is hauled away (as hazardous chemicals containing ash) to a county dump.

2. A typical coal-burning plant (1GW nameplate) generates:
   a. 3,500,000 tons of CO_2 per annum, which, combined with the waste from thousands of similar plants around the world, goes up in the atmosphere and contributes to the greenhouse effect, feeding global warming trends.
   b. 10,000 tons of sulfur dioxide (SO_2), causing acid rain that damages forests, lakes, and buildings, and forming small airborne particles that can penetrate deep into the lungs.
   c. 500 tons of small airborne particles, which can cause chronic bronchitis, aggravated asthma, and premature death, as well as haze-obstructed visibility.
   d. 10,000 tons of nitrogen oxide (NO_x), as much as would be emitted by half a million late-model cars. NO_x leads to formation of ozone (smog), which inflames the lungs, burning through lung tissue and making people more susceptible to respiratory illness.
   e. 700 tons of carbon monoxide (CO), which causes headaches and place additional stress on people with heart disease.
   f. 200 tons of hydrocarbons, volatile organic compounds (VOC) which form ozone.
   g. 170 pounds of mercury; just 1/70th of a teaspoon deposited on a 25-acre lake can make the fish unsafe to eat.
   h. 200 pounds of arsenic, which will cause cancer in one out of 100 people who drink water containing 50 parts per billion.
   i. 110 pounds of lead, 4 pounds of cadmium, other toxic heavy metals, and trace amounts of uranium. These are extremely toxic and deadly in very small quantities.

3. One MT (310 gal) of crude oil emits 584.5 kg. of carbon as CO_2 and similar gases. Oil burning power plants—tens of thousands around the world—have similar problems to those described for coal plants. The emitted gas types and quantities are different, but the end effects are the same—heavy environmental pollution and serious threat to human health.

4. 1 gal. of gasoline emits ~ 2.77 kg. of carbon as CO_2 and similar gases harmful to human life and the environment.

5. 1 cubic meter natural gas emits 0.5 kg. of carbon as CO_2, as well as similar gases.

6. Biofuels (trees and bushes) emit 50% of their weight in carbon-based gases during burning. The majority of the African population still uses these biofuels for their daily needs, so we need to find a way to replace them with renewables; solar is the fastest, cheapest, and cleanest way to do this.

7. 1 ton of natural gas floating free in the atmosphere traps as much global warming-causing radiation as 20 tons of CO_2 in the same place.

8. Fine-particle pollution from US power plants cuts short the lives of over 30,000 people each year. In more polluted areas, fine-particle pollution can shave several years off its victims' lives. Hundreds of thousands of Americans suffer from asthma attacks, cardiac problems, and upper and lower respiratory problems associated with fine particles from power plants. The elderly, children, and those with respiratory disease are most severely impacted by fine-particle pollution from power plants. Metropolitan areas with large populations near coal-fired power plants feel their impacts most acutely; their attributable death rates are much higher than in areas with few or no coal-fired power plants.

Power plants outstrip all other polluters as the largest source of sulfates, the major component of fine-particle pollution in the US. Approximately two-thirds (over 18,000) of the deaths due to fine-particle pollution from power plants could be avoided by implementing policies that cut power plant pollution containing sulfur dioxide and nitrogen oxide. Some progress in this area has been made already, and more activities and regulations are planned for the near future in the US, but the rest of the world is not even close to this; the grand scale pollution will continue for the foreseeable future.

Environmental and Global Warming Issues

Pollution from energy generation accounts for over 50% of all air, soil, and water pollution today. Electric power production from coal and oil, the largest contributor to global warming, is blamed for environmental damages. EPA estimates that fossil fuel-based power generation has an environmental health cost of 10.5 cents per kilowatt hour—almost as much as it's actual cost. This means that its actual cost is double its present value, half of which is adverse health effects. And we, the consumers, are the ones who end up paying for it with our taxes and our health.

Environmental buzzwords are all around us: anoxic waters, ocean deoxygenation, climate change, global warming, global dimming, fossil fuels, rise in sea level, greenhouse gas, ocean acidification, shutdown of thermohaline circulation, conservation, species extinction, pollinator decline, coral bleaching, coral reefs extinction, holocene extinction, invasive species, poaching, endangered species, dam impacts, environmental degradation, eutrophication, habitat destruction, environmental health dependency, air quality, asthma, electromagnetic fields, electromagnetic radiation, indoor air quality, lead poisoning, asbestos poisoning, sick building syndrome, genetic engineering, genetic pollution, genetically-modified food controversies, intensive farming, overgrazing, irrigation, water depletion, irrigation-water pollution, water pollution, eonoculture, environmental effects of meat production, slash and burn, pesticide drift, plasticulture, land degradation, land pollution, desertification, soil conservation, soil erosion, soil contamination, soil salination, land misuse, urban sprawl, habitat fragmentation, habitat destruction, ozone depletion, CFC, light pollution, noise pollution, visual pollution, nonpoint source pollution, point source pollution, rain water pollution, acid rain, marine pollution, ocean pollution, ocean dumping, oil spills, thermal pollution, urban runoff, water crisis, marine debris, ocean acidification, ship pollution, wastewater contamination, fish kill, algal bloom, mercury in fish, air pollution, smog, tropospheric ozone, volatile organic compound, particulate matter, sulphur oxide, nanotechnology, nanotoxicology, nanopollution, nuclear issues, nuclear fallout, nuclear meltdown, nuclear power, radioactive waste, overpopulation, burial, water crisis, overpopulation in companion animals, tragedy of the commons, consumerism, consumer capitalism, planned obsolescence, over-consumption, overfishing, blast fishing, bottom trawling, cyanide fishing, ghost nets, Illegal fishing, unreported and unregulated fishing, shark finning, whaling, logging, clearcutting, deforestation, Illegal logging, mining, acid mine drainage, mountaintop removal mining, slurry impoundments, toxins, chlorofluorocarbons, DDT, endocrine disruptors, dioxin, toxic heavy metals, herbicides, pesticides, toxic waste, PCB, bioaccumulation, biomagnification, waste, e-waste, litter, waste disposal incidents, marine debris, medical waste, landfill, leach rate, recycling, incineration, Great Pacific Garbage Patch, resource depletion, exploitation of natural resources, overdrafting, energy conservation, renewable energy, efficient energy use, renewable energy commercialization.

Quite an array of words, conditions, and situations, isn't it? So how do we know what is good and what is bad behind each of these? Where do we start, once we have determined that there is a problem? What are the most important issues, and how are these interconnected? What should we do to help the environment, ourselves, our way of life, and that of future generations? Do we care enough to do the right thing ASAP, or should we procrastinate some more, hoping that the issues will just go away?

"You can always count on Americans to do the right thing—after they've tried everything else," Winston Churchill reportedly said a long time ago. We believe that Americans will not fail this time—and one can even see the signs of their trying to do the right thing today—but everything else has not been tried as yet. It might take a generation or two, but Americans will do the right thing. Let's just hope that it is not too late by then.

Environmental Damage Examples

On December 5, 1952, the residents of London, England, awoke to the dawn of a five-day reign of death. A temperature inversion had trapped the coal smoke from the city's furnaces, fireplaces, and industrial smokestacks, creating a "killer fog" that hovered near the ground. People began to die from respiratory and cardiopulmonary failure. Not until the weather system that had trapped London's pollution finally loosened its grip, and the soot-filled air cleared out, did death rates return to normal. The end of the episode saw more than 3000 dead, and many more people suffered the coal smoke aftermath. It is not hard to imagine such a scenario developing again, and over larger-populated centers around the world, which is something we are just not prepared for.

Driving North from Hong Kong into mainland China, you'll notice dozens, if not hundreds, of large factories bellowing black smoke into the air and discharging fizzling, green-brownish liquids into the ground. Walking down the streets of major Asian cities, you'll notice most people wearing masks over their mouths and noses. This unprecedented environmental pollution on such a large scale started some 20-30 years ago, and it is getting worse as we speak. It coincides with the tremendous growth of the Asian industrial complex at the expense of local and world environments and the health of the people working and living close to the polluters.

But you don't have to go to Asia to see examples of man-made hazmat dumps and generators of poison gasses. Drive by any large paper, cement, or chemical factory, or any coal-burning power plant, in the US or Europe and you'll see the signs of pollution: heavy smoke in the air, dead trees, dusty fields, green and brown bubbling lagoons, and a number of other signs of man's intervention and recklessness. Yes, this is the mark of the fast-growing capitalist society, which we all must agree has to change before it is too late.

Just mentioning the word Chernobyl sends chills through the spine of every Russian or European who lived through that nightmare. Of course, many did not survive the disaster, while many more are still suffering, and many generations will go on bearing the tell-tell signs of this larger-than-life nuclear disaster. With hundreds of nuclear plants operating well past their life expectancy, we won't be surprised if we witness more such accidents in the near future.

And let's not forget the *Exxon Valdez* and Deepwater Horizon nightmares, other forms of man-made pollution related to energy production which had detrimental effects to the environment and life forms in it, including people.

The 2010 collapse of the coal mine in Chile, where 30 miners made the world news surviving for many days buried underground, and the Japanese nuclear disaster where many people were not that lucky, are yet other serious reminders of the dangers in the conventional energy field. The people involved in these events will never forget the day and hour when the Earth shook and took its victims. We shouldn't forget either and should consider replacing these dangerous technologies with safer and friendly wind and solar power generation.

Each of these events might be a small thing in itself in the grand scheme of things, but connected together they have a profound and most negative effect on the environment and human life in the long run. Making this connection, however, is a very complex undertaking, with many variables and unknowns, thus the split among scientists.

Solar cells and modules manufacturing, transport, and recycling processes also emit large amounts of CO_2 and other gasses into the atmosphere. They also fill the county dumps with rusting and harmful materials, so we will take a closer look at these phenomena.

Figure 8-2 is an example of a failed solar power plant in the US, one of several that were supposed to bring us a bright energy future way back in the 1980s and 1990s but instead created a new phenomenon, which we call "solar junk yards." There is also the abandoned

Figure 8-2. Carrizo Plain PV Power Plant

Deker Lane PV power plant in Austin Texas, which was supposed to be a demonstration plant. It did not survive the test of time and the demonstration failed, with the site shutting down some time ago. Why did these large and high-visibility installations fail? Was it the technology, the installation, or the operation that failed? Did the failed equipment cause any damage to the surrounding environment beyond the visual pollution and the rusting metal parts? Several other solar projects failed within 10 years of installation in the 80s and 90s, either because the technology was not efficient and reliable enough, or because of poor planning and management. Readily available and cheaper conventional energy sources did not help the matter much either.

There is no question that all industries and energy-generating technologies use natural resources and generate waste by-products. Renewable energies (wind, solar, etc.) are no exception. The only question is how to estimate, control, and reduce the negative effects. We don't have the answers to most of these questions either, but hopefully this text will address some of the issues and bring them out of the closet for discussion by all interested and affected parties.

CO_2 Generation and Issues

Of all environmental culprits, CO_2 is the most notorious. It is the evil of all evils, and the reason for everything bad happening to the environment—or at least this is the present-day consensus. So let's take a look at CO_2 as the all-encompassing and most urgent concern as far as the global warming goes. We need to examine it also because PV installations are measured in terms of tons of CO_2 emissions saved, as compared with those generated by fossil fuel plants of the same size during a certain period of time.

A gallon of gasoline, which weighs about 6.3 pounds, could produce 20 pounds of carbon dioxide (CO_2) when burned. Impossible? Yes, possible! The C+O2 combination, expressed in CO_2 units, is a wicked one with some very special properties. When gasoline or other carbon-containing fuels burn the carbon and hydrogen separate, the hydrogen combines with oxygen to form water (H2O), and carbon combines with oxygen from the surrounding air to form carbon dioxide (CO_2).

A carbon atom (see Figure 8-3) has an atomic weight of 12, and each oxygen atom has atomic weight of 16, giving each single molecule of CO_2 an atomic weight of 44 (12 from one carbon atom and 32 from 2 oxygen atoms). So, to calculate the amount of CO_2 produced from a gallon of gasoline, the weight of the carbon in the gasoline is multiplied by 3.7 (44/12). Since gasoline is about 87% carbon and 13% hydrogen by weight, the carbon in a gallon of gasoline weighs 5.5 pounds (6.3 lbs. x .87). Then we multiply the weight of the carbon (5.5 pounds) by 3.7, which equals 20 pounds of CO_2 produced by a gallon of gasoline. Amazing!

Thus, consider the case of an average car that burns one gallon of gasoline every 10-15 miles (or every 5 miles if we drive a Hummer, SUV, or RV when driving around the city). Every 10-15 miles, we leave 20 lbs. of CO_2 behind us, which goes up in the atmosphere to accelerate the global warming process. Every 100 miles driven in

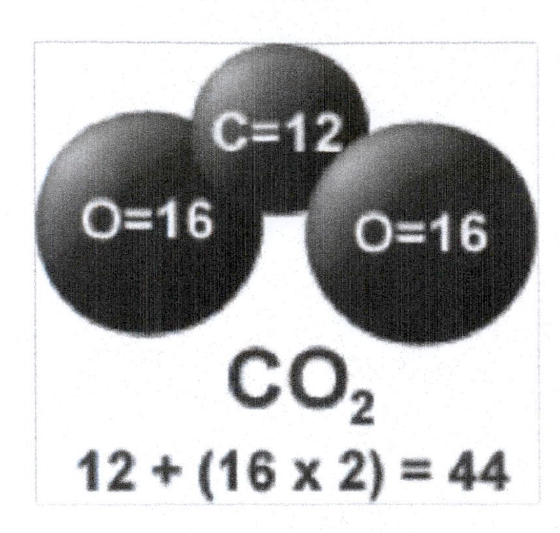

Figure 8-3. CO_2 molecule

the Hummer or the family RV generates 400 lbs. of CO_2, not a commendable footprint. A summer vacation trip of 1,000 miles will leave 4,000 lbs. (over 2 tons) of CO_2 footprint behind our happy family's SUV. And 100,000 families with similar summer trip plans will load the atmosphere with an additional 400,000,000 lbs. (200,000 tons) of CO_2 within weeks... 1 million families...you get the picture.

How about a 2 GW coal-burning power plant bellowing dense clouds of smoke all day long, or a cruise ship navigating aimlessly around the Gulf of Mexico burning thousands gallons of diesel? Multiply that by thousands of such activities and you'll see that millions of pounds of harmful CO_2 gas are emitted into the atmosphere every day. Not much has changed, even after all the talk about the environmental disasters around us. How long could this continue before we run out of fossils and suffocate ourselves to death?

Environmental and Safety Concerns of PV Manufacturing and Use

Environmental damage in its broad sense is a misnomer, because the environment will survive in one form or another just like it has on several occasions in the past. Every time it has recovered and evolved even better at the expense of the living things in it at the time. We, however, may not like the upcoming changes and may not even survive them this time around if this pattern continues. Since the serious environmental changes we are presently experiencing might have devastating effects on the human race, we need to consider the preservation of the environment (in its present form and shape) as a prerequisite for sustaining life on Earth, and a vehicle to our wellbeing.

Human activities have been blamed for recent negative environmental changes and, since there are no human activities that are absolutely pollution-free and totally environmentally friendly, there will always be some environmental effects and damages brought about by them. We just must be fully aware of consequences and take measures to balance the good and bad activities.

The manufacturing of PV cells and modules, as well as their transport, installation, operation, and subsequent end-of-life decommissioning, transport, and recycling are no exception. The safety and environmental concerns here start from the moment we dig a shovel full of sand to make the silicon material from which we will then make the solar cells and modules. Having a better understanding of the processes and their effects will bring us closer to understanding and evaluating the related environmental damages, which will allow us to find ways to control them.

MG Silicon Production

Let's take a brief look at the environmental effects of manufacturing and using solar cells and modules, starting with silicon-based technologies. The silicon solar cells manufacturing process starts when a mountain of sand somewhere in the world is bulldozed, transported, melted, refined, and solidified as chunks of silicon. The huge amounts of sand excavated from these locations, as well as the heavy traffic in and around these areas, have caused extensive land erosion, excessive CO_2, and dust pollution, as well as other serious damage to environments and wild life in the affected areas. Millions of gallons of diesel and gasoline are burnt in order to get the sand out, load it on trucks or train cars, and transport it to the melting facility, where huge amounts of electric energy and millions of tons of coal and other carbon-containing materials are burnt in the huge furnaces to melt and purify the silicon in the sand and convert it into metallurgical grade (MG) silicon. And this is only the beginning.

The melting process, although relatively well controlled, uses crude equipment, uses a lot of energy, and is very expensive due to the huge amounts energy and material to be processed before, during, and after the melting process. Tons upon tons of different additional materials (chemical additives) are also bulldozed into the furnace, melted, and then removed and disposed of later. These additives were also dug out, melted, boiled, refined, packaged, and transported to the melting facility before being shoved into the furnace. When the silicon melting process is complete, after several hours (even days), the melt is discharged, cooled, and crushed into small chunks.

When the dust settles, which happens only after the operations have been discontinued, we have sent millions of tons of CO_2, SO_2, NO_x and dust particles into the atmosphere. Day after day, the non-stop process goes. Many such excavating operations and melting facilities are in full swing around the world as we speak, while many others are planned for the near future, when they'll add many more millions of tons of harmful gases to the atmosphere and many tons of solid waste to the local dump sites.

Silicon Wafers and Solar Cells Production

Thus produced, MG Si is loaded on trucks and railroad cars and delivered to poly silicon production plants to be purified and converted into solar grade silicon (SG Si). The SG Si refining plants are actually huge chemical factories where a number of solid, liquid, and gaseous chemicals (most of which are hazardous, toxic, and poisonous) require special handling, processing, transport, and disposal. Here, all these materials, chemicals, and gasses are mixed, boiled, baked, sifted, crushed, liquefied, gasified, and solidified in a never-ending action. There a number of liquid chemicals that are quite expensive in this process, so they are recycled and otherwise reused via complex distillation, filtration, and other chemical process, all of which use huge amounts of electric power, cooling water, and additives. Millions of cubic tons of CO_2 and other toxic liquids and gasses are the by-products of these processes. Some are difficult to recycle or capture, so they are just freely exhausted or disposed of into the environment without any treatment, as in many Asian polysilicon plants.

Organic chemicals such as silane, dichlorosilane (DCS), trichlorosilane (TCS), silicon tetrachloride (STC), and many others are mixed, heated, evaporated, condensed, and transported, along with many inorganic liquids and gasses such as HCL, HF, HNO_3, H_2O_2, etc. Some of these are also dumped in the soil or vented into the air, where they mix within the soup of other gasses. The resulting mixture of organic and inorganic compounds sometimes stagnates over population centers as it becomes part of the atmosphere to accelerate global warming.

Hazardous liquid by-products, some dangerously corrosive and even pyrophoric (self-igniting), are created, transported, and processed along the complex process sequence. All of these require special handling, placing chemical and fire safety on the top of management's list of priorities in most facilities. The personnel are trained in the proper handling of these chemicals, including emergency procedures, for all eventualities. In addition, proper equipment and building designs/ procedures are used throughout the facilities, which include ventilation, electrical system safety, static electricity control, control of all ignition sources, personal self-contained breathing apparatus use, and, of course, a no smoking ban (which is not fully enforced in most Asian facilities).

At the end of the SG Si purification process, the silicon rods are crushed again for ease of transport and sent to different facilities for the manufacture of solar wafers, cells, and modules. At these facilities the SG Si chunks are melted in special, high-temperature furnaces in the presence of different gasses, where they are shaped into long cylindrical ingots (single crystal silicon) or square blocks (poly crystalline silicon). The rods and blocks are sliced into thin wafers on special saws, after which they are ready for processing into solar cells.

Large amounts of electric power, chemical additives, slurries, liquids, and gasses are used during the melting, shaping, cutting, slicing, and cleaning operations. The slurries and cleaning liquids have to be processed and disposed of eventually, which is a hazardous undertaking as well.

Thus produced, wafers enter the solar cells manufacturing process where they undergo a number of thermal and chemical operations, and where a lot of electric energy and thousands of gallons of different chemicals, liquids, and gases are used and disposed of. These amounts and expenses also need to be taken into consideration in the overall wafers-cells-modules manufacturing budget.

The work safety in these facilities, where hi-tech equipment and procedures are used and where strict processing and safety standards are currently enforced, is usually acceptable. The disposal of gases and chemicals in many cases, however, is problematic.

The actual solar cell manufacturing and modules assembly process is executed in clean areas where, in most cases, safety and environmental standards are the norm. Nevertheless, there is still a lot of energy and chemicals used in this process, which also needs to be considered in the energy balance estimates.

Thin-film Modules Production

Thin-film PV (TFPV) technologies leave their signature mark on the environment as well. The TFPV manufacturing process starts in a dark, dusty, or muddy, mine somewhere in Asia or Africa. Low-paid men and women work in deplorable conditions, digging, dragging, loading, processing, and transporting the precious and often toxic metals. Health problems and even death are common occurrences among these workers, but the actual facts are not widely publicized.

The freshly mined raw materials are transported to refining facilities to be refined into useful form and shape. The raw materials are exposed to a number of complex mechanical, chemical, and electro-chemical operations in order to separate and purify the different metals. Dangerous gasses and chemicals are used in these processes too, and some of the resulting equally dangerous and toxic gases and chemicals are vented into the atmosphere or dumped close by. Some of the mining and refining operations look, feel, and smell like they have just jumped out from the pages of Dante's *Inferno*. Humans working there, in some of the worst cases, look like they belong on the same pages—not a pretty picture.

Thus refined metals are shipped to TFPV module manufacturing facilities for processing into modules. In sharp contrast with the above described "holes in the ground" mining and refining operations, the TFPV manufacturing facilities are sparkling clean, modern, semiconductor-type fabrication plants, equipped with the latest state-of-the-art equipment. Well-paid engineers and technicians follow well-defined processes executed in clean room environments under the strictest of process controls. The safety, efficiency, and productivity of these facilities is the envy of the PV industry, and we have only praises for their setup, operation, and quality control.

Here, the refined and still toxic materials are placed in special chambers, where they are deposited in the form of very thin films onto glass substrates (panes) under near-ideal process conditions. The deposited films on the first glass pane are then covered with a second pane and encapsulated between them. The resulting TFPV modules are flash-tested and packed for shipping. This is a most efficient, cost effective, and high-volume production process that is leading the PV industry. These facilities are glimpse of the future. It is how the PV industry of the 22 century will be set up and operated, which is a good thing.

Nevertheless, although the metals used in TFPV modules manufacturing are by-products of the mining and production of other metals, they are still scarce, in addition to being toxic. These are serious issues that the TFPV manufacturers are not willing to discuss openly, so we hope that we can bring the issues out of the closet for an honest discussion.

Another issue of environmental concern worth mentioning here is the fact that most TFPV modules are frameless. That is, the active thin film structure (CdTe or CIGS) is deposited on a glass pane, covered by encapsulant and another glass pane. So, the thin film structure is protected from environmental attacks by a thin film of plastic, exposed to the elements at the open edges of the modules. Since no plastic material can last very long under harsh desert conditions, the thin edge seal would deteriorate under the relentless IR and UV bombardment and would allow moisture and harmful environmental gasses to attack and destroy the thin films. This will not only shorten the useful life of the TFPV modules, but might also result in environmental contamination, damaging the air, soil, and water table.

While this might not be a serious problem in small size installations, the millions of TFPV modules installed on thousands of acres of desert land in the newly proposed large-scale PV power plants pose a potential danger to the environment and life in the affected areas. More preliminary work must be done before a full proliferation of these unproven PV technologies in the US deserts occurs.

It was discussed in previous chapters that frameless modules are more prone to degradation due to ingress of moisture and environmental gases into the glass/glass modules (front and rear glass covers) which also retain excess heat more than modules with plastic or metal back covers. The excess heat could result in a number of unwanted effects (delamination, overheating, hot spots etc.).

Some TFPV modules are both frameless and of glass/glass construction; therefore, the active thin films inside are more prone to degradation and power loss. The frameless structure also facilitates the escape of toxic materials from the module in case of mechanical disintegration or chemical decomposition.

As discussed in the previous chapter, there are ways to improve the edge sealing of the TFPV modules. Some manufacturers actually wrap up the thin film structure in a sheet of protective material, similar to that used in c-Si PV modules. This completely seals the thin film structure and the module edges in one continuous envelope, thus preventing moisture from entering the module and attacking the thin film structure. On top of that, some manufacturers add a metal frame around the module edges which is "glued" to the glass panes by edge sealer, thus providing a solid, several-layer barrier which protects thin films for a long time regardless of environmental attacks.

Most TFPV manufacturers, however, continue making frameless all glass modules in order to keep costs down. Price vs. healthy environment? This doesn't sound very safe nor green, does it?

Renewable, Green and Safe?

Speaking of safe and green, solar energy proponents and manufacturers claim their PV products to be "renewable," "green" and "safe." This means that from an environmental and safety point of view the related technologies and their components are nearly perfectly clean to manufacture and use. But how much of that is really so?

Here is a brief look into the subject:

1. c-Si silicon materials.

Basically speaking, silicon is one of the most abundant materials on Earth—sand. There is more sand around us than anything else, so why even bother bringing it into this discussion? Simply because not all sand is suitable for solar cells manufacturing, and because a lot of pollution is generated during the mining, transport, and processing of silicon. In one estimate, over 7.0 tons of CO_2 are generated in producing just one ton of SG silicon. Then that much, or more, is produced during the subsequent steps of refining the silicon and converting it into PV cells and modules.

Most sand contains so much dirt and other unwanted impurities that using it for such a purpose would be like making a potato soup with one potato mixed with a bucket of dirt. Around 25% of the Earth's crust consists of silicon, which is not pure but is mostly in the form of "clay" and other alumino-silicate materials. Pure silicon dioxide (SiO2), a.k.a. silica, from the Latin "silex," is a mineral best suited for making SG silicon. Silica is also quite abundant, making up ~12% of the Earth's crust. So the best type of sand for our purpose is "silica sand," which contains a high percentage of silica. Silica sand is formed from the weathering of silicate minerals and rocks, as part of a natural cycle. Naturally occurring silicate materials in contact with CO_2 and water are eroded over time into Silica and CaCO3. All we have to do then is find the place, dig out the silica sand, and extract it from the CaCO3 and other ingredients with which it is mixed. The purer the sand, the less effort and energy it takes to convert it into useful SG silicon. There are a number of "pure" silica sand mines around the world, but the purity varies significantly from place to place.

Desert "sand" is often misunderstood to be the kind of Silica sand that one finds on beaches, and the kind we can use for solar cells. In fact it is not even close, for it is nothing more than dried earth (clay). One way to tell them apart is to spit into the palm of your hand, add a small quantity of the "sand" and then rub it. If it is clay, it will turn into brown mud, but if it is silica, it will just become damp and will clump together. Thus, as the Sahara Desert advances because of lack of rainfall, more sand appears, but really it is just earth turning into dust.

We've seen a number of plans to convert the Sahara's desert sand into solar cells, so we wonder what these engineers and scientists are thinking about, or if they know something we don't and will be having the last laugh. We will just have to wait and see. The few places on Earth where pure silica sand can be found are often in isolated areas, meaning that unless the silicon foundries are built nearby, the sand has to be transported, another great expense.

Silica and silica sands are also widely used for the manufacturing of many everyday products such as glass, optical fibers, diatomaceous earth, cement, and ceramics. They are also used as additives in foods, not to mention the use of silica sand in making millions of semiconductor-type silicon wafers. There is a lot of competition for it, and yet we don't foresee any major shortages any time soon, although prices will certainly fluctuate with overall demand and energy costs going up and down.

So how renewable, green, and safe is silicon? Renewable? Yes, to a large extent. Green? Yes, except for the large quantity of materials and energy used to produce it and the accompanying bi-products emitted and released. Safe? Yes, if properly manufactured and used.

2. Silver, copper and other metals.

Huge quantities of silver, aluminum, copper and other metals, as well as plastics and many chemicals are needed to produce millions of PV modules. Large amounts of silver metal are used to provide good ohmic contact between the metal grid on top of the cell and the interconnecting wires. Silver is also used for reflective backing for mirrors in thermal solar plants. According to the VM Group in London, over 1,000 tons of silver have been projected for making PV modules in 2011. This is over the 1.0 million kilograms of silver used today, and the amount is projected to triple by 2016, to nearly 3,000 metric tons of silver used for making PV modules worldwide every year. If we add that much more silver metal for coating heliostat mirrors, we end up with some very large numbers. So the question here is, "How much silver will be left in the world in 10, 50 and 100 years if we use it at this pace?"

Silver is a precious metal, the price having gone from $4.00/ounce several years ago to over $40.00 today. Prices are expected to go higher with time. What will that do to PV module prices? Similarly, prices of copper and aluminum metals have sky rocketed lately, and although there are large deposits of these left on Earth, the increased prices will play a significant role in PV modules' use on the world's energy markets. Therefore, although there are significant amounts of these metals around the world, they cannot be considered "renewable." They are, however, non-toxic, so they could be considered "green," and "safe."

3. Cadmium

Cadmium (Cd) is generally recovered as a by-product of zinc concentrates. Zinc-to-cadmium ratios in typical zinc ores range from 200:1 to 400:1. It is used for making NiCd batteries, electroplating, lasers, electronics, paints, and most recently in thin film PV modules.

In January 2010 USGS estimated ~600,000 tons of cadmium reserves worldwide (calculated as a percentage of available zinc reserves), of which the world mines and uses ~19,000 tons annually. If USGS is correct, in 32 years there will be no more cadmium.

Cadmium is a toxic heavy metal, which displaces zinc in many metallo-enzymes in the body, so cadmium toxicity can be traced to a cadmium-induced zinc deficiency. It concentrates in the kidneys, liver and other organs, and is 10 times more toxic than lead or mercury. Inhaling cadmium-laden dust leads to respiratory tract and kidney problems which can be fatal. Ingestion of significant amounts of cadmium causes immediate and irreversible damage to the liver and the kidneys. Japanese agricultural communities consuming Cd-contaminated rice developed itai-itai disease and renal abnormalities, including proteinuria and glucosuria.

Cadmium is one of several substances listed by the European Union's Restriction on Hazardous Substances (RoHS) directive, which bans certain hazardous substances in electrical and electronic equipment but allows certain exemptions and exclusions from the scope of the law. In February 2010 cadmium was found in an entire line of Wal-Mart jewelry, which was subsequently removed from the shelves. In June 2010 cadmium was detected in paint used on McDonald's tumblers, resulting in a recall of 12 million glasses.

Compared to other serious toxins, cadmium is more dangerous as it accumulates, and is not disipated. Even a negligible dose of cadmium in the air or water, inhaled or ingested daily, will accumulate to eventually reach toxic levels, causing cancer or organ failure.

USGS says, "Concern over cadmium's toxicity has spurred various recent legislative efforts, especially in the European Union, to restrict its use in most end-use applications. If recent legislation involving cadmium dramatically reduces long-term demand, a situation could arise, such as has been recently seen with mercury, where an accumulating oversupply of by-product cadmium will need to be permanently stockpiled." (12)

So how "renewable," "green" and "safe" are the cadmium-based PV technologies? From the above facts we see that cadmium is not a "renewable" commodity *per se*, but we don't foresee a shortage during the next 3 decades. It is, however, far from "green" or safe."

4. Tellurium

Tellurium (Te) metal is produced by refining blister copper from deposits that contain recoverable amounts of tellurium. Relatively large quantities of tellurium are also found in some gold, lead, coal, and lower-grade copper deposits, but the recovery cost from these deposits is too high to be worthwhile. Tellurium is used mostly in making steel and copper alloys to improve machinability, in the petroleum and rubber industries, and for making catalysts and some chemicals. One of the rarest elements in the Earth's crust, it is found in considerable quantities as a secondary metal in mining operations.

The world produces 100-200 tons of tellurium annually, and while its total availability is uncertain, we do not foresee a shortage anytime soon. Tellurium is used as cadmium telluride in manufacturing thin film PV modules. It is mildly toxic material; nevertheless, utmost precaution must be taken when handling its pure form or its basic compounds as contained in the PV modules.

Tellurium and its compounds are known to cause sterility in men working with tellurium-containing materials, even under strict monitoring conditions such as in semiconductor fabs and hard disk manufacturing operations. Although there is significant amount of tellurium around the world and it cannot be considered "renewable," we do not foresee a shortage anytime soon. Due to its toxic properties it is not "green," or "safe."

5. Selenium

Selenium (Se) is a non-metal that is chemically related to sulfur and tellurium. It is obtained by mining sulfide ores, and is used in glassmaking, metallurgy, and pigments. While toxic in large amounts, trace amounts of selenium are needed for cellular function in most animals. Selenium toxicity was noticed first by doctors who found increased sickness among people working with it. A dose as small as 5 mg per day can be lethal, causing selenosis. Symptoms include a garlic odor on the breath, gastrointestinal disorders, hair loss, sloughing of nails, fatigue, irritability, and neurological damage. A number of cases of selenium poisoning of water systems were attributed to agricultural runoff through normally dry lands.

Selenium quantity and price depend on mining operations of other metals and minerals, but we don't foresee a shortage anytime soon. Its use is estimated at 1,500-2,000 tons annually. Though there is a significant amount of selenium worldwide, it cannot be considered renewable, green, or safe.

6. Arsenic

There are over 1 million tons of arsenic (As) worldwide, of which 54,000 tons are extracted annually, mostly in the form of arsenic sulfur compounds. Another 11 million tons might be recovered from copper and gold ores. The main use of metallic arsenic is for strengthening alloys of copper and especially lead used in automotive batteries. Arsenic has proven toxic and poisonous qualities. Although there is significant amount of arsenic around the world it cannot be considered "renewable." Due to its toxic properties it is not "green," or "safe."

7. Gallium and Indium

These and other hard-to-find-and-isolate mildly toxic metals are also presently used in significant amounts in thin-film PV modules manufacturing processes. These elements are rare, so they cannot be considered "renewable." They are mildly toxic but could be qualified as "green," or "safe," if properly produced and used.

8. EVA and other plastics

A number of organic materials (plastics) are used throughout the entire PV module manufacturing process. They are too varied and complex to qualify or quantify in this text. Since their production is based mostly on fossils (extracts from crude oil, coal and such) they are not "renewable," but we don't foresee shortage anytime soon. Because they are manufactured with the help of poisonous solvents and other toxic materials, we'd have a hard time classifying them as green or safe. They are, however, safe enough to work with, following basic precautions like wearing gloves and masks. Due to outgassing, their safety in long-term operations has been questioned and needs more research—especially in light of the large-scale PV installations in the deserts, where organics are most vulnerable and unpredictable.

One can easily see that the complications here are endless, and interwoven so much so that we could write an entire book on the properties and effects of the different combinations and permutations of the availability, mining dangers, toxicity, adverse short- and long-term environmental and health effects, improper field use, etc. of the materials and components used in making and using PV cells and modules.

Many questions remain to be answered in these areas, so we must insist that these issues are brought out for honest discussion by all parties involved, including manufacturers and proponents of TFPV technologies.

How do we justify the claim that PV technologies are renewable, green, and safe, if some of the materials used to manufacture them are not? Should we not qualify and use the different PV technologies according to the type and amount of non-renewable, not green and unsafe materials used during their manufacture and use?

Note: At the time of this writing, several manufacturers have applied for "non-toxic" certification of their PV modules. The trend will continue until it becomes an industry-wide standard, and we envision different types and levels of "toxicity" assigned to different types of PV modules in the not-so-distant future.

This process will take awhile, but it is unavoidable, because the consumers must be fully aware of how renewable, green, and safe the products they use are and what to expect in the long run.

Energy Generation and the Environment

Comparing conventional energy sources with the solar technologies of today is not a straightforward process, and there is no line that can be drawn between them, but there are definite differences, plus and minuses. Although our objective is not to compare the different technologies from an environmental point of view, we will discuss the characteristics and draw some conclusions, to open the subject to an honest discussion. This may not happen soon, but it will happen eventually, make no mistake about it!

Conventional Energy Sources

The activities related to electric power production (coal mining, crushing, oil drilling, pumping, transport, and power-generation processes) no doubt contribute to an extent to environmental pollution, global warming, acid rain, and disease migration. Increased production and use of fossil fuels, especially coal, has severe local and regional impacts. Locally, air pollution already takes a significant toll on human health. Acid rain precipitation and other forms of air pollution degrade downwind habitats—especially lakes, streams, and forests—and they are damaging crops and buildings.

One recent study warns that in the absence of sulfur abatement measures, acid depositions in parts of South Asia could eventually exceed the critical toxicity load for major agricultural crops by a factor of 10. Without the use of the best available technology and practices, coal mining leads to land degradation and water pollution, as does the disposal of hazardous coal ash.

On a global level, increased burning of fossil fuels will mean an accompanying rise in greenhouse gas emissions, along with the potential adverse impacts of global warming and other climate changes. Nuclear fuel, too, has obvious environmental costs associated with materials production and disposal, although nuclear power produces virtually none of the air pollution and carbon dioxide discharges of fossil fuels during power generation. Its drawback is the thousands of tons of waste materials stored at all nuclear plants worldwide. That amount (70,000 tons in the US alone) increases daily and begs the question, "How long can we keep piling such large quantities of deadly materials in our backyards?" And, of course, the Japanese nuclear disaster brings another set of questions, which affect human life in a most direct, serious, and permanent way.

Natural gas has been hailed as a "clean" energy source, but a recent study (April 2011) by a prestigious scientific group claims (although there were no scientific data to support the claim) that the production of natural gas creates a much larger quantity of harmful gases (mostly due to raw methane) than any other industry. We already know that raw methane gas released in the atmosphere is worse than CO_2 in accelerating the green house effect, but we didn't know that it is released constantly in such large quantities as part of the natural gas production process. It appears that the natural gas industry is trying to discredit the findings without giving the scientific community a chance to verify the data and take the necessary measures. We are quite familiar with this trend; we have seen it before. We only hope that the scientific community and US regulators will step in to take a closer look at this potential disaster. Judging by other such incidents in the past, however, we cannot be sure that this will happen any time soon, if ever.

Shale gas is also raising our hopes for cheap energy independence. It, too, brings several dangers and creates a serious environmental impact. The drilling and high pressure water pumping (fracking) are so extensive that some geologists fear that these activities might destabilize the ground in affected areas. When drills penetrate through aquifers the boreholes must be well sealed, so water doesn't leak in and waste doesn't leak out. If this isn't done right, there's trouble. Also, some of the fracking water injected into the newly drilled well gets absorbed by the shale, but some "burps" back contaminated with toxic chemicals and must be disposed of as hazardous waste. These are very complex and dangerous activities, even when done right. When they are not done right, they simply contaminate and devastate the area—Deep Horizon type fiascos on land.

All these activities are conducted by people and are close to population centers. Needless to say, through the years, the lives of millions of people have been seriously affected by coal, natural gas, diesel, and other forms of energy generation. Basically, we die to make life easier—and we are not even close to done yet. On the contrary, energy demands are growing and related activities are increasing accordingly.

According to the U.S. Department of Energy, America needs 20,000 MW (or 20 GW) of new power generation every year for the next 20 years to meet projected demand. An additional 2,000,000 MW are needed around the world over the same period. This represents the need for adding 15-20 conventional power plants per year in the US and 1,500-2,000 additional power plants world-wide.

The natural resources will be depleted by the end of this century if we continue digging and pumping at this rate, and our health will deteriorate even faster as well if we don't find and implement cleaner power sources in the very near future. So, the overwhelming conclusion is that we need wind and solar. Yes, but solar, although called "green" and "clean" energy, also has its own problems.

PV Installations

Above we looked at the environmental impact of different energy-generating technologies. Now we will take a close look at the impact of solar power technologies and installations. Just like any other place where people and equipment operate, the PV installations have some impact on the local environment—and quite negative in some cases. As the land area used for these plants increases, so do potential negative effects, some of which are as follow:

Environmental impact studies and analysis show that large amounts of CO_2 and other environmentally unfriendly gasses and toxic liquid by-products are generated during the manufacturing, transport, and recycling stages of PV modules and other components. Hopefully, this is much less than the amounts generated by fossil-fuel power generation of similar size during the 20-30 years operation of such components. Even if we assume the least advantageous 1:6 ratio used by some for energy payback, PV systems will still prevent many tons of pollution from entering the atmosphere. The larger the PV plant is, the more harmful gasses will be kept unreleased vs. conventional energy generation.

Various sources use different calculation methods and come up with different estimates, so we are taking a middle road, assuming that each kilowatt hour (kWh) generated by burning coal produces ~1.5 lbs. CO_2 and other harmful gasses (CO, SO_2, NO_2 etc.). This means that a small, commercial PV system of 10kWp operating in Arizona will produce 20,000kWh electric power annually, or total of 500,000kWh (considering the losses) during its 30-year lifetime. This, then, means that our small 10kW commercial PV system will save the environment from absorbing 750,000 lbs. of harmful gasses during its lifetime.

Wait—not that fast! While the above is true, we must consider how much CO_2 was generated during the production of the PV modules and other components in this example. PV systems are estimated to have created (during their manufacturing, transport, installation, operation*, and recycling cycles) the average of 0.25 lbs. of CO_2 per kWh produced, or, to put it another way, another 125,000 lbs. of CO_2 just went up in the air. So in the end, our small 10kWp system will still save 625,000 lbs. (750,000 – 125,000) of poisonous gasses from entering the atmosphere during 30 years of operation. Still, that's not bad for our humble PV system.

Now, let's look at our larger 100kWp system. Using the same logic, we see that it will prevent 6,250,000 lbs. of harmful gases from entering the environment every year. How about a 1.0MWp system? Now we are talking… 62,500,000 lbs. poisonous gasses kept back from reaching us. A very large-scale 100MWp system will keep a total of 6,250,000,000 lbs. from changing the climate and causing global warming during its 30 years of operation. A 500MWp plant will reduce the CO_2 emissions by 5 times this amount… and on it goes…

Of course these are only estimates, with a significant margin of error and enough variables to move the decimal point in either direction. Nevertheless, the numbers are significant and leave no shadow of a doubt that we are talking about major forces at play—forces that must be well understood and controlled if humans are to live a normal life on Earth for another millennia, or more.

NOTE: No large quantity of CO_2 gas is produced during the actual on-sun operation of PV modules, but there is some measurable outgassing from them and the peripheral components in the power plants. Also, as new technologies (thin-film modules containing poisonous elements as Cd, Te, As, etc.) are being installed in ever-increasing numbers on thousands of acres of desert lands, we need to consider and reconsider where we install these and how we use them. Also, although the emitted gas amounts are very small, the power fields get larger and larger in size, so we need to pay more attention to the outgassing and its negative effects. Since some of the new PV technologies have no history, no precedent, and no proof or any other data to show their safety for the duration, we must be careful how we apply them.

So let's take a look at the different solar technologies from environmental and safety point of view:

1. CSP systems

In general, CSP technology is considered to be environmentally friendly, except for the environmental impact during the manufacturing of the different components (mirrors, support structures, steam turbine, cooling towers etc.). A lot of energy is used during the manufacturing process of CSP equipment. Tons of metals and other materials are consumed too, the production of which requires a lot of energy, and generates large amounts of harmful liquids and gasses. Production of mirrors, when compared to the overall PV cells and modules manufacturing process, is less energy-intensive and more environmentally friendly, yet uses lots of energy and toxic materials. The types and quantities of these activities and their byproducts must be entered into the equation, if we are to get a good idea of the overall energy balance.

Like all power plants, CSP power generation has some impact on the local area:

- CSP equipment uses a lot of water which is not abundant in desert areas where most CSP plants are located, creating another set of problems for the locals.
- Since the technology is based on use of mirrors, the terrain before installation needs to be equalized (leveled and developed), which requires seriously disturbing thousands of acres of virgin desert land. The impact of this intrusion is significant, but the future will tell how good or bad this activity is.
- Mirrors do not require much maintenance. Once installed, however they must be cleaned at a small cost (water and workforce) and a lot of mess—muddy terrain and run-off which causes concerns with water table contamination in US deserts.
- The plants are almost neutral for landscape except for the solar power towers, which stick very high above the ground and are at times considered a nuisance or visual pollution.
- The noisiest parts of the CSP systems are the steam turbines and the Stirling engines (in dish Stirling case), but the plants as a whole are quiet. To our knowledge, there are no documented reports of noise complaints from locals.
- CSP equipment may have occasional spilling of oil or coolant, but this is negligible and preventable.

2. PV systems

PV power plants are basically environmentally friendly, but many of the above mentioned conditions apply here too. As PV technology was developed, it was not until the 1980s that researchers began to consider its environmental implications. As its use expands daily, pressure is applied by the scientific community and administrations that all costs of the creation and use of energy systems be taken into account.

The oil crisis had demonstrated the need to reduce the world's dependence on imported oil, while the Chernobyl and Japan nuclear accidents alerted the world to the hazards of nuclear power. Larger environmental problems such as global warming and acid rain also reiterated the need to decrease our dependence on fossil fuels.

The first German-American workshop was organized in Ladenburg, Germany, in October 1990 to discuss the "External Environmental Costs of Electric Power: Analysis and Internalization." Papers at this workshop considered environmental damage and ways in which environmental costs may be internalized. PV was not discussed specifically at this workshop; however, the impacts of external costs on wind energy were covered and can be easily related to those of the PV industry. (13)

Ottinger's group at Pace University, Centre for Environmental and Legal Studies, published an extensive review of environmental costs/risks and covered all electricity generation technologies, including photovoltaic installations. Various environmental costing models were also discussed, and this has formed the basis of many of the studies on external/environmental costs that have been published (14). Knut Sorensen, Statistics Norway, published an extensive report on life cycle analysis with regard to energy systems. This report outlines the principles of life cycle assessment and how they may be applied to energy systems, particularly renewable energy technologies. (15)

The summary of a 1997 report, "Environmental Aspects of PV Power Systems" (10) indicates that the immediate risks to human health and the environment from production and operation of PV modules seem to be relatively small and manageable.

The methodology developed was applied to multicrystalline and amorphous silicon cells and gives a comprehensive breakdown of the technologies involved and how the life cycle analysis is structured. The results are divided into environmental, social, economic, and other impacts, showing a significant impact on its respective area of influence. Impact type and size vary from case to case, but the overall conclusion is that the environmental impact of all PV technologies must be analyzed and

Table 8-1. Immediate and long-term environmental and health issues

TASK	IMMEDIATE CONCERNS	LONG TERM CONCERNS
SUPPLY CHAIN OPTIMIZATION	~Investigate the availability and short term constant supply of rare and exotic materials such as Cd, Te, In, Ga, As, Ge and Ag now and in the future ~Optimize the methods and improve the efficiency and safety of rare, scarce and exotic materials mining, manufacturing operations and field use.	~Investigate solutions for supply chain shortages ~Develop thinner active film layers in TFPV modules ~Optimize efficiency of rare and toxic materials use ~Optimize EOL module decommissioning procedures ~Design new materials and products for complete recycling and safe waste disposal systems
EFFICIENT ENERGY USE	~Develop methods to reduce energy use during silicon material, cells and modules production ~Optimize energy use for module frames and BOS	~Optimize energy consumption of Si processes ~Optimize energy consumption of recycling processes ~Optimize energy-efficient frame and BOS designs
IMPROVEMENT OF CLIMATE CHANGE CONDITIONS	~Optimize CO2 mitigation potential of PV technologies ~Investigate release of fluorinated (FFCs) and other toxic and unsafe compounds during manufacturing ~Investigate air and land contamination from PV modules operating in extreme climates ~Design & suggest hybrid energy generation options	~Investigate the CO2 mitigation potential of autonomous PV systems ~Develop FFC alternatives for use in PV production ~Optimize gas and liquid release methods ~Investigate the role and impact of dynamic assessment methods
HEALTH AND SAFETY CONCERNS	~Optimize safe use of compressed and explosive gases in the manufacturing processes ~Investigate and develop procedures for safe use of "black list" materials such as Cd, Te and As in manufacturing and long term field operations	~Develop safer materials and safer alternatives ~Investigate using thinner active cell layers ~Develop more efficient material utilization methods ~Investigate the long-term risks from (low-level) releases of "black list" materials in large scale fields
RECYCLING AND WASTE CONCERNS	~Investigate leaching of heavy metals from modules in long term landfills ~Optimize module & BOS waste management options	~Investigate the environmental aspects of relevant recycling and waste management methods ~ ~Investigate and optimize long term landfill safety

considered.

The PV industry has progressed since 1997 and many things have changed. They must be taken into account when discussing environmental aspects of PV power generation:

a. PV power fields are growing in size. From the 5-10 MWp maximum size in the 1990s, we are now witnessing 500 MWp PV power plants under development. This represents millions of PV modules installed on thousands of acres, so the expansion in size means larger quantities of harmful effects which need to be accounted for and taken care of before it is too late, and
b. The largest PV power plants are to be installed in the US deserts, where failure rates are many times higher than in moderate climates. This is also a new phenomena, the negative consequences of which could be too great with time, and should not be ignored.

Environmental and Health Impact of Solar Power Generating Activities

Assessing the environmental impact and health and safety issues related to manufacturing and operation of solar energy generating systems is very complex undertaking, which would require an entire new book dedicated to the subject to thoroughly address and analyze. The actual impact is only "suspected" in some cases, without proven qualitative value, so we dare only summarize the environmental impact and health issues due to solar energy activities (equipment manufacturing and use) below and in Table 8-2.

1. Heavy metals, benzene compounds, diesel fumes and radionuclides contribute to reduction in life expectancy and diseases such as lung and other cancers, osteroporosia, ataxia, and renal dysfunction.

2. O_3, SO_2, PM10, PM25, CO, and O_3. are responsible for respiratory illnesses, congestive heart failure, chronic bronchitis, restricted activity days, asthma attacks, and frequent hospital admissions.

3. Mercury is proven responsible for loss of IQ in young children.

4. NO_2, NO_x, SO_2 and O_3 cause yield reduction in wheat, rye, barley, oats, potatoes, sugar beets and sunflowers.

5. CO_2, CH_4 and N_2O influence world-wide morbidity and mortality rates, coastal and agricultural impact, and economic impacts due to temperature change and sea level rise.

Table 8-2. Environmental and health impact of solar industry activities

Energy and CO2 payback	2-4 years are needed to replace the energy used, and compensate for the CO_2, emitted during the materials, cells and modules manufacturing processes
Resource depletion	The availability of some metals used in c-Si and TFPV modules is limited
Land use	Environmental impact varies between PV technologies and world's locations
Manufacturing safety	Safer materials and processes must be considered and developed Emphasis must be on prevention of accident-initiating events Capturing accidental releases and preventing human exposure is a priority #1
Field operation safety	Thorough measurements of contaminating species must be implemented
End-of-life recycling	The EOL recycling process must be well designed, executed and enforced
Waste handling	Safe waste handling during manufacturing and at EOL is an absolute must Flammable/explosive gases like silane, phosphine, germane, and toxic metals like cadmium in TFPV and lead in c-Si PV need to be thoroughly controlled Efficient and controlled recycling facilities and landfills must be implemented The PV industry should learn HSE mgmt. from the semiconductor industry
Climate change	PV technologies provide clean energy and reduce greenhouse gas emissions, which is their greatest contribution to improving quality of life.

6. Acid rain, SO_2, NO_x, and NH_3 cause elevated acidity, eutrophication, and 'PDF' of species.

7. Fatalities from traffic, and workplace related accidents (mining and refining are prime suspects).

8. Long-time exposure to noise can be blamed for life expectancy reduction, and some operations are so noisy that one must shout in another's ear to be heard.

These chemicals and activities are found at different stages of solar equipment manufacturing and operations. Remember that the solar industry consists of a myriad of operations, executed in different production plants and factories around the world. The manufacturing process of c-Si and TFPV modules includes some very large industrial operations—metals mining and refining, sand melting and silicon refining are huge (in size and volume) enterprises that use a lot of energy and chemicals, in order to make and transport tons of materials and products. These activities sometimes cause significant damage to air, soil, and the water table in an area, including damage to human health and life as we know it.

The facts herein are related mostly to the manufacturing process, because the operation of PV power plants is a much cleaner undertaking. There is a small amount of outgassing measured during field tests of c-Si PV modules, but the amounts are too small to cause any concern at this point.

Thin film (TFPV) modules are somewhat different in that:

1. Many contain toxic chemicals (Cd, Te, As, Se and such).
2. They are a relatively new product that have not been proven safe for large-scale desert use.
3. We must be aware of potential dangers because of the rapid escalation of TFPV modules deployment in the deserts, with the PV power plants getting very large in size. This concern might grow into a much larger issue.

Because of that we'll take a closer look at TFPV modules from an environmental and safety point of view.

Environmental and Safety Assessment of Thin-film PV Modules

Thin-film PV cells and modules are the "new kid on the block," and have not been thoroughly tested and proven safe, especially when used in large-scale power fields in the deserts. There are a number of issues which we will address only briefly herein, but we would like to encourage the reader to take a closer look at the available literature and make an independent decision on the subject.

CdTe and CIGS thin-film PV modules have seen a quick rise and have been deployed successfully in large numbers around the world lately. The low efficiency and other issues are obviously not hindering the TFPV technology deployment (most likely due to favorable incentives and subsidies), so they are growing faster than the other PV technologies currently available. And amazingly enough, they are the first PV technologies planned to be deployed in gigantic 250MW and 550MW power fields in the US deserts. This is quite unusual, even abnormal, because CdTe and CIGS thin-film PV modules are a fairly new development and have neither been thoroughly tested nor proven practical safe for some applications—especially for use in the harsh US deserts, which is exactly the location planned for these gigantic power fields.

The key issues here are:

1. Some thin film PV modules contain compounds of cadmium (Cd), tellurium (Te), selenium (Se), arsenic (As), and other metals and chemical compounds

2. The active thin film layers in the TFPV modules are mechanically fragile structures, which could easily be disintegrated by heat-freeze action, and/or mechanical friction, bending or impact upon the modules.

3. The thin films are also, under the right conditions, chemically active compounds. They can easily be decomposed, or reacted by moisture, rain water (via weak acids and other chemicals containing in it), as well as by almost anything else that enters the modules.

4. Presently most thin film modules consist of two glass panes within which the thin films are encapsulated and sealed within a plastic layer. This is not enough protection, so damage to the active structure during the long term (30+ years) non-stop operation is expected.

5. The active thin films in some TFPV modules are encapsulated by layers of organic polymers, which could easily break-down, disintegrate and decompose under intense IR and UV radiation in the deserts with time. Once that happens, the thin film

structures are vulnerable to the elements, which will eventually penetrate the module and destroy them.

6. Millions upon millions of thin film PV modules have been planned for installation onto thousands upon thousands acres in the US deserts, the immense scale of which undertaking is where the real danger lies.

7. Toxic materials based thin film PV modules used in the desert, where temperatures within the modules could reach 180°F, and where summer storms, rain and hail, would aggravate the above mentioned mechanical and chemical degradation to the point where major changes in the modules and the thin film structures are to be expected with time:

   a. The never ending heat-freeze action of the desert environment will stress and eventually damage and stress the thin film structure,

   b. The encapsulating layers will eventually break down with time, thus allowing the elements to penetrate the module, react with the toxic thin films and decompose them

   c. The resulting from the above actions particulate, liquid and gaseous contamination might damage air, soil and water table in the local area, and might even pose threat to human health and threaten all life forms in the area.

A number of scientists have looked at the toxicity issues of PV modules and have published papers on the subject, as follow:

1. Scientists at Brookhaven National Laboratory (BNL) in the USA, have been publishing health and environmental information of interest to the PV industry since the 1970s; mostly funded by the United States Department of Energy. The work was started by Moskowitz and then continued until the present day by Fthenakis. It was the early studies by Moskowitz that first alerted the PV industry and environmental scientists to the possible hazards associated with the manufacture, use and disposal of PV modules containing toxic materials.

2. In 1995, BNL and other scientists published a comprehensive study on the health and environmental issues of the manufacturing, use and disposal of thin film modules. They examined the hazards associated with producing and using thin film modules, focusing on the potential of workers in manufacturing facilities to be exposed to chronic, low levels of Cd. They also review regulations and control options that may minimize the risks to workers and discuss recycling and disposal options for spend modules.

3. Hynes, Baumann and co-workers in the UK have published several papers on environmental aspects of many thin film deposition processes. These include environmental risk assessment and hazard assessment of the manufacture of CIGS based thin film PV cells, the chemical bath deposition of CdS and the deposition of alternative window materials to CdS for use in thin film modules. Steinberger from the Fraunhofer Institute in Munich has investigated such risks of Cd containing thin film modules in the operation phase, considering hazards that may occur due to fire, weather influences or damage of the module due to mishandling.

4. Steinberger firstly identifies the amount of material in the modules and then investigates the concentrations of selenium, cadmium and tellurium in the air due to a fire lasting about 1 h. His results show that there is no acute danger posed by such fires, and that releases of cadmium or selenium from a burning PV module are less than those given out by a coal fired power station in normal operation.

5. Dr. Fthenakis has considered the cadmium emissions from cadmium telluride thin film cells also. He has taken a cradle to grave approach and comes to some interesting conclusions. The Cd present in a NiCd battery is elemental Cd and not CdTe, a much more stable and insoluble form. Coal and oil burning power plants routinely produce Cd emissions, whilst a PV cell does not produce any Cd emissions during its operation. Cd can either be used or discharged into the environment, where it is normally "cemented* and buried or landfilled as hazardous waste.

In conclusion, the above works have several things in common:

1. They all recognize the serious implications of using cadmium and other potentially toxic metals in PV products but offer no concrete solutions to avoid or minimize the obvious dangers.

2. They are all old, outdated, and incomplete. Due to the seriousness of toxic exposure (and to avoid liability and further responsibility), they always recommend more studies and further investigations of the related harmful phenomena and effects.

3. They always refer to tests done under "standard," or "normal" conditions, which usually means dry air and 25°C operating temperature. These conditions, however, are very far from the extreme heat and humidity in the world's deserts and other areas where most of these products are planned for installation.
4. The actual lab tests, although properly executed, offer no real solutions. Their conclusions and recommendations fail in the attempt to somehow extrapolate the results of a small lab bench test to the behavior of the millions of modules in a mega PV power field in the desert.
5. They offer no actual test data, nor any kind of scientific proof for 10, 20, or 30 years' exposure to potentially toxic metals containing PV modules to a harsh desert environment. And we have not seen any long-term tests to date, done under actual extreme-desert or excessively humid climatic conditions, which is our main and only concern.

So, what will happen to these untested and unproven-for-this-application PV modules, containing heavily toxic and carcinogenic materials and operating non-stop during 30 years under the desert bliss? How would we explain large-scale air, soil, or water table contamination in these fields, which is likely to happen during the 30 years of non-stop bake-freeze cycles? Who will be responsible for the damages to the environment and peoples' health? Who is going to pay for the cleanup, medical treatment, and resulting law suits?

Some of these issues have been addressed presently by the manufacturers, but they are far from being fully resolved. Instead, regardless of the potential dangers, mega-fields consisting of millions upon millions of Cd-toxic compounds containing PV modules are planned for installation on thousands of acres in extreme climates. There is a lot of money to be made before the window of opportunity closes, so the race is on. The manufacturers and the responsible parties—politicians, regulators, scientists, owners, installers, operators, investors, and utilities—are, due to ignorance or negligence, simply closing their eyes to the dangers, hoping that it will be OK somehow. We hope so too, for their sake and for that of the future generations, because if something goes wrong in those mega-fields we all might end up with the greatest environmental disaster even known to man.

The responsible parties' responses are quite inadequate. See some examples below.

Case 1. A government scientist claims in defense of cadmium toxicity that, "...one nickel-cadmium flashlight battery has about as much as a square meter of PV module using current technology." True, but we are yet to see 10,000 acres of desert land evenly covered with NiCd batteries, and we'd be truly alarmed if we saw such plans. However, the work to cover many times that much desert land with CdTe and SIGS cadmium-containing modules is already underway.

Case 2. A government scientist conducted lab bench tests with small (25 x 3cm) pieces of CdTe modules and found that 99.5% of the cadmium was encapsulated in the molten glass, thus proclaiming the fire safety of CdTe modules. It is not possible, however, to extrapolate this to several million 3'x4' modules to also encapsulate 99.5% of the cadmium in the molten glass. But even if they did, 0.5% of millions of CdTe modules in AZ and CA deserts would still be way too many modules outgassing and leaking cadmium compounds. Would this be good or bad? Nobody knows, since there is no such precedent anywhere in the world to date. And yet, the hasty conclusion of this test is one of the main arguments used for assuring the safety of the CdTe TFPV modules in large-scale installations today.

Another test by government scientists shows that Cd and Te are readily leached out of the PV modules in small amounts of dilute acids and other chemicals, which process is used to recover these metals during recycling of old CdTe modules. Yet, nobody is worried about similar TFPV modules in the mega-fields during the long term exposure to harsh environmental conditions, accidents, natural disasters, and such unwelcome and unplanned events for the duration.

Case 3. In an email from October, 2009, a NIOSH toxicologist writes to a colleague, "We had a call not too long ago from the Health and Safety guy from a company making the (CdTe) panels. He was pretty cool. Based on the low bioavailabilty of CdTe in the form it exists in these solar panels, I think I'd take my chances. In the event of a meteor strike or the like sufficient to vaporize the metal, I think Cd tox is the least of my worries."

How superficial is this? The issue here is many thousands of acres of desert land covered with millions of cadmium-containing modules broiling under the hot desert sun—unproven for this use. But the NIOSH scientist, while flirting with the "cool" guy from the company that makes CdTe modules, dismisses the issues at hand by considering the "low bioavailability" in the brand new modules (duuhh) and the improbability of "melting the metal" in case of a meteor strike, none of which are relevant to the real dangers of long-term exposure of millions of CdTe modules to the desert elements.

Case 4. In a response to a letter about the potential dangers of CdTe long-term exposure to the desert elements, a local official responsible for the land proposed for a large CdTe power plant says, "I feel comfortable that I have (done my job properly) as, after all, it is my responsibility to do so. I am satisfied with the answers I have received to your questions and to my own."

As for this equally relaxed and superficial response, void of any scientific or practical proof, consider the following scenarios:

a. The CdTe modules operate 30 years with no problems; the local official is right, the concerns are unfounded, and he has done his job.
b. Severe cadmium contamination invades the area, shuts down the power plant, and forces the detoxification of thousands of acres after 15-20 years of on-sun operation. What is the local official going to say or do in that case? The inevitable follow-up question to all preliminary planning here must be, "What if the unthinkable happens?"

These are not easy questions, are they? Unfortunately, due to the novelty of the TFPV technology, the immensity of its deployment, and the uncertainties around the related issues, most of these questions will go unanswered for now. So millions of potentially toxic PV modules will cover California and Arizona deserts in the near future, and all we can do is wait and see.

Air monitoring programs recently mandated by BLM for large-scale TFPV power plants are a step in the right direction. Air quality monitoring stations will be deployed to measure particulate, heavy metals and other emissions.

Weighted averages of hourly results will be reported every 24 hrs. and published on a website during 3 years post-construction. Locals will be given advanced notice of elevated emissions to take measures as needed.

External Impact and Cost of PV Power Plants

The "external" costs of PV plants' setup and operation is defined by monetary quantification of the socio-economic and environmental effects and damage they do (and will) as they inflict a local area, a nation, and the world as a whole during their cradle-to-grave life cycle. These effects and the related damages could be expressed in $ per kWh generated, for lack of better way, and should account for all materials, procedures, and events from cradle to grave.

Included in these calculations are: environmental effects; human health; materials production; effects of use and disposal (materials, gasses, chemicals) on agriculture; noise; audio and visual pollution; ecosystem effects (acidification, CO_2 damage); and all other effects. Thusly obtained numbers could be used to provide a scientific basis for legislative and regulatory policies, energy taxes and incentives, global warming policy adjustments, etc.

The "ExternE (5) project," addressing these issues in detail, was introduced and funded by the European Commission in order to develop a methodology to calculate the external costs caused by energy consumption and production. This is another far-thinking program of the EU community, which, along with their RoHS (7) program, makes them far superior in this area when compared to the fragmentation of the issues in the US and other developed and developing countries.

The ExternE project defines costs as the monetary quantification of the socio-environmental damage, expressed in eurocents/kWh, with the possibility of providing a scientific basis for policy decisions and legislative proposals such as subsidizing cleaner technologies and energy taxes to internalize the external costs. It looks at all energy-production technologies using a methodology developed for this project that allows the various fuel cycles to be compared. An outline of the initial results for the PV fuel cycle, starting with a very small sample of PV systems, shows that the results are not consistent and require more work.

There has been pressure from the PV industry for the PV fuel cycle to be re-done using a larger sample of more representative systems, but this is still in the making. So, one of the earliest publications from this project by Baumann et al. is still valid. In this paper the author outlines the basic assumptions and requirements of the methodological framework for the quantification of external costs and compares the environmental effects of different energy technologies, including renewables and the conventional technologies of coal, oil, nuclear, and natural gas.

Basically, PV systems must be designed and manufactured with long-term environmental concerns and considerations in mind, which must include:

1. Manufacturing materials and procedures, including the production and use of MG and SG Si, thin-film metals and chemicals, glass panes, metals frames, gasses, chemicals, etc. process supplies
2. Solar wafers and cells manufacturing processes
3. Module encapsulation and framing processes
4. Evaluation of direct and indirect processing energy (including transport, storage, etc.)

5. Gross energy requirements of input materials (supply chain and internally generated products and byproducts)
6. Allocation schemes used in the calculations
7. Separate thermal energy, electrical energy, and "material energy" calculations
8. End-of-life recycling and disposal calculations (including transport and storage)

There are a number of dislocated efforts in the above areas, but no uniform, standardized method presently exists that is capable of capturing all environmental factors into one, all-encompassing methodology. Such a methodology is needed to account for the effects of the above concerns on the PV manufacturing processes and long-.term use of the related PV products.

Lifetime Energy Balance

Life time energy balance (LEB) is the energy used during the PV products manufacturing, transport, installation, and operation, compared with the power they generate during their useful lifetime. LEB is an important factor, which needs to be kept in mind when analyzing and calculating the energy benefits from a PV power plant. LEB basically tell us how much energy we save by using PV energy generating sources.

The solar cells and modules manufacturing process starts with melting and purifying sand in huge, dirty, dangerous, and energy-guzzling furnaces. The produced metallurgical grade (MG) silicon is crushed in huge mills and further purified in a complex network of chemical and electric equipment, again using enormous amounts of natural resources and energy. The product at this point is solar grade (SG) silicon of varying purity, the quality of which depends on the raw materials and process quality. The SG silicon is crushed again and melted again in large furnaces at high temperatures for a very long time. Thus produced ingots of mono or poly silicon are sliced into thin wafers onto which the solar cells will be built. The wafers are cleaned, baked, fired several times, and coated several more times until finally a solar cell emerges at the end of the line. They are then sorted, lined up, and vacuum-bake sealed into the module frames, then transported to the location. The next time these PV modules are transported will be at the end of their life cycle, going to the crusher.

All this requires energy, materials, and resources. Fortunately, the production equipment and processes are very efficient these days, but even so, making a solar cell and module is an energy-consuming undertaking. Consider this equation for lifetime energy balance (LEB):

$$LEB = \frac{Eprod + Etrans + Einst + Euse + Edecom}{Egen}$$

Where:

Eprod = energy used during production of materials, wafers, cells, and modules
Etrans = energy used to transport materials, modules, and BoS to PV site
Einst = energy used to assemble and install the PV power plant
Euse = energy used to operate the PV power plant
Erecyc = energy used to decommission, transport, and recycle the PV field
Egen = energy generated during the life of the PV power plant

In all cases, Egen must be much higher than the sum of the other sources of energy used in order for the system to be an effective energy source.

The term "energy payback" describes this "energy in-energy out" ratio and is what we will have to consider when designing, pricing, and justifying a PV system. Or in other words, how long do we need to operate a PV system before we recover the energy used, and pollution generated, during its manufacture, transport, and operation before it is decommissioned?

There are energy payback estimates ranging from 1 to 4 years for different technologies; more specifically:

- 3-4 years for systems using current multi-crystalline silicon PV modules
- 2-3 years for current thin-film PV modules

These estimates, however, vary from product to product and from manufacturer to manufacturer. Cost of energy (crude oil in particular) also has a great effect on the estimates, so these numbers must be adjusted periodically. The ever changing socio-economic situation in different countries is another great factor and we expect that the present-day economic slow-down and worldwide financial difficulties will drastically reshape these and most other estimates as well.

Lifetime CO_2 Balance

Similarly, a system's lifetime CO_2 balance (LCB) is a factor that takes into consideration the CO_2 used during the manufacturing and use of PV components vs. the amount of CO_2 saved by using the PV components instead of coal- or oil-fired power generation.

The solar wafers, cells, and modules manufacturing processes generate significant amounts of CO_2 and other harmful gasses, which must be taken into consideration when talking about the advantages of PV technologies over the conventional energy sources. Consider this equation for lifetime CO_2 balance (LCB):

$$LCB = \frac{Cprod + Ctrans + Cinst + Cuse + Cdecom}{Csave}$$

Where:

Cprod = energy used during production of materials, wafers, cells, and modules
Ctrans = energy used to transport materials, modules, and BoS to PV site
Cinst = energy used to assemble and install the PV power plant
Cuse = energy used to operate the PV power plant
Crecyc = energy used to decommission, transport, and recycle the PV field
Csave = energy generated during the life of the PV power plant

In all cases, Csave must be much higher than the sum of the CO_2 generation sources in order for the system to be an effective energy source. Nevertheless, and regardless of the LCB ratio, the PV plant will receive carbon credits for the CO_2 -free power generated during its lifetime.

Here again, we have seen estimates that the total quantity of CO_2 generated during the cradle-to-grave cycle of PV components is compensated within 2-4 years of CO_2-free PV power generation, by reducing CO_2 emissions as compared to conventional power generators.

PV Modules Recycling

PV companies worldwide are actively moving towards the recycling of their products, both for manufacturing waste and end-of-life modules. EU companies have proposed a voluntary take back system capable of meeting the future waste recycling demands. The new recycling process lines must be capable of processing crystalline silicon and thin-film modules alike. The silicon recycling is now at the pilot stage, and reuse of silicon material will be timely for the industry. US companies have been reclaiming solar cells and semiconductor process wafers for many years, so the expertise and equipment are available. Scaling up to accommodate the large demand of the future is a key to success here.

Recycling and reclaiming reduce the energy payback time by a factor of 4 in the best case scenario, but this depends on the available insolation at the PV power site and the particular PV technology used.

The forecast is that 40,000 MT of PV components will be ready for decommissioning and recycling by 2020. This number will double and triple during the following decade. This includes silicon and thin-film based PV components recycling. The thin-film component will be ~ 20% by then. CdTe thin-film PV modules, or 8,000

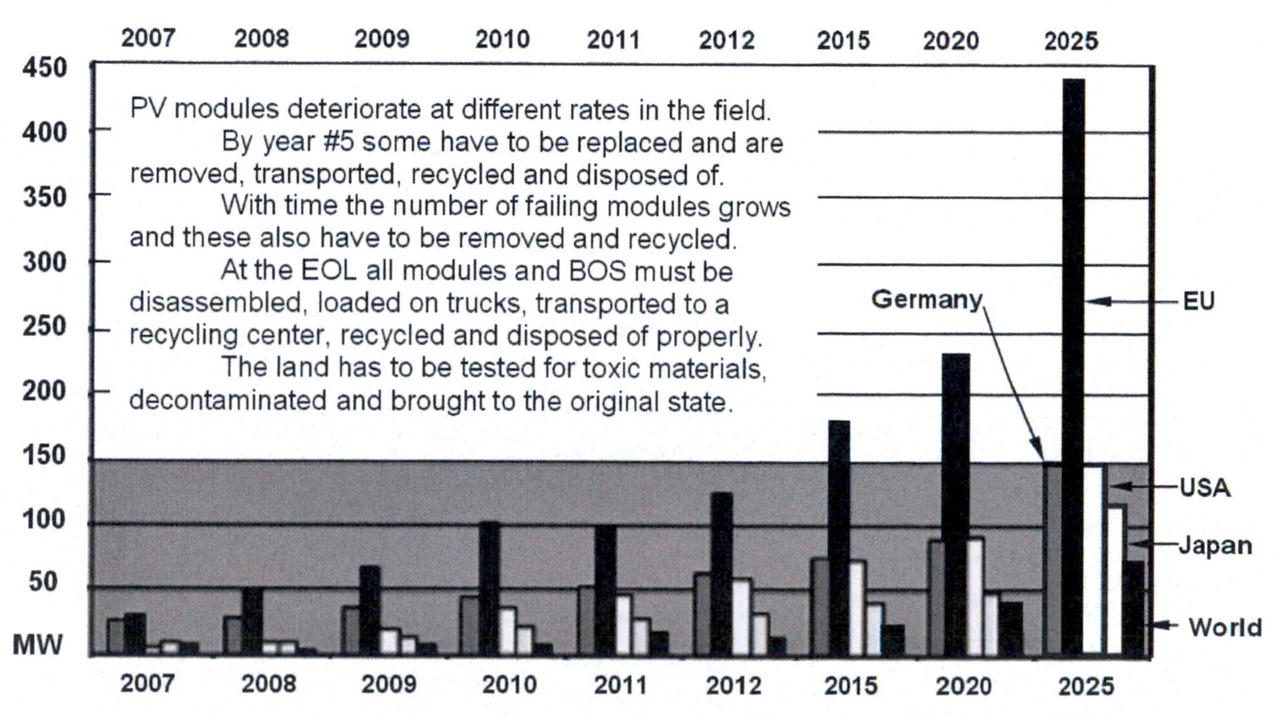

Figure 8-4. Forecast of PV waste recycling in the decades to come

MT, will be the majority of the recycled products in this category. 8,000 MT CdTe PV modules is a lot of modules which, with an average of 8-9 grams of cadmium* in each module, could create a serious hazmat debacle if not handled and processed properly.

NOTE: Cadmium is one of the six most toxic carcinogenic heavy metals on Earth, which together with tellurium, selenium, arsenic and other toxic metals and their compounds must be handled with utmost caution during the uninstall, transport, crushing, and disposal. Disposal of some components as hazardous waste is another major issue to be addressed and fully resolved in the years to come, because special handling, transport, processing, and containment would be required.

The EU already has directives for voluntary and extended manufacturer responsibility, where the decommissioning, transport, storage, processing, and disposal of the modules are the ultimate responsibility of the original manufacturer. The directives have been integrated into the legal system of several member states. Different paragraphs outline the registration, packaging, transport, waste disposal, documentation, legal responsibilities, etc. These components of the overall effort to protect the environment and life during all stages of the manufacture, use, and recycling of PV products clearly place the responsibility on the manufacturers' shoulders. Customers and users are required to be aware of the issues at hand and to observe and complete their obligations as well.

This means that whoever made the PV products is legally responsible for them and obligated to dispose of them at end of their useful life. The manufacturer is obliged to assure the proper execution of this process in all cases—even if no longer in business. Special arrangements, such as insurance or delegation of responsibility must be provided at the time of installation and be well-documented prior to starting the actual work on each project. The cost of the decommissioning process is to be agreed upon and reflected in the installed cost of the power plant.

Every year around the world, a large quantity of PV modules will reach their end-of-life state, become obsolete, or fail. From all these old modules, structural steel, aluminum, inverters, wiring, and other hardware will have to be disassembled, sorted, loaded on trucks, and transported long distances for recycling and / or disposal. This is a great effort, which in many cases is neither planned nor taken into consideration during the PV plants' design and financing stages. If this is not done properly, or if it is hidden from view, then we have to assume irresponsible behavior, something that we have to watch out for.

The PV plants designers, installers, investors, and operators, in addition to the responsibility to ensure the safe plant operation, are also fully responsible for the proper decommissioning and disposal of the plant's components, as well as for final cleaning and bringing the land to its original state.

As can be clearly seen in Table 8-3, the manufacturer is legally responsible for the safety and proper execution of all steps from the beginning to the end of the useful life of their products. The retailers, installers, and plant operators are also responsible in some stages

Table 8-3. End-of-life outline of the participants' responsibilities

EOL TREATMENT OF TFPV MODULES	DECISION AND EOL* NOTICE ⇒	REMOVAL ⇒	TRANSPORT ⇒	RECOVERY & DISPOSAL
GENERAL REQUIREMENTS	Notify proper authorities of EOL	Proper and safe procedures use	Authorized Hazmat carrier use	Authorized facilities use
ORGANIZATIONAL RESPONSIBILITY	Plant operator	Manufacturer and operator	Manufacturer	Manufacturer
TECHNICAL RESPONSIBILITY	Plant operator	Manufacturer and operator	Manufacturer	Manufacturer
FINANCIAL RESPONSIBILITY	Plant operator	Manufacturer and operator	Manufacturer	Manufacturer
LEGAL RESPONSIBILITY	Plant operator	Manufacturer and operator	Manufacturer	Manufacturer

Table 8-4. PV modules recycling steps and methods

RECYCLING OF PV MODULES AT EOL

c-SI MODULES			TFPV MODULES		
RECYCLABLE	kg/m²	%	RECYCLABLE	kg/m²	%
Glass	10.0	90	Glass	15.0	90
Aluminum	1.4	100	Aluminum	0	0
Solar cells	0.5	90	Solar cells	0.1	90
EVA, Tedlar	1.4	0	EVA, Tedlar	0	0
Ribbons	0.1	95	Ribbons	0	0
Adhesives	0.2	0	Adhesives	0.2	0

RECYCLING AND DISPOSAL METHODS

Mechanical	Crushing, attrition, density separation, flotation, adsorption, radiation, laser beam, metal separation, and other methods.
Chemical	Acid / base treatment, extraction with solvents, dissolving, precipitation, slagging, and other methods.
Thermal	Incineration, burn-out, pyrolysis, melting, slagging, and other similar methods.
EOL disposal	Recycling into the same product, recycling into another product, recovery of energy from thermal treatment of organic layers, utilization of mineral fractions (e.g., concrete, road material, etc.), and landfill disposal.

of the product life. Overall, all entities who have been involved in the power plant's planning, design, installation, and operation are responsible to one degree or another for the proper execution of the respective steps, including the final decommissioning, recycling, and disposal of the plant components and land clean-up.

Table 8-4 is a list of operations that have to be considered, planned, and properly executed during recycling and/or disposal procedures. Some of these operations are accompanied by toxic gas and liquids generation, so special caution must be exercised during their execution as well. Also, the proper final disposal of waste materials and related chemicals and gases is absolutely necessary. That too is the responsibility of the project owners and products manufacturers.

Recognizing and embracing cradle-to-grave responsibility for products and land safety, as well as proper handling and disposal, must be one of the conditions to participate in any PV project in the US. We have no reason to ignore this critical aspect of PV industry development, regardless of what other countries do or do not do. The future generations' well being depends on our decisions.

PV Installations on Contaminated Land Sites

Identifying and using land located in areas with high quality, renewable energy resources will be an essential component of developing electricity from renewable energy sources. As the number of large PV projects increases, the demand for large pieces of land will increase as well. Land that has been previously used for activities such as mining and waste disposal will become increasingly important for this application.

The U.S. Environmental Protection Agency (EPA) estimates that there are approximately 490,000 sites and almost 15 million acres of potentially contaminated properties across the United States tracked by EPA. This estimate includes Superfund, Resource Conservation and Recovery Act (RCRA), Brownfields, and abandoned mine lands.

Cleanup goals have been achieved and controls put in place to ensure long-term protection for more than 917,000 acres. Through coordination and partnerships among federal, state, tribal, and other government agencies, as well as utilities, communities, and the private sector, many new renewable energy facilities can be developed on these contaminated properties.

The EPA Office of Solid Waste and Emergency Response (OSWER) Center for Program Analysis (CPA) is seeking opportunities to facilitate the reuse of contaminated properties and active and abandoned mine sites for renewable energy generation.

These lands are environmentally and economically beneficial for siting renewable energy facilities because they:

1. Offer thousands of acres of land with few site owners.
2. Often have critical infrastructure in place, including electric transmission lines, roads, and on-site water, and they are adequately zoned for such development.
3. Provide an economically viable reuse for sites with significant cleanup costs or low real estate development demand.
4. Take the stress off undeveloped lands for construction of new energy facilities, preserving the land carbon sink.
5. Provide job opportunities in urban and rural communities.
6. Advance cleaner and more cost effective energy technologies, reducing the environmental impacts of energy systems (e.g., reduce greenhouse gas emissions).

Here is the $64 trillion question: Are we killing two birds with one stone, or are we just killing all the birds because we can, and for the sake of killing? What are the advantages and disadvantages of installing PV power plants on contaminated land? Are there risks associated with such a move? How do we separate the past, present, and future issues, be they environmental, health, etc.? Who is ultimately responsible for the environmental and health safety of these potentially dangerous fields, used in an untested, unproven, and unprecedented manner?

What are US regulators saying on the matter? Are they aware of this development, or did it slip between their fingers? This is another set of important questions awaiting an answer, demonstrating again how young and immature the PV industry is. As with any young person or new thing, we must protect from abuse and misuse, which is what we trying to do with this book—bringing the issues to the attention of the US public, the scientific community, and the regulators.

Safety and Environmental Issues in the Desert

The desert areas is where the maximum amount of sun energy is concentrated and where we expect most of the efficient and profitable future large-scale power generating plans to be located. This brings a number of issues into play, some of which are very serious and critical to the future of the PV and energy industries. Work and operation in remote desert areas, however, is difficult, Paved roads and other infrastructure are few and far between, and PV modules and people alike must endure the climate extremes—heat and freezing, dust,

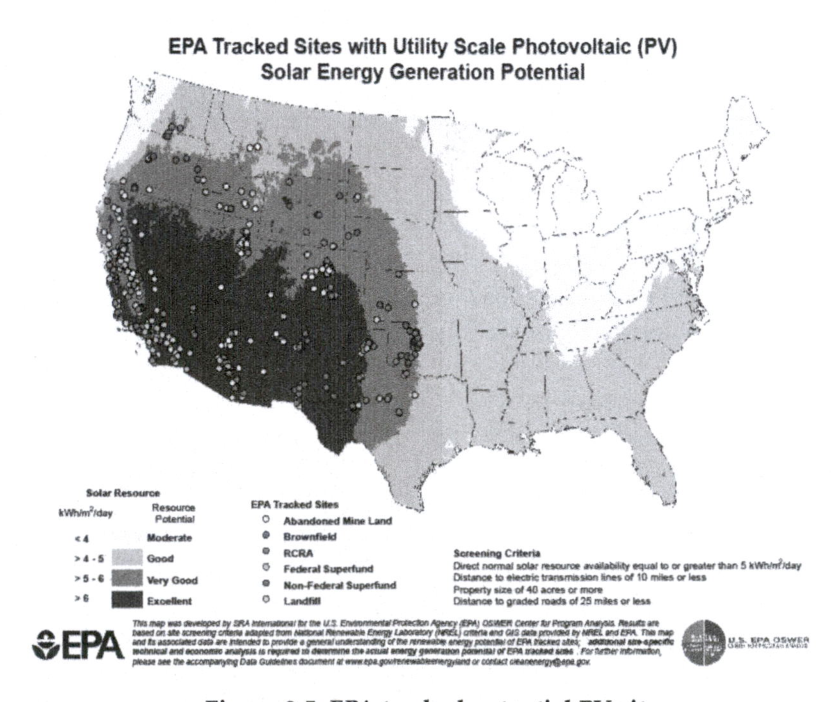

Figure 8-5. EPA tracked potential PV sites

storms, etc. The IR and UV radiation in the desert is extreme, and the PV modules suffer and undergo serious changes, deterioration, and failures with time.

Desert-based PV installations are exposed to the whims of Mother Nature, which is quite cruel in these areas, so some foreseen and unforeseen damage has to be expected—and in more exaggerated form than in any location around the world. One of the most serious and unavoidable dangers, in addition to unrelenting extreme heat, is that of sand storms and lightning strikes, which can damage the structures mechanically or electrically. They can also cause additional damage by sand-blasting the PV modules' optical elements, thus reducing their efficiency or eventually rendering them useless. Of course, frequent and serious surface soiling is an unavoidable damage and another expense we must face in the desert.

There is no known method of preventing wind and sand damage, but some precautions could and should be taken during the design and installation stages, as follows:

1. Gravel-type deserts are best suited for large-scale PV installations, because they contain less free sand and are less likely to cause damage by sand blasting critical components.
2. Build protective wind barriers around the edges of the power plants to reduce the wind speed and minimize the sand-blasting effect.
3. Incorporate in the overall system design an efficient way to stow trackers, or to turn them in the direction of the wind, thus protecting their optical side from sand-blasting damage.

We need to remember that work in the desert is always difficult for a number of reasons, some of which we discussed above. Large scale PV installations are a new phenomena and we don't have enough experience to be able to take all precautions needed for safe and efficient operation. So no doubt that the first large scale installations will be the "guinea pigs" of the utility scale PV power generation in the US.

The Stakeholders

Who are the key players in today's solar energy market? Who is responsible for assuring the proper execution of the steps in the different areas of the PV energy field and the related environmental areas?

Government agencies such as BLM, EPA, CPUC, AZCC, and others are fully engaged in solar projects and are well aware of the existing and potential problems of the different PV technologies. They are tracking the developments in the areas where PV is to be deployed in large quantities, and they are controlling the safety...or are they?

As can be seen in Figure 8-5, the most likely areas for large-scale PV installations are those of higher sunlight intensity, especially the desert areas of Utah, Colorado, California, Arizona, and Texas. As discussed in the previous chapters, these areas—the deserts—have the harshest climate on Earth and will provide a number of serious challenges to any and all PV technologies operating there.

There are a lot of players in any PV installation, and a lot of responsibilities to be shared, but there are no official, uniform standards and procedures outlining the individual and collective responsibilities. So for now there are a lot of improvisations, stepping on toes, gaps, and finger-pointing when things go wrong. We do hope that this chaotic situation will change soon, and that we will end up with a well-regulated and controlled PV industry—especially when utility-scale projects are involved.

Here is a list of major players in each of the PV installations with varying responsibilities from project to project:

1. Manufacturers, suppliers. and dealers
2. Independent system operators
3. Utilities
4. Public utility commissions
5. Developers and installers
6. Investors and financiers
7. Public and private land owners
8. Mining industry executives/managers
9. Transport companies
10. Recycling facility operators
11. EPA regions/headquarters
12. BLM and other federal agencies
13. Federal and state entities and regulators
14. Tribal governments and communities
15. Local governments and chambers of commerce
16. Local and county planning commissions
17. Environmental organizations
18. Other public and private parties

All these entities are in one way or another involved in, and responsible for, the solar power plants' planning, finance, design, installation, operation, decommissioning, safety, and regulation—and everything in between. Each has some responsibilities and accountabilities, but since this is a fairly new game, the rules of which change from time to time, there are number of major gaps, delays, and failures in the PV energy markets.

One of the major problems we see these days is that many new players are entering the field just because it is growing so fast. New solar companies and many existing ones that have recently switched to solar are the new specialists. The governing authorities and decision-makers are also quite new at this game, and most of them, just like their private counterparts, don't have the necessary background and experience to understand the issues at hand needed to make the right decisions. The lack of standards and uniform procedures complicates things further. Things are getting clearer as time goes by, but it will be a long time before the solar industry is fully standardized, optimized, regulated, and running smoothly. Until then there will be lots of improvisations and corrections.

The major issues that we propose for debate and resolution by users, investors, and regulators are:

a. The evaluation and implementation of PV technologies coming from new and low-cost manufacturers, in light of their efficiency, reliability, and longevity in large-scale PV power fields.
b. The evaluation and implementation of PV technologies containing toxic materials, and their long-term use in large-scale PV power fields in the US deserts; this must be open for discussion and resolved by the responsible parties—manufacturers, users, and regulators.
c. Manufacturing, permitting, and regulatory procedures must be optimized and standardized to ensure a quick and efficient development of PV power plants in the US.

In summary: The PV industry is currently going in the right direction, but there are still a number of unresolved issues, as spelled out above. Thus, the near future of large-scale PV development is not clear enough, and we know even less what to expect in the long run. We just have to hope that wisdom and professionalism will prevail in all our decisions in order to allow the quick and trouble-free implementation of large-scale PV power generation in the US and around the world.

Notes and References

1. http://www.aproundtable.org/tps30info/globalwarmup.html against global warming.
2. What is the energy payback for PV? US DOE.
3. R.W. Miles et al. Progress in Crystal Growth and Characterization of Materials (2005).
4. RE-Powering America's Land: Renewable Energy on Potentially Contaminated Land and Mine Sites, EPA, 2008.
5. ExternE web site ExternE web site. http://www.externe.info/
6. Energy Pay-Back Time (EPBT) and CO_2 mitigation potential, Evert Nieuwlaar, Erik Alsema.
7. RoHS web page. http://www.rohs.gov.uk/
8. Department of Energy (DOE) cancer death case. http://www.dol.gov/owcp/energy/regs/compliance/PolicyandProcedures/proceduremanualhtml/unifiedpm/Unifiedpm_part2/Chapter2-1800Exhibit10.htm
9. The World's Worst Environmental Disasters Caused by Companies. http://www.businesspundit.com/the-worlds-worst-environmental-disasters-caused-by-companies/
10. Environmental Aspects of PV Power Systems, http://www.energycrisis.org/apollo2/pvenv1997.pdf
11. ExterneE, Impact pathways of health and environmental effects, http://www.externe.info/
12. USGS: http://minerals.usgs.gov/minerals/pubs/commodity/cadmium/mcs-2010-cadmi.pdf
13. http://www.abebooks.com/servlet/BookDetailsPL?bi=1108560583&searchurl=ds%3D30%26isbn%3D9783540541844%26sortby%3D17
14. http://www.fas.org/ota/reports/9344.pdf
15. http://www.iiasa.ac.at/Admin/PUB/Documents/IR-02-073
16. Environmental risks regarding the use and end-of-life disposal of CdTe PV modules. http://www.dtsc.ca.gov/LawsRegsPolicies/upload/Norwegian-Geotechnical-Institute-Study.pdf
17. Health, safety and environmental risks from the operation of CdTe and CIS thin film modules, Steinberger, 1998 http://onlinelibrary.wiley.com/doi/10.1002/(SICI)1099-159X(199803/04)6:2%3C99::AID-PIP211%3E3.0.CO;2-Q/abstract

Chapter 9

PV Energy Markets Now and in the Future

"Sometimes one pays most for the things one gets for nothing."
Albert Einstein

BACKGROUND

We don't know exactly what (or who) Einstein had in mind when he wrote the above quote, but it perfectly fits today's solar energy generation scenario. We get lots of energy absolutely free from the sun every day and then pay dearly for this same energy, converted in one form or another, for use in our daily lives.

Our energy-hungry world gobbles as much electricity as it produces, and at a high price. As a matter of fact, it is one of the key drivers of our economies and personal lives. It is also far from enough, with nearly 1/3 of the world's population still living in total darkness and another 1/3 barely able to afford it. As energy prices go up constantly, due to rapidly increasing energy demand in developing countries and dwindling energy resources, we can't help but wonder how long the other 1/3 of the world will be able to afford to keep the lights on...

Figure 9-1 is a nighttime satellite picture that reflects the difference between the abundantly lit populated centers of the developed countries and the darkness of the rest of the world. Guess where the largest energy markets are now and will be in the near future? Yes, there is a lot of talk about providing power to all people in the world, but in the end only those who have money to pay for it and resources to implement it will get it. Make no mistake, the lit areas of the world in Figure 9-1 show exactly where most of the future energy power production, including alternative energy, is going to be generated and used at least for the next several decades.

The countries that use most power need more energy and new fuel sources, because the old fossil-based sources are almost depleted, are getting prohibitively expensive, and are polluting. Oil is fluctuating around $100/bbl, and there it could get much higher with time, which could create havoc in the world's economies. Coal and natural gas are very dirty; nuclear is too dangerous; hydro is limited (and environmentally damaging), so in order to keep the lights on and cars and economies running, the developed and many developing countries (some of which have no other choice) will need a lot of renewable energy in the near future. The race is on.

Government subsidies and incentives created a solar frenzy in the world energy markets lately, which can be summarized with one sentence: "Produce as many PV modules as quickly and cheaply as possible, and install them as quickly as possible before the subsidies dry up." It's a simple capitalist approach to maximizing profits, while the opportunity lasts. No time is allotted for concerns with quality, reliability, longevity and such. Nope, not now, maybe later. Now we are busy installing millions of cheap PV modules as quickly as possible anywhere we can—even in extreme climate regions where many of these modules may not survive.

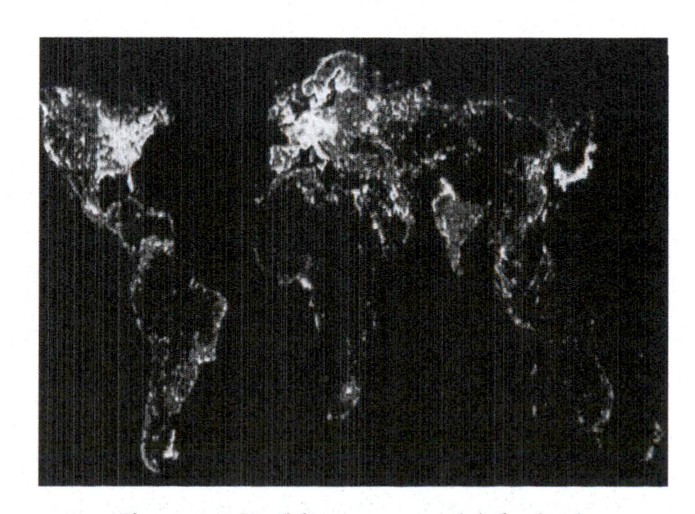

Figure 9-1. World's energy use (night time)

THE ENERGY BALANCE

Energy use has nearly doubled during the last 40 years, from 1,200 kg oil equivalent per capita in 1970 to 2,000 kg in 2010, with most of the world's electricity produced mostly by the conventional energy sources, oil, coal, natural gas, nuclear and hydro power generation. Millions of tons of fossils remains were pumped up, dug out, transported, burned and refined, while millions of tons of toxic byproducts were constantly dumped somewhere in the soil and water, or exhausted into the air.

Coal, oil and natural gas are equally polluting (yes, natural gas production process is a major polluter), and as their consumption increases, so does the level of environmental contamination in its many forms. It goes without saying that this is how we have been generating energy for the last 100 years—by heavily and indiscriminately contaminating the very place we live in, and harming humans in the process. This is how we will continue doing it until we poison and choke ourselves to death—unless we wake up and change our ways.

Note: Do you know that the US government pays over $5.0 billion every year to oil companies in the form of subsidies, tax breaks and other perks, and that this payoff has existed since 1916 when the government instituted income tax incentives to encourage individuals and corporations to drill for oil?

Yes, the same large companies, most of which are multi-national conglomerates with billion-dollar gains are paid billions of US taxpayer dollars every year. This is a drop in the bucket for these companies and a great loss for the taxpayers. So why is the government wasting our money? Your guess is as good as ours, but it is one of those things that makes governments what they are—detached from reality and unfair in many ways.

During the 1930s, federally financed dams created hydroelectric power. We don't know how much money was spent to prop up the hydro industry, but just think of the enormity of the Colorado river dams built in the mid-20th century. This was an immense effort at an equally immense price; billions upon billions of dollars for sure. Thousands of workers built world-class structures which are still up and operating and are the pride of our nation. Could solar follow in the steps of the hydro industry? Not likely any time soon, and at least not until all technical, logistical and financial issues are properly addressed.

From the 1950s onward, the federal government also financed extensive research into nuclear power. Studies of the subsidies to the nuclear-power industry over the decades indicate aggregate subsidization at well over $150 billion, and a subsidy intensity (government support per kWh output) normally exceeding 30% of the market value of the energy produced. This support continues with millions of dollars spent on further research and improvements.

These are big numbers, which we cannot even dream of applying to the solar industry today for a number of reasons we discussed previously. Plus, most of this is history. Getting back to the present, we wonder what would happen if the $5.0 billion in subsidies paid to the oil companies were offered to the US solar industry instead. This annual influx of billions of dollars, in our opinion, would put solar on its feet in no time. This alone represents over 1.0 GWp of guaranteed PV installations (as much as installed in 2010 in the entire USA) and thousands of jobs every year. Most importantly, it would mean that the solar industry finally has the *full and unwavering support of the US government* (which is not the case today), taking the government's reputation and credibility to much higher levels. Imagine for a moment the US government committing to an ongoing $150 billion support for the solar industry. This is a sure way to solving our energy problems. Far fetched? Yes, no doubt! Ain't gonna happen anytime soon—not until we find ourselves in total distress, like after WWII when the great infrastructure construction in the US occurred.

Full and sustained government support is something that has been lacking in the past and is very important for future progress of solar energy in the US. If that happens, the solar industry will no longer be looked at as a step child, or as an afterthought, but rather as a rightful member of the US energy sector. Wishful thinking yes, but it is arguably the best way to put the solar energy on its feet quickly.

Nuclear power has been providing significant amounts of energy since the 1960s. It is clean power, but it is quite dangerous as demonstrated by the Three Mile Island, Chernobyl, and Japan 2011 disasters. Since most nuclear plants are old, with major parts rusting and deteriorating over time, we can expect more serious incidents in the future.

That's not all. We still have not resolved the radioactive waste storage issue. There are presently 70,000 tons of nuclear waste stored *in situ* (in the backyards of the 104 nuclear power reactors in the US alone). Soon the number will grow to 140,000 tons here, with that much or more around the world, and so on, until something happens—something we don't want to think about, but something that's unavoidable.

The now infamous Yucca mountain nuclear storage

silo was hailed as the solution to the nuclear waste nightmare, only to be mothballed after spending $14 billion on its development, thus making it the most expensive high tech cave in the world. This means that we spent 20 years thinking that we have a long-term nuclear waste disposal solution, just to find out that we don't have any solutions. Going back to the drawing board, most likely without success, new efforts will create more "gifts" for future generations—gifts bestowed in our drive for bigger and better things for ourselves.

Who would've thought that natural gas, which we have always considered "clean" and safe happens to be the biggest polluter of them all? The complex and expensive shale natural gas "fracking" process creates large amounts of methane, gases and liquid waste products, which contaminate the ground, water table and air in large areas of the US and the world. And drilling is expanding by the day.

Methane gas escaping fracking sites causes 20% more environmental damage than the dirtiest of them—coal mining and burning—and is 30% more damaging to air, soil, and water than natural gas obtained via conventional means. Since fracking is the way of the future for natural gas production, its pollution can only increase.

Natural gas power plants are significant air pollution sources as well, releasing hazardous air pollutants and fine particulate matter. The methane they release traps 20-25 times more heat than carbon dioxide. We don't know the quantity of greenhouse gas (GHG) emissions from natural gas because producers are not required to report those emissions.

The Obama Administration took steps to mandate reporting of heat-trapping gases, but the oil and gas industries were exempted. Imagine that: the biggest polluters are not required to report the damage they cause. All other GHG-generating industries are required to report—a penalty for not being able to afford expensive lobbying.

Fuel (grain) ethanol must be mentioned here as one of the most promising renewable energy sources. The world's fuel ethanol industry has been around for half a century and the production equipment and processes are well established and optimized.

Ethanol is used in most countries as an additive to gasoline and for many other purposes; thus, it is an essential link in world energy markets. Yet, the US fuel ethanol industry has been receiving substantial government (federal and state) subsidies since 1978 at the rate of 51¢ per gallon of ethanol used. This several billion dollars per annum of taxpayer money is going into the pockets of the multi-national energy giants, because the subsidies are in the form of tax credits to the energy companies, not to farmers who produce the grains, or the producers who supply the ethanol.

The PV subsidy equivalent of 51¢ per gallon (not counting additional subsidies paid to farmers and producers, which could be quite significant) would represent ~4.25 million added to the income of our 100 MWp PV power plant operating in the Arizona desert. But the solar industry has no chance of getting that lucky.

Ethanol fuel is one of the few truly renewable, green and safe technologies available today, so one would think that after so many years of non-stop operation it should be able to make it on its own. It has its own problems, however, and as a matter of fact went through a major scale-down recently. So it remains to be seen how it will hold up in the future and especially after upcoming subsidy cuts planned by US politicians.

Hydro, while clean and safe during power generation, is creating some incredible environmental damage and causing disasters on an unheard-of scale. From dams collapsing and wiping out entire populations, to flooding millions of acres of land, hydro power is leaving its irreversible imprint on the Earth's surface. Just look at the controversy surrounding the Three Gorges Dam in China, which flooded close to 1,000 km^2, including archeological and cultural sites and populated centers, and displaced 1.3 million people. "The end justifies the means" the decision makers probably thought then, but now they are forced into re-evaluating the entire situation, because there are so many problems. See Wikipedia for full details.(16)

Itaipu Dam in South America is even bigger than the Three Gorges and is considered the most expensive structure ever built. It flooded over 1,500km^2 of populated land and displaced over 10,000 indigenous families. An electrical failure at the dam in 2009 shut down power to the entire country of Paraguay and parts of Brazil, affecting over 50 million people and the local economies. (17)

Is this how we secure an energy future? Not many new dams can be built, even if we decided that this is the only way to energy independence.

During four years in the first decade of this century, the world population increased 5%, CO_2 emissions increased 10% and energy production increased 10%. If this trend continues until 2050, the world population will double, requiring the production of 100% more energy, accompanied by a 100% increase of CO_2 gasses. Actually, the trend has been slowing down since 2008 due to economic stresses and other fiascoes, forcing the world into a "survival" mode, which is less wasteful and is making us take a closer look at our limited resources. This period of slowing down and awareness is even contributing to the present energy and environmental awakening, which resulted in expansion of alternative energy generation and use. Coincidence? Nope, it is the inevitable reminder that things are bad and could quickly become desperate.

Where does all this energy come from? 34.80% of energy used today comes from crude oil, 29.4% from coal, 23.8% from natural gas and the rest is split equally between nuclear and hydro power. Wind and solar contribute a minuscule <0.2%, which doesn't even show on the energy graphs.

Note: This breakdown of energy sources does not include the energy used in developing countries, where open wood and coal fires are used by millions of people on all continents (yes, North America too) for heating, food preparation and other household needs. It is also hard to estimate the CO_2 and other harmful gasses emitted from these fires, millions of which are burning every day.

Just think: most of the world's energy resources originate from the sun's rays. Some of that energy has been preserved as coal, crude oil, natural gas, and trees. So our God-given solar energy has been around for millions of years, and some of it was stored under the ground for us to use as we see fit, and boy, did we use it. In less than one century we've pumped out and burned most of the fossil fuels that were stored underground over thousands of years.

Solar power generation has been mostly on the back burner until now, so we must conclude that comparatively speaking it is not in its infancy, as many people say. It is rather an *embryo,* which is still going through metamorphosis. This embryo was conceived in the 1970s from the desperation and frustration of the western world with OPEC's maneuvers, but it went back into hibernation when oil prices went down. Now it is beginning to reawaken during a worldwide movement for energy independence and environmental clean-up. Solar developments of late are good, albeit far from what we expect solar to be 100 years from now, when the solar energy industry will be fully matured and competing head-to-head with conventional energy generators.

Embryo or not, solar is here to stay. This time it cannot go back in stand-by mode as it did several times since the '70s, but its progress won't be as smooth as we want it to be. It is, and will be, hindered by a number of technical and other issues which we discuss and analyze in this text. Like it or not, solar is the world's future energy source and we must keep in mind that whatever we have now is just the beginning, not the final solution. Far from it!

The inevitable questions persist: If we are in such desperate need for new energy and a clean environment, why is it that although there are so many solar projects planned and/or applied for in the US and Europe, so very few get under construction? Why are the completed PV projects in the US fewer than those in the rest of the world—below the estimates and far from our expectations? Why is there so much talk about global warming, energy crisis, green energy, energy independence etc., and so little action? Why is there so much money floating around and so few PV projects started? Where should we look for solutions?

We will take a close look at these and many other energy-market issues, considering technical, administrative, logistical and financial advantages and disadvantages, and the issues and barriers in the solar industry, while staying focused on large-scale PV power generation. We will take a look at the key factors contributing to the growth and stagnation of the PV industry and the variations in the PV markets in the US and abroad.

INTRODUCTION

IEA issued a report in 2008, called, "Time is Running Out," which correctly stated: *"The world's energy system is at a crossroads. Current global trends in energy supply and consumption are patently unsustainable—environmentally, economically, socially. But that can—and must—be altered; there's still time to change the road we're on. It is not an exaggeration to claim that the future of human prosperity depends on how successfully we tackle the two central energy challenges facing us today: securing the supply of reliable and affordable energy; and effecting a rapid transformation to a low-carbon, efficient and environmentally benign system of energy supply. What is needed is nothing short of an energy revolution."*

The US energy revolution actually started in the 1970s sparked by the OPEC embargo. US taxpayers were so fed up with the never-ending games of the Arabs, and their pain was so real and the cries so loud that that they were heard in Washington. New energy policies were adapted and millions of dollars used for R&D and energy subsidies. Solar benefited directly from the very start of the revolution, with many private companies and government labs working on new solar technologies and markets.

Unfortunately, the fast-paced revolution slowed down several years later and went dormant soon after. Efforts were made to revive the revolutionary spirit several times during the 1980s and 1990s but cheap oil prices and spectacular gains of the conventional energy generators on Wall Street did not allow significant progress.

The IEA report goes on to explain how the existing world oil supplies are getting depleted, and that the solution is in finding new energy sources and new ways to generate energy. While we agree with the report's conclusions and suggestions for "an energy revolution," we'd emphasize that the revolution should start by stepping up the implementation of renewable energy sources in a safe, sensible and efficient manner. Any delays will have negative effect on the world's population, due to insufficient and very expensive energy supplies, and the deadly pollution most of these create. On the other hand, hasty deployment of cheap but unproven energy sources may be nothing short of wasteful and maybe even dangerous in the long run.

Nevertheless, talk is cheap. There is nothing revolutionary in continuing with the empty talk of "what could be," while the status quo of "Drill, baby drill" persists. It will only extend the agony, and won't solve the world's energy or environmental crisis.

"Revolution is an event," someone said, to which we must add "a successful revolution is a process." It is a work in progress which could last an eternity. The second, or third phase of today's energy revolutionary process started again in 1997-1998, when oil prices over $150 caught the world unprepared.

The fear and the energy crisis were especially pronounced in Europe, because the old continent is poor on energy resources. Europe depends heavily on Arab oil and Russian natural gas—not a particularly favorable combination and a very shaky partnership at a time when energy demand is increasing significantly and energy prices are going up dramatically.

So, 1997-1998 marked the next phase of the ongoing energy revolution. Lots of solar installations went up in Europe—more than anywhere else—mostly financed by government subsidies. Talk about high hopes for a new and brighter solar future were the norm and can still be heard among the media.

Lately, however, we are noticing a familiar dampening of the revolutionary spirit. Technological, logistic, regulatory and financial problems are threatening to put an end to this latest phase of the energy revolution process and send it into a slumber...again.

Spain, France, the UK, and even Germany are drastically reducing solar subsidies and incentives, and we see a significant slowdown of solar activities there and elsewhere. Would this be another failed attempt to bring a bright energy future to the world?

We need to emphasize here also that, regardless of the development in one or the other energy sector, one single energy source, or one single company cannot solve all energy problems. A joint effort, consisting of many different technologies and many different companies is needed to provide the energy markets with efficient, reliable, safe and profitable energy generating sources.

Large-scale PV power generation in the US is one of the key elements on the road to energy independence. It is, however, still marginal, although California utilities are actively promoting PV installations and that might help sustain the growth of this most important segment of the solar industry at least in that geographic area. Nevertheless, the existing and planned large-scale PV installations are a drop in the energy market's bucket. And as with every new beginning, the first steps are the hardest and slowest. Considering the serious technical and financial difficulties solar is experiencing today, the California utilities are headed in the right direction. We hope that they will be successful in reaching their goals to set a good precedent for the world's PV industry to follow.

THE PV ENERGY MARKETS

A clarification is needed here. Energy markets are those areas in the US and around the world where electricity is produced, distributed, sold and used. For example, Los Angeles, California, is a huge and complex energy market with ever increasing needs, served by a number of utilities and energy service companies. Most small towns and villages, on the other hand, are served by a private local entity (a wires-only utility in most cases) which provides all local energy needs. At the same time, most of Central Africa and many other areas of the world are not even part of the energy market (see Figure 1 above) simply because their infrastructure and economies cannot support significant and profitable energy generation, distribution and use. There are some notable exceptions, as we'll see in the text below.

PV power generation, although still quite new and un-established as a meaningful power generating source, is nevertheless part of the energy market and has the same goals as conventional power sources: to provide cheap and reliable energy. Thus far, however, the PV energy supply has not been cheap or reliable.

We have discussed the different issues and drivers in this text, as well as the need to find ways to reduce PV energy prices and increase the efficiency, reliability and durability of PV technologies. These are challenging but "must-do" tasks, if PV is to become an integral part of the energy market.

Supply and Demand

Supply and demand is a tricky variable, especially where immature business such as PV power generation as we know it today is concerned. And we can say confidently that the present-day solar business (and PV in particular) is as immature as they come. We insist that it is in its embryonic state, and although it is showing some signs of maturity, these are not a sure indication that the trend towards maturity has even started.

Supply and demand picture is a reflection of the PV industry's immaturity, but represents only one facet of the complex picture of technical, political and financial issues it faces at present. It is exposed to the growing trend of uncertainty, lack of consistency and any type of market stability or controls in the solar industry.

Note the higher levels of (past and future) global supply, vs. demand in Figure 9-2. Actually we believe that the ongoing world's economic crisis will bring some even more serious surprises in this area during 2011-2012, with surplus hitting record levels—unless China and other Asian countries find a way to use the surplus in their own territories.

It's a complex picture to be sure, but overwhelmingly with supply consistently much higher than demand. This, and the related price drops, allowed serious growth of PV installations in 2010, where global PV cell and module production peaked at ~24 GW and 20 GW respectively. This represented an annual increase of 110% over the previous year. Thus world PV installations reached 16.2 vs. 7.5 GWp in 2009, or a staggering 146% increase.

This is due to several factors:

1. The strong European FiT markets, mostly in Germany (7.2 GW) and Italy (2.2 GWp),
2. Lower PV module prices that have kept pace with subsidy cuts; i.e., Chinese c-Si modules dropped from \$2.35/Wp in 2009 to less than \$1.80/Wp in 2010, and finally but very importantly
3. Surplus of polysilicon material, wafers, cells, and modules due to unprecedented expansion of manufacturing capacities, with China and Taiwan leading the pack of low-cost manufacturers which is bringing prices down to \$1.10/Wp in 2011.

The demand during several periods (especially 2nd and 4th quarters) of 2010, however, challenged the ability of suppliers to keep up despite the significant increase of production capacities in that part of the world. This makes us believe that PV installations could've exceeded the 20 GW mark in 2010 if there were no temporary and unpredictble shortages.

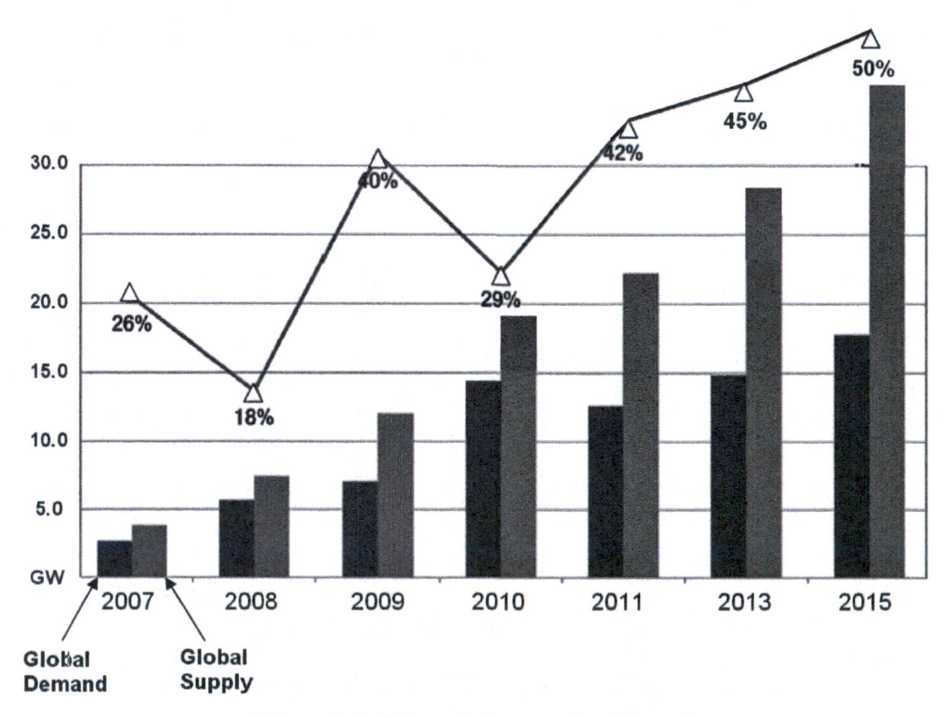

Figure 9-2. PV modules supply-demand

The breakdown of PV technologies manufactured in 2010 was approximately as follow:

c-Si modules	19,800 MWp
CdTe modules	1,450 MWp
a-Si modules	1,350 MWp
Super m-Si modules	950 MWp
SIGS/CIS modules	450 MWp

No question, silicon based technologies were well ahead of the rest, and we believe they will remain so until new, much more efficient technologies make their way onto the energy markets. China was responsible for nearly 50% of the PV modules manufacturing capacity in 2010 (mostly silicon based) and is planning to increase its share on the world PV market in the years to come. This, in the light of the new world-wide economic situation remains to be seen, however, and we would not be surprised if many Chinese PV module manufacturers go out of business, or go back to making LEDs or whatever else they were doing before getting subsidized into the PV modules making business.

The picture of significant surplus of PV modules observed in the past will most likely dominate the solar industry in the future. But there are other reasons for the unexpectedly slow growth of PV projects in the US, with land permitting, regulatory changes, interconnection and transmission issues contributing heavily.

Another very important factor affecting large-scale PV projects is lack of confidence in the quality, reliability and longevity of the PV modules—especially those coming from low-cost and newly hatched manufacturers. Potential customers and investors are not convinced in the reliability and profitability of these PV technologies (modules and inverters in particular) for long-term use in large-scale PV installations in the desert.

One doesn't have to be a specialist to figure out why these are such big issues. Lack of confidence in cheaply made PV modules is a main subject of every meeting of PV professionals, installers and investors. Some take it seriously, while some take advantage of the situation for the sake of making a quick buck. Many books have been written on the matter and entire meetings and conferences dedicated to it. The results thus far, however, are spotty for lack of better words, and simply do not meet the expectations.

While the numbers of manufactured modules are impressive and quite obvious, the quality of the modules from the different Asian manufacturers is neither. Just recalling that 3-4 years ago most of these manufacturers didn't even know what a PV module was, makes us wonder how they became so quickly world class specialists and reliable manufacturers of one of the world's longest-lasting (or expected to last) products? Even more importantly, how, when, and where did they test their modules for long-term reliable operation? How do they calculate the reliability and longevity of their modules? And since most of them haven't even seen a desert, let alone understand the conditions in one, how did they make sure that their modules would be able to withstand the blistering desert heat and UV radiation during 30 years non-stop operation? The answers, or lack thereof, are quite unsettling.

So, customers and investors need assurance and insurance in order to put their trust and money into these products, unless they have money to burn. Annual power loss, temperature degradation, and failure rates of PV products during 30 years non-stop operation in the deserts are the Trojan horse and the best kept secret of the PV industry. But the public is not easily misled into thinking that just because a PV module looks nice and shiny, it will stay that way and operate trouble-free for 30 years under the torture of the elements.

Note: A major solar modules and services provider, using high-quality, US-made PV modules issues the following long-term power output warranty: "The System's electrical output during the first ten (10) years of the Lease Term shall not decrease by more than fifteen percent (15%)."

This, however, is 1.5% annual power loss for US-made modules in 2011. Keep this in mind as a "best case" practical scenario when designing or purchasing power-generating systems. Many other manufacturers are making promises (some that sound much better than the above) that cannot be verified, so we have a choice of closing our eyes and taking the plunge, or taking a very close look at what we are buying.

So supply is expected to be high, demand low and many of the large-scale PV projects which could solve the demand problem are going nowhere for a number of reasons. Due to the complexity and high cost of large-scale PV installations, their large size and long duration, there is an obvious need to make sure that the quality of all components in these large fields is acceptable. We need to ensure that modules, inverters etc. equipment do not degrade excessively or fail during the long exposure to the harsh elements.

There are ways to do this, but a standardized team effort is needed, something that is not available, nor easy to do presently. Nevertheless, developers and investors should open the communication channels and start serious discussions with the manufacturers on the above issues. Additional actions, such as plant visits, QC system and documentation inspections, test and other data collection would help to get an idea of the manufacturer's readiness to provide quality product and stand behind it for the duration. But the best and most successful approach would be for manufacturers to participate in the projects using their products as part owners. This way their unwavering cooperation would be secured, thus ensuring the long-term success of each project.

Background of the Energy Markets

Many countries are far ahead of the US in the alternative and renewable energy game, with:

#1. Iceland, 100% hydro and geothermal power
#2. Norway, 99% hydropower
#3. Brazil, 83% hydropower
#4. New Zealand, 68% hydro and geothermal power
#5. Canada, 58% hydropower
#6. China, 82% fossil and nuclear power

While:

— Half of South Asia has no access to electric power at all
— Most Sub Saharan African people live without electric power
— Only Australia and the UK are ahead of the US in the use of conventional fuels with 92% and 96% respectively fossil or nuclear power.
— The UK and many other countries have a good excuse; they just don't have a Mojave Desert where sunlight abounds, but we don't know what the problem in Australia is, for they have more sunshine and deserts per capita than the all others put together. Actually we do know; they are having a governmental solar energy policy meltdown. See details below.

The progress of late in the alternative energy area, however, is undeniable. 300 GW of new electricity generating capacity were added globally in 2008 and 2009 from which 140 GW, or almost half, came from renewable energy sources. Solar was a significant part of it, and the trend continues.

Some estimates point to an irreversible move towards using renewable energy, with close to 80% of the worldwide energy in 2050 coming from renewable energy sources. We are not sure if this is achievable, but we know for sure that solar—one way or another—will play a major role in the upcoming alternative energy boom.

One third of the planet's landmass is covered by deserts, which receive intensive solar radiation every day, so there is plenty of energy...we just have to get it somehow. Several expert studies have estimated that using just 4% of the total desert area for solar systems is sufficient to supply all the electrical energy requirements of the world. The Gobi desert alone receives ~20 times the total world primary energy supply. The solar radiation falling on just 4% of the world's desert is sufficient to meet all world electrical energy requirements today. Of course, having the energy is an important prerequisite, but we also have to find a way to capture and deliver it where it can be properly used. Not a small task!

The theoretical potential of solar energy is enormous—far beyond what the world could ever use—even in the face of increased energy demand by developing countries. Greenpeace estimates 1,884 GWp PV energy will be generated by year 2030 around the world. This is possible, but only with a serious commitment by many governments and companies. A well coordinated effort and standardized procedures are needed if we are to get even close to achieving that goal.

A great number of solar thermal and PV projects have been planned and some approved for development in the Southwest, since it is where the sun shines most abundantly. Some of the thermal solar projects were converted into PV power plants lately, so we need to wait and see if this is a trend that we'll see in the years to come.

Nevertheless, there is a lot of sunshine in the Southwest, which is why most US and many world PV products manufacturers are gearing up for work in the US energy markets in the future.

Recently, PV prices (as indicated in Figure 9-3) have been falling steadily while energy markets were hot for a number of reasons. Prices will continue falling, nearing $1.25/Wp average in 2011 and even lower in 2012, mostly due to oversupply resulting from sig-

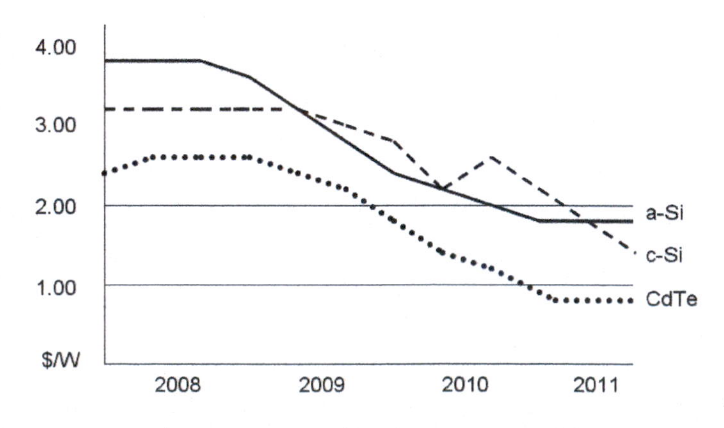

Figure 9-3. Module prices 2008-2010 (1)

nificant cooling of the world's PV markets. The solar installations slowdown during 2011-2012 will affect all PV components and production equipment manufacturers, resulting in lower PV modules prices and significant production equipment order cancellations.

Figure 9-4 reflects the difference of estimates made mid-2010 and mid-2011. The top of the dark column shows the 2010 estimates for PV installations around the world, while the lighter column is the revised estimates made in 2011. Recent changes in the energy and economic policies of the major European countries, and increasing financial difficulties around the world forced cumulative caps, FiT cuts and overall slowdown of solar activities. Depending on the financial situation in the EU and Asia, projections might have to be adjusted again next year, but the solar boom is over for now and the foreseeable future.

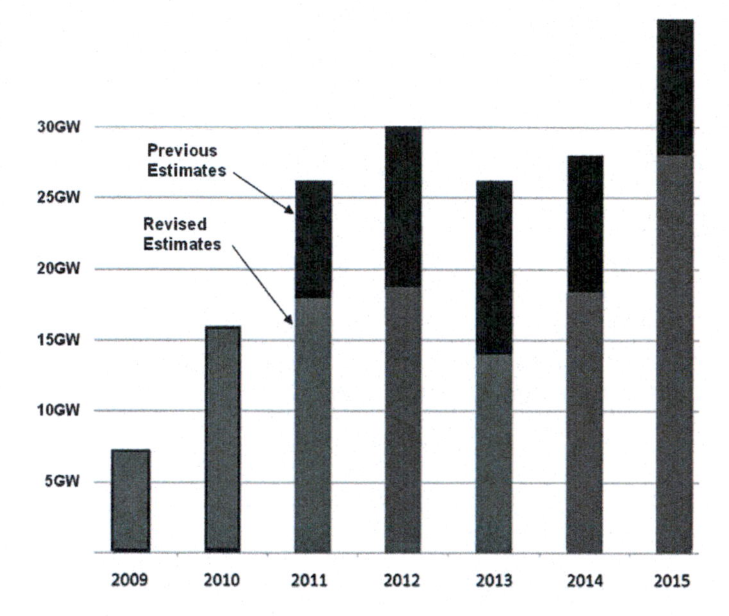

Figure 9-4. Estimates of world-wide PV installations.

Thermal solar looks promising and is quite appropriate for large-scale installations in some locations in the US deserts. The manufacturers and contractors (see Table 9-1) are experienced enough and are doing a good job thus far in delivering quality systems at the right price. The issue with using water for cooling must be resolved, however, in order to preserve dwindling water resources. Using air-conditioning units instead of water cooling is a possibility which, although not as efficient, might be the most appropriate approach in the long run.

Table 9-2 is a partial list of the major thin film PV modules and equipment manufacturers during 2010. We didn't even bother to list the c-Si PV cells and modules manufacturers, because even a partial list will result in dozens of entries.

No doubt, there are a great number of PV products manufacturers already, and new are popping up almost

Table 9-1. Solar thermal equipment manufacturers and costs

TECHNOLOGY TYPE	PLANT SIZE	PRODUCTION COST	INDUSTRY PLAYERS
Power Tower	50-200MWp	$0.12-0.15/kWh	eSolar
			Solar Reserve
			Bright Source
			Bechtel
			Abengoa
Updraft Tower	100-200MWp	$0.12-0.15/kWh	Enviro Mission
Stirling Dish	10-50MWp	$0.15-0.25/kWh	Infinia
			SES
Parabolic Trough	50-500MWp	$0.15-0.25/kWh	Abengoa
			Sky Fuel
			Solar Millennium
			Solel
			FPL Energy
			Acciona
Linear Fresnel	10-100MWp	$0.10-0.15/kWh	Ausura
			HelioDynamics
			MAN Ferrostaal

every day. Some are involved in manufacturing established technologies via established processes, while others are trying new materials, products and processes.

Note in Table 9-3 the large number of European PV plants. Europe is a good place for PV modules, both c-Si and thin film. Weather conditions and solar insolation levels are optimum for the present-day state of the art. Unlike the scorching heat and unforgiving UV of the deserts, most of Europe offers moderate climate which is favorable for PV modules' normal operation and longevity. Thin film modules offer some advantages over silicon based technologies in the "darker" partially cloudy climates and will continue to expand their share in these geographic areas.

Table 9-2. Thin film PV equipment manufacturers per type

Company	Product	Company	Product
Abound Solar	CdTe	Nano PV	a-Si
Ascent Solar	SIGS	Nanosolar	SIGS
Ascentool	CdTe	Opti Solar	a-Si
AVA Solar	CdTe	Primestar Solar	CdTe
Avancis	SIGS	Qingdao New Energy	a-Si
Bekar Europe Gmbh	a-Si	RESI	CIS
Daystar Tech.	SIGS	Shell Solar	SIGS
Dow Chemical	CIS	Signet Solar	a-Si
Energy PV	a-Si	Solar Field	CdTe
EPV	SIGS	Solexant	a-Si
First Solar	CdTe	Solopower	SIGS
Global Solar	SIGS	Soltaix	a-Si
Heliovolt	SIGS	Solyndra	CIS
Innovalight	a-Si	Stion	CIS
ISET	SIGS	Uni-Solar	a-Si
ITNIES	CIS	Wurth Solar	SIGS
Light Solar	CIS	XsunX	a-Si
Miasole	SIGS	Zia Watt Solar	CdTe

The US Energy Markets

Contrary to common belief, the huge overall US energy market—the oldest, largest and most advanced in the world by far—is not a single unit. It is instead a compilation of components, some fragmented, and which simply lacks the overall coordination and standardization needed for efficient operation of such a large entity. It has evolved into a conglomeration of a large number of different markets operating under federal, state and local rules and regulations, executed by the local utilities under the supervision and guidance of central and local authorities.

So what we have in the 21st century USA is 50 well-defined state markets with hundreds of different (much less defined and ever changing) local market variations within each state energy market. Or as one analyst put it, "We have over 3,000 utility-driven energy markets in the US."

Feudalism operated this way, with the local landlords making the rules and calling the shots. It worked for a while, but then it stopped

Table 9-3. World's largest PV power plants, 2009

97 MW	Canada,	Sarnia PV power plant
84.2 MW	Italy,	Montalto di Castro PV power plant
80.7 MW	Germany,	Solarpark Finsterwalde I,II,III
70 MW	Italy,	Rovigo PV power plant
60 MW	Spain,	Parque Fotovoltaico Olmedilla de Alarcón
54 MW	Germany,	Solarpark Straßkirchen
53 MW	Germany,	Solarpark Lieberose
50 MW	Spain,	Parque Fotovoltaico Puertollano
48 MW	USA,	Copper Mountain Solar Facility
46 MW	Portugal,	Moura photovoltaic power plant
45 MW	Germany,	Solarpark Köthen
40 MW	Germany,	Solarpark Waldpolenz
36 MW	Germany,	Solarpark Reckahn I,II
35 MW	Czech Republic,	FVE Veprek
34.5 MW	Spain,	Planta Solar La Magascona & La Magasquila
34 MW	Spain,	Planta Solar Arnedo
31.8 MW	Spain,	Planta Solar Dulcinea
31 MW	Germany,	Solarpark Tutow I,II
30 MW	Spain,	Parque Solar "SPEX" Merida/Don Alvaro
26 MW	France,	Gabardan PV power plant
26 MW	Spain,	Planta solar Fuente Álamo
25.7 MW	Germany,	Solarpark Helmeringen
25 MW	USA,	DeSoto Next Generation Solar Energy Center
24.5 MW	Germany,	Solarpark Finow
24 MW	France,	Parc Solaire Les Mees
24 MW	Korea,	Sinan power plant
23.4 MW	Canada,	Arnprior PV power plant
23.2 MW	Spain,	Planta fotovoltaica de Lucainena de las Torres
23 MW	Spain,	Parque Solar Hoya de Los Vincentes, Jumilla
22.068 MW	Spain,	Huerta Solar Almaraz
21.78 MW	Germany,	Solarpark Mengkofen
21.47 MW	Spain,	Parque Solar Parque solar El Coronil I + II
21.2 MW	Spain,	Solarpark Calavéron
21 MW	USA,	Solar electric power plant, Blythe
20 MW	China,	Jiming Hill, Xuzhou City PV power plant
20 MW	Germany,	Solarpark Rothenburg
20 MW	Korea,	Seoul power plant
20 MW	Spain,	Parque Fotovoltaico SOLTEN I+II
20 MW	Spain,	Planta solar fotovoltaico Calasparra
20 MW	Spain,	Solarpark Beneixama
20 MW	Spain,	Parque Solar El Bonillo

working. Equally so, until now, having utilities and local regulators dominate energy markets has worked, but they are becoming obsolete and some of their actions (or lack thereof) are even harmful to the development and progress of alternative power generation.

The US energy market needs a complete refurbishment...from the top down. Alternative energies have to be given a priority for our sake and for that of future generations. It is unreasonable and selfish to think that we can continue digging and using dinosaur remains indefinitely in order to run the world's largest economy. This line of thinking is not new. It was clearly identified, followed, and later on ignored in the 1970s, 1980s and then in the 1990s. This time things are somewhat different and much more serious, so we hope that the unprecedented trend of solar energy generation growth of late is here to stay and that federal, state, and local authorities will *stay* behind it this time.

One key element of that growth is the anticipated increase in commercial and utility scale systems installations. Because of the importance and urgency of these installations, we focus on them and the related complexities in this book. Nearly 7 GW of utility scale projects are under contract for implementation by 2015 in the US. Nearly double that (14 GW) are announced but have no signed PPA as yet. The solar energy field is changing as we speak in favor of PV power generation, and we foresee drastic increase during the next several years.

As with any other beginning, PV installations are not growing fast enough. Even the most aggressive and dedicated California utilities—PG&E and SCE—offer fairly small contribution to that growth. They have plans for adding less than 1.0 GW PV power by 2015 which is a good step ahead and a good example to follow, but is still just a drop in the energy bucket. So, PV power generation is still a very small part of the US energy complex with a total of ~2 GWp solar generated capacity thus far, which is equivalent to the power generated by a single nuclear plant. At the same time, the US hydro and nuclear power plants generate over 100 GW of electric power. Coal and oil fired plants generate even more power than all others put together and are also expected to grow.

The US PV industry is growing slowly in the shadow of its much larger cousins and it will be a long time before the PV embryo grows enough to be even noticed by the energy giants. It will take even longer before it walks on its own, which is not going to happen any time soon, so it will be something that future generations will have the honor and pleasure to turn into reality.

We are absolutely confident that 100 years from today at least 80-90% of the energy produced in the US will come from renewable sources—wind and solar mostly. That will be the only choice then, because there will be no oil or coal left, and nuclear waste will be piled sky-high by then too. We hope we can move the development of alternative energy sources far enough for future generations to build upon and recover quickly from the energy and environmental problems they will certainly inherit from us.

The US PV Markets

World PV installations peaked during 2007-2010 in the US and around the world. While Germany installed over 7.0 GWp in 2010, the US managed to squeeze less than 1.0 MWp PV installations on its territory. Large portions of these installations were roof-mounted PV systems, heavily subsidized by government, state and local entities. This was not a brilliant revival of the solar revolution in the US, but the solar embryo is alive and well, and US solar markets are still, no doubt, the most promising for its development in the long run. Just in time, because with 91% of our power coming from fossils (coal, oil, natural gas, etc.) and nuclear power sources we are still one of the most energy *inefficient* and polluting nations in the world.

Global solar energy growth is expected to remain below 20% per annum for the next several years, and will most likely slow down some, due to the financial problems in Europe and Asia. As a consequence, the US share of the global market might double and even triple, with global PV companies targeting the US for expansion in this huge energy market.

During 2009 the US PV market grew 36% mostly due to federal incentives and subsidies. This growth was accompanied by the introduction of foreign-made low-cost PV modules and products, new marketing strategies of PV vendors, utilities and investors. The growth was sustained despite the challenging economic difficulties in the country and the ever changing political and regulatory policies.

Residential installations represented the bulk of the gains in the PV market expansion. But how did ordinary people pay for the solar installations in the midst of the financial difficulties we face? According to some analysis, the major payment methods in 2007-2009 were: 68% of the new solar installation owners paid cash, 20% paid with home equity loans, 10% with home mortgage refinance, and 2% used signature loans. The numbers today are somewhat different and suggest third-party financing as the major money source for residential and small commercial systems (less than 10 kW). A number of different financing sources are used for larger residential and commercial systems (greater than 10 kW).

Solar installers have come up with variety of schemes to reduce the amount of initial payment for a new system and spread the cost over time. This way a home or small business owner could have a PV system installed on their roof with minimal or no down payment. The financial benefits vary with location but in most cases they are quite small. The systems division of SunPower was the leading company in terms of PV installations in California during 2009; Chevron Energy and SPG Solar performed strongly in 2009 too and moved up to the #2 position. Among residential and small commercial installers in California, REC Solar, SolarCity and Real Goods Solar led the field in those sectors. This line-up changed somewhat in 2010, but these companies are still major players in US residential and small commercial PV markets.

Although not as explosive as in Germany, 2009 was a year of transformation for the US solar market. According to Solarbuzz summary, "Changes in the roles of utility companies, new market entrants, lower cost PV modules from Asia and new direct-to-market approaches became more prevalent. As a result, solar companies doing business in the States will need to adapt quickly to these challenges while also being responsive to frequent adjustments in the fragmented incentive and regulatory environment." And this is just the beginning.

California's 2009 edge as a critical lead and base load state market for the US was obvious, with 53% of US's PV on-grid installations (Tables 9-4 and 9-5). Despite the slowdown in demand from the corporate sector across the US, government, residential and utility growth more than offset the slowdown. Price cuts in residential installations provided the foundation for steady growth across the country. A wide range of start-up markets in other states are well underway as new PV incentives were launched.

Currently PV installations in the US are slowly increasing and the US share in the global energy market is nearly 6% over the past decade, although it fell down to 5% in 2010, mostly due to the rapid ramp up of PV installations in Europe. California contributed 49% of the PV installations in 2010 and the indications are that California, with its flagships PG&E and SCE, will continue to be the leader in solar installations—small and large—for the foreseeable future.

Ten states accounted for over 85% of the solar energy generation in 2010 with California still leading the pack. New Jersey is number two with 15% market share, while Nevada, Arizona and Colorado follow with 6%

Table 9-4. Grid-tied PV installations in the US, 2009

States	*MW_{DC}*	*Market Share*
1. California	768	61%
2. New Jersey	128	10%
3. Colorado	59	5%
4. Arizona	46	4%
5. Florida	39	3%
6. Nevada	36	3%
7. New York	34	3%
8. Hawaii	26	2%
9. Connecticut	20	2%
10. Massachusetts	18	7%
All Other States	83	7%
Total	**1,256**	—

Table 9-5. 2010 US solar installations

U.S. utilities	2010	Cumulative
PG&E (CA)	157.3	476.5 MW
FPLC (FL)	87.2	117.4 "
PSE&G (NJ)	74.7	117.4 "
SCE (CA)	68.4	578.3 "
Xcel Energy (CO)	42.0	85.6 "
Tri-State G&T (CO)	30.2	- "
APS (AZ)	29.9	52.5 "
SDG&E (CA)	27.1	89.5 "
JCP&L (NJ)	22.9	51.1 "
DEC (NC)	20.8	- "
CPS Energy (TX)	15.5	- "
ACEC (NJ)	15.2	35.1 "
JEA (FL)	12.2	- "
TEP (AZ)	11.9	- "
Ohio Power (OH)	10.4	- "
SMUD (CA)	10.7	- "
HEC (HI)	9.8	- "
LADW&P (CA)	9.8	- "
LIPA (NY)	8.7	- "
PECO Energy (PA)	7.6	- "
NSTAR Electric (MA)	6.2	- "
MEC (PA)	5.0	- "
Oncor Electric (TX)	5.0	- "
Nevada Energy (NV)	-	90.0 "

each. We do foresee US market growth outpacing the global by doubling in size annually for the foreseeable future, but that, in light of a number of political and financial changes, remains to be seen.

Solar installations are growing slowly, but surely, though not as smoothly as desirable, and a number of changes are expected to rearrange the entire PV industry during the next several years. Nevertheless, the potential of the US solar market is the envy of the world—with plenty of sun (which many of the world's PV leaders just don't have) and abundant quantities of desert lands, which are critical for efficient implementation of large-scale PV power plants. Not to mention the infrastructure which is ready, albeit under duress, to accept the additional amounts of generated power. This makes the US energy market the most suitable and desirable for solar development in the near future. For this reason alone many US and foreign companies are starting to position themselves for a place in the US energy market, and we foresee the field getting even more crowded.

While there are a number of barriers and issues to be resolved, the central policy thrust in the US over the past several years remains positive. Sixteen states and Washington, DC, have enacted a Renewable Portfolio Standard with solar set-asides to promote PV installations. The dispersed funding sources mean the US market does not carry the same level of risk compared to countries driven by a single national policy. Nonetheless, federal incentives are playing a much larger role in stimulating demand.

Within the next five years, industry watch dogs forecast the market will grow to between 4.5 to 5.5 GW depending on the political and financial scenarios. This is around ten times the size of the 2009 market, an average annual growth rate of 30% per annum. The key drivers of this outcome will be more aggressive positioning in the utility segment based on the need to meet their Renewable Portfolio Standard obligations, the development of new state markets together with the return of the corporate segment, and steady growth in residential demand stimulated by cuts in end-market pricing. The US order book for photovoltaic systems currently stands at 12.0 GWp. This represents the total of solar set-aside RPSs, projections of demand from multi-year funded incentive programs, stimulus-funded projects and other large utility identified projects.

The US energy market is accepting the PV growth, albeit somewhat unwilingly, with utility scale projects growing proportionally. (See Figure 9-5.) If everything goes well, the trend will expand during

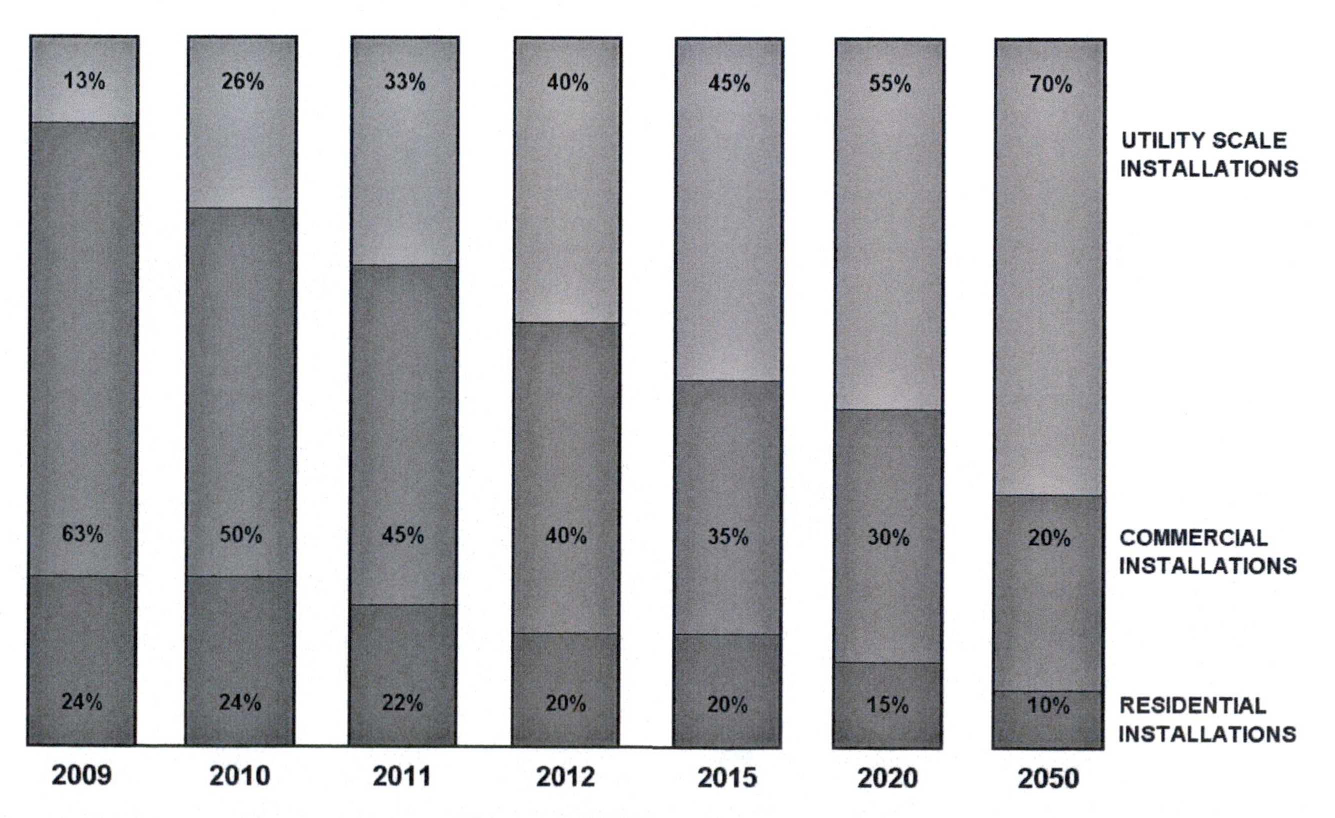

Figure 9-5. US PV markets 2009-2050

the next several years, bringing great gains to the US solar industry.

Note in Figure 9-6 the humble ~2.2 GWp actual contribution of solar power to the multi-Giga watt conventional power industry in the US. The story of David and Goliath comes to mind. What are the chances of another miracle? David is not going to give up, but all things considered, a long and difficult road awaits him.

The future looks good for solar so far, but the new political and socio-economic conditions in the US might bring some changes which we cannot qualify and quantify precisely. What will happen when present-day incentives and subsidies are decreased or eliminated? Would customers, utilities and investors continue their interest in solar power if it were less profitable and more difficult to implement than the conventional power sources?

Solar energy generation is here to stay for now, but a number of critical questions still need to be answered, as follow:

1. Which state energy markets offer the greatest development opportunity?
2. Which technology is most appropriate for the particular locations?
3. Which utilities will procure the most PV over the next five years?
4. How are acquisitions altering the project developer landscape?
5. Which developers will lead the market in coming years?
6. When will PV become cost-competitive with natural gas?
7. How can PV suppliers serve the utility PV market?
8. What new technologies and approaches will bring new developments?

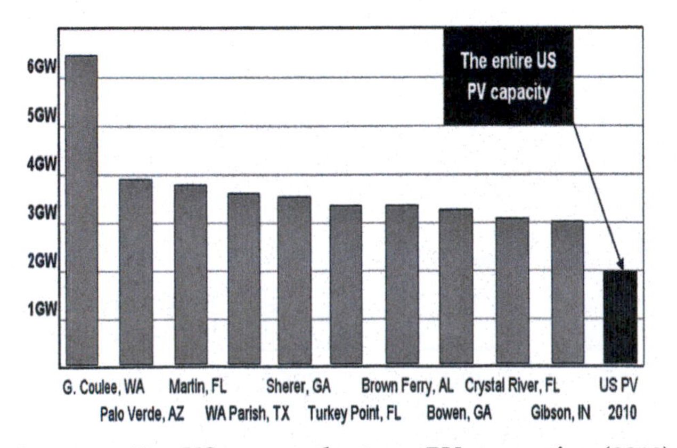

Figure 9-6. Top US power plants vs. PV generation (2010)

9. How would the different states and utilities deal with reduced incentives?
10. What lessons can we draw from other countries' PV experience?

2010-2011 PV Market Update

The US solar power sector grew 67% in 2010, reaching $6 billion, up from $3.6 billion in 2009. Solar electric installations reached 956 MW, including 878 MW of PV systems. In 2011 PV installations are expected to double from the 2010 level, while the global market will experience slower growth due to subsidy cuts in Europe.

The largest PV power plants in the US in 2011 are shown in Table 9-6:

Table 9-6. Largest US PV projects 2010

US Utilities	*Project*	*Installed*
PG&E	Copper Mountain	48.0 MW
TSG&T (CO)	Cimarron 1 Solar	30.0 MW
FP&L (FL)	DeSoto Solar	25.0 MW
SCE (CA)	FSE Blythe	21.0 MW
Xcel Energy (NC)	Greater Sandhill	16.0 MW
DEC (NC)	Davidson 1 & 2	15.5 MW
CPS (TX)	Blue Wing Solar	14.5 MW
DEC (NC)	Jacksonville PV	12.0 MW
Ohio Power (OH)	Wyandot Solar	10.1 MW
FP&L (FL)	Space City Solar	8.0 MW
Exelon (IL)	Exelon City Solar	8.0 MW
PG&E (CA)	CalRenew-1	5.0 MW

PV is taking the lead, no doubt. Industry sources quote 2,700 MWp of large-scale PV installations are in the pipeline (permitting and pre-development activities) for development in 2011 mostly in the California and Arizona deserts. Most, if not all are under PG&E and SCE contracts and/or supplying power to these and other California based utilities.

Still, the total capacity of the fully operational PV plants in the US presently is several times lower than the PV installations in Germany alone. The US share of worldwide photovoltaic solar installations fell to 5% from 6.0%, mostly due to booming growth of PV in Germany propped by generous government incentives.

Contracted PV projects, with a signed PPA in 2010 were over 5,000 MW, while announced projects were over 10,000 MW. Several projects were completed, and a large number were cancelled, or are still pending and in negotiation or litigation mode. There are a lot of good developments to report on the PV front, especially in California, but we need to also look into what is hindering more intensive growth of solar energy markets.

As we saw in the previous chapters, a number of issues (political, regulatory and legal) stopped several solar projects in 2010, and are presently stopping the development of other major solar projects too. We must point out also that while in the past having a signed PPA meant that the project is 100% "shovel ready," today this is no longer the case. There are a number of solar projects (some of them very large) with PPA and yet still on standby for a number of unresolved regulatory and environmental issues and/or are in litigation. So, our conclusion is that a project is not "shovel ready" until the first shovel is in the ground and no red flags are anywhere to be seen.

At a broader level, demand remains driven by the Section 1603 Treasury Cash Grant program, state-level incentives, and improved project economics following the module price crash amidst the global financial crisis. The nagging question of what will happen to the US solar energy market in the post-Section 1603 era persists, so we must be prepared for the consequence of the inevitable changes it will bring. We saw the drastic changes in Europe and can't help but wonder who or what is next.

Finally, a number of unresolved technical issues, as discussed in the previous chapters, are still looming over the fledgling PV industry. Quality, reliability, and longevity of PV modules depend on a number of factors, most of which the users have no control over. Materials, manufacturing processes, and the related quality issues constitute a "black box" in PV installations design, implementation and operation schemes.

New and unproven manufacturers are pushing millions of cheap PV modules on the US energy markets, where ignorant and/or greedy buyers don't hesitate to install them on any vacant roof or lot, hoping that the unproven PV modules will operate and last as needed. This, however, is gambling, and while we wish all gamblers out there good luck, we would warn them that just because a module looks new and shiny, it won't necessarily perform or last as specified.

Keep in mind that saving a penny by buying cheap and unproven PV modules from new and unproven manufacturers today, might cost you a buck or more tomorrow. Please take this free advice, coming from 35 years experience with these products: think twice before buying cheap and unproven modules. They are not disposable toys, but serious equipment expected to last 30 years or more. Any compromise now might mean a disaster later on.

The US Solar Future

Nobody knows what the future will bring with regulatory and financial changes, as well as legal challenges, more of which are expected in the near future. Anything is possible with the serious political and economic changes the US and the world are going through. We have seen this scenario play several times before, when things looked so good one day, only to wake up the next day in a different political and economic environment which did not support solar energy development. And so the solar embryo had to go back into hibernation several times since the oil embargo of the mid-1970s. Is this another cycle? We don't know yet, but recent developments in Spain, Germany, and most other EU countries point in that direction, so we have to be on our toes for awhile yet.

By mid 2010 solar markets in the US and EU looked so good that the forecasts were mind boggling. There were daily reports of increased production capacities, new PV projects planned and going into construction. Some predicted 60% increase in solar installations in the US during 2011 over 2010, vs. 20% percent forecasted previously, attributed mostly to commercial and utility projects as many companies transition to solar energy driven by favorable incentives and attractive prices.

In total, 878 megawatts (MW) of photovoltaic (PV) capacity and 78 MW of concentrating solar power (CSP) were installed in the US in 2010, enough to power roughly 200,000 homes. In addition, more than 65,000 homes and businesses added solar water heating (SWH) or solar pool heating (SPH) systems. This growth was driven mostly by the federal Section 1603 Treasury program, completion of significant utility-scale projects, expansion of new state markets and declining technology costs.

Compared to many countries where subsidies and location variables drive a certain segment more than others, activity in PV solar energy market segments in the US is spread out evenly which leads to market growth stability. Residential, commercial, and utility PV installations account for approximately 1/3 each. In contrast, we have seen in Germany, Italy, and France, countries with large solar energy subsidy programs tend to have emphasis on particular market segments and have provisions which can overheat installation activity leading to boom and bust cycles and sometimes a complete market collapse.

Sure enough, the EU, with some exceptions, is going through a serious "bust" cycle—the most serious in Spain, where all government support was withdrawn with the stoke of a pen. The Spanish solar embryo will have to go into hibernation now... Are any others of its foreign cousins going to follow?

Combined with steady federal and state subsidies, strong incoming solar radiation, large numbers of roofs and available land, rising cost of fossil fuel generation and high electricity demand, most industry observers believe the US solar energy market will double again in 2011 and will continue growing through the next several years. We have the resources, the infrastructure and a growing need, so who or what is going to stop us?

When viewing the rising cost of coal and natural gas prices as a result of increasing economic activity combined with growing cooperation on Capitol Hill regarding energy policy and rapidly decreasing PV system cost, we will not be surprised if the US PV market doubles and triples during 2012 and beyond. See the estimates in Figure 9-7. Large commercial and utility-scale PV projects are the most prominent and will contribute most to the increase of solar energy in the US.

PV manufacturing operations in the United States increased the production of solar components substantially in 2010. Production of solar modules rose 62%, while wafer production grew 97% and cell manufacturing rose 81%. Although impressive, these numbers represent only a small part of worldwide PV products manufacturing capacity. Sure enough, stiff competition from low-cost regions such as China forced three domestic PV facilities to close last year. The BP Solar plant in Maryland, Intel-backed SpectraWatt's New York facility, and Evergreen Solar's factory in Massachusetts all shut down due to rising costs and not being able to compete with the cheap imports.

Additional plant closures are expected in 2011-2012 for the same reasons, and worldwide slowdown in the sector. New plants are planned to be built during the same time period by Wacker Chemie AG, Flextronics, and Stion. The outcome of the US PV products manufacturing sector is uncertain at this point, with increasing materials and labor costs, and falling product prices being the main reasons.

We cannot solve the world's solar woes, so we should take care of the situation at home and close to home. The international mandates in Table 9-7 reflect the plans of Mexico, the US, and Canada to expand the solar energy contribution to their energy portfolios. The US will lead the pack, and we expect international cooperation to grow and many joint projects to be develop on the continent.

The US government is doing its part, albeit somewhat hesitantly and even clumsily, in trying to help the solar industry overcome the problems it is facing. Recently, the DOE allocated $27 million to eliminate solar installations' red tape by bringing down the "non-hardware installation" costs. Of this sum, $15 million goes to "streamlining the installation process for local governments." It's pocket change used to solve a mega-problem, and it won't even touch some of the key issues plaguing the broader permitting processes of large-scale

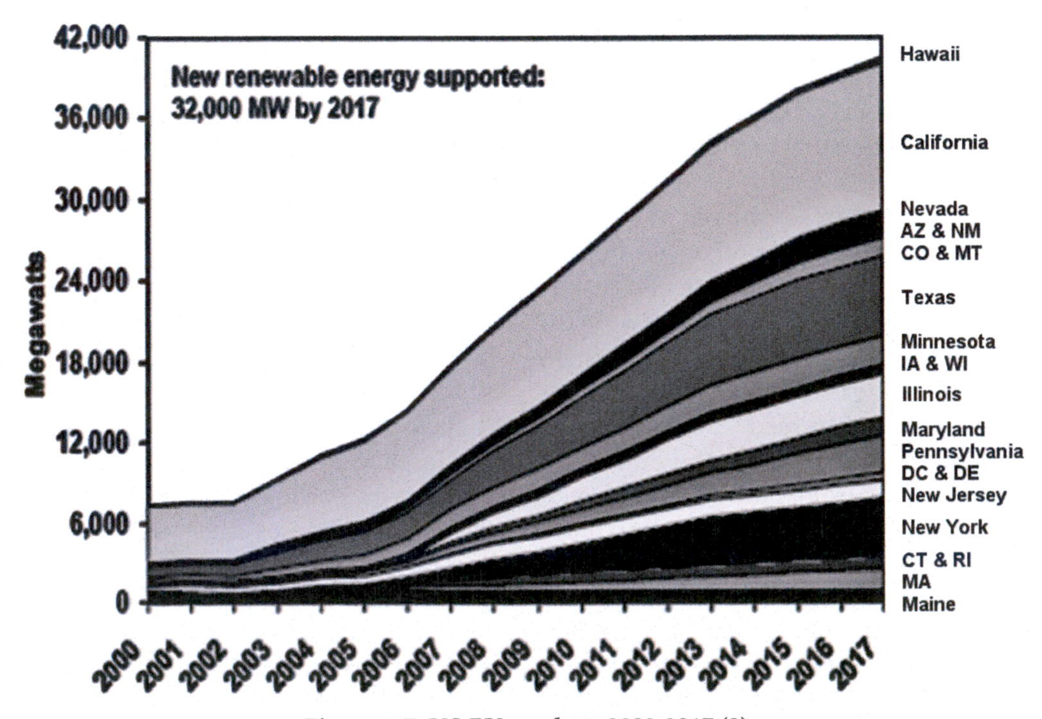

Figure 9-7. US PV markets 2000-2017 (2)

Table 9-7. Renewable energy mandates 2006-2017 (2)

Year	Mexico (MW)	Canada (MW)	US (MW)
2006	1271	395	14381
2007	1356	1745	16967
2008	1356	2406	19520
2009	1457	2406	21866
2010	1559	4006	24143
2011	1710	4006	26450
2012	1812	4079	28761
2013	1913	4079	31171
2014	1913	8760	32949
2015	1913	8760	34854
2016	N/A	9140	36094
2017	N/A	9140	37175

installations. It won't even address the critical land and interconnection issues, and the related project delays and cancellations.

"Innovations in IT and local business processes, such as online permit applications, can deliver significant savings for solar energy systems and will help America to compete globally in this growing market," says Dr. Chu. That's a good "Introduction to Permitting Costs 101," Mr. Secretary, but where is the beef? Lots of words and little money won't rebuild the US solar industry. Billions of dollars, thousands of people and several years of total dedication and hard work are needed to address the key issues and remove the major obstacles. Maybe it is a good first step, this $15 million morsel, but many, many such steps and much bigger morsels are needed to get where we are headed. Many!

Finally, we must remember the fact that most PV modules used in the new US PV installations are Chinese-made. Suntech, Yingli, and Trina heavily subsidized by the Chinese government, are invading the US energy markets and filling their pockets with taxpayers' money. What do we know about these manufacturers, some of whom have 2-3 years production experience? What do we know about the quality of their products—remember the durability and longevity issues discussed previously? When and how did they test their newly designed PV modules?

How do we know that these modules won't fail, thus converting the US PV power fields into junkyards, 10-20 years down the road? The Chinese companies will go away as quickly as they came into existence, and then what? Would the US PV energy program become a lesson on how *not* to do things?

Why don't we bring the PV products manufacturing home, where we can control the pricing to begin with and the products' quality in the long run? Why do we still need to depend on unstable foreign governments to supply our energy needs. Did we forget the lessons of the past? Isn't it time to switch from a "consumer" to the "Made in the USA" mode so well respected worldwide?

The California Case

California is leading the way, no doubt, and we believe that the trend will continue. A quick daily reality check is provided by the California Independent System Operator Corporation (CAISO) which launched a new tool in 2011, "Today's Outlook" (9). The charts show the actual real-time solar energy being generated on the CAISO wholesale power grid, which serves 80% of the state. The 24-hour real-time feature provides transparency to the variability of solar production during each hour, day, week, and season. This includes minute-by-minute plots of the MW of power being generated by solar at any given time. See Figure 9-8.

The overwhelming impression one gets from these charts is that solar power generation is inconsistent and of short duration. The largest amount and most useful solar energy produced on this particular day was from 10:00AM until 2:00PM, or total of 4 hours. The total load demand, however, was many times larger than the alternative energy can provide; i.e., CAISO power graphs (9) show nearly 30 GW total demand, while only 300-375 MW were generated by solar, and only at peak hours (around noon) during this particular day.

Today's Outlook charts also show that the state

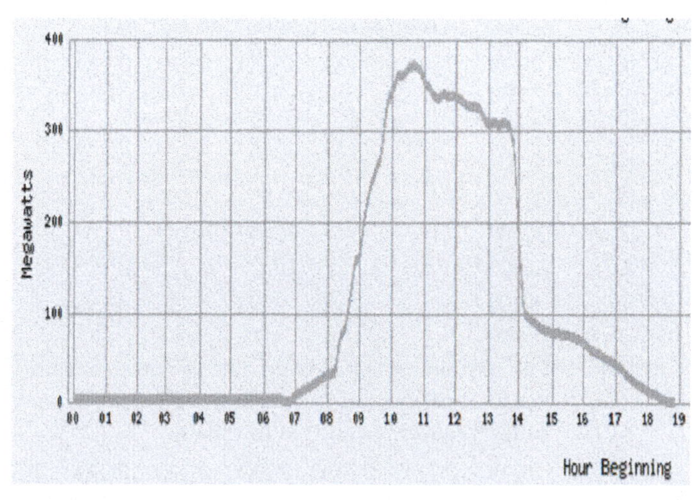

Figure 9-8. California solar power generation, daily chart, April 2011 (9)

(the CAISO domain) had more than 40 GW of available resources (with some power plants on standby) on that day, which means that the power industry at its present state is fully capable of taking care of the power demand fluctuations. This also means that the state is not desperate for solar and wind, and that a long uphill battle is ahead for alternative energy sources as they try to replace conventional generators.

In summary:

1. 300-375 MW solar power is produced for several hours around noon in Figure 9-8, and much less during the rest of the day, so basically the total power produced by solar is much less during early morning and late afternoon hours, and none at night.
2. There are more full-power generating hours in summer, less in winter, and none on cloudy or foggy days.
3. PV power generation is only part of the total solar power generation in the state, and we cannot tell what portion of the 300-375 MW in CAISO's graphs is contributed by PV.

No matter how you look at it, solar energy generation is a small drop in the sea of energy produced in the state, and even smaller on a national level. Although the share of solar energy generation is steadily increasing, it will still take a long time before it makes a significant difference on the US energy markets.

Still, California is way ahead of the other states. Just take a look at the numbers of total solar installations in different states compared with those in California:

CA	57,000	solar installations through 2010
NJ	5,300	"
PA	2,500	"
HI	2,300	"
MA	1,900	"
NY	1,700	"
AZ	1,500	"
CO	1,500	"
CT	1,200	"
WI	1,100	"
Other	The remaining states were able to squeeze from 1; yes total of one and two solar installations in some states to several hundred in others.	

Surprising numbers: the mighty Texas has total of 300 solar installations and only 35 MW on its entire territory, sunny Florida 400 installations and 74 MW PV power, mostly flat desert New Mexico 300 installations and only 43 MW PV power generation. And the other surprise is New Jersey. How was a state with half cloudy and half snowy weather able to install several times more solar than ever-sunny Arizona? Politics? Finances? Probably both.

We must give New Jersey due credit for defying the odds and just like Germany proving that although location and climate matter, political will and money rule the game sometimes. Just like Germany, however, this growth (in our opinion) is temporary because it is not sustainable. The basic rule, "Solar energy generation requires a lot of sunshine" cannot be changed to accommodate poor weather conditions. Politics and money can delay the inevitable, but sunshine locations have much more chance to remain in and win the solar race in the long run.

These are signs of the times and clear indication that solar has not only unpredictably variable power output, but that it also has unpredictably varying acceptance as a power source in different parts of the country.

Things are still moving in the right direction, and even accelerating in California, since it is increasing the renewable portfolio standard (RPS) from 20% in 2010 to 33% in 2020. This means that 33% of all electric power in California must be generated by alternative power generators. So, let's see:

1. Assuming that California's entire electric power generation capacity is 70 GW AC, and that
2. 14% (~10 GW AC or 12 GW DC) of the energy mix today comes from renewables (biomass, geothermal, small hydro, wind and solar), then
3. By 2020 California must add at least another 15 GW AC alternative energy in order to meet the 33% RPS mandate.

Considering the part-time operation and hourly variability of solar and wind generation, adding energy storage capacities, and converting DC to AC increases the total to ~25 GW DC alternative energy to be added to the California energy mix by 2020, which will cost ~$150 billion for the solar installations alone. Replacement of old and outdated technologies from existing power fields (bio and hydro generators, steam turbines, coolers, CSP trackers, liquid transfer lines, pumps, heliostats, inverters, PV modules, etc.), and addition of transmission lines, substations etc. would cost, let's say, another $50 billion. So a minimum of $200 billion (in our estimate) is needed for this undertaking to be completed by 2020, if adding solar and wind energy was the only alternative.

These are staggering numbers under any circumstances, but especially during the current financial slump, the end of which is not in sight. Granted, lots of this money would come from private and institutional investors, government loan guarantees, grants etc., but even if California and the US government have all the money they need (which they simply do not have now), how is this huge undertaking going to be accomplished? Is this even possible, technically and logistically speaking? What technologies will be used? Where will the hardware come from? Are we going to cover the deserts with cheap China-made PV modules and the mountain tops with cheap China-made wind generators? How about the land ownership, permitting and right-of-way issues? How are we going to solve these in such a short period of time, if it takes years to go through the processes of each individual installation? And let's not forget the interconnection points and transmission lines, which are one of the big show-stoppers today. How fast can we build substations and run transmission lines across the deserts? With so many questions, we hope there are people who have answers, because our energy future depends on that.

How much of this added energy is going to be generated by solar, and how much of the solar component is going to be PV is another unknown at this point. We have seen plans for adding ~1.0 GW of PV power generation by 2015, plus several GW of other alternative energy technologies (CSP and wind mostly), but the total is still very far from the 25 GW DC needed to meet the RPS goals. CSP generation is going through some major changes too, so we are not sure how this will affect solar power development in general and PV in particular.

There is still time, though, and we should not discount American ingenuity and drive, which are especially effective and productive under duress. So we keep our fingers crossed for California, hoping it will show us and the world that alternative energy is achievable and that it is here to stay.

A Glimpse into California's Energy Future

The electric power industry and the responsible state agencies are planning the operational, technological, and market changes needed in the power sector to reach the 33% RPS mandate. The California Energy Commission (CEC), the California Public Utilities Commission (CPUC), the California Independent System Operator (CAISO), and the local regulatory authorities (municipal utilities and irrigation districts) are actively participating in the implementation of the RPS program.

The CEC role is principally to certify renewable resources (mostly solar thermal [CSP] plants of up to 50 MWp) for RPS eligibility and to develop a tracking and verification system to ensure that renewable energy output is properly accounted for and within RPS compliance. The CEC, in coordination with the Western Governors Association, established the Western Renewable Energy Generation Information System (WREGIS) as a tool in the compliance with its mandate to develop a renewable energy registry and tracking system. WREGIS tracks generation from power plants using renewable certificates in order to verify compliance with state and local regulatory requirements throughout the Western power grid. Several additional market programs have been included on a voluntary basis.

The CPUC and local regulatory authorities focus on procurement activities of their respective jurisdictional retail sellers of electricity, review and approve renewable energy procurement plans (and long-term procurement plans that include renewable integration capabilities). These also review contracts for RPS eligibility, establish standard terms and conditions for renewable energy contracts and establish compliance rules and procedures for investor-owned utilities (IOU) and other electricity service providers.

The CAISO is responsible for providing transmission services in the state and minimizing grid congestion. Optimizing grid management is the goal, and one aspect of this approach is using a mathematical approach to analyze and solve problems, instead of the present less efficient "zonal" approach. Clear price signals and "day-ahead" scheduling are expected to improve the power flow and optimize its real time management. These actions are critical for maximizing grid reliability and are expected to provide efficient electric power generation, transmission and utilization within the California energy markets to accommodate the new (variable) power sources, including wind and solar.

Renewable energy certificates (RECs) and energy from qualified renewable resources procured together as "bundled" commodities are currently eligible for California RPS compliance purposes as well. RECs sold separately from the underlying energy are "tradable" and are not presently eligible for RPS compliance, although the CPUC has statutory authority to permit the use of tradable RECs.

Major California utilities, PG&E, SCE, SDG&E, LAWD and others, are proceeding with plans to add significant amounts of alternative energy power to their power generation mix. Due to a number of recent developments, there is uncertainty as of how much of the planned solar power will be generated by CSP and how much by PV sources.

In addition, and very importantly, California's legislation is driven by serious environmental concerns, which forced the state to adapt Assembly Bill 32 (AB 32), the California Global Warming Solutions Act of 2006 which requires the reduction of greenhouse gas emissions state-wide in order to reduce potential climate change impacts, reduce dependence on oil, diversify energy sources, save energy, create new jobs, and enhance public health.

The California Air Resources Board (CARB) estimates that sources within the state emitted just over 460 million metric tons CO_2 equivalent (MMT CO.e) in 2010. The state aims to reduce emissions to 1990 levels, or 427 MMT CO.e by 2020. In 2008, greenhouse gas emissions from electrical generation for California markets amounted to 116 MMT CO.e, or about one quarter of the total California greenhouse gas emissions, and the emissions from these must be reduced proportionally as well. A cap and trade system via the Western Climate Initiative will assign emission caps to all emitting generators in California, including the imports (the rules for these are still on the drawing board).

Thus far, the 33% RPS has a solid foundation and backup on all levels, so we must believe that it will drive the solar energy markets in California and lead the country into a bright solar future by 2020. The only thing left to do now is to start the installation of the 25 GW DC alternative energy power plants around the state as needed to reach the RPS goals. How many of the new solar generators will be PV power generators is another question and will be hotly debated among customers, utilities, investors, politicians and regulators in the near future.

California Energy Update

As of mid-2011 all above mentioned activities are on track, and the major utilities and regulators are actively involved in the alternative power programs. There are some clear plans and actual work underway to add several GWs of solar and wind power by 2015. Plans beyond that are not clear. Putting all known and planned projects together and considering the best possible scenarios, however, we still don't get even close to achieving the 25 GW DC power estimated to achieve the 33% RPS by 2020. Viable known solar projects, which are estimated to start production by 2015, total only ~2.0 GWp. So let's give due credit to, and thank the California politicians, regulators and utilities for their hard work, and hope for the best.

Summary

All utilities are basically proud of their residential installations, using them as an example of their involvement in the solar revolution. What is not open for discussion, however, is that the money supporting these installations comes from ALL customers, including the unemployed, pensioners, and the disabled. They are ALL obligated to pay into the program so that the few who can afford the $4-5,000 initial payment can have solar on their roofs. Fair, efficient, and sustainable? Not really; just playing the game.

In the best of cases, residential installations cannot solve the energy problem. Large-scale installations are needed to make the difference, and the utilities control their development. Ultimately, until the utilities get fully involved in solar—just like they are fully involved in installing and operating conventional power generators—solar will not get very far. Sun or no sun, without the utilities, solar will be just a slow trickle from house roofs and the big dream of a few enthusiasts.

The new trend in 2011 in that some utilities only allow "load offset" for commercial PV installations. In other words, commercial customers are not allowed to send power generated by their PV systems into the grid. They can only use it on the spot at their location. This does not encourage PV power generation in Arizona.

So most utilities project a "positive attitude" towards solar and are making an effort, mostly in residential PV installations. But this is nothing more than a "feel good" bandage on the national energy wound. And it will be so, until utilities undertake massive, Gigawatt type, large-scale PV installations.

The Future of Large-scale Solar Installations in the US

The US Department of Energy (DOE) and the Bureau of Land Management (BLM) (the agencies) have completed a Draft Programmatic Environmental Impact Statement for Solar Energy Development in Six Southwestern States (Solar PEIS), which evaluates utility-scale solar energy development in Arizona, California, Colorado, Nevada, New Mexico, and Utah. The agencies are preparing the Solar PEIS to reach goals established by Congress to make suitable BLM-administered lands available for solar energy development. These are BLM-administered lands that would be available for right-of-way (ROW) application under each of the alternatives evaluated by the BLM in the Draft Solar PEIS.

The preliminary list goes something like this:

1. **In Arizona**, approximately 9,218,009 acres (37,304 km^2) of land would be available for ROW applica-

tion under the no action alternative, and 4,485,944 acres (18,154 km^2) of land would be available under the solar energy development program alternative.

Three special energy zones (SEZs) would be identified: Brenda (3,878 acres [16 km^2]), Bullard Wash (7,239 acres [29 km^2]), and Gillespie (2,618 acres [11 km^2]).

2. **In California**, approximately 11,067,366 acres (44,788 km^2) of land would be available for ROW application under the no action alternative, and 1,766,543 acres (7,149 km^2) of land would be available under the solar energy development program alternative.

   Four SEZs would be identified: Imperial East (5,722 acres [23 km^2]), Iron Mountain (106,522 acres [431 km2]), Pisgah (23,950 acres [97 km^2]), and Riverside East (202,896 acres [821 km^2]).

3. **In Colorado**, approximately 7,282,061 acres (29,469 km^2) of land would be available for ROW application under the no action alternative, and 148,072 acres (599 km^2) of land would be available under the solar energy development program alternative.

   Four SEZs would be identified: Antonito Southeast (9,729 acres [39 km^2]), DeTilla Gulch (1,522 acres [6 km^2]), Fourmile East (3,882 acres [16 km^2]), and Los Mogotes East (5,918 acres [24 km^2]).

4. **In Nevada**, approximately 40,794,055 acres (165,088 km^2) of land would be available for ROW application under the no action alternative, and 9,587,828 acres (38,801 km^2) of land would be available under the solar energy development program alternative.

   Seven SEZs would be identified: Amargosa Valley (31,625 acres [128 km^2]), Delamar Valley (16,552 acres [67 km^2]), Dry Lake (15,649 acres [63 km^2]), Dry Lake Valley North (76,874 acres [311 km^2]), East Mormon Mountain (8,968 acres [36 km^2]), Gold Point (4,810 acres [19 km^2]), and Millers (16,787 acres [68 km^2]).

5. **In New Mexico**, approximately 12,188,361 acres (49,325 km^2) of land would be available for ROW application under the no action alternative, and 4,068,324 acres (16,464 km^2) of land would be available under the solar energy development program alternative.

   Three SEZs would be identified: Afton (77,623 acres [314 km^2]), Mason Draw (12,909 acres [52 km^2]), and Red Sands (22,520 acres [91 km^2], and Millers (16,787 acres [68 km^2]).

6. **In Utah**, approximately 18,182,368 acres (73,581 km^2) of land would be available for ROW application under the no action alternative, and 2,028,222 acres (8,208 km^2) of land would be available under the solar energy development program alternative.

   Three SEZs would be identified: Escalante Valley (6,614 acres [27 km^2]), Milford Flats South (6,480 acres [26 km^2]), and Wah Wah Valley (6,097 acres [25 km^2]).

Wow, this land mass could power the US many times over, and provide energy to the entire world. Only time will tell how much of this land will be actually used for solar power generation, but the US government is making a giant step in the right direction. It is now up to the states, the regulators, the utilities and the US solar industry to take advantage of the situation.

Most importantly, we must find a way to reduce and eventually eliminate the dependence on foreign energy imports—be it oil or PV modules. Importing mass quantities of strategic energy products (which PV modules are) from any foreign country is not a solution to our energy problems; it will only complicate the situation and prolong the agony!

2011 update: In Q1 2011 over 250 MW of grid-connected PV was installed in the US, according to media reports. This is a 65% increase over installations during the same period in 2010. PV capacity grew across all three PV market sectors—residential, commercial, and utility, with commercial installations showing the strongest growth.

The US is set to increase its global market share, due to the newly developed financial and energy situations at home and abroad. In addition, the solar energy sector is now a major employer, where the total number of green jobs growing steadily are outnumbering those in oil and gas.

The US PV market is still concentrated in a half dozen states which have over 85% of all installations (a 5% increase from 2010), and this situation will not change anytime soon.

US module manufacturing also increased by 30% during the same time. It is not clear, however, if these are US manufacturers, or foreign companies setting up shop in the US.

The CSP market is falling behind for now, with no CSP projects completed in Q1 2011, although over 1.0

GW CSP plants are under construction, some built by foreign companies benefiting from generous US government loan guarantees.

Overall, the US PV market, as fragmented and uncoordinated as it might be, has the best chance to survive the test of time. We only hope that the US government, the regulators, and the scientific and business communities will start working together on expanding the "made in America" brand of solar energy components. We have the know-how, the resources and the need. What is stopping us?

EU Solar Energy Markets

With some small exceptions, Europe doesn't have much sunlight—not nearly as much as some countries blessed with deserts and large areas of direct, uninterrupted solar radiation. Nevertheless, Germany, Spain, Italy and others have done very well during the past several years by installing gigawatts of solar energy—mostly grid-tied, rooftop residential installations or small commercial plants. A number of small to medium PV plants were completed and several are in the making.

This trend is/was driven by the rising needs and decreasing supply of energy, and supported by extremely favorable government policies and subsidies—i.e., in the fall of 2008 several European utilities paid up to $0.75/kW for solar residential power generation. There was a boom of solar installations never seen before or since, mostly driven by the extremely generous subsidies. It was obvious even then, however, that this is an unsustainable situation. Sure enough, the trend is slowing down even faster than anticipated due to the accelerate financial problems in EU.

Solar in EU is not dead yet. It is just going through major changes and needs some time to reorient itself in the new situation. The upward movement will likely continue in the near future with some fluctuations, mostly due to political and regulatory policy changes, and varying incentives and subsidies.

The European PV markets started developing in 2000 with ~100 MWp total installed capacity, which grew slowly to ~1,000 MWp in 2004, and then accelerated quickly to reach 3,400 MWp in 2006, 4,900 MWp in 2007. It escalated to 10,400 in 2008, and capped all expectations at 15,800 MWp installed capacity in 2009. Europe presently generates 73% of the PV power in the world. Cloudy, foggy Europe, mind you, is leading the US and the world in solar energy production.

Remember: A small corner of the Arizona deserts sees more kilowatts of sunlight in one day than the entire country of Germany sees in a full year. What an incredible story the German PV industry is! It's a confirmation that solar energy belongs to those who are determined to use it—regardless of where they are located—and that solar can and will make the difference in the future, when no more fossils are left to burn. For fairness sake, however, we must also say that such rapid growth of solar energy generation under unfavorable climate conditions is unsustainable in the long run. Germany must come up with other approaches and alternative energy sources if it is to achieve **true** energy independence without going broke in the process.

In addition to the favorable policies, generous subsidies and exceptional FiTs there is one major technical reason for the success of PV power in Europe—the fact that c-Si and thin film PV modules do actually like colder, cloudy climates, where they produce a decent amount of energy and last much longer than in the deserts. They function better under cloudy skies, because they don't have to deal with excess temperatures, which cause serious power drop and annual power degradation. Harmful UV radiation which causes additional power loss and premature failures due to EVA degradation and active layer deterioration is also reduced under cloudy skies and cool climate, so the modules perform better and last much longer.

Due to their specific structure and higher sensitivity to a broader range of the sunlight spectrum, thin film PV modules perform exceptionally well—comparatively speaking—under cloudy skies and in cooler climates. This makes them a good choice for most European locations. The encapsulation and the active layers in thin film modules are exposed to much less stress under cloudy skies, limiting the degree of their disintegration and decomposition, thus reducing the level of annual power degradation. The fact that TFPV modules are the cheapest PV type available today also helps in their successful implementation in great numbers in Europe.

We need to mention, again, our concern with the fact that most TFPV modules are all glass, frameless structures with only a thin layer of encapsulation separating the active thin film from the elements. It would take much longer for the elements to damage the protective edge seal under the cooler and darker European skies, but if and when they do, the excessive moisture in some locations would attack and decompose the active thin film. This in turn would result in fast power degradation and might even contribute to environmental contamination from TFPV modules containing toxic materials.

Europe leads the world in utility-scale PV installations, with Germany in the lead, as shown in Table 9-8. While this trend of unsustainable growth, fueled by government subsidies and incentives is slowing down in most EU countries, US PV commercial and utility installations are expected to surpass European solar power generation during the 2011-2015 period.

The European PV Champions

Germany

Germany leads the number and size of PV installations with Spain as the distant second. Germany alone

Table 9-8. Europe's largest PV plants

80.7MWp	Finsterwalde Solarpark in Germany consists of three different PV projects, installed by Q-Cell SE, which also installed the Strasskirchen Solar Park. The entire project can power over 20,000 homes.
60.0MWp	Parque Fotovoltaico Olmedilla de Alarcón in Spain consists of 270,000 PV modules producing 87.5GWh annually, which is enough to power 40,000 homes. It was completed in 2008 at a cost of €384 million.
54.0MWp	Strasskirchen Solar Park in Strasskirchen, Germany contains 225,000 PV modules installed on 135 hectares, capable of producing enough power for roughly 15,000 homes. It abates ~35,000 tons CO2 yearly.
53.0MWp	Solarpark Lieberose in Germany was the third largest operational PV power plant in 2010. It is operated by Jiwi on 163 hectares and can power 15,000 homes at a total cost of €180 million.
47.6zMWp	Parque Fotovoltaico Puertollano in Spain consists of 231,653 PV modules. Part I, producing 82.5MWh, powered 39,000 homes in 2010. The cost of the entire project (70MWp when finished) is €346 million.
46.0MWp	Moura Photovoltaic Power Plant in Portugal produces 93GW/h power annually. The modules are mounted on trackers, which produce more more power. Operating in one of the sunniest places in Europe helps.
45.0MWp	Köthen Solar Park in Köthen, Germany consists of 205,000 PV modules installed on 115 hectares former air field. It was built within 12 months by RGE Energy AG at a cost of €133 million.
40.0MWp	Waldpolenz Solar Park in Saxony, Germany consists of 550,000 thin film PV modules producing 40GWh electric power per annum. It was built in 2008 on 220 hectares at a cost of €130 million.
34.5MWp	Planta Solar La Magascona & La Magasquila in Trujillo, Spain covers 100 hectares and produces 46GWh electricity per annum. The PV modules are mounted on single axis trackers. It was completed in 2008.
34.0MWp	Planta Solar Arnedo in Arnedo, Spain consists of 172,000 PV modules installed on 70 hectares, producing 50GWh power annually to power 11,500 homes. It cost € 181 million to build.

* Note: The difference in producing different amounts of power above is due to the difference in solar insolation levels at the different geographic areas.

installed 1,500 MWp PV in 2008, 3,800 Wp in 2009 and 7,200 MWp in 2010. Total installed PV capacity in 2009 was ~10,000 MWp, exceeding 16,000 MWp in 2010 and projected to reach 25,000 MWp by 2015. This energy revolution has a humble beginning, with an unimpressive 12 MWp installed in 1997, when the futuristic for that time "1,000,000 solar roofs" program was introduced.

Note: Amazingly, this is exactly when the US Department of Energy pronounced solar dead, or unneeded and put it on the back burner. "Drill, baby, drill," the government dictated, and focused on spending billions on burying nuclear waste in mountain caves, a fiasco resulted in the most expensive cave in the world—the $14 billion Yucca mountain silo—which sits idle now as a reminder of the real dangers of nuclear fuel and our futile efforts to control its evil powers.

The German solar boom was characterized and driven by large government and local subsidies and incentives, and grew very fast so that Germany's PV power generation now represents almost 10-15% of the total electricity production in the country at peak time. Note the AT PEAK designation, which actually represents ~1% of the total power generation throughout the year.

Every boom is inevitably followed by a bust, and sure enough an amendment voted by the German Bundestag in March 2011 states that the FiT will be lowered by up to 15%, depending on the growth of the PV market until May 2011. This will then accelerate the flexible component of the legally prescribed gradual FiT decrease. At the end of the year, there will also be an additional reduction, so that PV FiTs could drop by as much as 24% by January 1, 2012. Further reductions are expected thereafter as well, and we see the cooling of the overheated German solar markets already.

German ingenuity will kick in and there is still a good chance that Germany will maintain its leadership in the EU solar energy markets, especially since most other EU countries are experiencing the same (or worse) problems, accompanied by additional FiT reductions and installation caps.

Germany's solar industry got an unexpected boost early 2011 when seven old nuclear plants were shut down following Japan's Fukushima nuclear disaster. This development might override the damage inflicted by the FiT reduction, at least temporarily. It might also serve as an example to other countries who follow in the nuclear race. Unfortunately, the US will not be one of them anytime soon, due to the government's focus on nuclear and coal power generation. Let's hope the next administration will be more willing to look beyond the immediate benefits.

This incident got the attention of Germany and other world powers, and sure enough, it was followed shortly thereafter by an announcement in June 2011 that all German nuclear plants will be shut down by 2022. Other EU and Asian countries are considering similar measures, while others are looking into spending additional billions to make the old, decrepit nuclear facilities safer in order to milk them a while longer.

Reality check: The incredible progress of PV power generation in Germany comes at a high price. Real net cost of the solar PV installed between 2000 and 2008 was €35 billion (~$50 billion). An additional €18 billion in 2009 and 2010 brings the total to €53 billion (~$70 billion) in ten years. The utilities also pay heavy premiums for the generated power as well, ~$0.43/kWh, which is higher than the normal rate they charge their customers. This additional expense is spread among all customers, including these who chose not to install solar on their roofs and those who cannot afford the increases.

An estimated 65,000 jobs were created by the PV boom, though most were temporary and low paid. Extrapolated over the overall expense, each newly created job costs ~$225,000 in government subsidies, which is almost 10 times more than most workers make. Germany's PV ambitions continue, albeit in a somewhat different direction, with $15 billion planned for expanding the manufacturing capacity and R&D of new technologies during the next 3 years.

And at the end, in 2010 solar PV is producing a grand total of 1.0% of Germany's electricity at a cost of $70 billion—a great portion of which money went to Asia for the purchase of cheap PV modules. The Asian modules are untested and unproven for long-term operation, which increases the owners' risks and liabilities. Is this a good investment? We will let the future generations answer this question.

In conclusion, and as a justification and clarification of the German PV experience, we must note that Germany, as most European countries, doesn't have many energy choices. With Russia keeping its finger on the energy spigot, Europeans are in a desperate situation and are forced to make a choice. Coal, oil and nuclear power are limited in their usefulness for a number of reasons we've discussed in this text, and since there is not enough sunlight for solar thermal development (which requires large areas and intense desert-like sunlight) they have chosen PV as an urgent path to energy independence.

Although cost was not the major driving factor, the recent economic crisis is bringing changes which override all other considerations. The reduction in FiT and caps in most EU countries is slowing down the solar boom, which simply means that the solar industry is not capable to stand on its own feet, so the Europeans just have to find a way around it.

A long winter and weak demand following FiT tariff cuts implemented at the beginning of 2011 could lead to additional cuts in the 3-6% range on top of everything we've discussed thus far. Official German installations figures for January 2011 highlight a 77% decline from installation figures for December 2010. Also, over 90% of system installations were small—below the 200 kW class which, together with the other changes, points to a solar future that is quickly changing and is entering somewhat uncertain territory.

We here in the US, however, have to thank Germany for showing us and the world that PV can be a significant energy source regardless of climatic and financial conditions. Our situation is much different, so we should chose a different path to achieving energy independence. We have a lot of excellent bright sunshine in many locations and an infrastructure that can support adding large quantity of solar power.

We just need to develop our own PV products manufacturing capabilities, introduce better regulated energy finance systems, and solve several other logistics issues. That will put the solar industry on its feet and bring us quickly to energy independence.

Spain

Spain's PV beginning was even humbler, with 0.0 MWp in 2000. Now Spain isn't too far behind Germany, and has shown even more impressive relative solar installations growth lately, due mostly to the availability of sunshine regions in southern Spain and excessive government support. Starting with 60 MWp installed capacity in 2005, it jumped to 700 MWp in 2007, a whopping 3,400 MWp in 2008, and slowing down in both 2009 and 2010.

The slowing trend of PV installations in Spain during 2009 and 2010 was due mostly to drastic changes in the regulatory system, FiT reductions, etc. Long-term projections are still quite optimistic, calling for over 7,000 MWp installed capacity by 2015. This may not happen, however, due to recent drastic FiT and cap changes. The Spanish solar boom decline will continue for the foreseeable future and until Spain and Europe pull out of the economic slump. PV is here to stay, but in moderation, which might be a good thing as time will allow careful analysis of historical events and better-informed planning of future endeavors.

In an emergency session, Spain's Ministry of Industry approved over 900 new solar projects for 2011. 116 MWp of new projects will benefit from the current FiT rates just in time to avoid the drastic subsidy reductions planned by the Spanish government from Q2, 2011 onward. 80% of the 4,718 subsidy applications filed in the first three months of 2011 were turned down by the government. Of the approved bids, 74.6 MW were for small (total of 7.4 MW) and large-scale (total of 67.2 MW) rooftop installations. Only 41.6 MW were for ground-mounted systems. Q2's 2011 tariff reductions were ratified in 2010 and will see reductions to €0.29 (an 8% drop) for rooftop installations under 20 kW, €0.20 (27%) for larger rooftop systems and €0.13 (47%) reduction for ground-mounted projects. The timeframe to apply for the new licenses ended on April 6, 2011.

The latest development in Spain is the law RDL 14/2010, approved by the Spanish Congreso de los Deputados, which introduces heavy restrictions on new installations, and even totally unexpected reduction in the remuneration of existing PV installations in the country. Solar industry representatives and activists are protesting, and plans are underway to file lawsuits with the Supreme Court and the European Commission, to provoke EU proceedings against Spain's anti-solar measures. This is an unexpected and unwelcomed development for the entire solar community, the consequences of which are reverberating through Spain, Europe and the world. It is, however, only a confirmation of the fact that the solar industry, like any other for-profit business, must be able to stand on its own two feet, instead of relying on government subsidies and other handouts.

Hastily put together solar farms, using low-quality PV products and unskilled labor have often proved inefficient and problematic. Many farmers and industrialists who enthusiastically invested their own savings into the "new energy" future during the 2007-2009 solar boom find themselves in a desperate situation today. Due to the slimmed down FiTs and other government restrictions, their solar farms cannot generate profit and are taking them into financial ruin. Will Spain's promise of cheap energy and solar prosperity end up in a large number of junk yards popping up on the Spanish landmass in the near future?

If this were not enough, Spain's energy sector regulator, CNE, suspended FiT subsidies for 350 existing solar power plants which are operating under full power, after it was suggested that they were providing fraudulent electricity generation figures.

Note: This is a new one—fraudulent energy data reporting—the likes of which we've never seen nor dreamt of before. Maybe this is an indication of things to come to our shores too. 350 fraudulent operations sounds like organized crime is coming to the solar energy sector too. Would these operations survive without the FiT payments which are major part of their budget? This would be a good test of the readiness of the solar industry to manage without external support. The other remaining question is will any new installations be undertaken in Spain under RDL 14/2010?

Still, the Spanish government is proposing a 300% increase in solar installations by 2020, bringing Spain's installed solar-power generating capacity to 70 GW. We just have to wait and see how they'd do that in light of the above changes and fiascos in the midst of the economic crisis that will surely last awhile longer.

United Kingdom

And speaking of humble beginnings, there were only 100 MW of PV power installations added in the UK in 2010. This is a relatively small number for one of the world's strongest economies, and a slow beginning for a brilliant alternative energy future—a future getting slower and dimmer with time.

Like Spain, the UK government is slashing guaranteed prices paid for electricity from solar projects that are larger than 50 kilowatts. That's because it wants to prioritize roof-top projects over large-scale solar farms. Climate Change minister Greg Barker said in Parliament that he still wants bigger roof-top community plans to go ahead. "Solar panel makers need to bring down prices so that communities can build bigger sun-powered projects," Barker said, six days after proposing cuts in incentives. "Community-based projects that are larger than 50 kilowatts, equivalent to 2 tennis courts, will still get a tariff which is comparable to that paid in Germany," he told lawmakers. "We are hoping that many community projects, in particular around the 100-kilowatt size, will still go ahead but the pressure must be on manufacturers to bring down prices."

So the UK will rely on small solar installations (roof mounted) and lower prices (most likely dirt cheap PV modules coming from China) to achieve energy independence. Cheap is good, but not for long-term application, so we'll be watching the developments in the UK closely.

The review of feed-in-tariffs comes less than a year after the program was started. Solar power companies including module maker Sharp Corp., Solar Century Holdings Ltd. and project developer Low Carbon Solar have said the government's plans threaten to kill off an industry that was just beginning to take off in the UK. So the UK solar embryo is threatened to be sent back in hibernation.

Under the plans announced in March 2011 by the government, PV projects greater than 50 kilowatts would have their tariffs cut by at least 42 percent. Those larger than 250 kilowatts would be slashed by 72 percent. "Those investors who were looking to invest in larger schemes were disappointed," Greg Barker said. "We were absolutely convinced it was the right thing to do. We've taken measures which will avoid the boom and bust seen in other countries across Europe." We know what the minister means, but have to conclude that solar is not a priority for the UK government. This is probably not a great loss, and in part justifiable, because the UK is not exactly the solar capital of the world.

As a matter of fact, the Brits might be much better off betting on Desertec (the large EU energy project scheduled to bring wind and solar electric power from the North African deserts to Europe) instead of relying on the foggy and soggy rooftops of the UK countryside. Also, tourists would not appreciate looking at shiny PV modules on top of the picturesque, centuries-old roofs, so be careful with what you wish for, Mr. Barker.

In either case, we are glad the UK has plans for solar in its energy portfolio and only hope they get more serious about it in the future, for alternative energy is the only thing their future generations could rely on.

Wind might be a better energy source for the UK, but an interesting fact emerged in March 2011 in the report by Stewart Young Consulting on wind power in the UK (7). The report made several striking, but not overly surprising, conclusions as follow: "The nature of wind output has been obscured by reliance on "average output" figures. Analysis of hard data from National Grid shows that wind behaves in a quite different manner from that suggested by study of average output derived from the Renewable Obligation Certificates (ROCs) record, or from wind speed records which in themselves are averaged. It is clear from this analysis that *wind cannot be relied upon to provide any significant level of generation at any defined time in the future.* There is an urgent need to re-evaluate the implications of reliance on wind for any significant proportion of our energy requirement."

These are remarkable statements, in our opinion, which confirm several facts we have been talking about, and which the scientific community should pay especial attention to:

1. No single source of energy can be counted on for supplying a large quantity of reliable and efficient power. Instead, a number of sources—in some cases a wind and solar mix might be a good match—are needed for serious and constant energy supply.

2. No single information source—including gov-

ernment—can be trusted and used as an ultimate source of reliable information for decision making.

Note: We have noticed similar disinformation trends in the PV sector in the US as well, where companies, and scientific and government bodies provide biased information on technologies and projects they favor, or are working on. While such behavior is typical for the capitalistic "Make a quick buck" approach, it is damaging to the overall goal of developing and implementing reliable, safe and efficient alternative energy sources.

France

720 MW of PV installations were connected to the grid in 2010, which is almost three times the amount installed in 2009. This is a total of 13,000 in 2010, compared to 4,430 PV installations in 2009. 29 PV installations above 250 kW were grid connected in 2010, providing a total of 130 MW, compared to 24 installations providing 33 MW in 2009. Not a bad progress for the "nuclear power capital of the world."

However, France also got hit hard by the economic crisis, and the solar energy market was capped at 500 MW for 2011 and beyond, including the projects that have already received permits by the end of 2010, according to French media reports. The French government estimates that 3.4 GW of projects have received construction permits in 2010 and roughly 2 GW of these projects could ultimately get built in the 18-month timeframe which stretches to the first quarter of 2012. If these projects are not completed, the government may consider increasing the installation cap from 500 MW to 800 MW. The new 2011 FiT could mean that future permits will depend on project size (for projects above 100 kW, the FiT would be determined by an auction process), while for the residential market, tariffs could get adjusted by 20% in 2011 and 10% in 2012. The revised regulatory outcome allows for an additional 3.4 GW of projects to be completed potentially in 2011 or 2012 (in the case of supply constraints) and thus improves the 2011 demand expectations.

In December 2010 the French government also announced that no new PV projects larger than 3.0 kWp will be approved for the next three months. The decision will remain in force until the government submits a new framework for PV promotion, according to Prime Minister, François Fillon. Installations up to 3 kWp will not be affected. The reason cited is the unexpected growth of the PV market in France, due to high FiTs, which increases the costs to the government and consumers. "In order to work against the speculative backlash, a new equilibrium has to be established," according to Mr. Fillon. So France is also trying to cover their centuries-old, picturesque roofs with brand new, shiny PV modules. We can't picture the Louvre roof covered by cheap China made PV modules, but it might be the next thing to get used to.

One of President Nicolas Sarkozy's goals for 2011-2012 is to limit the development of renewable energies to 500 MWp per year. Since the number of projects already registered is far above this goal, the definition of the new conditions is considered justified by the French government. All this sounds like an echo from the UK's Mr. Barker—focusing on smaller PV installations and limiting their volume as well.

Our advice for France, like the UK, is to preserve the exotic look of the old roofs that tourists love so much and rely on the sun bathed North African deserts to provide power to its citizens.

Update: The French government is working on regulations geared to boosting solar energy industry deployment, and hoping to stimulate the development of innovative technologies, while at the same time promoting made-in-France products and services.

As of the summer of 2011, all PV installations larger than 1,000 m2 will be subject to two different types of tenders, the conditions for which are still not finalized.

The first type—installations between 1,000 and 2,500 m2—will be offered simplified conditions and favorable prices for the generated electricity to be sold to the utility. The government launched the first lot for 120 MWp in August, following the recommendation of the French energy regulator CRE. Later on (date and conditions unspecified) France will call six additional tenders of 30 MWp each.

The second type of tender will affect installations larger than 2,500 m2 with estimated total capacity of 450 MWp. The power sale price and other criteria will be defined at a later date, but will include considerations for industrial projects, contribution to R&D, the related environmental impact and other factors.

As an example, government officials have expressed an opinion that they would give priority to solar plants in industrial zones or in abandoned commercial sites (mines and quarries), rather than those located on agricultural land.

Italy

Italy also showed impressive gains recently, when its installed PV capacity jumped from 79 MWp installations in 2007 to 417 MWp in 2008. It grew again in 2009 and 2010 and is expected to total over 5,500 MWp by 2015.

In May 2011 the Italian Council of Ministers signed the long-awaited Conto Energia IV effective June 1, 2011. Due to legislative uncertainties the Italian solar market was on standby since the beginning of 2011. This effectively stopped commercial and utility-scale projects in their tracks and caused the PV supply chain to become saturated with PV module inventory estimated at 2-3 GW. The module surplus was exacerbated by the slow start in PV installations in Germany this year.

Industry specialists estimate that delays in creating a new FiT had put 10,000 jobs at risk. This also caused over €8 billion in orders to be blocked, while €20 billion in PV contracts were on hold.

Key changes incorporated into the Conto Energia IV include a monthly degression of the tariff system and a funding cap for large systems. Interestingly, the new Conto Energia provides a 5% tariff bonus to system operators, should 60% of the investment cost (modules and inverters), rather than installations costs be sourced from companies making such products actually in the EU. Importantly, existing permitted projects under previous tariff would be allowed to continue through August 31, overlapping with the new FiT.

A total cap on installations was feared, but according to the new rules there is a cap (€580 million) imposed for large systems through 2012, which would enable new installed capacity of 2.8 GW based on current prices. Funding from 2013 to 2016 has been limited to €1,361 billion, which would allow for 10 GW of installed capacity under the same conditions.

A half-year reduction of tariffs will take place from 2012, but in 2013 an increase of between 5 and 10 cents is possible. All other forms of funding such as tax breaks and investment subsidies, however, are to be abolished for good. Funding is then to be adjusted on a half-year basis starting mid-2013.

There are several positive aspects to the new FiT for the PV industry. The major one is that projects under 1 MW are classified as small projects, and no cap is placed on roof system installations. With this, the majority of the Italian market remains intact. There is also no cap on ground systems of less than 200 kW, which eliminated the fear of the previous draft which called for a cap on all ground-based systems regardless of size. With these changes and with permitted systems now eligible for incentives without a cap up to August 31, the Italian PV market is expected to add 2 GW of installations in 2011.

The large-system cap lowers previously estimated installations and the priority ranking system for large systems, and these factors need further clarification. The Conto Energia IV establishes new provisions for connection to the net, which basically allows only 30 days for a completed system to connect to the electricity grid. Delayed connections would result in penalties, which are also to be clarified. Nevertheless, the new Italian FiT system is much better than originally thought or even expected, so barring some unforeseen disasters, Italy is on its way to a bright solar future.

Update: FIT schemes in Italy are also undergoing constant modifications and restructuring. The latest decision from Rome is to create a special list and register all large PV installations by August 1, 2011. All such projects have to register with GSE (government network agency) in order to get on the list and get an approval. Then special criteria apply to different cases.

All PV installations connected to the grid by August 1 will continue to receive the current FIT rate, with no further changes. 172 photovoltaic installations were connected to the grid in July, 2011 and will receive the current FIT rate.

A total of 947 photovoltaic installations are presently on the GSE list and are entitled to receive some type of subsidies IF connected to the grid by December 31, 2011. This includes 252 PV installations that were already completed by the end of June, but have not yet been connected to the grid. An additional 695 projects are under construction presently or have obtained the required permits. The race is on!

The last official approval for new PV installation was given in February, 2010. Any plant with a permit issued after that will not be entitled to claim government support. However, some of them may be qualified and able to get on the present GSE list if one of the listed projects is cancelled. These latecomers can also apply for inclusion in the new list in November, 2011, for implementation in 2012.

GSE will also publish a list with the possible alternative candidates for solar subsidy in 2011. Another list will detail the PV projects that have not submitted all of the required documents and therefore may not be able to benefit from the subsidy.

Other countries in the region are also showing encouraging results, but in much smaller size, relatively speaking. The economic crisis is affecting all EU countries, and most EU27 countries are taking measures to reduce spending. Solar incentives and subsidies will be drastically cut across the board. There are many examples of this trend in the other EU27 countries (and the world), which have disrupted the upward PV dynamics set in motion during the 2007-2010 PV sector boom. The direction and shape of the PV future is as unclear as ever.

Long-term European PV Programs

Europe's (EU27 countries) total electricity generating capacity (including conventional energy generation sources) is estimated at ~800 GW, with 58% oil and coal, 18% hydro, 17% nuclear, and 7% wind. Solar power generation is estimated at ~2% presently. Though not a great number, it's a great beginning of a trend that Europe intends to follow. The EU community is looking at solar (and PV in particular) as a key solution to its very survival, so serious plans are underway for its implementation in even larger quantities in the future. The "Resource-efficient Europe" initiative calls for PV to contribute over 10% of the total generated electric capacity and 30% reduction of greenhouse gas emissions in Europe by 2020. It also requires completing the internal energy market and implementing the European Strategic Energy Technology Plan (SET-Plan). Those are very ambitious plans, is all we can say, wishing the Europeans the best of luck in their implementation.

An example of this drive is the European Photovoltaic Technology Platform (EPVTP) comment on "Towards a new Energy Strategy for Europe 2011-2020" (5). In it, they "welcome the public consultation on the European Energy Strategy and propose EPVTP as a qualified stakeholder group that intends to engage in a constructive dialogue with the EU institutions on this and other topics of strategic importance that fall within its scope.

"The European Commission, the International Energy Agency, the US Department of Energy, and Institutions and Policy Makers worldwide recognize Photovoltaic Solar Energy (PV) as a key technology to address environmental and climate change challenges, in particular energy safety, security, sustainability, access and affordability for all. The sector's track record has provided evidence of continuous innovation and cost reduction, resulting in a fast learning curve that allows photovoltaics to make a substantial contribution to the EU electricity supply already by 2020. Under favorable conditions this contribution may be more than 10%, as an important first step towards an even much larger share. Already in 2009, PV represented the third largest net new energy generating capacity installed in the EU27, after Wind Energy and Gas."

Having said that, achieving these results has only been possible thanks to continued and predictable policy support—and significant incentives and subsidies. The dialog in Europe continues on all levels, and we clearly foresee great things happening there in the PV area in the future.

The biggest problem we see in the European solar future is the lack of intensive direct beam sunlight, as needed for serious utility type power generation. Solar power provided by thousands of 3 kW rooftop installations, as planned in some EU countries, won't provide energy independence anytime soon, if ever. It will only ruin the esthetic appearance of lots of old roofs, in addition to creating lots of leaks in them.

The greatest and most exciting solar project in Europe, and maybe the world, is the concept proposed by the Desertec Foundation for making use of solar and wind energy of the North African deserts. This, in our humble opinion, is the most plausible and fastest road to energy independence and environmental cleanup of Europe. Some plans calling for this concept will be implemented in North Africa and the Middle East by the consortium DII GmbH formed by a group of European companies and the Desertec Foundation. The effort was initiated under the auspices of the Club of Rome and the German Trans-Mediterranean Renewable Energy Cooperation (TREC), and aims at promoting the usage of the generation of electricity in the deserts using solar power plants and the transmission of this electricity to the consumption centers—in this case via underwater cables transporting the generated DC power from North Africa to Europe.

The original and first region for the assessment and application of this concept is the EU-ME-NA region (Europe, Middle East, Northern Africa). The realization of the Desertec concept in this region is pursued by using wind, concentrating solar power and PV systems, located on 6,500 square miles (17,000 km2) in the Sahara Desert areas of North Africa.

Thus produced electricity would be transmitted to European and African countries by a super grid of high-voltage DC cables. It would provide a considerable part of the electricity demand of the MENA countries and also provide continental Europe with 15% of its electricity needs. By 2050, investments in solar plants and transmission lines would total €400-600 billion. The exact plan, including technical and financial requirements, will be designed by 2012, with the implementation beginning soon after.

The German experiment

The solar energy developments in Germany from 2007 through 2010 have been hailed as an example of successful mass deployment of PV energy. While the alternative energy gains are undeniable, there are gaps and weak points in the overall program, as explored fully in a paper issued by Ruhr University team, titled, "Economic Impacts from the Promotion of Renewable Energy Technologies. The German Experience." (6).

The Ruhr team's conclusions are strikingly simple and the arguments within quite convincing. "We argue that German renewable energy policy, and in particular the adopted feed-in tariff scheme, has failed to harness the market incentives needed to ensure a viable and cost-effective introduction of renewable energies into the country's energy portfolio. To the contrary, the government's support mechanisms have in many respects subverted these incentives, resulting in massive expenditures that show little long-term promise for stimulating the economy, protecting the environment, or increasing energy security."

When the German solar and wind programs began in 2000, PV was in a class of its own. It offered index-linked payments of €0.51 for every kWh of electricity produced by solar PV, which were guaranteed for 20 years. France had the highest payment in 2008 of €0.75/kWh, at least for a short while. Other EU countries started similar programs too; i.e., the UK's PV subsidy of 41p. The solar subsidies were, and remain, much greater than those for other forms of renewable technology.

The net cost of the solar PV installed in Germany between 2000 and 2008 was €35 billion, and €18 billion in 2009 and 2010 respectively. Or a total cost to the taxpayers of ~€53 billion in ten years. These investments make sense for those who could afford to install PV, since the returns are guaranteed by taxing the rest of Germany's electricity users.

The shocking revelation is that these great expenditures have achieved very little in real world terms! By 2008, solar PV energy generated only 0.6% of Germany's electricity—a 0.6% gain at a cost of €35 billion to be paid by the customers. Is this a good investment? After 10 years of large subsidies, saving 1,000 kg. CO_2 from using PV still costs €716. IEA estimates €1000 per ton, which shows that PV is not the solution for environmental cleanup in high-latitude climates like Germany, since the net saving is zero.

The Ruhr team estimates that every solar PV job in Germany cost €175,000, or much more than workers usually earn in any occupation. Also, most of these people (skilled workers) have been drawn out of other industries where they could be more useful. Keep in mind that most of the PV products installed in Germany and the US are made in Asia, meaning that no additional PV products manufacturing jobs have been created, and lots of money was spent on cheapo Asian exports.

The authors of the Ruhr paper conclude, "Hence, although Germany's promotion of renewable energies is commonly portrayed in the media as setting a "shining example in providing a harvest for the world" (The Guardian 2007), we would instead regard the country's experience as a cautionary tale of massively expensive environmental and energy policy that is devoid of economic and environmental benefits. As other European governments emulate Germany by ramping up their promotion of renewables, policy makers should scrutinize the logic of supporting energy sources that cannot compete on the market in the absence of government assistance."

This author would like to add that a full sunshine day in Germany and most EU countries is a rare occasion. The question, then, is how much power can PV modules generate on a cloudy day? What marketing and financial formulas does the German government use to justify spending billions on solar power, when there is not much sun to speak of in Germany? We do suggest that the money would be much better spent on other energy initiatives, like the Desertec project, where CSP and PV power plants in the Sahara desert would generate power and send it to Germany and the other EU countries. We will leave these decisions to the German people, and will concentrate on developing PV power plants in the deserts, where solar power is abundant.

Nevertheless, official 2010 PV installation figures for Germany show a booming market in 2010 with almost 7.4 GW installed, an increase of 75% compared to 2009. Once again Germany was by far the PV market leader. However, for the market to prosper in 2011, against a backdrop of lower feed-in tariffs, prices much drop fast.

According to PV analysts, the ground-mounted market typically requires an investor ROI of approximately 10% to make that sector of the German market attractive to kick-start projects. Based on the German FiT, the cost per watt would now need to be at least €2/W to make investment viable. The residential sector requires an ROI of 8% for homeowners to see the payback benefits of installing PV on roofs. This would require a €2.7/W price to ensure market development for small <30 kW residential systems. Prices, however, are not falling fast enough for the market to follow the installation trend seen in 2010. So, it remains to be seen if Germany will remain in the lead of the solar revolution.

Note: Amazingly enough, the German experiment goes almost against everything we know (and preach in this book) about solar energy generation and use.

Sorry, Germany, but look at this:

1. The location, which is a primary concern in solar power generation, due to mostly cloudy skies, is very far from a best-case scenario. Is the pattern

of developing "solar under cloudy skies" sustainable?

2. Small installations as used in Germany are too expensive to install and maintain; thus, they are not the solution to mass power generation.
3. Significant subsidies are usually temporary and realistically speaking unsustainable, so they cannot be the solution to energy independence in the long run.
4. Using foreign-made energy products (cheap PV modules) is counterintuitive to achieving national energy independence. Jumping from Arab and Russian oil into Asian PV products is like jumping from the frying pan into the fire.
5. The quality of the imported PV modules provides an additional uncertainty and a major risk factor in the long-term benefit of the hastily put together installations. The questionable quality of materials and processes with cheap imported products are issues that Germans, more than anyone else, should be quite sensitive to, but they've have chosen to close their eyes.

Our conclusion is that Germany will do much better in the long run if it focuses its attention on and funnels its money to more feasible and sustainable solar and wind projects, such as the planned Desertec project which will generate power in the North African deserts and transport it to Germany via DC cables. We do believe that this is the long-term direction Germany and the major EU countries will choose in the end.

The German industrial machine, however, did not hesitate to jump into the solar race even before anyone was aware that the solar revolution had started, and it has not ceased to impress us thus far. Many German companies were and still are on the front lines of solar production equipment and products manufacturing. Some of these are:

German Polysilicon Companies

Wacker Chemie is a chemicals conglomerate which increased polysilicon production capacities rapidly in Germany and is expanding in the USA as well. Wacker has the majority of its profits coming from its nearly 25,000 ton/annum polysilicon production. It is one of the world's major producers of semiconductor wafers as well, so it uses some of the poly in-house while selling most of the rest to Asian customers. Wacker is expanding in Germany and the USA to keep its No. 2 global position.

German Solar Panel Manufacturers

Solarworld is the largest German producer of solar modules. The company is one of the few to still have operations in Europe and the US. The company has been battered by low-cost competition; however, it is strongly expanding in US as growth slows down in Germany. Solarworld is also entering the polysilicon sector through a JV with Qatar. It is the only Western company to not yet have a major factory in Asia.

Q-Cells was the largest solar producer of cells in 2008, but it faced horrendous 2009 losses of over €1 billion. After restructuring, it moved its factories to Malaysia and has diversified into solar modules and systems. The company has seen its stock price drop more than 95% from the peak levels and is now using Flextronics as an outsourced module producer for its cells. It has expanded into making PV modules and systems, including thin film modules through its Solibro subsidiary.

Bosch, the German auto giant, has expanded into solar energy by buying up small German companies like Ersol and Aleo Solar. Bosch is now vertically integrated with operations in solar wafers, cells and modules manufacturing.

Schott, the German glass maker is a big component supplier to the CSP industry but does not provide turnkey solutions. Schott is also involved in PV technologies by producing both crystalline silicon and thin film modules. It recently established a JV with the Chinese company Hareon and plans to build 700 MW of solar panel capacity in emerging markets around the world. It has won a contract to supply 20 MW of thin film modules to the Indian solar company Premier Solar with 10 MW of supply in 2010 and 10 MW in 2011

There are a number of other German solar companies which have some capacities in making solar cells and solar panels, like Solon, Solar Fabrik, Centrosolar.

German Inverter and Production Equipment Manufacturers

Uniquely so Germany still holds the leading position in manufacture of solar production equipment—inverters and cells and modules manufacturing equipment. The high-tech nature of the inverter sector has made it harder for other companies to compete with well established and successful German manufacturers, some of which are:

SMA Solar is the big daddy of solar inverters with a 40% market share of the global market. SMA has become the most valued solar company in Germany. It has maintained its market share and grown as fast as the global solar demand, unlike some of its competitors. SMA Solar faces tough times ahead as the growing inverter industry attracts numerous competitors, and it remains to be seen whether it can avoid the fate of some of its competitors (i.e. Q-Cells).

Kaco New Energy also benefited from the strong growth of the worldwide solar market in 2009 and 2010 to become a large player. It has not grown as fast as SMA but is still a big player.

Fronius International is very similar to Kaco in terms of its growth. Fronius was the #3 player in 2009, but has seen startups like Power-One take more market share and threaten its position.

Schieder Electric is a European electrical equipment giant who got into the solar inverter market by buying the Xantrex, Canada. It is now fighting for its spot in the marketplace too. The battle is on, and the winners are still to be determined, but these German companies will do well in all cases.

The major German production equipment makers are part of the largest and most advanced solar production equipment manufacturing machine in the world; supplying a majority of the tools and machinery needed for manufacturing solar cells, wafers and modules. While the US has almost caught up with Germany in supplying machinery to the low-cost Chinese solar producers, German equipment makers are still heavily present in the top 10 manufacturers.

The major solar production equipment manufacturers and exporters in Germany are:

1. Centrotherm Photovoltaic AG
2. Roth and Rau
3. Meyer Berger
4. Manz Automation
5. ALD Vacuum Technologies

While low-cost Chinese PV modules are having an impact on the solar industry in Germany, other sectors like solar inverters, and production equipment manufacturing have continued to flourish. SMA Solar is the largest solar inverter manufacturer and exporter in the world, while Centrotherm is the 2nd largest solar production equipment maker.

The German solar market has been the biggest reason for the growth of the global solar industry in the past decade, where solar energy subsidies have fostered both the development of robust renewable energy and green industries and jobs. This is a remarkable achievement, compared with the sluggish development of the solar industry in other countries, including the US, most of whom have delayed their entry in this new market or have missed the trend altogether.

What we should learn from the European solar energy experience, and especially the PV installations boom from 2007-2010 and the factors that contributed to the impressive gains in PV installations there, is that the key countries had:

1. Full government and industry commitment (during the period)
2. Emphasis on stimulating national market and the solar industry
3. Well designed and implemented support systems, which are:
    a. Sustainable (for the time)
    b. Forecastable
    c. Reasonable
    d. Guaranteed in time
    e. Encouraging price reduction
4. Low administrative barriers:
    a. Permits, support schemes, etc.
    b. Grid-connection
5. Regional subsidies for PV companies
6. FiT of $0.40-0.84.5/kW was a key driver (compared with $0.10-0.15 in the US)

A careful look at this list leads to the immediate conclusion that the EU situation is much different than that in the US. While this is so, and many of the above factors cannot be changed, or adapted for use in the US, we should at least analyze them thoroughly to avoid repeating the mistakes of the EU countries. The areas where a lot of work is needed in the US are obvious. Our experience shows a lack of government commitment, and our national alternative energy market is not a priority, as a major part of the subsidies go to foreign companies or risky projects. The support systems are not sustainable, forecastable, or guaranteed to last, and no encouraging price reductions are in sight. Administrative (regulatory, utility, legal, etc.) barriers are controlling the PV energy market and stifling its growth. Grid connection is too expensive and simply unreachable for most smaller companies. Subsidies and FiT levels are fluctuating and unreliable.

Any of these factors alone is a major barrier. Together, they create a wall that few companies can climb successfully, try as they may.

Asia and the Rest of the World

The solar future in Asia looks good for PV sales. New demand for cheaper energy, driven by the record-high oil prices of 2008-2009, falling prices and increased reliability of solar technologies in 2009 and the growing concerns over global warming, are forcing governments throughout the region to look to the sun as an increasingly feasible source of renewable energy.

China, India, Japan, Taiwan, and Thailand have introduced subsidies to promote the use of solar panels on rooftops and as small power producers.

China

China, the world's leading PV exporter, had only 1.0 GW of installed PV power generating capacity in 2010. So, China, one of the largest and most rapidly developing countries in the world, has less PV power generation than Japan, the US, and most West European countries. Things are about to change, though. The Chinese government is promoting a sharp increase in domestic use of PV power. The government aims to increase China's solar power generation capacity from 50 MW in 2008 to 10 GW in 2015 and up to 60 GW in 2020. Note the incredible several hundred-fold increase in solar energy installations and use.

A newly released report, "China's Solar Future—A Recommended China PV Policy Roadmap 2.0," uses the IEA Solar PV Roadmap as a benchmark for China, so that the country, which is well known as an exporter of PV products, can become a large producer and user of PV power as well. The report speculates that 1.3% of the China's electricity can be generated by PV power plants by 2020, while by 2030 China can reach 4.6% PV generation. To achieve the 2020 and 2030 PV electricity goals, the Chinese government will have to support the PV products manufacturing and PV projects development of up to 60 GW of installed PV capacity by 2020 and 270 GW by 2030.

By following the detailed PV installation roadmap with annual targets up to 2030, China can immerse itself in PV power consumption and reach the goals, according to international energy specialists. No doubt, China can do this and more, but it won't be easy. China's populated centers are predominantly foggy, cloudy, and smoggy and basically not suitable for solar power generation. The areas of higher direct insolation are far from populated centers, and China's electric grid and basic infrastructure in most areas is mostly a rag-tag of partial solutions that need to be upgraded and expanded to add solar power sources. So basically, making lots of PV modules is one thing; adding significant amounts of PV power generators is another that requires significant additional effort and money.

China is still the dominant solar products manufacturer, so Europe, the US, and Australia will continue to rely on Chinese-made PV modules if they want to be price-competitive. Solar "Made in China" is becoming less and less a qualitative risk but rather an opportunity to make solar PV investments at reasonable prices. This is a counterintuitive approach, however, because installing cheap products today can mean paying dearly tomorrow.

China-based solar companies will continue to dominate the PV market for the foreseeable future. Backed by loans from local banks, they continue expanding production and cutting unit costs. This allows them to offer PV modules at lower costs, which in turn helps them to gain an unprecedented market share. The trend will continue with 50% of the world's total PV cells and modules capacity manufactured in China. PV module prices are expected to fall disproportionately because of decreased demand. Average prices of slightly over $1.00/Wp are expected in 2011-2012, causing manufacturers' gross margins to drop significantly, but they hope that increased volume will compensate for the price drop.

At the same time, US and European companies are focusing more on project development and installations activities to stay in the game. The industry's high financial risk will persist in the future and debt levels will continue to rise as China-based solar makers expand production capacity and extend global sales exposure.

The Chinese solar industry began to expand the production of cheap inverters, batteries and other components in 2010 as well. In 2011, the expanded production of Chinese PV companies is expected to result in an excess capacity situation which will lead to a decrease in prices but also in decline of the profit margin up to 20%. Due to the decreasing demand from Europe, the demand side will be very advantageous while Chinese suppliers will be fiercely competing for clients in 2011.

The current heavy dependence of the Chinese solar industry on exports will, following the declining demand from Europe, be shifting towards more investments in technological innovations and quality improvement. The Chinese government will be increasing investments in its solar sector, not only in order to save the many export-dependent domestic solar companies, but also in order to use the vast potential of the country and decrease China's overall oil, gas and coal dependency. China is thus still a major factor to consider in the solar PV market of 2011.

After the Japanese nuclear disaster, the Chinese government started looking into replacing some of its nuclear power plants with solar installations. This is not an easy thing to do, but keeping in mind the enormous PV manufacturing capability and ever increasing energy needs, China might just become the new leader in the PV revolution, so we'll be watching closely.

2011 update: China's government is to introduce a national photovoltaic feed-in tariff (FIT) scheme to be applied at two basic rates. The first will be 1.15 Rmb/kWh (1.0 Rmb=US$0.18) to be paid for projects approved before July 1, 2011 and completed by the end of the year. The second rate is 1.0 Rmb/kWh (US$.156/kWh). The Chinese government's Renewable Energy Development Fund (REDF) will be the FIT manager and funding source.

Sounds good, right? Yes, but this is China—land ruled by an outdated, clumsy political system with a record of one-sided decisions. The way to get something done in China is ONLY through good contacts with and bribes to regional communist government officials. One must also go through a complex permission process based on similar connections and bribes.

Chinese companies are used to this and are very good at it. They are also very proud of their connections and successes in this area. Any foreigners thinking about getting contracts in China will have to learn the system and be willing to play by the rules without any hesitation or remorse.

Japan

In Japan, no newcomer to sun power, subsidies and people's growing awareness of climate change helped solar sales to more than double in 2009, when capacity surged by 484 megawatts, according to the Japan Photovoltaic Energy Association. An estimated 50 million Japanese households now have solar panels on their roofs—many for hot water use. Solar homes account for 88.7 percent of the total output. In January 2011, Japan reinstated subsidies for homeowners who purchase solar power systems, reducing costs by almost half, and introduced a feed-in tariff, similar to Germany's, which requires utilities to buy excess solar power supplied to the electricity grid.

Following the nuclear disaster in Japan, the potential now exists to install an additional 100 GW of PV by 2015, according to some analysts. A Swiss-based bank has raised its global forecasts for newly installed PV capacity from 166 GW to 296 GW until 2015, some of which will benefit Japan. The nuclear disaster might encourage politicians to put an end to the continuing trend of cutting subsides. New safety standards will also contribute to making existing and new nuclear power plants unattractive from safety and financial points of view.

Japan's solar FIT will be extended to geothermal, wind, small hydro, and biogas energy generation as of April 2012. A host of restrictions and serious limitations have been forced onto the energy sector over the objections of experts, many local communities and other interests that sought a more robust, comprehensive FIT.

The Fukushima disaster obliterated 1/5 of Japan's nuclear power production—a heavy price to pay. Energy shortages are causing serious technical problems and personal hardships around the region. The Japanese government is looking seriously into all options at its disposal to be implemented in the near future, and solar is one of them. We will watch closely as Japan struggles to recover from the disaster and regain its electric energy power generation capability.

Taiwan

Taiwan, another leader in PV exports, has allocated US$280 million to subsidize a "100,000 solar roofs program." The government hopes to install 1,000 megawatts in solar-power capacity annually by 2025, compared with 6 megawatt operational now.

India

India has announced ambitious plans to boost solar output almost 1,000-fold to 20,000 megawatt by 2022. The average number of sunny days in India averages around 300 per year, and due to the high sun radiation particularly in the south and the center of the country, the potential for solar power is extremely high and makes India a favorite destination for domestic and international solar PV investments. India plans to install 20 GW of solar power by 2022, or 12% of the country's total energy share then, and 200 GW by 2050 (making it the biggest solar PV market at that time). There is still much to do in order to accomplish all this, since the installed capacity stood at about 12 MW at the end of 2010, and the national infrastructure needs a major overhaul to handle this new energy source.

Wind power and biomass have the main share right now, with a total installed capacity of renewable energy sources at about 17 GW by the end of 2010. The PPA agreement between the large Indian conglomerate TATA and the state-run power distributor Gujarat Urja Vikas Nigram (GUVNL) to the construction of a 25 MW solar power plant is clear sign that India's domestic companies have recognized the governmental encouragement and promising business opportunities with regard to India's solar PV market in 2011. The government is eager to increase the solar share by concurrently improving infrastructural conditions in the country. Another important business for 2011 and beyond will be solar home systems.

More than 40% of households in India, mainly in rural areas, still lack access to electricity and are bound to be benefiting from the Indian government's efforts to promote electrification and increase the share of solar power.

Korea

The South Korean PV market is expected to grow substantially throughout 2011 and beyond and is thus very lucrative for investments from overseas. The reason lies in the enhanced governmental support of solar and other renewable technologies. In March 2010, the South Korean National Assembly passed the introduction of a Renewable Portfolio Standard (RPS), effective from 2012. The RPS requires a 4% renewable share in energy generation by 2015 and 10% by 2022. Solar companies will be increasing their investments and expanding business activities in Korea to meet the expected mounting demand in solar PV technologies there.

Malaysia

Malaysia's parliament has approved a sophisticated system of feed-in tariffs to develop its renewable energy resources. It passed a Renewable Energy Bill creating the feed-in tariff policy and the Bill for the Sustainable Energy Development Authority which will go into effect in mid 2011. The 2011 Malaysian quota for solar PV is 29 MW, and in 2012 the target is an additional 46 MW. One-third of the solar PV capacity is set aside for projects of less than 1 MW.

By 2020, Malaysia expects to have installed more than 3,000 MW of new renewables of which about one-third (1,250 MW) will be from PV and one-third from biomass (1,065 MW). Malaysia's feed-in tariffs are divided into multiple tranches, with PV divided into six branches, not including Malaysia's four separate bonus branches for locally manufactured components. It's a complex picture, that will get much clearer with time.

Thailand

Thailand took the lead in promoting solar use in southeast Asia in 2010 when the government introduced a feed-in tariff to encourage companies and homeowners to invest in renewable energy. Corporate interest has been surprisingly strong, with proposals already on the table to invest in a crop of solar farms with capacity of 30 to 90 megawatts.

Australia

The Australian government has set a target of generating 20% of its electricity from renewable energy by 2020, starting with a $1.8 billion funding program for the development of solar projects. The government is also working on setting a price on carbon emissions starting in July 2012 in preparation for a trading system to begin as early as 2015.

A number of government policies encourage the development of renewable energy projects and are expected to attract close to $40 billion of investment by 2020, according to industry specialists. $30 billion will be spent on utility-scale wind farms and $10 billion will be invested in smaller rooftop PV systems. In 2010 Australia was one of the top 10 nations in attracting funding for "large-scale" renewable energy projects.

BP Plc, AGL Energy Ltd., TRUenergy Holdings Pty and many other are vying for government solar grants to develop PV projects. A joint proposal from Suntech Power and Infigen Energy is expected to bring more alternative power to the continent.

During 2011, however, renewable energy investment in Australia is expected to fall to $3.5 billion from more than $4.5 billion in 2010. It is expected to rise again close to $4.5 billion in 2012 and remain higher than that for most of the decade. At least these are the expectations.

2011 update: the Australian Minister for Climate Change and Energy Efficiency, Greg Combet is planning to reduce the Solar Credits support multiplier effective July 1, 2011. In addition, the solar support program will be ended all together by July 2013. As a result, New South Wales government has permanently closed the solar bonus scheme to new applicants. On top of that (as if this were not enough) all existing FiT beneficiaries will also lose out as AUD$0.60/kWh rate is reduced.

New South Wales Minister for Resources and Energy, Chris Hartcher, has announced that reduced the existing AUD$0.60/kWh tariff will be reduced to AUD$0.40/kWh as of July 1, 2011 for an unspecified time period. The AUD$0.20/kWh tariff will not be changed at least for now.

This new situation is causing the Australian solar industry to sputter and stall, worrying about the future of feed-in tariff rates countrywide. It is simply unclear how this new situation will affect the solar industry in the country and the region, but at the very least it will slow it down.

Middle East and North Africa (MENA)

UAE and Saudi Arabia, along with Jordan and Egypt, appear to be leading the show for thin film and notably are considering CPV as one of the main alternatives. Saudi Arabia is planning to add 5 GW solar power by 2020 as part of its $100 billion targeted to development of renewable energy.

The demand for energy is expected to peak at 120 GW by 2012 according to industry analysts and to 250 GW by 2032. "Saudi Arabia is currently looking at various alternatives that can help boost its power generation to meet rising demand," Mohammad Al Hussaini from Riyadh Exhibitions Company, said recently. Masdar is one of the examples in the MENA region's drive towards solar future.

More than 70 solar companies are operating in the UAE, but the regulations are still a work in progress. The financial and real estate crisis, which hit the region in 2008 has slowed the progress, as had the recent turbulence which has brought additional uncertainties in the region.

"The Abu Dhabi government is working on a policy which will entail a regulatory and financial framework. The policy development is led by the Executive Affairs Authority," the spokesman said. He adds that because of the long-term and stable power purchase agreements, Abu Dhabi is becoming very attractive for investors.

The coastal emirate of Ras Al Khaimah (RAK) has also shown interest in using PV energy, especially for solutions related to cooling systems and desalination. The Swiss Centre for Electronics and Microtechnology (CSEM) in RAK is making significant efforts in research and development. It is building an open-air laboratory on 100,000 square meters featuring renewable energy R&D infrastructure, including the prototype of Solar Islands. The facility will house the Solar Cool Centre, which will produce up to 100 TR (tons of refrigeration) using solar power by 2012, besides the PV Test Centre Lab and Solar Outdoor Calorie Metre which will test the efficiency of solar insulating materials.

UAE, however, has decided to continue to subsidize the conventional electricity generation and water sectors. In a statement, Mohammed Saleh, director general of the Federal Electricity and Water Authority (FEWA) said the country will still "shoulder the gap between sale prices and product costs, despite the sharp increase in diesel and gas prices in global markets." What all this means for the solar future of the region remains to be seen, but we are convinced that solar energy will become an ever increasing part of it.

Jordan also shows progress in this area. The Kawar Energy Group and CH2M HILL launched a United States Trade Development Agency (USTDA), which funded a feasibility study to explore and develop utility-scale solar power generation in the kingdom. Jordanian clean energy developer Kawar Energy selected global consultants CH2M HILL to perform a feasibility study for its Shams Ma'an Photovoltaic Solar Power Plant, which will generate 100 MW of PV/CPV in the near future.

Jordan currently imports 96% of its energy, but the competitive PV prices and other incentives offer a lucrative long-term opportunity for Jordan's government.

Israel

The Israeli government came up with a new long-term, clean energy plan in mid-2011 calling for 10% of its total energy needs to be generated by renewable energy sources by 2020. This effort is geared to both ensure energy dependence and to reduce pollution from conventional power generators and users.

The new energy program calls for generating 2.75 GW electric power in Israel via renewable sources by 2020. Experts in the field claim it's ambitious, but achievable. The government is fully committed to subsidize the sector as planned and as needed. An inter-ministerial panel will oversee the evaluation, the design, development and implementation of innovative technologies by Israeli companies.

Not very long ago, the Israeli government pledged over $600 million for the development of green energy technologies and an additional $3 million will be used immediately to develop model facilities for the new green technologies.

Presently a 250 MW solar complex in Ashelim is in the bidding stages and promises to be the largest such facility in the country in the near future. Israel is going solar! What other choice do they have? And why did they wait until now?

In conclusion, the solar boom in these regions is due to a number of reasons, driven by increasing oil prices and the sharp decline in PV prices lately, due to oversupply in a market hard hit—especially among Western buyers—by the global recession. That glut was in part due to Asian PV suppliers that had beefed up production to meet demand in affluent markets such as Germany, which has been heavily subsidizing investments in solar usage for more than a decade.

Use of solar energy is largely about pricing for now (and somewhat about environment). To encourage investment in solar systems, governments introduce feed-in tariffs (with rates that are not always acceptable) and other strategies which work some of the time.

The bigger the market for solar systems, the greater the supply and, eventually, prices come down because of economies of scale and new technologies. But there are other socio-economic and price-factors involved.

In Thailand, for example, solar roofs haven't taken off partly because most roofs are built for a hot climate, and cannot take the extra weight without added investment in roof reinforcements.

Solar farms are land-intensive too. It takes about 5-6 acres of land to install a 1-megawatt electric power generator. And not all land is suitable for solar; i.e., the Philippines' Department of Energy deems the country's position just above the equator (with hazy, cloudy skies) not suitable for solar energy applications, as compared to other sources of renewable energy such as wind and geothermal.

Then there is the question, of just how long can governments afford to subsidize renewable energy such as solar, before passing the cost on to the consumers. Because the customers in most Asian and African countries are among the poorest in the world, they simply cannot afford solar, regardless of how useful it might be.

The MENA countries are in an altogether different class. With an abundance of sunshine and money to burn, most of them are well positioned to supply their own energy needs via solar and wind power, and even help Europe to get a significant amount of solar energy (via Masdar and similar projects). Does energy coming from foreign lands constitute energy independence? Europeans are trying to answer the question now, and until a proper solution is found, the controversy will deepen.

Note: Unlike CSP, which requires a large, centralized generation process, PV energy has proven to be well-suited for small and large; off-grid and utility applications. The cost of solar power has been steadily decreasing thanks to developments in technology and supply-demand factors, with solar cell production capacity forecasted to increase by 39% in 2011, as a result of which the average selling price of solar panels is expected to drop by 15-25%. This, combined with improvement in the region's economies will contribute to an increase in solar energy generation.

Bright Ray of Hope in Africa

It is obvious from Figure 9-9 that there is enough sunshine to generate several times the electricity demand of most countries. Africa is exceptionally "blessed" with the largest area of desert lands and with the highest quality sunshine and weather suitable for solar power generation. The economic status of most countries in the continent, combined with total lack of suitable infrastructure, however, will not allow development of significant large-scale PV plants for local consumption anytime soon.

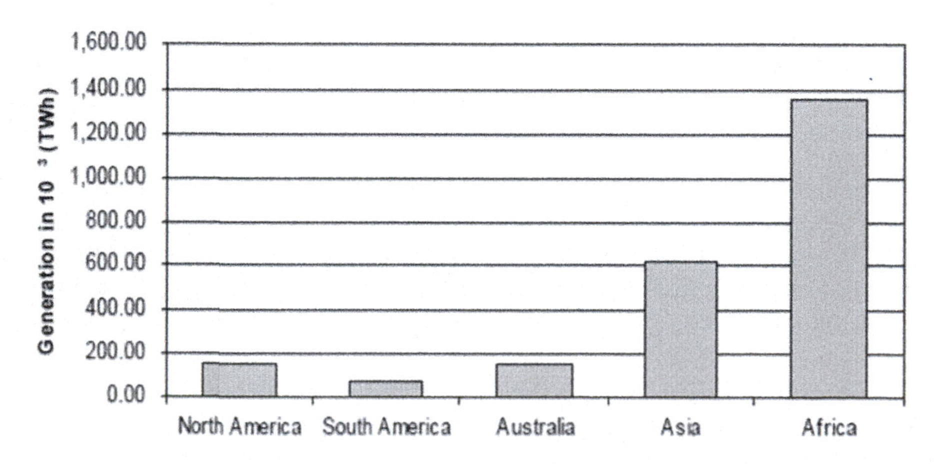

Figure 9-9. Potential solar power generation capacity

Nevertheless, North Africa will benefit from its proximity to Europe, because it meets the requirements of the grandiose plans of the EU community to install several TWp of wind, CSP and PV power plants in the North African deserts, close to the southern shores of the Mediterranean Sea in the near future. The actual work is going slowly now, and will continue so until the European beast emerges healthy from the economic debacle it is presently in. When those plans for the projects named Desertec are funded and executed, North Africa's deserts could provide a lot of cheap power to Europe. The complete details are still unclear, but here are some facts and preliminary plans that make sense and that promise a bright solar energy future for the region.

Europe has already established a Trans Mediterranean Energy Renewable Cooperation (TREC) between the countries of Europe, the Middle East and North Africa (EU-MENA). The construction of solar thermal power plants is already in progress in the Sahara. Germany and Spain lead in the drive for clean energy from the desert; other countries are now warming up to the idea of desert electricity. Even in Great Britain, a country that is rightly considered as being rather too far away from the Sahara desert, people are excited about the prospect of generating and transmitting clean, affordable sun-electricity from the Sahara to the UK.

Studies by Greenpeace, confirmed by a study commissioned by Germany's Federal Ministry for the Environment, Nature Conversation and Nuclear Safety (BMU) and executed by the German Aerospace Centre (DLF), showed that solar electricity imports from solar thermal plants in North Africa and the Middle East are likely to become one of the cheapest sources of electricity in Europe, and that includes the cost of transmitting it to Europe. Europe sees the generation of electricity from the sun energy in Africa's Sahara desert as a source of its future energy supplies and as security from the whims of Arab and Russian-controlled energy sources.

With the exception of several North African countries, which are part of the EU-MENA Desertec project, the rest of Africa appears to be either not aware of EU-MENA Desertec project currently in progress in its backyard, or is not interested in it. Worse, many African countries do not understand the ramifications of being left out of Africa's Sahara Desertec project.

While most countries see the need to switch from nuclear to renewable energy as demonstrated by current plans to generate electricity from the sun in Africa, Nigeria has recently indicated interest in nuclear energy instead. Nigeria, like the rest of Africa, is obviously oblivious of what it has—enormous amounts of sun energy, falling as sunshine on African deserts, arid Sahel region and savanna grassland begging to be harnessed by anyone who has the resources and an appreciation for sustainable, clean, affordable and renewable electricity and, of course, money is an obstacle.

Sustainable power in Africa can be based to a great extent on utility-scale renewable energy generation from the sun, as well as energies from wind and micro scale water turbines (instead of giant water dams). A well balanced mix of renewable energy sources with fossil fuel backup can provide affordable power capacity on demand throughout Africa. Africa is the only region in the world with many sites which are considered as either good (Sahara, Kalahari and Namib deserts) or suitable for solar power installations (arid Sahel steppes and savanna Grassland which stretch from East to West Africa).

Note: A site or region is deemed suitable for solar if it receives at least 2,000 kilowatt hours (kWh) of sunlight radiation per m2 annually, while the best site locations receive more than 2,800 kWh/m2/year. Ideal site regions, where the climate and vegetation do not produce high levels of atmospheric humidity, dust and fumes, include steppes, bush, savannas, semi-deserts and true deserts, located within less than 40 degrees of latitude north or south.

In conclusion:

— Availability of affordable and clean electricity the world over, and Africa in particular, would have a significant impact in cutting worldwide emissions of CO_2, and lead to significant improvement in the quality of life.
— North Africa will eventually benefit from the EU plans to generate electricity there and to transport it to the European countries.
— The shaded areas under solar mirrors have the potential to support local horticulture. Thus, a solar field can be used as a frontline installation for reforestation of the deserts.
— CSP plants can feed waste steam into the steam turbines of existing conventional steam power plants in North African counties thus putting back to use power plants that have become non-operational due to supply difficulties or price increase with oil or gas. This and other creative use of solar energy would go a long way in solving Africa's energy need in the long term.
— In Africa, the dominant source of fuel in low-income African homes is wood, which women and children spend many hours daily in search of, and which generates significant amounts of CO_2. Availability of affordable electricity would extend study hours for these school-children, free up time for other activities for women and reduce the CO_2 generation. In addition, deforestation with associated land erosion and desertification would be controlled as there would be no need to cut down trees for fire wood.
— Drinking water could be produced either by the desalination of sea water using the waste heat from CSP plants, or by harvesting dew and fog from the sky using thermal cooling from CSP plants to cool tentacles for dew harvest.
— Every year, each square kilometer of desert receives the solar energy equivalent of 1.5 million barrels of oil. Multiplying by the area of deserts in Africa or in the world, this is several hundred times as much energy as the world uses in a year. Countries like Nigeria can move directly from being oil-rich to being solar-rich.
— Electricity from solar thermal power plants in northern Nigeria's arid Sahel zone would be sold both in Africa and exported to Europe, generating more revenue than Nigeria currently earns from petroleum exports.

For Africa, solar technologies could offer more than electricity or drinking water. As a matter of fact, solar will eventually go on record as a singular case of technology that ushered in a rapid transformation of Africa. With a sustainable electric power supply as a major development catalyst, many African countries endowed with natural resources would have enough electricity to develop new kinds of processing industries. Many natural resources previously exported unprocessed, would be processed and exported as value added intermediates or end products.

This won't be a quick or easy process. There are many hoops to jump through, such as poor infrastructure, harsh (extreme heat) climate conditions, lack of qualified labor, etc. There is also a lot of money needed for the planned undertakings, so we can only hope and pray that they will materialize soon enough for the benefit of all involved in Africa and the entire world.

Note: Socio-political events, such as those in 2011, in several North African countries, however, might delay the development of solar energy (and everything else) in the region.

Obviously PV is growing, see Table 9-9, and despite the expected obstacles and delays, it will continue to grow. The solar revolution is on and cannot be stopped.

There are still unresolved issues, however, which will hinder the expansion of the PV energy markets. Let's take a close look at some of these.

PV ENERGY MARKET CONCERNS

A number of factors and problems determine the energy markets' behavior and choices. Using PV products is only one of these choices and because of that we must know what the markets need and the risks that must be considered when discussing use of PV products for large-scale applications.

c-Si and thin films PV modules are presently the key building blocks of the PV industry. While thin film and other PV cells and modules will continue to find application in residential and some large-scale PV markets, there are more c-Si PV cells and modules sold than all other technologies put together for all kinds and sizes of installations. Although there are some changes on the horizon, c-Si PV modules will dominate the residential and large-scale PV markets for some time to come. But they have problems, which we must be well acquainted with, in order to present a proper picture to the interested parties—owners, investors, etc.—for their consideration and decision making.

Some of the major concerns with the progress of the PV energy markets are discussed below.

Cost

We start the discussion on the PV energy market's concerns with PV products' cost because it is one of the most obvious barriers to the expansion of PV projects in the US and abroad. PV is currently at a disadvantage because of the high cost of PV modules and inverters. Painfully aware of the high cost of PV products, customers rush to build projects with cheapo PV modules from new, untested and unproven Asian manufacturers. How these modules would perform in the long run and how long they would last is of no concern to the hasty solar projects developers. As long as the modules work when they are first installed, they are given a stamp of approval. "We'll worry about the rest later," the common wisdom goes.

PV has a place in the energy market because it is flexible and has the most desirable set of attributes. The potential of a PV system to generate profit varies widely, however, and depends on a variety of factors, including the PV technology's quality and efficiency, the power plant's location, grid connectivity, and a number of technical requirements.

The major factor that has significant effect on costs

Table 9-9. World PV markets 2007-2020

	INSTALLED PV CAPACITY (MWp)					(PROJECTED MWp)		
Country	2007	2008	2009	2010	2011	2012	2015	2020
Germany	1,100	1,500	3,800	7,200	5,500	3,500	4,000	5,000
Spain	560	2,500	75	850	500	650	1,200	1,500
Italy	75	420	550	2,200	3,200	1,500	2,500	3,000
France	55	150	250	700	1,100	1,000	1,750	2,500
Japan	201	230	480	975	1,500	1,500	2,200	4,500
China	70	110	175	380	800	1,500	5,000	15,000
India	10	15	30	80	150	300	900	4,500
USA	207	342	480	900	2,100	3,750	6,000	15,000
World	Balance	Balance	Balance	Balance	Balance	Balance	Balance	Balance
Annually	**2,800**	**5,600**	**7,500**	**16,250**	**~18,000**	**~20,000**	**~30,000**	**~60,000**
Total	**9,550**	**15,150**	**22,650**	**39,000**	**~57000**	**~77,000**	**~150,000**	**~300,000**

today is the presence of market stimulation measures, which can have dramatic effects on demand for equipment. But the stimulus varies with the changes in the political wind's direction, so no solid predictions and forecasts can be made until the subsidies and incentives subside to a constant and manageable level. Lots of stimulus packages are available, but they are changing fast with political currents—so fast in fact that many potential owners and investors have dropped PV projects from their portfolios due to the uncertainty of the situation and the growing risk factors. 2011-2012 will be critical years in determining the future of the subsidies and incentives in the US solar markets.

The installation of PV systems for grid-connected applications is increasing yearly, but the grid-connected energy market also still depends heavily on government incentive programs and the related utilities' conditions. The installed cost of grid-connected systems varies widely from place to place depending on national and local support programs, equipment, and labor costs. Conditions are ripe for an overall drop in PV products' prices and the development of production capacities that could support rapid and quality deployment of large-scale PV plants in the US and some advanced countries. Their progress, however, depends on too many political and social-economic factors, which are driven directly and indirectly by political and economic developments.

One way to accelerate the process is to step up the economies of scale for the PV products manufacturers. Today a typical PV modules manufacturing facility has a capacity in the range of 100-200 MW/year. It would take such a plant 10-20 years to produce enough PV equipment (modules and such) to match the power-generating capacity of one large coal fired power plant. Because of that, we need many of these manufacturing facilities, and/or an increase in production capacity of existing factories, in order to support the growth of PV installations. Very importantly, these facilities must be located in the US proper, and not imported from the other end of the world, as is the case these days.

Another path towards radically lower cost is changing manufacturing processes. Today's technology of choice is based on crystalline silicon (c-Si) modules. It is the most understood and most reliable PV technology. However, it is inherently material and labor intensive, using batch production methods which are mature but still quite inefficient and expensive. The great hope for the future lies with technologies (or manufacturing processes) similar to today's thin film PV modules, and other such technologies, that are much less material and labor intensive, since they can be automated and are sustainable for continuous production processes. They would offer the potential to shift to lower cost mass production methods and provide higher output. The problem with today's thin film PV technologies, of course, is their low efficiency, and questionable field reliability and safety. So basically, we need to find a compromise between the efficiency and reliability of c-Si products, and the cheap mass production of thin film modules.

Manufacturing PV modules represents approximately half the cost of PV installations today. Other cost elements include the mounting hardware and electrical components such as inverters and frames, and labor for installation. All those elements add up to a typical "installed cost" of around $5.00-6.00 or more per watt today. A major part of this cost is the expense of purchasing, transport, additional warranty, insurance, decommissioning, and waste disposal—all components of the sand-to-grave life cycle of PV modules. Manufacturers usually prefer not to disclose all these costs at the beginning of negotiations, so when they quote you $1.25/Watt, you must ask what is included. Only then will you find the real value of the product in question.

Basically, the lowest PV project cost is obtained by optimizing the watts per square foot, the production cost per square foot, the overall product efficiency and reliability, and reducing the O&M costs. There are countless efforts, and we can testify to many successes in these areas. The longevity of PV products is an exception to the rule. It is hidden behind insufficient test data, misinterpretations of accelerated test data, silence, and no convincing long-term (15-20 years) field test data of the products sold on the energy markets in the past—data which could be translated to today's products as well. Longevity, however, is a critical (if not decisive) parameter of the final ROI calculations, so it cannot be ignored or shoved in the closet.

When negotiating a large purchase, show the manufacturer's rep the field test results in this book and ask for comparative test data (even if it is from accelerated tests) to see how their product compares, in order to verify their claim of product performance and longevity.

The Asia PV Syndrome

We cannot discuss cost or longevity of PV products without mentioning the developments of late in Asia and their effect on the world's energy markets. Asian countries are presently the largest producer of PV modules and other energy products. There are a number of reasons for this, but the main one is cost. In addition to labor costs that can be less than $200/month per worker, many Asian solar companies also benefit from huge government subsidies and much lower SG&A (sales, general and administration or 'cost of doing business'), research & development, peripheral costs, and tax rates.

There is also an expanding production equipment manufacturing industry in Asia that provides equipment at a fraction of the cost of equipment made overseas, although most of it is copied (some illegally) from Western models. The current generation of Asian manufactured solar equipment includes module laminators, wafer etch/bath, mono-crystalline wafer pullers and polycrystalline batch growers. Also there is a growing list of lower cost consumables suppliers (slurry, aluminum and silver pastes) which also helps to ensure low manufacturing costs. These activities result in a low-cost final product, the quality of which is impossible to verify or test for long-term efficient performance and endurance.

Asian manufacturing lines tend to be more labor intensive and use more domestic materials and equipment which require substantially lower capital expenditure. Most Asian PV module manufacturing lines use an all-manual labor approach, reducing the initial investment significantly.

Greatly increased production of cheap raw poly silicon material is another reason for the low c-Si PV module costs. There were three major poly silicon manufacturers dominating the domestic landscape in China just 3-4 years ago. Sichuan Xinguang with initial production in 2007 and 2008 capacity of 1,250 metric tons. LSCS with capacity of ~1,000 metric tons, and ESM producing ~1,000 tons. Now there are over 30 large poly silicon producers in China alone, and more planned to come on line in the near future. What is the product quality of these quickly put together enterprises? We have no access to production data so we must assume that it is acceptable. We all know how to spell ass-u-me, right, and many of us have paid the consequences of blind ass-u-me-ing?

Asia presently leads the world in production of many types and sizes of PV modules and other energy products, which are exported cheaply around the world (Table 9-10). The top 20 module suppliers (most of which are Asian) shipped ~12 GWp PV modules in 2010 (~80% of worldwide installations), and are ramping up production for over 18 GWp shipments in 2011. Including other suppliers, the PV industry is poised to ship 20-22 GWp PV modules in 2011, and there are projections for further increase of exports.

This extraordinary and familiarly abnormal growth is creating controversial Asian domination of the energy markets, which is causing friction on several business and government levels, and which must be corrected in order to balance the complex solar industry and the related energy markets situation. We here in the US also need to consider the damage this invasion of foreign-made energy products is causing our economy. We need to beware of the fact that importing such great quantities of energy products will only shift our energy dependence from Arab oil to Asian-made PV. How independent is that?

The worst part of the Asian invasion is the attitude of some Asian manufacturers. Emboldened by the rapid growth of the solar industry, and thinking their product is absolutely indispensable, many of them act like used cars salesmen, "This is a nice car, but you cannot take it for a test drive. And please don't open the hood, because there are things we don't want you to see."

Amazingly enough, US customers (designers, installers and investors alike) have accepted this abnormality only looking at, and arguing about, the price tag. If the modules are shiny enough, they are good for 30 years torture in the desert, the thinking goes. How wrong and unfair to the US solar industry and the unfortunate owners is this? And who will be held respon-

Table 9-10. PV cells production totals

PV CELLS PRODUCTION TOTALS, 2010

BY TYPE		
Technology	**MW DC**	**%**
c-Si	19,800	83%
CdTe	1,400	6%
a-Si	1,350	5%
Super m-Si	920	4%
SIGS	430	2%
TOTAL	**23,900**	

BY REGION		
Country	**MW DC**	**%**
China/Taiwan	14,200	59%
World	3,300	14%
Europe	3,100	13%
Japan	2,200	9%
N. America	1,100	5%
TOTAL	**23,900**	

sible for abnormalities and failures of these unproven modules, 10-15 years down the road?

Raw Materials Supply, Quality and Prices

Volume, cost and quality fluctuations of raw materials and consumables in the solar industry is a major subject that needs another book if it is to be fully explored, so we'll only touch on the basics.

1. Polysilicon is needed to manufacture c-Si solar cells, and although it is an abundant material in nature, it has been and still is a bottleneck in the solar photovoltaic energy industry explosion. The high initial capital and technical requirements of polysilicon production plants means that only large size (over 1,000 tons per annum) production can be economical, effective and competitive. This also means that only large enterprises with government help (which is the case in Asia) can undertake the huge risks and compete in this area.

   The gap between demand and supply has been and still is huge in Asia. The total demand in 2005 was 2,825 tons while the supply was only 130 tons. A number of new polysilicon production plants closed the gap somewhat, but PV has been growing about 30% per year lately, and since this growth is faster than anticipated, a temporary shortage in silicon feedstock has occurred. With the projected slowdown of the PV industry during 2011-2013, the gap might even be closed. In the midst of this, polysilicon prices dropped significantly from $400/kg in the 2008 spot-market to $30/kg in 2010 and even down to $20.

   The quality of the silicon feedstock material is also of major concern. Just think how the 130 tons of silicon produced in 2005 has grown to over 30,000 tons today, and the plan is to reach 90,000 tons in the near future. Some of the silicon production plants still use outdated Russian equipment and processes, while others are copying Western technology, so production is growing and prices are going down, but what about product quality? Are all production facilities equipped with proper equipment, processes and qualified labor? Who controls the process and product quality? How do they do it? This is another black box in the arsenal of the Asian PV industry, and we have no access to the details; we just have to assume that what we are buying is of good quality.

On top of that, if solar cells manufacturers rush the production process, not paying attention to the proper and thorough inspection and tests of the incoming Si materials and wafers, then the outcome would be questionable and in some cases disastrous.

Processing untested and unsorted Si wafers, for example, usually increases the rejection rates of the finished product, which translates into lost income. Most importantly from our point of view, cells and modules made out of lower quality Si wafers will have reduced efficiency and will be much more likely to fail after several months or years of field operation. This is the problem we see with PV modules exposed to merciless desert heat and is what we need to consider, watch for, and prevent one way or another. See Figure 9-10.

With millions upon millions of PV modules coming from all over the world, we are faced with a serious large-scale problem which will become more pronounced as time goes on—and especially when the PV fields have been 10-15 years in operation and the original manufacturers are out of business or moved on to the next technological gold rush.

In his article, "An Eye on Quality" in July 2011 (15), Ian Latchford of Intevac (US equipment manufacturer) addresses this problem and suggests strict quality control, especially at the beginning of the manufacturing process, by using 100% photoluminescence (PL) testing of the wafers' bulk material to identify low-quality or contaminated Si wafers. PL is one of the most effective and reliable non-destructive processes suitable for mass production operations.

Sounds like a no brainer, right? Good quality in, good quality out! Yes, but not nearly enough PV cells manufacturers are doing PL, or any other 100% initial bulk wafer material quality testing. The reality is that a 10% visual and surface resistivity check is the accepted maximum, and even that is cut significantly during rush production periods.

The industry needs "total" quality control, including PL or similar non-destructive qualitative testing of the integrity and quality of wafers and cells after key steps of the manufacturing process.

The key tests, in addition to the initial PL bulk material tests described above, must include:

a. Initial and in-between-steps wafers surface cleanliness test, where both wafer surfaces must be visually inspected and then scanned or otherwise tested for organic and inorganic contamination and mechanical defects.

Building a solar cell on a chemically contaminated

or mechanically damaged surface is like building a house on a sandy hill—it will collapse after the first rain.

b. Diffusion length and concentration testing is absolutely necessary to determine any process inadequacies or material imperfections. The diffusion layer is the engine of the PV power generation and not checking it is like entering a race without checking your car engine.

c. Metal fingers depth of penetration into the wafer surface and quality of the alloyed area. Imperfections in the metal-silicon boundaries will result in latent problems which will affect the cells' field performance and longevity.

Not easy or cheap, this is the only way to ensure the quality of each cell and module. The rule of thumb is that the lowest quality solar cell will determine the highest quality of the entire module (and string). So if one cell in a 100-cells module is of low quality and/or is underperforming, the entire module will respond in kind. And if the cell fails, so will the module. What if this failure occurs in our large-scale power field after only 10 years of operation? Can we afford this? Can we afford not knowing EXACTLY the quality of the product we are using?

Remember: solar cells are expected to last and perform for 30 years in the most inhospitable environments. We even put our well-being and reputation on the line that this will happen. Or should we...?

What we have on the solar cells' mass production lines today is a "partially closed eye on quality abnormality" (no pun intended), where some inspections and tests are done on some of the product, some of the time. The final result is "some garbage in, some garbage out." This issue is complicated to the point that even the production line engineers would not be able to tell you what to expect, simply because they are working blindfolded some of the time too.

The technical specifications, however, will show only good final product, which in the short run might be OK. In the long run (20-30 years in the desert), however, any bulk quality problems, contamination, or process imperfections will only aggravate the situation. Latent problems not detected during the manufacturing process will eventually cause accelerated annual degradation, higher temperature coefficient and/or premature field failures.

Manufacturers cannot assure us that their product will last 30 years IF they have skipped some of the steps of the solar cell production QC/QA sequence for the sake of faster, cheaper mass production. And critical QC steps are normally skipped or poorly performed.

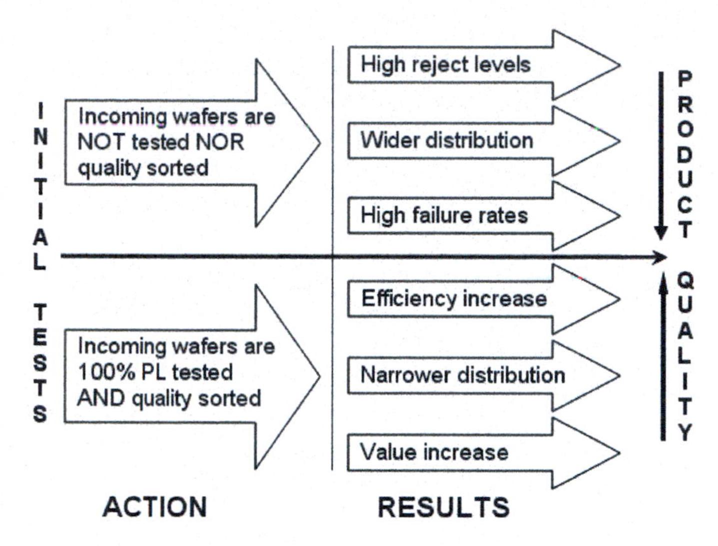

Figure 9-10. Wafers test and sort

Note: Thin film PV modules manufacturing, on the other hand, is quite different from a process/quality control point of view, because the process is quite sophisticated and the quality control procedures are built in. Automated production equipment and clean-room type facilities are state-of-the-art in most cases, and the production process is run by highly qualified engineers and technicians. Elaborate QC/QA procedures are built into the process too, ensuring superior quality and performance. Mistakes can happen, yes, but they are the exception, unlike the case with c-Si production where quality gaps can be found at every step of the unsophisticated, labor-intensive process.

2. Huge quantities of silver, aluminum, copper and other metals, as well as plastics and many chemicals are needed to produce millions upon millions of PV modules as demanded by the solar markets. Large amounts of silver metal are used for providing good ohmic contact between the inside of the cell and the outside circuitry of the PV modules and for reflective backing for mirrors in thermal solar plants. According to the VM Group in London over 1,000 tons of silver have been projected for making PV modules in 2011. This is over 1.0 million kilograms of silver used today, an amount projected to triple by 2016. Nearly 3,000 metric tons of silver will be used for making PV modules worldwide every year at that time. If we add that much more silver metal for coating heliostat mirrors and use in the semiconductor and other industries, we end up with some very large numbers. So the question here is, "How much silver will be left in the world in 10, 50 and 100 years if we use it at this pace?"

Silver is a precious metal, the price having gone sky high during the last 10 years from $4.00/ounce several years ago to nearly $40.00/oz. today. As things are going, prices will continue to rise. What will that do to the price of PV modules?

Similarly, the price of copper and aluminum metals has sky rocketed lately, and although there are large deposits of these left on Earth, the increased prices will play a significant role in PV modules' use on the world's energy markets.

3. Thin film modules manufacturing requires a number of rare and not-so-abundant raw materials, such as cadmium, tellurium, arsenic, indium gallium, selenium, etc. Some of these are already in limited supply and it won't be long before they are depleted altogether. Their prices will go up and down according to the supply and demand ratios, but as the demand increases sharp price increases are expected during the next several years.

Solar Industry Standardization

Every beginning is difficult and the developments in the solar industry today are no exception. There are a large number of manufacturers, installers, regulators, products, processes, regulations and laws, operating in a fragmented manner. There is no comprehensive and all-encompassing standardization in any of these areas and activities. Instead, the different parties make their own rules, which have resulted in a patchwork of products and services, some of which are hard to classify and qualify.

Table 9-11 clearly identifies some of the major areas in the PV industry that have limited, or no, standardization. Full standardization of these key elements must be implemented as we expand solar energy use, but they are complex and expensive to tackle. The lack of standardized supply chain quality, equipment reliability, and manufacturing and quality control processes integrity is quite obvious. New companies pop up daily and offer product that has been manufactured any way possible, with a level of quality that remains to be seen. In many cases, the quality of the shiny, cheap, modules we buy for our new PV power plant is a dark secret.

A number of fragmented efforts in different areas are helping, but an all-encompassing standardization effort has to be started from scratch. This is not an easy task, especially during this frenzied period of frantic growth, where new companies pop up every day in Asia and use whatever materials, equipment, processes and procedures they find suitable or profitable. This situation reminds us of the chaos in the semiconductor industry several decades back and how hard it was to introduce, establish, and enforce the materials and manufacturing standards. Some Asian semiconductor manufacturers are still far from complete standardization and total quality control even today. We believe that it will be equally difficult to introduce standards in the international solar manufacturing industry too, and that it will take decades to report serious progress in this area, much less complete standardization.

The supply chain (purchase of raw materials, chemicals and gasses from a third-party vendor) is a wild card in the present-day PV industry's deck. It, too, holds secrets which we, the customer and buyer, do not have access to and which are not discussed openly by most of the participants. The chain is at least partially invisible in most cases, while complete details won't be provided in all cases. And we wonder if there are complete details available even to the manufacturers in some cases. We know from experience that the quality of the supply chain, especially in Asian countries, has a lot to be desired. This is another large subject, which requires its own book, so we'll just touch the surface.

The overall process of verifying the product quality, permitting, installation, O&M and decommissioning of solar power fields is also without standards today. These key elements vary in structure and quality from state to state and from company to company. If there is some standardization in some of the above processes, the enforcement usually is not in place, which simply means that standardization is far from being a uniform and practical tool used by all manufacturers, installers, and operators around the world.

On the administrative side, the lack of enterprise, view, and demand planning are other signs of the immaturity of the solar industry. In our opinion, it is not even possible to discuss these parameters in the chaos of the present situation, where manufacturers are pushing whatever product they can squeeze through the lines, while customers are only interested in the lowest prices. Why would any manufacturer care about anything but volume under these circumstances? At least for now and until US customers and investors get more educated about the long-term effect of the issues at hand, the immature solar industry and its products depend on decisions that might be less than beneficial to its long-term health. See a list of needed standards in Table 9-12.

Power Transmission, the Essential Link

The efficient use of the transmission networks and the need to accommodate renewable energy sources are

Table 9-11. Key PV manufacturing standards needed

Supply Chain	*Mfg. Process*	*QC/QA*
Raw materials	Equipment	Initial inspection
Process chemicals	Process specs	In-process inspect
Consumables	O&M procedures	Final inspection
Waste treatment	Enforcement	Long-term tests

Table 9-12. Key PV power plants standards needed

PV products	*Permitting*	*Logistics*
PV modules	Federal policies	Land preparation
Inverters	State policies	Interconnection
BOS	Local policies	Installation
Misc.	Utilities policies	O&M

key elements in the future of power generation and distribution in the US and the world. Electrical transmission lines have to be used because electric power is usually generated in a fixed area, and then the power must be transported to a point of use (POU) which might be many miles away. Furthermore, the solar energy source is often located in remote areas such as solar energy generated in the desert. Since it is not practical to store electric energy in large quantities, at least for the time being, we need transmission lines to transport produced energy to the users in the POU centers immediately upon generation.

From a technical point of view, a large amount of power can be transported on AC and DC transmission lines over very long distances. Additionally, between areas separated by water, the use of a DC system allows a utility to transmit the energy by submarine cables. This has been successfully done by running such cables under the Mediterranean Sea, connecting Africa with Europe, where several additional solar installations and underwater connections are planned. Research studies show only small power losses and insignificant adverse health effects on humans or aquatic life in and around these areas.

The type of transmission used depends on the type of generation and distances involved. High-voltage transmission is defined today as grids with voltages from 69 kV and above. Most of today's interconnected and meshed networks use three-phase alternating current, AC, with a frequency of 50 or 60 Hz as the commonly used technique, taking advantage of the easy transformation between voltage levels. Direct voltage, DC, is used especially for long transmission lines where it gives the advantage of the same power level being transmitted via just two conductors and with lesser overall power losses.

Higher AC voltages result in lower losses and provide economic advantages. But, HVDC (high voltage DC) can also be used for special applications when it is possible or when it is otherwise difficult to connect the two networks, e.g. for stability reasons. The capacity of an 800 kV AC line is around 2000 MW, while an anticipated figure for the future for 1200 kV lines is 5000 MW.

A realistic maximum distance for an uninterrupted AC transmission is around 800 miles. For transmission lines moving electricity, this means that there is the potential of using long transmission lines to a larger extent than is the case today. So the proportion of renewable and cleaner types of electricity generation would be increased, reducing the amount of greenhouse gas emissions. Such bulk power transmissions over long distances can be built at a reasonable cost. The transmission of 2000 MW over 600 miles would cost less than 1 cent/kWh.

All this is fine and doable, but adding transmission lines and substations to transport solar power generated far out in the deserts is a complex and expensive proposition. There are plans, however, for expanding and upgrading the power grid at key locations (mostly in California) which will allow the installation of additional PV power plants in the deserts.

The Real-life Obstacles

As with any developing sector of the world's economy, solar energy has its own problems. There are some clouds on the bright energy future's horizon, which we need to be aware of, because it might throw the world PV industry into a spin, as it did several times before. The major obstacles in the broad implementation of solar energy in the US and world's energy markets are:

Location

Actual PV power plant location is the single most important consideration for owners and investors. Maximum sunlight and minimum sky obstructions (clouds, fog, smog, shade etc.) are the preferred conditions. The weather is an uncontrollable variable, but is more consistent in some places than others; i.e., the deserts are much more predictable than the shoreline or the mountains. In all cases, good research is needed to determine the available solar radiation during the different seasons.

NASA's solar charts (11) could be used as a source of long-term solar and weather information for most areas of the world and is a good start when choosing location.

In addition, NREL has an excellent tool called Solar Advisor Model (SAM) which allows thorough performance and financial modeling of any type and size solar projects. It is location specific and uses the latest climate data available for the particular location, which could be quite useful for decision making solar professionals and investors. (12)

Infrastructure

Solar energy is most useful when installed in large-scale PV power fields and plugged into the grid for mass distribution and use. Not many countries have an efficient distribution system that is accessible and capable of handling mass solar power influx. The US is an exception, so we are looking forward to a rapid increase of solar energy generating projects in the near future.

Political and Regulatory Issues

Permitting the installation and operation of solar power plants in the US and other countries is a difficult, slow, and expensive process. Government bureaucracy, environmental issues, utility companies' inefficiencies and other factors are hindering the quick development of solar installations. Increased awareness of the problems with conventional energy producers around the world is helping solar energy development and we expect it to become a primary energy source in the US in the very near future.

Toxicity

Some PV technologies contain toxic materials in their structures (i.e., thin film modules contain cadmium, tellurium, indium, arsenic etc.), or use such in their operation (i.e., thermal liquids and lubricants in CSP plants). While in small quantities these are not so dangerous, the deployment of large quantities on very large land areas presents a concern of large-scale contamination. Toxic materials could outgas and otherwise leak harmful substances, which could cause serious environmental damage in the affected areas. The effects of these materials on the local environment must be evaluated and the risks calculated prior to starting work on the project.

RoHS Compliancy Certification

With the increase in the size and scale of PV installations in US deserts, the toxicity issue is becoming an important concern to the locals since it directly affects their lifestyle and well-being in perceived and real ways. It is of utmost importance in cases where millions of toxic materials containing PV modules are installed on thousands of desert-land acres. Even small amounts of toxic chemicals in the individual modules might prove detrimental to the environment and all life forms in and around these mega-fields. We have discussed the details of potential problems in this text and are convinced that the danger is real and that something must be done to eliminate the possibility of converting large desert areas into hazmat dumps—no matter how remote the possibility might seem at first glance!

A significant step in the right direction is the recent trend (mid-2011) where BLM mandates the use of air monitoring stations in large-scale PV power fields and submitting hourly data of the type and quantity of particulate, heavy metals and other pollutants contained in the air samples. In this way, the locals will be alerted ahead of time to any air contamination and be able to take the necessary precautions. Another, and maybe even more important step is taken by some world PV industry leaders. A number of responsible manufacturers are voluntarily submitting their PV modules to RoHS (Restriction of Hazardous Substances) testing and certification by qualified facilities in the US and EU. Thus certified modules are labeled "environmentally friendly" and are given green light for use anywhere and in any climate.

For example, Inventux Technologies AG was awarded such a certificate in mid 2011 by TUV Rheinland (a testing and certification lab in the U.S. and Germany). The certification means that Inventux manufactures PV modules free of toxic substances, such as cadmium, arsenic, lead and other toxic chemicals. The modules are therefore classified as "totally environmentally friendly" and can be used anywhere and in any climatic, and can be easily recycled and disposed of as non-toxic waste. According to Inventux, their thin film modules are classified as "conventional building glass," and can be safely used, disposed of and recycled without elaborate chemical processing. The company had its modules checked and certified on a voluntary basis for conformity with the EU RoHS Directive and with the German Electrical and Electronic Equipment Act (ElektroG).

Another world class PV modules manufacturer to follow this example in 2011 and voluntarily obtain RoHS certification for its CIS modules is Solar Frontier, Japan. While solar modules (including TFPV modules of any type and size) are currently exempt from RoHS compliance, Solar Frontier is one of the few manufacturers claiming a toxicity-free product. It is one of the few to voluntarily submit their product for testing by TÜV Rheinland, in order to make the claim official.

The "Restriction of Hazardous Substances" (RoHS)

adopted by the European Union in 2003 sets a 0.01% concentration limit on cadmium, and a 0.1% limit on lead, mercury and other hazardous substances that can be found in any component of a tested PV device. To ensure the long-term safety, including recycling and disposal, PV modules are tested and certified and officially recognized as RoHS compliant.

We strongly encourage this trend of responsible treatment of a critically issue, and would like to see it expand to all sectors of the solar industry—all products and installations. A worldwide program to categorize PV modules used in large-scale US power fields on the basis of the type, amount and seriousness of their toxic content during long-term use is urgently needed. We do hope that until RoHS certification becomes a standard, all manufacturers, regulators, installers and owners involved with PV products containing toxic materials will be aware of the potential dangers, consider the consequences, and act responsibly. For now, using PV modules without RoHS certification in large-scale power fields—especially in the US and world's deserts—must be done with utmost caution. It is the responsible thing to do!

Cost

Everything considered, solar energy generation requires complex and expensive technologies. The trend now is to use the cheapest possible equipment, regardless of its quality. This is a dangerous situation, which might soon contribute to the failure of a number of PV plants.

Many people around the world cannot afford the expensive PV technologies, even with the generous incentives and subsidies. This situation won't change anytime soon, but progress is made slowly, so we expect PV to be at LCOE (comparable to the conventional energy sources cost) within the next 15-20 years.

Efficiency

Some of the existing solar technologies—including PV—are just not efficient enough to stand on their own. A lot of land and equipment is needed to convert relatively small parts of the incoming sunlight into useful energy, which increases the overall costs. More efficient PV technologies are needed, to provide more bang for the buck. Efforts in these areas continue and we have no doubt that PV module manufacturers will succeed in this challenge.

Longevity

According to MIL-STD-721C, "reliability" is the duration or probability of failure-free performance under stated conditions, while "durability" is a measure of useful life (a special case of reliability). We lump these two qualities together and call the resulting factor "longevity" which is directly related to the length of the useful life cycle of PV modules, other components and the related power fields.

The useful life cycle (or longevity) has different values for different situations, according to agreed terms. If, for example, a PV power plant is designed to operate 20 years at a very minimum of 80% efficiency, but is down to 75% efficiency by year #20 (due to PV modules degradation), it might be decided to shut down and decommission the plant. In this case, the longevity of the modules (and the power plant) is considered to be less than 20 years, and represent the end of their useful life cycle. Thus, the longevity (reliability and durability) of the PV plant in this case is also less than 20 years.

In the same case, plant owners might come to an agreement with a manufacturer to continue operating the facility even though the power output would be well below the warranty levels. Under this scenario, the PV modules would be operating beyond their useful life cycle, but their longevity has been extended with the understanding that the power output would be much less than the previously agreed upon, unless the discrepancy is corrected somehow. Replacing excessively degraded and failed modules is one way, but the manufacturers must take the responsibility and rectify the situation by providing replacements. Here is where performance warranties and performance insurance would play a critical role in the plant's survival.

While all other obstacles can be worked out and improved one way or another, longevity—or the time a PV module will last while producing electricity per spec, or before failing all together—is an uncertainty which is not well understood, or widely publicized. Because of its importance, however, we need to address it every time we discuss a solar module or project. Because longevity determines the bottom line, which affects the energy markets, we will take a closer look at the elusive longevity factor.

For discussion and analysis purposes, we divide the PV modules life cycle, or their longevity, into several different and distinct stages:

a. *Early mortality* is associated with rapid power loss, or total failure, shortly after installation and field operation, as discussed in the previous chapters and hopefully before the manufacturer's warranty has expired. In this case the "longevity" is not calculable, because the modules fail prematurely.

Nevertheless, the longevity period starts at the day when the power field is activated.

b. *Useful life* is the period from the time we decide that the modules are no longer victims of early mortality, until they reach a certain predetermined loss of power. Twenty years of field operation with a maximum of 20% power loss for the duration is presently accepted as "normal" longevity. We argue, however, that this is just not enough time, and that 30 years, and even more, must be considered as normal longevity.

c. *End-of-life* is the period close to the end of useful life, which is characterized by accelerated power loss and/or abnormally high failure rates. The day when the power field is switched off for good marks the end of the modules' longevity period.

The objective of longevity optimization is expressed in terms of prolonging the useful life of the PV modules as much as possible by all means available. This effort must be in the foundation of all PV modules manufacturing operations, because once the modules leave the manufacturing process line there is very little we can do to optimize their longevity. Keeping track of the life cycle stages and recognizing the signs of degradation and failure is one of the few tools we have for dealing with longevity issues once the modules are installed in the field.

Once passed the early mortality stage, most well-made PV modules settle down into a normal production cycle with the expected 1% annual power loss until they enter the end-of-life stage, which could be at any time after 30-40 years operation in Germany, or Washington, DC, but only 10-15 years of non-stop operation in the desert.

Longevity is one of the major, and most unpredictable, issues plaguing the PV industry today, which increases the uncertainties and risks of long-term operation, especially in excessive heat and moisture regions. All PV modules are doomed to fail slowly or suddenly in the field. They all start their useful life by losing ~1% efficiency every year, due to material and process instabilities. The trend could accelerate quickly and cause catastrophic failure with time—especially under desert or other harsh climates. This uncertainty is great enough to cause serious concerns to customers and investors alike, and must be solved to ensure efficient long-term operation of PV installations.

A number of disturbing facts and data on the subject were presented in the previous chapters, so it suffices to say that the PV technologies—with no exception—have unresolved issues and have not proven fully reliable for the long run—30+ years of non-stop, in-sun reliable operation in the deserts. To be fair, we must also note that no PV modules have been given a chance to operate 30 years, to be fully tested and proven reliable.

Lab Tests and Certifications

PV modules go through serious tests during certification for use in the US and EU. These tests consist of a series of visual and electro-mechanical procedures designed to evaluate the ability of the modules to operate properly under normal conditions and to withstand some abnormal conditions. See Figure 9-11 for details on the types of tests the PV modules generally undergo.

These tests, however, do not come even close to the types and magnitudes of climate extremes in deserts and humid areas. So, very often, modules that have passed the initial certification tests can deteriorate quickly or fail in the field, as we see in the examples herein. This is not surprising, because no test can accurately reproduce 20 years of daily blistering heat and UV bombardment, midnight freezes, sand and monsoon storms, and never-ending thermal and electro-mechanical stress.

Long-term Desert Field Tests

As a confirmation of the facts and issues discussed in the previous chapters, a research study published in the fall of 2010 (8) confirms the presence of major issues in PV modules operating under harsh desert conditions. The study describes the results from long-term scientific tests done with PV modules operating in an official desert test field for a number of years (ranging from 10 to 17 years).

Note: To our knowledge, this study is the largest-ever evaluation of the long-term performance of production-grade PV modules under desert conditions.

The study was done at one of the most prestigious PV test sites in the world; APS's STAR solar test facility in Phoenix, Arizona, where most world-class PV modules manufacturers send their modules for long-term testing, so we would not be surprised if the tests include some famous names.

The work was supervised by one of the world's most reputable PV specialists, Dr. Govindasamy Tamizhmani (Mani) from the Arizona College of Technology and Innovation, and President of TUV Rheinland, Tempe, Arizona, at Arizona State University (ASU).

Dr. Mani's team carefully and thoroughly visu-

Figure 9-11. PV modules certification tests(14)

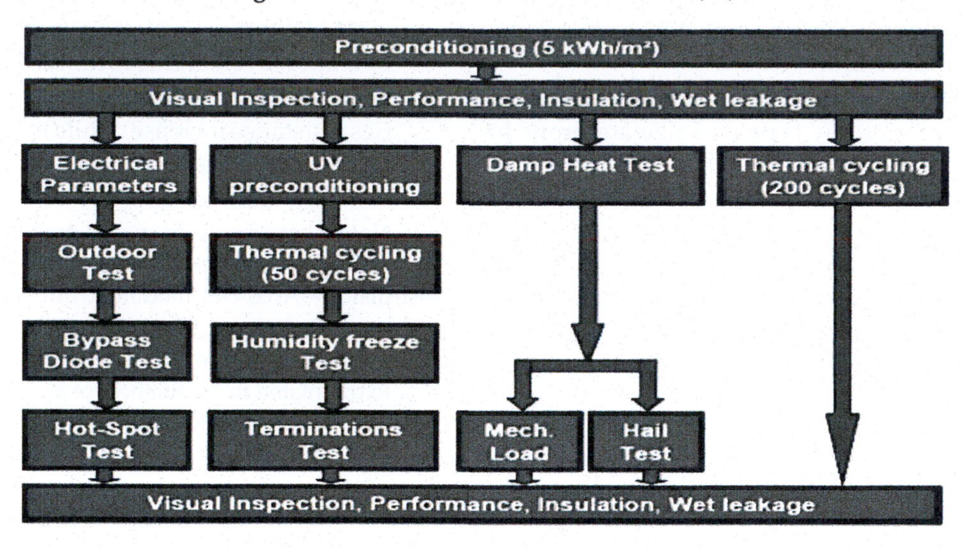

ally inspected, tested electrically and IR measured each module in the nearly 1,900 PV modules batch.

The final results from the long-term tests in the Arizona desert are shown in Table 9-13.

Table 9-13. Long-term field tests in Arizona desert (8)

Module	*Module Count*	*Modules Affected*
A—Mono-Si (17 years; glass/polymer	384	
Browning in cell center	384	100.00%
Frame seal deterioration	384	100.00%
Hot spot (IR Scan)	4	1.0%
B—Mono-Si (12.3 years; glass/polymer)	1092	
Encapsulated delamination	2	0.2%
Browning in cell center	1092	100.0%
Hot spot (IR Scan)	2	0.6%
C—Poly-Si (10.7 years; glass/glass)	171	
Broken cells	47	27.5%
Encapsulated delamination	55	32.2%
Hot spot (IR Scan)	26	15.2%
White material near edge cells	2	1.2%
White material browning	56	32.8%
D—Poly-Si (10.7 years; glass/polymer)	48	
Browning in cell center	37	77.1%
Hot spot (IR Scan)	4	8.3%
E—Mono-Si (10.7 years; glass/polymer)	50	
Bubbling substrate	33	66.0%
Browned substrate near Jbox	50	100.0%
F—Poly-Si (10.7 years; glass/polymer)	120	
Discolored cell patches	6	5.0%
Backsheet bubbling	1	0.8%
Browned spots on backsheet	2	1.7%
Metalization discoloration	22	18.3%
Frame seal deterioration	15	12.5%
Hot spot (IR Scan)	4	3.3%
Total Modules:	**1865**	

Amazingly, almost 100% of the nearly 1,900 PV modules tested in that study showed some sort of visual deterioration. Some were more impacted than others, but the overall conclusion was that a large percent of PV modules exposed to constant sunlight and heat in the deserts deteriorate beyond the manufacturer's warranty within 10-17 years of operation. Many of the tested modules show average 1.3 to 1.9% annual power degradation, which translates into 23-33% total power loss for the duration (some only 10 years). This, extrapolated for the extended time of use, is well above the manufacturer's warranty of maximum 20% degradation in 20 years.

Interestingly enough, significant EVA encapsulation discoloration and delamination was observed on many modules, even those that were exposed to the sun only 10 years. Solar cell discoloring (due to a number of factors) was observed on most cells too. Since the oldest modules in this study were made in the mid-1990s we must assume that encapsulation and cell manufacturing processes have improved some during the last 10-15 years to reduce the defect ratio. However, we have no proof of such progress, so this becomes another unknown that needs our attention when dealing with PV modules destined for desert applications.

In summary, browning in cell center was almost 100%, encapsulation issues followed as a distant second, followed by frame seal deterioration, hot spots, broken cells and glass, etc. Many of these defects are serious enough to cause power loss and eventually lead to total failure, be it due to optical interference, electrical resistance, overheating, mechanical disintegration, or any combination of these. This is simply not supposed to happen; not in such large numbers at such an important test site, so we must ask, "What caused the problems? Was it the materials quality (silicon, metals, EVA, glass), or the manufacturing process (diffusion, metallization, encapsulation), or was it some special combination of these that was primarily responsible for such a large number of defects and failures?"

In any case, this study confirms the unfortunate reality that PV modules of any type and size are prone to rapid deterioration and failure under harsh desert conditions. Although we cannot predict future performance and overall condition of modules during years 20, 25 and 30, the logical assumption is that the deterioration processes will continue at the same rate, and even faster in some cases. Because of that, we must conclude that each module type and size must be very well designed and properly manufactured with high quality materials and proper processes in order to be given a chance to survive the desert elements.

Note: The reasons for the above described defects and the related explanations and justifications are too numerous to launch into at this point, but references and explanation to many of them can be found in the previous chapters.

Another set of long-term field tests, albeit with a reduced number of modules and under different climatic conditions, was conducted by the Institute for Energy, Ispra, Italy. The test field is located in northern Italy, so climate conditions are much different from those in the Arizona desert, causing the field inspection and test results to be somewhat different as well:

# of tested Modules	Measured power loss
32%	1-5%
25%	10-20%
24%	5-10%
8%	20-30%
6%	30-50%
3%	75-100%
2%	50-75%

Summary:

- 50% of the modules show yellowing after 10 years, and 98% after 20 years of on-sun operation, with 78% showing excessive yellowing.
- 74% of the modules show delamination after 10 years, and 92% after 20 years.
- 76% of the modules show sealant infiltration and 22% have noticeable signs of hot spots.
- The average annual degradation of connected modules is ~1.0%, while that of modules left in open circuit mode was only 0.6%, showing the negative influence of current flow and additional temperature increases.
- Glass-glass modules also had a greater degradation, most likely due to excess temperature levels during operation even at this location,
- High level power losses (>20%) are attributed to series resistance increase, while moderate power losses (<20%) to optical properties degradation.

Note: These results are not as dramatic as some other field tests we've seen (probably due to the moderate climate and good construction—Arco c-Si modules), but are still significant and another confirmation of the fact that PV modules will perform as well as they are made.

Extrapolated to the finances of our 100 MWp power plant, the average power degradation and failures represent a major income loss from year 10 on. And the question remains, "Can we afford even moderate power and product losses like this?"

Catastrophic Failures

Those of us who live in the desert are well aware of its destructive nature and the forces that are in play nonstop day after day, year after year. We know that the desert has no friends or favorites and shows no mercy. Anything left in it for a long period of time will be damaged or destroyed sooner or later. There are few exceptions to this reality, but unfortunately PV modules are not one of them. Because of that, we must be prepared for the worst when installing PV fields in the desert. Excess stifling heat, freezes, highly destructive UV radiation, severe sand storms, strong winds, golf ball-size hail, and fires just to name a few must be expected at any time.

These conditions, added to the high voltage current flowing through the panels create scenarios on the extreme end of most materials tolerance limits. Plastics crack and disintegrate with time, the active structures inside the modules experience excessive thermal and mechanical abuse, shrinking and expanding with the temperature changes, and all this can create defects in the structure and lead to failure.

The defects caused by mechanical, electric or chemical abnormalities in the modules could lead to partial or total failures. Partial failures result in power degradation which results in power loss, but the modules are still operational. This is the case with the modules in the field tests; they show damage, defects and power loss, but are still functioning. Partial failure could also be caused by sand storms that temporarily disable the power production by soiling the modules, or damaging their support structures. After proper maintenance, the field could be up and running within hours.

Total failures result in permanent damage, be it mechanical damage to the modules or electrical failure causing open or closed circuits, or chemical disintegration of the active structure. These abnormalities could result in total shutdown of modules and strings of modules, and in some cases even cause fires. These are then classified as catastrophic failures.

Catastrophic failures are not an everyday occurrence in PV power plants, but when they happen they cause a lot of damage. The most dangerous catastrophic module failure, DC arcing, occurs when the modules lose their dielectric properties after long-time operation under the elements (due to lamination break-down) and start leaking electric current across the module elements. DC arcing is an example of premature module failure and is the worst catastrophic field failure.

The problem with DC arcing is that it cannot be easily foreseen or prevented, and once started it cannot be stopped. The arc will burn out only when there is no more material to burn, and in worst cases it could ignite the module, burning it and the entire array. This is a catastrophic failure, which could cause damage to the entire project. Proper O&M procedures (including using IR sensors to detect hot spots and overheated modules), albeit expensive and time-consuming might be able to detect the precursors of DC arc creation, so they are highly recommended.

There is simply no way to predict, let alone prevent, the desert elements and related surprises, but properly designed and manufactured modules, using highest quality materials and processes are a good start in that direction. Proper installation, again using highest quality materials and procedures, is the next best thing we can do to extend the useful life of the modules. Using proper O&M procedures for the duration will close the circle of our responsibilities.

In addition, an adequate manufacturers' warranty, including long-term performance warranty, is needed to provide protection against product defects. And finally, insurance against natural disasters, complete with long-term performance insurance, will bridge the gaps between the manufacturers' warranties and what Mother Nature has in store for our modules during their 30 years field operation. This way nothing is left to chance, and only now are we ready for anything the desert is going to send our way—for the duration.

Module Construction

Module construction—materials and type of construction—is of utmost importance for a number of reasons and might mean the difference between success and failure, as seen in the above long-term desert test. One useful observation in the ASU-APS test study (8) was that many modules are of the glass/glass type construction, which is basically two glass panes joined together with encapsulation with the thin film structure in between. This configuration is preferred by many manufacturers these days (TFPV modules especially) in order to reduce costs. We insist, however, that these initial cost savings are going to cost dearly in the long run, and many PV projects might suffer long-term consequences.

An interesting conclusion in the study was that glass/glass construction of some PV modules contributes to heat retention which causes additional increase of temperature in the modules and which could result in excess delamination and performance deterioration. "It is immediately obvious that Model C has the highest percentage of hot spots. It is believed that the thermal stress of the glass/glass construction and large size could be primary causes for this failure type. Again, seeing as how this study was conducted in a hot and dry climate, it is not to say that the glass/glass module construction would not succeed elsewhere."

In addition, since most of the modules in the study (65%) were frameless (no metal side frame to cover the exposed edges) the high defect rate must be re-evaluated from that point of view as well. See Figure 9-12. Intense IR and UV radiation could quickly degrade the plastic edge seal, which would cause excessive degradation and failure of the module.

So the ASU tests confirm that due to the above issues, frameless glass/glass PV module construction is incompatible with desert applications. This should be a warning to all manufacturers that frameless glass/glass construction has its place in energy markets, but that deserts are not the best choice.

Also, it was noted in the study, that some manufacturers have replaced the metal back covers of their modules with plastic sheets, which would allow moisture penetration with time. Plastic materials have some

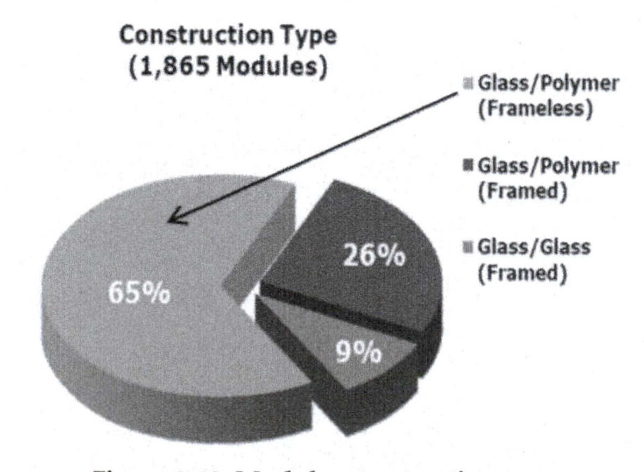

Figure 9-12. Modules construction type

advantages (mainly lighter weight and breathability) so these plusses and minuses must be considered in all cases. Our recommendation would be to use modules with metal back covers for long-term operation in the desert, because plastics just don't last too long there—regardless of their benefits.

Connecting many modules in strings causes additional overcurrent, overheating and other issues, or as the ASU-APS field study (8) concludes, "The higher degradation rates of grid-tied PV modules as compared to individually exposed modules are attributed to the system voltage related corrosion, module mismatch and insolation levels." This also confirms that the modules' temperature coefficients in the field are very different from those measured in the lab, so the study recommends using field measurements instead of taking them from the manufacturers' spec sheets. This is also in line with our conclusions and recommendations on the matter.

In addition to being disappointing, the excess encapsulation damage and delamination of frameless modules (some due to glass/glass construction overheating) is also quite dangerous because some all-glass, frameless TFPV modules contain toxic materials. The IR and UV accelerated destructive processes would easily damage the encapsulation in the open edges, where cracks and voids would allow moisture and environmental gases to enter the modules. The unwanted penetrants would then attack and slowly overstress and decompose the thin film structure, which has the potential of contaminating the surrounding area. This is especially dangerous when considering the huge number of toxic materials containing TFPV modules planned to be installed on thousands of acres of desert in the near future.

So the question here is how to predict and prevent quick delamination, moisture penetration, and premature failures of the protective encapsulation layers and the subsequent destruction of the active module components. What are the variables that we should keep in mind, and what should be done to ensure proper protection in preserving optimum performance of PV cells and modules in large-scale PV plants? What kind of quality control is needed to provide maximum quality of the final product? How do we attract the manufacturers' attention in a positive way, to obtain the relevant data and cooperation needed for analysis of the quality of their products?

Better yet, shouldn't we start looking for modules that offer better solutions to edge sealing and protection, similar to that provided by some TFPV manufacturers as Solar Frontiers, who put a special effort in ensuring the integrity of the thin film in the module by protecting it with several moisture ingress barriers, including a frame all around the module edges? See Figure 9-13.

Note the carefully designed redundant seal of the module edges in Figure 9-13, where the moisture has to penetrate: a) thick layer of frame sealing material, b) specially designed moisture resistant MVTR film and c) a thick layer of encapsulation, before reaching the active thin film structure.

This module construction is very similar to that of framed c-Si PV modules, with the silicon cells replaced by CIS thin film, thus preserving the "cover glass/encapsulation/cell/encapsulation/backsheet/frame all around" configuration. Based on proven PV industry technology, this module design would provide excellent moisture ingress prevention, and protect the active thin film structure inside much better than the frameless glass/glass PV modules, where only a thin layer of encapsulation separates the thin films from the elements and eventual decomposition.

Would these modules operate flawlessly 30+ years in the desert? We actually don't know the exact answer to this question, because no module has been in field test for that long, but common sense and years of hands-on experience tell us that framed TFPV modules with triple edge seal barrier will outlast any frame-less module installed in any part of the world by far. This is even more true for modules operating under extreme desert heat or

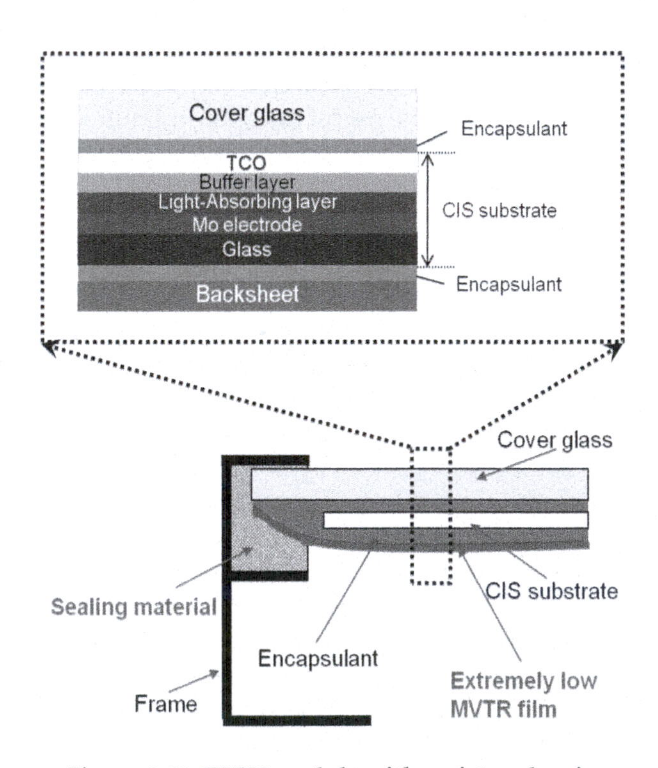

Figure 9-13. TFPV module with moisture barrier

humid areas conditions. This is an important consideration that has to be taken into account by customers, designers, installers and investors alike, because it might mean the difference between success and failure of our large-scale PV project in the Arizona desert

Note: Thin film PV installations depend on low cost and optimum efficiencies, which are driving the increased use of TFPV modules in the world-wide energy market—expected to grow from $3,406 million in 2009 to $19,422 million in 2015. CIS/CIGS and CdTe TFPV module installations are expected to grow 43.9% during the same period with a-Si PV modules following closely. Grid-connected power generation via TFPV modules is expected to have a 39.1% increase from 2010 to 2015.

Success is good, but TFPV modules are a new product, untested and unproven for long-term desert operation, so TFPV module manufacturers need to study the issues and decide if their TFPV modules are ready for 30 years of desert torture? What is the possibility of excess power degradation, total failure and/or environmental damage from the millions of TFPV modules installed in the deserts?

Desert Phenomena

In order to wrap-up this discussion and put it in the right perspective, we need to take a long and serious look at the actual exposure of PV modules to long-term desert operation. This leads us to emphasize again and again the importance of well designed and constructed modules, always keeping in mind that they are not disposable toys to be stored in a dark closet after two weeks of use. Instead, they are high-tech devices destined to experience 30 years of non-stop torture in the world's harshest deserts.

The desert does not pick and choose; it attacks decisively and violently anything in its path. It destroys anything that dares challenge its reputation as an unforgiving and cruel tyrant. Blistering heat, freezing cold, hail, sand and monsoon storms (accompanied by high humidity and violent lightning strikes) are some of its tricks, which we who live in the desert know well and have learned to respect. We wonder, though, how many module manufacturers are even aware of the extreme conditions their modules are expected to endure in the deserts. We wonder if they have taken all this into consideration in their design process and manufacturing operations.

PV modules are exposed to this harsh treatment non-stop, day after day, and for very long time. They heat up in the morning and overheat at noon, which not only causes a drop in efficiency, but actually creates a minuscule damage in the cell and modules' structures (micro-cracks, electrical contact deterioration, etc.) with every cycle. This daily minuscule damage is compounded for the duration and can at anytime grow larger and more noticeable.

In many cases, when the damage becomes noticeable, by visual inspection or electrical measurements (as in the ASU-APS field test study mentioned above) it means that a) the modules have deteriorated significantly, b) their power output is reduced above the specified values, c) there is a good chance of further power reduction, and d) there is increased chance of total failure with time.

No type or amount of lab testing can accurately reproduce the different types of field failures. There is no test available today that can reproduce 30 years of non-stop desert torture, because no one can accurately predict the types, level and duration of the different cycles of torture the desert will choose to put the modules through for the duration. Underestimating the power of the desert, however, will surely lead to certain field failure.

On top of that, the modules in our large-scale power field are connected together in long strings that are continuously exposed to the internal torture of several hundred DC volts of electric current running through them—non-stop, day after day, all day long. The electric current also generates significant amounts of heat in the cells and modules in its path to the grid. The sun heats the modules from the outside, while the produced electricity heats them from inside.

Any slight flaw in the cells' or modules' structure will eventually create even more heat due to increased resistance or sunlight (IR and UV) absorption and things only get worse from here. Again, due to the complexity of the external and internal factors at play, no type or amount of lab testing can reproduce and predict accurately the effects of internal heat generation under desert conditions.

During many years of PV cells and module testing we have seen different variations of field behaviors, but one thing remains constant in the desert: internal overheating of the modules that peaks around the noon hour. Internal mid-summer temperatures of up to 185°F have been measured in the Arizona and Nevada deserts. At these temperatures the abnormalities caused by any small defect in the solar cell/module design or construction are amplified and could lead to a disaster.

In conclusion, the proper PV module design and choice for large-scale desert operation requires:

1. Understanding and respect of the harsh desert conditions and related behavioral abnormalities of PV materials and structures.

   Warning: Remember: the desert kills anything

left there for a long time, and PV modules (especially the plastic materials in them) are no exception.

2. Understanding the long-term behavior of solar cells and active structures, and choosing the best and most appropriate materials and manufacturing processes to fit the extreme operating conditions.
3. Understanding the long-term behavior of PV modules operating in the desert.

   Caution: Remember: every new module is expected to last a long time under hellish conditions, so it better be made properly and with high quality materials. Any mistake or compromise might bring doom to it, the batch of modules and the entire PV project.
4. Designing and executing module construction that is suitable for desert operation by
   a. Providing best possible edge sealing to protect the active layers
   b. Providing well designed side frame to cover the exposed module edges
   c. Avoiding glass/glass module construction
5. Proper and careful field installation as needed to avoid corrosion and overheating.
6. Proper operation and thorough maintenance, starting with careful watch for signs of early mortality and power fluctuations. This is needed to identify failing modules or strings and take immediate action, hopefully before the warranty period has expired.
7. There is no way to predict with any certainty what will happen, when and where in the desert, so we need to have contingency plans for all possible circumstances, so
   a. Manufacturer's and long-term performance warranties must be well designed to account for, and take care of, any materials, labor and performance deviations,
   b. Performance insurance must be well designed to take care of excess power drop, module failures, and any accidents and incidents—man-made or caused by Mother Nature.
8. And finally, pray. Pray unceasingly, because the desert is one place where bad things happen on a regular basis; sometimes very quickly and in an awesome fashion.

Practical Applications

Taking a careful look at the ASU-APS field test study (8) makes us pause, take a deep breath and take another, closer look at the results. An almost 100% defect rate and excessive annual deterioration (much higher than expected) after only 10-17 years is something to think about. How would that reflect on the bottom line of our 100 MWp PV project in the Arizona desert? How would we justify the annual power losses and make up for the power expected by, and contracted with, the utility?

The study is another, albeit shocking, confirmation that many of the unresolved and unaddressed issues presented in this text are real, measurable, and quite significant, so they must be addressed and resolved soon—preferably before we enter into an irreversible large-scale PV nightmare. These test results are shocking even to us, who have been fully aware of the problems, and who have been issuing warnings for years about the quality (or lack thereof) and related performance issues of PV modules.

Our 35+ years experience with engineering issues and quality control matters cannot process the near 100% defect ratio suggested in the study, let alone put it into the right perspective. The practical advice we have for perspective PV customers or investors is to make sure they know EXACTLY what they are dealing with and getting into:

1. Don't let brand new, shiny PV modules mislead you into thinking that they will perform as well as they look for the duration. These are not disposable toys, but serious high tech devices, which will be exposed to the most ferocious climate on Earth and endless torture for many years, so they better be made right,
2. Don't let the manufacturers' promises mislead you into believing that they know what their modules will do under excess desert heat, unless they have long-term test data to show,
3. Don't let financial gains shown on paper for the first 3-5 years blind you to the fact that what comes during years 10-15 might not be as wonderful, and
4. Do your homework thoroughly; do not leave anything to chance or to unconfirmed promises:
   a. Know who the module manufacturer is, and how long they have been making these modules (check if these particular modules are a newly designed product without any desert operation history).
   b. Learn the production process details: equipment, process steps, materials, labor conditions and everything else that goes into the manufacturing process.

c. Get as much quality-control documentation as you can: initial inspection, in-process QC and test procedures, final test procedures and results, and long-term test procedures and results.
d. Get information on the supply chain (the quality of materials and chemicals made by a third party).
e. Learn all you can about the modules: where they were made, where and how they were tested and certified, any long-term test results done—when, by whom, and how.
f. Become familiar with the details of the technical information available, describing the modules' structure, function and performance.
g. Discuss and negotiate warranty conditions, in detail.
h. Parts and labor warranty conditions: What is considered a failure? By whom, when, and how will the failed modules be replaced? How would the lost time be compensated?
    ii. Long-term performance conditions: what is a failure for every year after year one? Over 1.0% annual power loss, or >20% total power loss should be considered failure. How would the replacement be done and lost time compensated?
    iii. Temperature coefficient: discuss and agree on the modules' performance under desert conditions, where >180°F in-module temperatures have been measured in the desert. If 0.5% per °C is the maximum accepted under the warranty, then what could be done if it gets higher with time? How would the replacement be done, and lost time compensated?

This is a list of the key basics, but there are a large number of secondary issues to be addressed and questions to be answered in order to get a complete picture and reduce the risks to a manageable level. Since some of the issues are quite complex, and since no one person can resolve them all, the intervention of a number of specialists in the different areas of the project is a must.

A Practical Example

In the previous chapters we reviewed all aspects of present-day PV technologies and their applications. Now we will apply our knowledge to design, finance, or purchase a PV power plant using real-world conditions. The PV plant is 100 MWp and is to be installed in an appropriate location in a US desert. This particular location has 300 days of full sunshine and averages 6 hours per day full output. The average solar insolation is 900 W/m^2.

So we can expect:

Total solar insolation **per m^2** per day = 6 hrs/day x 900 W/m^2 = 5.4 kWh/m^2/day

Total solar insolation **per m^2** per year= 300 days x 6 hrs/day x 900 W/m^2 = 1,620 kWh/m^2/year

This means that we can expect ~1,620 kWh sunlight to fall on each square meter of that location during the entire year. Or, if we had 100% efficient technology we could produce 5.4 kWh from each square meter of land each day, and 1,620 kWh from each square meter during an entire year. This sounds quite good, BUT:

1. We don't have 100% efficient PV technologies—not yet! So we'll use 15% c-Si PV technology for this example. And we don't cover 100% of the land with modules either, so we need much more land than needed for the modules alone.

2. 300 days per year and 6 hours per day full sunlight are averages we use here, but these are major uncontrollable variables, which vary wildly from hour to hour and day to day even in the deserts. We will, however, use these averages, keeping in mind that they could go up and down (mostly down with bad weather) and there is nothing we can do to change that.

There are a number of variables in our PV power plant design and operation, in addition to the uncontrollable ones, so let's see how we can get the best out of this installation.

We start with the theoretical calculations, which go something like this: A 100-MWp (nameplate) power installation will require 100 MWp of PV modules, fixed-mount or on trackers. Assuming that the modules are 200 Wp each, we'll need 500,000 modules to generate 100 MWp DC electric power. If we are using trackers, the number of modules to obtain the 100 MWp nameplate rating would be the same, but the daily and annual energy output will be greater. Tracking modules produce more power during the rest of the day simply because they receive more sunlight than fixed-mount modules.

NOTE: The nameplate number of 100 MWp is arrived at by considering the maximum output of the modules at STC (25°C module temperature and 1000 W/m^2 insolation) as measured by the manufacturer. In reality this power output will be reached only around the noon hour (thus the designation of peak power) every day. Keep this in mind, because it is an important variable.

Fifteen percent efficiency means that each square meter covered with these PV modules will generate 150 Wp DC power (at STC). Since we are using STC numbers to design the plant, this means that we need ~667,000m^2 (165 acres) of active area (area covered by PV modules). Adding 50% space between the rows of fixed-mounted modules, we end up with ~330 acres of land needed to install a 100-MWp power plant. Usually, however, we need much more than that, to compensate for shade-free operation, access roads, abatement, etc.; so a typical plant size would be in the 400-600 acres range.

NOTE: Most utilities prefer using Watts AC in their calculations, instead of Watts DC, which changes the calculations. To do this, we need to track the power delivered to the grid back to the PV modules. In our case, 100 MW AC power would mean adding at least 20 MWp to the original 100 MWp modules to compensate for the losses from the modules to the grid (wires, inverters, transformers etc.).

With all of that figured out, we finalize the calculations, obtain pre-construction financing (several million dollars would be needed for pre-development financing to secure the land, obtain permits and PPA), make electro-mechanical drawings of the installation, and start the permitting process. We'd need to find land in an acceptable location and negotiate a purchase or lease with the owners. After months of meetings, stacks of forms, payments, etc. we get a construction permit and a PPA and construction finance ($400-500 million is needed to build the plant). Then we proceed with the installation of fixed-mounted modules (or on trackers) on the site. When everything is ready to go we flip the main switch to ON and start measuring the output during different times of the day.

This is what we'd expect to see during the 30 years:

Day 1

It is noon on a bright and sunny mid-summer day in the desert. Solar insolation 900 W/m^2, air temperature 120°F. Since the modules' output was measured at STC (25°C module temperature and 1000 W/m^2 insolation) we get our first surprises:

a. Because typical solar insolation at high noon in the desert is ~900 W/m^2 instead of the theoretical 1000 W/m^2 (which was used for the nameplate calculations of the modules and the power plant), we see that we have lost nearly 10% of the modules' output. Instead of 100 MWp, we are getting 90 MWp DC power output as a first measurement.

b. Then we notice that the output starts going down quickly. This cannot be. Yes, it can! The modules start heating up under power, and with the help of the hot desert noon sunshine the temperature inside the modules could reaches 150, 160, 170, even 180°F and higher. Under these conditions internal resistance increases and other things happen (remember the 0.5% power drop for each degree C increase of temperature), so that we could lose another 10-20% of the power output during the most productive hours of the day. Now our 100 MWp power system is producing only 80 MWp during the noon hour.

c. While we are busy making these measurements, DC power is running through the wires, combiner boxes, inverters and transformers and finally flowing into the electric grid. When we measure the actual electric power going into the grid, we find that it has dropped another 8-10% due to conversion and resistive losses in the equipment between the PV modules and the grid. The higher the ambient temperature, the higher the DC to AC conversion losses. So we are now down to 70-75 MW AC power generated during the noon hour and sent into the grid.

d. When the sun starts going down, the power output will go down proportionally until no more power is generated by the system when the sun goes down.

e. The cycle starts again the next day with a gradual increase of power output as the sun goes up, until the modules' temperatures reaches the maximum. Then the output is level until the sun starts going down again.

NOTE: Using trackers will complicate the comparative measurements, but we should assume that the same type PV modules installed on trackers, would generate the same amount of power at noon as the fixed-mount modules. During the hours before and after the noon hour, however, the tracking modules will generate 10-20% more power than their sitting fixed cousins, due to better orientation towards the sun and better absorption of sunlight.

Money Talk

Now let's look at the bottom line of that first day of operation. During the noon hour of day #1 we generated 75 MWp AC power, or 75 MWh (75 MW per hour) went into the grid. We actually generated that much power from 10:00AM until 4:00PM on that summer day in the desert. This is a full 6 hours of maximum power production. After that (from 4:00PM until 8:00PM) the generated power goes down with the sun. The same is true for the hours from 6:00AM until 10:00AM that morning, when the generated power went from zero to maximum as the sun was rising.

So, during that day we had 6 hours of full power generation and 8 hours of partial power generation. These 8 hours would average the equivalent to ~3 hours of full power generation. So that day we had a full 9 hours of full power generation (6 hours of full power and 3 hours average from the rest of the day)—or—9 hours x 75 MWh = 675 MWh AC power was sent into the grid. This is 675.000 kWh, for which the utility company will pay us $0.15/kWh, or a total of $101,250.00 for one day of power generation. Over 365 days, this is $36,956,250.00 gross annual income. Not bad, but we don't have 365 days of full sunshine even in the deserts, and most winter days don't produce even half this amount of energy. Because of that we usually average the annual power production to 300 days annually at 6-7 hours per day. So, this would bring our annual gross income to somewhere in the $20-25 million range.

Note: $0.15/kW payback is another variable, controlled by the utilities. We have seen PPAs from $0.09 to 0.23/kWh, depending on the utility (they have their own ways to determine energy costs). The utilities also pay different amounts for power generated during different times of day during different seasons. This complicates things even further, but basically it means that we can get a higher return during the peak demand hours of the summer (1-8PM) and much less in the winter when the peak hours are 5-9AM and 5-9PM, at which times there is not much sun.

Year #1

One year has gone by since the first day of operation. At noon of the anniversary, and under the same weather conditions, we notice that the power output is 1-1.5% lower than this time last year. We look at the manufacturer's spec sheet and see that it predicts ~1% drop of power each year. If the loss is less than 1%, we are good and operation continues. If the output loss is higher than 1%, however, we need to negotiate with the manufacturer and obtain a replacement for the failing modules. But the warranty is often not clear if any replacements can be done during year #1. Also, how do we justify a plant shutdown as needed for the replacement of the failing modules, if an agreement is reached at all? A day or two of shutdown will result in several hundred MWh loss in production, which is even greater than the power loss of over 1% that we are experiencing. Tough decision...

Money Talk

If we average our power generation (summer and winter months) to 6 hours/day and consider 300 full sunshine days per year, we get 300 days x 6 hours/day x 75 MW AC = 135 MWh/year. This is 135,000 MWh x $0.15 = $20.25 million in gross income expected during year #1. At this rate, the investment would be repaid in ~20 years. Adding the different subsidies, incentives, carbon credits, minus O&M and other expenses the plant might be repaid in 10 years and maybe less—if everything works as planned. This is a big IF. Hoping that everything goes this way for the duration is a bit of a stretch, because things usually do not go as planned. Read on.

Year #5

The power output from several hundred modules and several inverters has dropped significantly and something must be done. The inverters are under warranty, so all we have to worry about is the lost power generation during a 2-3 day maintenance shutdown for their upgrade. Unless there is a stipulation in the inverters' warranty agreement, we have to eat the loss. But the failing modules have to be removed and replaced, which will take several days and maybe weeks depending on the quantity. This is profit that will fall while operating expense remains constant. Did we include this loss in the initial plant O&M budget?

Using trackers would cause an additional headache to the plant managers, because some of the moving parts, such as bearings, motors, as well as sensors and controllers need to be upgraded or replaced. This is additional expense and more time of useful power generation lost. This activity and the related expense must have been anticipated during the power plant design stages. If proper maintenance and replacement of failing equipment is not done on time, then the trackers just might stop tracking and things will only get worse.

Money Talk

$20.25 million were generated during year 1, dropping down 1% annually every year thereafter, so now we are at least 5% short on power generation and gross income, in addition to the above mentioned loss of time. This means that the 75 MW AC peak power generation we got during year #1, is now down to 71 MW AC power generation at peak hour. That also affects the bottom line, so now we get ~$19.00 million annual gross income. And some other things have gone wrong; additional maintenance and replacement of hardware and equipment is needed, which will cost us at least another $1 million. Basically, our power generation and revenue are now down an additional 6-7%, putting the gross annual income at $18 million. Considering inflation, and that we are 5 years in the future, $18 million might be… even less.

Year #15

Many modules show greater than 20% loss of power output. Several inverters are failing and their long-term warranty has expired too. Critical decision time. What is better, shutting down the plant to replace the failed modules (provided the manufacturer is still around and will honor the warranty) or continuing to operate like this for awhile longer? Do we replace the inverters now or later? This is a major expense so, unless it was provided for in the plant operation budget, it just won't happen.

Money Talk

Now things get serious. The power output has been dropping 1% every year and we are getting 15% less than the first year of operation. Our 100 MWp DC power plant is pumping only 64 MW AC power into the grid. Our gross income for the year is also down to ~$17 million.

Also, this is the time when major equipment replacement and maintenance work are projected, the sum of which could vary from $5 million to $15 million (if a number of inverters are to be replaced). This might be catastrophic on its own, if proper budget and other measures were not taken in the beginning, because without these the power plant is faced with a major revenue loss.

Year #20

The scenario is even worse than that described above for year #15. A proper plant operations budget would make the difference between success and failure at this point. Have we thought of this beforehand? Were we aware of all possibilities and did we provide for each accordingly? The answers to these questions will determine if the PV power plant will remain in operation or go into bankruptcy proceedings, thus converting another 500 acres of desert land into a junkyard.

Money Talk

At this time the power output has dropped at least 20%, so we are down to ~60 MW AC power generation, and $16 million in gross income annually. Is this enough to pay the bills?

In addition, and even worse, all manufacturers' warranties have expired and we are totally depending on luck and good performance insurance. Any major failure of, or damage to, large numbers of PV modules or inverters means that we'll have a hard time recovering—even if there is adequate insurance coverage and modules/inverters replacement options (due to lost power generation).

Year #30

The power output is down over 30%, so our 100 MWp power plant is now producing less than 50 MWp, everything considered. All warranties have expired, maintenance costs are increasing, and it is time to shut it down for good.

The plant now goes into the final stage of its life cycle—decommissioning. Who will remove the PV modules and the supporting structure? Who will load the metal waste on trucks and pay for its transport, recycling and disposal? Who will pay for handling, recycling and disposal of the remains as hazardous waste (in case of toxic TFPV modules)? Who will return the land to its original state and decontaminate it as needed? These are very expensive processes that must be done properly and efficiently. Hopefully we have allowed for this effort in our preliminary planning, design, and budget calculations. If not, there will be a lot of toxic junkyards in the deserts 30-40 years from now.

Money Talk

This, in our opinion, is where many PV power plants would be in shambles. With power generation and revenues decreased by 30%, increasing failures and pending decommissioning, recycling and land restoration expenses, most managers and owners would prefer the easy way out. This is also when and where winners and losers will be separated and a precedent for future work established at the expense of failed installations and broken dreams. Here is where the surviving technologies and operators will be proven adequate and safe for this use, and where proper initial planning and design will be verified and used as a standard for the PV industry to follow.

The above scenario is neither the best nor the worst of cases. Things might go much better if we had a capable team consisting of experienced specialists in the different areas of the project development, and if proper planning, plant design, installation and operation (including provisions for the above mentioned issues and activities) have been specified and implemented.

Things could go much worse, if we rushed into it to beat the government subsidy deadline, and/or used technologies that were dirt cheap at the time, without knowing their long-term performance characteristics.

We have seen examples of both cases in the US and

around the world through the years, and since there are no long-term examples for comparison, we must wait for time to separate the winners from the losers.

Summary of Obstacles

We have seen a number of issues to keep in mind for cells, modules and power field design. These issues are partly responsible for the uncertainty in the PV industry and lack of investments for PV installations today.

Some of these reasons are:

1. PV module manufacturers usually guaranty 80% output by the 20th year and stop there. This means an average 1% reduction in produced power every year up to year 20, leaving uncertain losses after year 20—which is when PV modules are nearing their end-of-useful-life stage and when they start deteriorating and failing at even faster rates.

2. Some insurance coverage can be obtained to cover damage and performance deficiency of PV power plants, but we are not sure that adequate insurance can be purchased for the period after year 20, which is just around the time when many PV plants become profitable. Or, if such insurance exists, can we even justify or afford its price?

3. The yearly power decrease represents a minimum 1% annual loss of revenue—1% loss the first year, 2% the year after, 10% by the 10th year, 20% by the 20th year (in best of cases) and a whopping 30% loss or more during the 30th year of operation, if the plant is still operational at all. And this is big IF with no guarantee whatsoever.

   Note: Long-term field test results show that regardless of the manufacturers' claims, most PV modules lose ~1.3-1.5% or even more power annually. There is no way to prove this with new modules, so using the field tests as the most reliable long-term test numbers available, we calculate 13-15% power and revenue loss by year 10, 26-30% loss by year 20 and a staggering 39-45% power loss by year 30.

4. PV modules also lose 0.5% power per degree C above STC (25°C); i.e., PV modules exposed to the blistering Arizona desert sun will heat to over 85°C in the summer. This is 60°C over the STC value. The modules, therefore, will lose 30% of their power output during the 4-6 most productive hours of the day. This phenomenon is also a well kept secret that we don't hear many people talking about. This is a significant loss, however, which should be addressed during the power plant design stages.

5. An equally serious loss, which cannot be easily estimated and is often overlooked or underestimated, as discussed above, is that from total field failures. There are many reasons for PV modules to fail shortly after installation and manufacturers usually offer 2-5 years materials and labor warranty, which is good enough in most cases. Developers of large-scale PV power plants should, however, also consider negotiating a post-warranty failure warranty with the manufacturers. Such a warranty, or special performance insurance is needed, because there are a number of manufacturing conditions that might cause premature field failures, many of which were discussed in the previous chapters.

6. Designers, customers and investors must understand the issues we are discussing herein and obtain assurance from the manufacturers that the PV modules will not degrade excessively, and/or fail prematurely. Purchasing performance insurance might be another way to ensure a plant's survival.

We all need to know what exactly would happen in the field during the first year, year 5, 10, 20… and have a clear idea of the measures to be taken in every case. We, the technical personnel and consultants, should provide the most realistic theoretical analysis and best-worst scenarios estimates based on actual experience and scientific data, backed by field measurements and tests. This is a lot of work, some of which has not even been started yet, but it has to be done, and done right to provide efficient and trouble-free, long-term operation.

Performance Insurance

We are now looking at the newest branch of the US insurance industry, performance insurance of solar power plants, as the knight in shining armor coming to save the PV industry from its problems. This type of insurance is too new for us to assign it an exact place and role, but our field experience shows that it is one (if not the only) way to fill the gaps between quality, performance, and profits.

Thus far we discussed the major technical, administrative, and financial issues facing the PV industry. We determined that long-term performance and longevity are some of the most important, albeit least discussed, issues in any PV undertaking with any of the existing PV technologies. And looking at the data in the above tests we can't help but wonder how anyone could put a positive spin on the results and accept the risks associated with 20-30 years on-sun operation of these modules in the deserts. The long-term field tests are conclusively assuring us that the risks are real and are warning us of the potential and serious long-term physical deterioration, power degradation, failures and other dangers.

The lack of convincing test data, plus the data in the ASU field test study, are major obstacles in the early stages of the PV projects' financing, because investors are concerned with the lack of reliable scientific or test data that can convince them that the risks associated with long-term on-sun operation are well-known and manageable.

Enter the US insurance industry. A number of insurance companies do offer conditional performance guarantees designed to put the risk factors on a level ground which would make potential customers and investors more comfortable. PV performance insurance is a new field and like the PV industry it is equally unstandardized and unregulated thus far. In some cases, coverage is available to all participants: manufacturers and distributors of PV modules, integrators, utilities, customers and investors, while in other cases it covers only certain segments of the PV cycle, such as PV modules' efficiency and performance only.

Different insurance companies cover different types and percentages of the risks related to PV products and services. That might be just enough to eliminate the majority of risks, but we'd advise customers and investors to take a close look at the guidelines and definitions of the coverage in each policy before making a final decision. This is new thing and we all need to be careful.

Nevertheless, this is an encouraging development that might be just what is needed to bridge the gap of financing PV projects today and in the near future. Proper performance insurance fills the gap between quality, longevity and profitability of PV modules and projects, so it might be the key to providing acceptable levels of risk management, thus increasing the customers' and investors' confidence in solar installations.

One serious catch: the insurance companies would favor the more established manufacturers (at least for now) who produce better quality PV products. But there are a lot of PV project owners and investors who prefer to use the "cheap" Asian-made PV modules. So how would the insurers determine which of these have acceptable performance risk? How do insurers know what a better performance is to begin with? What do they know about longevity, let alone assessing the risks associated with it?

Because of its importance, these questions and the entire solar performance insurance sector needs special attention, so let's take a closer look at the characteristics and conditions of providing "performance" insurance.

In order to mitigate the performance and longevity risks of PV modules, insurers must have a good idea what these are. Since they are not specialists in the solar field, and since there are many risks in all areas of the cradle-to-grave life span of modules, they will need to create a "PV products performance insurance" technical team, which would consist of a number of technical specialists with varying areas and degrees of responsibilities. As an example, we'll suggest the following additions to the insurance team as staff members or consultants:

1. Materials specialist, in charge of evaluating the type and quality of the all process materials and chemicals used for making the modules in each of the insured projects.

2. Front end process engineer, in charge of determining the quality of each step of the solar cells manufacturing operations.

3. Back end process engineer, in charge of evaluating the PV modules assembly process.

4. Quality control engineer, in charge of determining the quality procedures of each process step and the final product.

5. In addition, a group of other specialists is also needed to ensure the integrity of the PV project's installation and operation. These specialists should be able to evaluate the project's design, structure, land, interconnect and other aspects and issues related to the modules' installation and operation.
   Note: Actually two separate technical teams are needed to cover all aspects of PV modules manufacturing and use—one team consisting of c-Si cells and modules manufacturing and use specialists, and the second of TFPV modules manufacturing and use specialists. This is needed because the materials, processes, installation and operation of these different types of PV technologies are quite different. Therefore, the different tasks and responsibilities are not interchangeable.

These specialists' and engineers' tasks would be a) to obtain the most recent and reliable information on the manufacturer, the materials and products to be used, and b) to analyze and summarize the risks in the respective areas.

The goal is to arrive to conclusions in the key areas of interest, as follow:

1. Supply chain management and quality
2. Manufacturing process execution
3. Quality control verification
4. Efficiency evaluation under local climate conditions
5. Performance evaluation under local climate conditions
   a. Temperature coefficient verification
   b. Annual degradation verification
6. Longevity under the local climate conditions
   a. Early mortality estimate
   b. Useful life estimate
   c. End-of-life measures
7. Toxicity and safety evaluation
8. Land and interconnect evaluation
9. Evaluation of political and regulatory issues in the local area

After completing these tasks, the technical team members would make a thorough and detailed report on each material and process step in the cradle-to-grave life span of the modules and estimate the level of quality, efficiency, performance and longevity of the final product and the entire project.

With all this done, the insurers will have a very good idea what they are dealing with and will have a better chance to calculate and mitigate their risks. It is possible that within several years of this thorough due diligence, a pattern would emerge that can lead to standardizing the entire insurance process. Actually, it is most likely that several patterns would emerge—patterns related to different manufacturers (established vs. new companies), different locations (hot desert vs. cloudy climates), different technologies (c-Si, vs. TFPV, vs. HCPV), and several combinations of the above.

Short of that, the insurers are working with their eyes closed, so rolling a die would be the only worse thing they could do. While the insurance business is full of uncertainties and risks, mitigating the risks in the chaotic solar frenzy is only logical. Sorting and sifting the low quality players and technologies would prevent disasters and save money for all involved. This is not an easy task and some serious knowledge, professionalism and hard work is needed before obtaining a good handle on the situation. We hope that the insurance companies involved in providing performance insurance to the solar products business will be able to understand and successfully tackle this fascinating but risky business venture.

PV project owners and investors would have a choice of insurance protection, which could vary from full, unconditional coverage to none.

1. Manufacturers who use their own PV modules to build PV power plants may not need or want performance insurance, although they still might choose a level of conventional insurance coverage to protect the plant from natural and manmade disasters.

2. Another class of customers using the highest quality PV modules available may also decide to skip insurance coverage altogether for obvious reasons.

3. Most of the other players, however, especially those using cheap PV modules will be able to choose a level of coverage as they see fit. This way the gap between uncertain quality, deteriorating performance and reduced longevity will be at least partially closed and the PV plant will survive unpredictable events, including excess performance degradation and failures.

In summary, performance insurance might be the solution to successfully using any type of PV modules and products, including those of unknown quality. Provided that the insurance companies are well aware of the risks and accept them (for a price, of course) most PV power plants with adequate performance insurance coverage will survive the test of time. Without adequate performance insurance, however, many PV plants of type 3 mentioned above may experience serious problems in the deserts and other harsh climate areas.

"The insurance industry doesn't work this way," some might say. Maybe, but the insurance industry has no experience with long-term PV installations either, and is far from being a specialist in all areas of PV products manufacturing and use, so it needs to take a very close look at all options (including the above suggestions) before jumping on the solar bandwagon. Else, it might end up in a mess of its own and up to its neck in risks, problematic issues and unresolved cases. Just saying…

The Real Competition

The old and tired conventional energy competitors, oil and coal will stay awhile as major energy sources, but their prices rise continuously, which will make them less competitive with time even with extensive government and private support. The dangers of mining and pumping coal and oil are well known, so we don't need to spend time on that. They are also well known polluters. The biggest surprise of late is the unveiling of the best kept secret of the energy industry—the fact that the clean and green natural gas creates very large environmental contamination. According to reliable scientific information, the complex and expensive shale natural gas "fracking" process creates large amounts of methane, gases, and waste products, which contaminate soil, water and the air. Methane gas escaping the fracking sites causes 20% more environmental damage than coal, and is 30% worse than natural gas obtain via conventional means, according to these reports. (10) DOE is launching an investigation into the matter, so we'll be witnessing never-ending debates for a long time to come, but no big changes are expected anytime soon. Mother Earth has to wait a bit longer for a breath of fresh air.

Nuclear energy is in a revival mode worldwide too. Over 440 nuclear reactors are in full operation around the world with over 370 GW energy production capacity, which accounts for ~15% percent of the total electric power generated around the world. As of the summer of 2010, 58 new reactors with total capacity of 60 GW worldwide are under construction, which will increase the total world nuclear capacity by an additional 16%.

Most of the nuclear power plants under construction or in planning stages are in developing countries. China alone is building 23 nuclear reactors with a total capacity of over 26 GW, which will nearly triple its nuclear energy production. Other countries, Russia and India leading the pack, are also increasing their nuclear power capabilities. On the whole, there are over 150 nuclear plants in planning stages worldwide, and intended to be in operation within a decade or so, nearly doubling the existing nuclear power capacity.

There are also estimates of a total of 340 nuclear plants under consideration for the long term, mostly in developing countries. This will in effect triple the total worldwide nuclear power production. Even if only half of these plants are built, it will still bring the electricity produced by nuclear power in the near future to nearly 1,000 GW. Just imagine how many solar cells and modules we need to manufacture to match this amount of power. It is impossible, of course!

Just think, though: one single nuclear accident 20 years ago, Chernobyl, killed 1 million people according to newly released reports, and contaminated half of Europe—people, animals, soil and water. There are thousands of people still sick and dying from the effects of the radiation in that part of the world, and many thousands of agricultural acres are still heavily contaminated and will be useless for years to come. A large quantity of food grown in parts of Europe and Asia is contaminated too. The medical cases of the Chernobyl nuclear disaster are considered by some to be the best hidden facts in the history of man, and we learn new facts daily.

The 2011 Japanese nuclear disaster is several times worse than Chernobyl because it affected several nuclear installations and waste storage areas, and because it took so long to get them under control. We don't know what the overall effect on Japan and the surrounding countries will be, but we know that it is serious. The questions are, "Which nuclear plant will be next on the list of nuclear disasters? And how close to home will it be?

So, the fragile solar energy industry embryo is growing under the shadow of its much bigger, much more powerful and dangerous cousins, and its progress is slow but steady. The tremendous solar revival during the last several years is a good example of how solar could be used successfully for residential, commercial and large-scale PV power generation. We can build efficient solar installations quickly and profitably. Adding more solar power plants will make us more energy independent, reduce environmental pollution, and reverse global warming.

The fact that solar power is much less polluting than conventional energies is a well known fact and will become even more important as we witness the mass devastation from man-made disasters such as the Deep Horizon oil spill, which polluted the Gulf with crude oil and oil dispersants. The oil dispersants used there is an example of a man-made disaster on top of a man-made disaster—a greed-driven decision followed by a hasty mistake which is causing even more damage than the original disaster. The Japanese nuclear radiation disaster is on an even larger scale, for it threatens not only the environment but also thousands of Japanese people and even more living in the surrounding countries.

Looking at all this through the eyes of future generations, we cannot help but ask the all-important questions, Why are we killing the environment, life forms, and humans for the sake of mass energy generation? How long will we continue this unsustainable and expensive process of using energy sources that contaminate and kill, with the full knowledge that there are better and cleaner ways to generate electric power?

US PV Roadmap

"PV Roadmap" (4) is an industry-led effort to assess the best mix of research and market development supports to accelerate PV development as needed to provide reliable and cheap PV energy sources. According to the Roadmap, with a reasonable set of incentives the solar photovoltaic market in the US could grow more than 30% per year over the next 20 years, from 340 MW of installed capacity to 9600 MW. An increase in PV installations of this magnitude will produce substantial economic benefits for the states and regions that build these installations.

Because PV technologies use more labor per MW installed than other renewable technologies, the direct job benefits to the regions that install systems are significant too. Producing the raw materials and final PV products locally must be our objective here as well, for it is the best, if not the only way to keep costs low and to ensure the highest quality and reliability of the final product, in addition to supporting local economies.

The economic benefits extend well beyond the immediate installation and even beyond the regions where the installations occur. A program of the size documented in the PV Roadmap will create a substantial, new demand for the components and sub-components that go into a PV installation. To fully document the extent of the economic benefits offered by the PV Roadmap plan the total economic stimulus must be mapped.

Manufacturing accounts for the largest portion of the cost of photovoltaics, and that manufacturing could occur in places other than the installation location, bringing economic benefits to other places in the country. But it all starts with a nationwide program for manufacturing and installing large quantities of PV products—a program which presently exists only on paper. No actual efforts have been made for its implementation.

Instead, we continue to import millions of questionable quality PV modules and other products from Asia, just because they are available and much cheaper than those produced in the West. PV Roadmap points the way to achieving our energy goals and we should support its development, but due to the present political and socio-economic conditions we expect it to be delayed significantly. In the meantime we will continue buying megawatts of cheap Asian PV products.

SOLAR ENERGY MARKETS REFLECTIONS

There is no question that solar energy has a number of benefits and advantages over any of the existing energy sources, and that it is shaping up as a worthy competitor on the energy market. Because of this we expect solar power generation to take its rightful place as the flexible and practical source of clean renewable energy that the world so desperately needs.

Advantages and Benefits

Greenpeace (2009) has a special model that could be used to generate future energy market predictions based on reference scenarios (business as usual) for different countries. Moderate and advanced reference scenarios based on realistic policies can be used to support development of clean, renewable technologies.

NOTE: Please note that as with any other predictions those below are subject to change, be it due to political or economic developments and abnormalities. Since we don't have a crystal ball, these are just guestimates coming from a respected international organization that is very involved in these affairs.

The Model

Under just a moderate scenario, the countries with the most sun resources could together:

- create €11.1 billion (USD 14.4) investment, peaking at €92.5 billion in 2050
- create more than 200,000 jobs by 2020, and about 1.187 million in 2050
- save 148 million tons of CO_2 annually in 2020, rising to 2.1 billion tons in 2050.

To put these figures into perspective, the CO_2 generated by Australia alone is 394 million tons a year; Germany has annual CO_2 emission of 823 million tons—equal to the CO_2 emissions of the whole African continent. So, if developed instead of new and decommissioned fossil fuel power plants, solar technologies could reduce global emissions significantly.

During the 1990s, global investment in energy infrastructure was around €158-186 billion each year; a realistic solar figure would represent approximately 5% of that total. This is a technology that, along with wind energy, can contribute to a 'New Green Deal' for the world.

Prices

The cost of solar electricity is coming down, and many developers say it will soon be cost-competitive with power generation from mid-sized gas plants. The factors affecting the cost of solar electricity are the solar resource, grid connection and local infrastructure, and project development costs. Power costs can be reduced by scaling-up plant size, research and development advances, increased market competition and production volumes for components. Government action can bring costs down further through preferential financing conditions and tax or investment incentives.

Policies and Support Needed

Since 2004, some key national government incentives have boosted SOLAR technology, creating massive growth in local installations. The measures that countries in the world's sun belt need in order to make solar work are:

- A guaranteed sale price for electricity. Feed-in tariffs have been successful incentives for development in Spain, France, and Italy, although things have been slowing down lately due to the world's economic crisis. We hope that solar will come back when the crisis is over.
- National targets and incentives, such as renewable portfolio standards or preferential loans programs that apply to solar thermal technologies.
- Schemes placing costs on carbon emissions either through cap-and-trade systems or carbon taxes.
- Installation of new electricity transfer options between nations and continents through the appropriate infrastructure and political and economic arrangements, so that solar energy can be transported to areas of high demand.
- Cooperation between Europe, the Middle East and North Africa for technology and economic development.
- Stable, long-term support for research and development to fully exploit the potential for further technology improvements and cost reduction.

With these key policy foundations in place, solar is set to take its place as an important part of the world's energy mix. Or at least so we hope.

Author's note: The energy dialog has been going on since the early 1970s. For 40 years we have been talking about the potential of solar and wind energy in providing clean and affordable power. But, as they say, talk is cheap. While we talk, the politicians and big business throw billions into conventional energy sources like there is no tomorrow—as if the amount of oil, coal and natural gas remaining in the ground is unlimited and getting cheaper—as if the sky-high piles of nuclear waste could disappear overnight—as if the serious disasters associated with the conventional technologies don't matter—as if future generations have a chance, while the fossils are depleted close to nil and the environment is choking on poisonous gasses and solid waste.

The most promising areas of the world include the southwestern United States, Central and South America, northern and southern Africa, the Mediterranean countries of Europe, the Near and Middle East, Iran, the desert plains of India, Pakistan, the former Soviet Union, China and Australia.

In these regions, 1 sq. km of land is enough to generate as much as 100-130 GWh of solar electricity per year using solar technology. This is the same as the power produced by a 50 MW conventional coal or gas-fired mid-load power plant.

As the efficiency of the different solar technologies increases continuously, we can expect that the above estimate will double or triple during the next decade or two. This will make solar power fully compatible with conventional energy sources, coal, diesel, and even nuclear power generation. The estimates of solar potential in Table 9-14 are quite convincing.

Advantages of Large-scale Installations

Large-scale PV power generation is the most efficient, and in this author's opinion, the only practical way of generating significant amounts of cheap, clean alternative energy that could make a difference on a worldwide scale. Installation and operation costs can be kept lowest by building large installations, because prices are always inversely proportional to large quantity orders. Materials and labor for construction and operation of large installations are local, thus benefiting local economies in a number of ways.

Large-scale PV projects require large quantities of PV products (modules, inverters, and BOS) most of which are imported. Building new production capacities for PV modules, inverters, mounting structures, trackers and other components in the US or in destination countries will bring material costs down and will generate a lot of additional permanent jobs for the local economies. The new local manufacturing facilities will also ensure the uniformity and quality of the PV products which cannot be expected from imported products.

Some of the advantages of implementing large-scale power installations to be built with US made PV products are:

1. Producing mass quantity solar power generating equipment will make it much cheaper and more efficient over time, to a point where no incentives or subsidies will be needed to justify its installation and long-term use.
2. It is relatively easy to find cheaper suitable land in the deserts, where most of the large-scale PV power plants will be located.
3. Desert lands are usually in unpopulated and oth-

Table 9-14. Forecast of world's economic and solar potential(13)

	Economic Potential	*Used until 2050*	*Direct Normal Irradiance**
	TWh/y	**TWh/y**	**kWh/m²/y**
Bahrain	33	4	2050
Cyprus	20	1	2200
Iran	20000	349	2200
Iraq	28647	190	2000
Israel	318	29	2400
Jordan	6429	40	2700
Kuwait	1525	13	2100
Lebanon	14	12	2000
Oman	19404	22	2200
Qatar	792	3	2000
Saudi Arabia	124560	135	2500
Syria	10210	117	2200
UAE	1988	10	2200
Yemen	5100	142	2200
Algeria	168972	165	2700
Egypt	73656	395	2800
Libya	139477	22	2700
Morocco	20146	150	2600
Tunisia	9244	43	2400
Greece	4	4	2000
Italy	7	5	2000
Malta	2	0	2000
Portugal	142	6	2200
Spain	1278	25	2250
Turkey	131	125	2000
Total	**1632099**	**2005**	**2000-2800**

erwise isolated areas, so their use for solar energy production causes less socio-economic damage.

4. Desert areas have huge potential and could supply most of the energy for the US and the world, even as the energy demand grows daily. Land that is useless today might become energy producing tomorrow.
5. Large-scale PV solar installations are constructed step by step, using modular approaches which make them easy to plan and implement.
6. Large PV installations could be designed and developed very quickly, thus meeting investors' needs and filling existing energy needs at a particular location.
7. Large-scale PV plants will reduce CO_2 emissions drastically, thus helping to repair some of the environmental damages and prevent future emissions. Thus-generated carbon credits can be used to reduce the plants' expenses.
8. Large solar installations offer positive economic impacts in the areas of their deployment, such as jobs and economic growth.

These advantages make large-scale PV solar plants a very attractive option for energy generation and environmental rectification during the 21st century. Cost and availability of large quantities of materials (and components) as well as the efficiency and reliability of the solar technologies to be used in the large plants are some of the major issues that must be resolved.

The Near-term Energy Future

Deutsche Bank Securities, Inc., and other energy specialists predict that the solar PV industry is very likely to shift into over-supply in 2011, forecasting ~53% increase in projected shipment output (>23GWp), offset by a modest ~3% y/y growth in end demand (~15.3GWp).

Installations in Europe are expected to contract ~18%, offset by growth outside of Europe. Germany, with half of the worldwide PV installations in 2009-2010 is expected to contract ~18% y/y to 5.8GWp on declining FiT rates and exclusion of farmlands. Italy and Belgium are expected to grow too. A recent cascade of FiT cuts, caps, and new taxes are weighing heavily on most EU countries, so it remains to be seen how countries like Germany and Spain will pull out from under this economic wreck.

The top 19 key module suppliers shipped ~10.8 GWp in 2010 (~75% of worldwide installations), and are ramping output to ship >16 GWp in 2011. Accounting for other module suppliers/new entrants, the industry is poised for >23 GWp of shipments in 2011. The industry is capable of shipping the 2011 estimate of ~15.3 GWp even if capacity ramp is halted today. Current pipeline of capacity ramp (backlog of cap equipment orders) suggests module shipment capacity will reach at least 20 GWp in 2011.

PV module prices are expected to drop below \$1.25/Wp in 2011, and have the potential to go much lower very quickly and without warning. With ~\$0.54/Wp in gross margins (at top tiered vertically integrated suppliers), prices could drop quickly when oversupply sets in, as it is expected to do in 2011-2012. If price cuts fail to stimulate demand, under-utilization of capacity could then drive production costs higher by pressuring gross margins from both sides (lower prices and rising absorption costs).

Several large c-Si and thin film companies have a strong pipeline of projects and will likely outperform in an over-supply state during 2011-2012. The level of insulation from over-supply is expected to diminish over time if module prices/installed costs continue to decline. Rooftop BIPV markets will be least affected from FiT cuts. Hence, some companies could potentially outperform with successful launch of lower cost, higher efficiency PV modules during this period.

The risks for PV markets in the near future include, but are not limited to:

1. Rapidly changing market conditions,
2. Changes to government subsidization policies,
3. The impact of many competing PV technologies, and
4. General economic risks.

The Mid-term Energy Future

Recently the UN's Intergovernmental Panel on Climate Change (IPCC) issued a report assuring us that by 2050 approximately 80% of the world's energy could be supplied by renewables. An estimated investment of US$2,850 billion to US$12,280 billion would be needed to fund this green energy boom.

Until recently the "green" industries were estimated to contribute only 13% to the global energy supply, but this is changing, according to the report. Solar, bioenergy, geothermal, hydropower, tidal and wind will also benefit the environment by cutting greenhouse gas emissions by 220 to 560 gigatonnes by 2050 as well. Four out of 164 scenarios of the long-term viability of renewables replacing fossil fuels were investigated in detail. The results show that the renewables sector could account for as much as 77%, or as little as 15%, of global energy output in 2050. These wildly disparate numbers were mirrored in findings on the decadal investment, with figures ranging from US$1,360 to $5,100 billion for the period up until 2020 and US$1,490 to 7,180 billion for 2021 to 2030.

The report also revealed that solar was one of the energy sources with the most potential. At present, the sector contributes only a fraction of a percent of the world's energy supply, but in some of the IPCC's more optimistic forecasts, this number could rise to as much as 33% by 2050.

While the majority of scenarios saw solar account for around 10% of total energy production, the scope for growth remains considerable. Nevertheless, actual deployment is still dependent on continued innovation, cost reductions, and supportive legislation.

According to the report, this unprecedented growth is heavily reliant on considerable government support. "The potential role of renewable energy technologies in meeting the needs of the poor and in powering the sustainable growth of developing and developed economies can trigger sharply polarized views," according to Youba Sokona, co-chair of the IPCC working group. "This IPCC report has brought some much-needed clarity to this debate in order to inform governments on the options and decisions that will needed if the world is to collectively realize a low carbon, far more resource efficient and equitable development path."

The report estimates a wide margin of 15 to 77% growth by 2050 so we must conclude that the uncertainty of attaining anywhere near 77% is over 50%. And since government support is needed to get even near that number, we should not hold our collective breath, especially if the worldwide economic crisis is not solved soon.

We have seen government support diminishing drastically in some countries, including the European leaders, so we have to settle somewhere in between, let's say 30-40% growth, as a more realistic level of growth by 2050. Even that is 2-3 times more than what was believed until recently and is good news for the solar industry, since it will be an active participant in this effort. We, however, cannot predict how much of that growth will be attributed to PV installations, although we must believe them to be a significant portion.

Emerging Energy Markets

Emerging markets are changing the way the world's energy is distributed by bringing energy production and use into their economies. See Table 9-16. There are six socio-economic and political drivers that determine the economic vitality of emerging markets today, which in the long run also paves the way to their energy independence. These growth drivers are reflected in Table 9-16.

Rapid Economic Growth

In the coming years, growth in emerging economies is expected to outpace that of the developed world. This growth is fueling an increase in household income in places like China and India, where nearly 60 million people—roughly the combined populations of Texas and California—are joining the ranks of the middle class *each year*.

Table 9-15.
Global solar development estimates for countries with most solar insolation.
(Greenpeace, 2009)(13)

	2015	*2020*	*2030*	*2050*
Reference				
Annual Installation (MW)	566	681	552	160
Cost €/kW	3,400	3,000	2,800	2,400
Investment billion €/year	1.924	2.043	1.546	0.383
Employment Job-year	9,611	13,739	17,736	19,296
Moderate				
Annual Installation (MW)	5,463	12,6602	19,895	40,557
Cost €/kW	3,230	2,860	2,660	2,280
Investment billion €/year	17.545	35.917	52.921	92.470
Employment Job-year	83,358	200,279	428,292	1,187,611
Advanced				
Annual Installation (MW)	6,814	14,697	35,462	80,827
Cost €/kW	3,060	2,700	2,520	2,160
Investment billion €/year	20.852	39.683	89.356	174.585
Employment Job-year	89,523	209,998	629,546	2,106,123

Table 9-16. Key drivers of emerging markets(3)

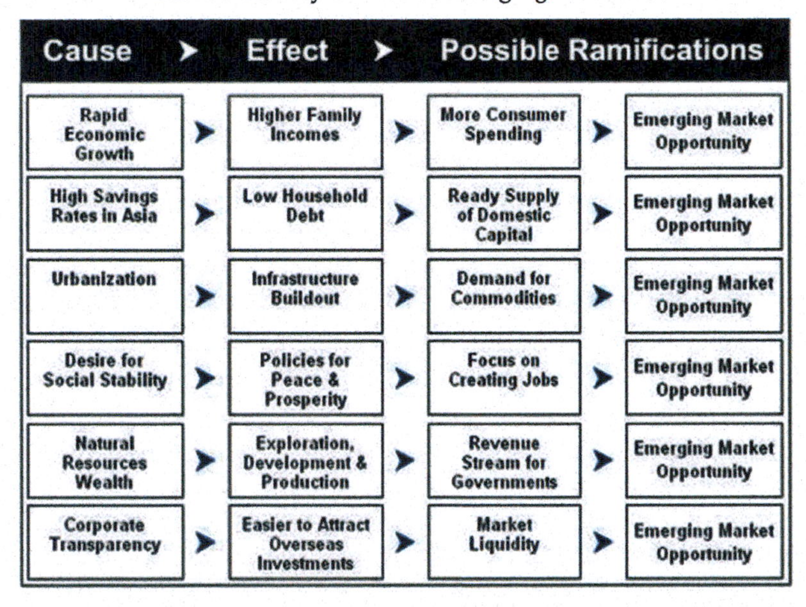

Cause	Effect	Possible Ramifications	
Rapid Economic Growth	Higher Family Incomes	More Consumer Spending	Emerging Market Opportunity
High Savings Rates in Asia	Low Household Debt	Ready Supply of Domestic Capital	Emerging Market Opportunity
Urbanization	Infrastructure Buildout	Demand for Commodities	Emerging Market Opportunity
Desire for Social Stability	Policies for Peace & Prosperity	Focus on Creating Jobs	Emerging Market Opportunity
Natural Resources Wealth	Exploration, Development & Production	Revenue Stream for Governments	Emerging Market Opportunity
Corporate Transparency	Easier to Attract Overseas Investments	Market Liquidity	Emerging Market Opportunity

High Savings Rates in Asia

Despite rising consumption, households in emerging Asia save 10-17 percent of disposable income—that's roughly four times what is saved in the US and much higher than the developed world. These high savings rates allow them to meet the higher requirements for home ownership—many require at least 20 percent down, and much more in some places—and have larger amounts of funds to invest in capital markets.

Urbanization

The world's urban population is growing by more than 70 million people each year. China already has over 100 cities with 1 million people and is expected to have over 150 of them by 2025. This urban migration has overwhelmed existing infrastructure like roads, sewers and electrical grids. The buildout of critical infrastructure will require vast amounts of copper and steel and increase demand for all commodities.

Desire for Social Stability

One main goal of emerging market governments is to remain in power and to keep the public happy. They are doing this by increasing personal freedoms for citizens and providing them with opportunities to increase their quality of life. Many governments have found the key to social stability is focusing on job creation which establishes a path of upward mobility for citizens.

Natural Resources Wealth

Many of today's most promising emerging nations sit atop some of the largest oil, metal and other valuable resource deposits in the world. Many of these nations have teamed up with private and/or foreign enterprises to bring these resources to market. Revenue generated through taxation and direct ownership allows these governments to build infrastructure, create jobs and pursue other economic opportunities.

Corporate Transparency

A history of corruption and political turmoil has given way to higher standards of corporate governance in today's globalized world. Though still far from perfect, the improved transparency and oversight has made important information available to investors and has reduced uncertainty. By aligning themselves with international business standards and requirements, emerging nations will attract more foreign capital and better integrate themselves into the global marketplace.

These six drivers have helped emerging economies weather the financial crisis and provided them with a blueprint for success as they continue to strive to build economic wealth. Many of them are becoming self-sustained and are making plans for significant wind and solar installations. Some have developed their own wind and solar equipment manufacturing industries, adding to their feasibility as a functioning economy.

Most Asian countries are actively participating in the "solar revolution" by becoming PV products manufacturers and/or users. Significant government support is helping the development of PV industries and power fields as well. This, combined with the quickly growing need for more and more energy (and its ever increasing cost) ensures the incremental growth of solar energy in that part of the world.

PV Technologies Review and Direction

Most solar cells and modules manufactured and sold on the world energy markets are based on the use of multicrystalline or monocrystalline silicon with cell and module efficiencies typically in the range of 16-18%. Higher efficiencies have been achieved in the laboratory, demonstrating that the production scale efficiency can be increased to >20%. Although these products have demonstrated their usefulness they are not cost competitive, yet, with other forms of power generation. This has led to intensified efforts to reduce production costs, by the introduction of different materials and manufacturing techniques.

Amorphous silicon based devices are becoming much cheaper to manufacture compared to the monocrystalline and multicrystalline devices. They are also more suitable for use in extreme climates, which is where most large-scale PV power plants of the future will be built. Stabilized efficiencies of commercial products remain <10%, even for double-junction, triple-junction and micro-morph devices, so more work needs to be done in this area for the a-Si PV modules to take their place in the sun.

A very interesting development is the HIT cell, which hybridizes amorphous silicon and monocrystalline/multicrystalline silicon technologies to lower production costs and produces high efficiency devices.

Results show that polycrystalline thin films have excellent potential for making solar cells, and it may be that other compound semiconductors without efficiency or toxicity problems can be developed in the near future.

The III-V compounds are excellent PV materials but are extremely expensive to produce and they contain either toxic and/or exotic and expensive elements. The costs can, however, be offset by using the III-V cells in concentrator systems, which have the highest efficiency to date and use very little material. The III-V compounds also offer advantages for space applications, improved power/weight ratios and better radiation resistances compared to silicon devices.

A most likely further development is that of tandem/multi-junction devices as these more fully use the solar spectrum to generate power. By using a stack of cells—a wide bandgap at the top to a progressively narrower energy bandgap at the bottom—it is possible to minimize both thermal and transmission losses. Multi-junction GaAs and GaSb cells (efficiency >36%) have been developed, so there is particularly good potential for using the tandem concept with thin film devices.

Excellent progress has been made in improving the efficiencies and production processes for thin film solar cells based on the use of CdTe and CIGS. The use of these products has grown sharply lately and they promise significant cost reductions compared to the crystalline silicon technologies. But there has been continued debate over the market acceptability of CdTe, due to its low efficiency and toxicity, and with respect to the scarcity of Cd, Te, Ga, and As, which might limit large-scale use of these materials in the long run.

The future also may include the development of solar cell structures that use organic materials such as dyes, semiconductor polymers and fullerenes or devices that incorporate nano-technology into their design.

One promising technology worth mentioning here is high concentration PV (HCPV) tracking. It was developed during the 1980s by several US companies with the financial assistance of the US Department of Energy and the technical assistance of several US government labs. This technology consists of very efficient solar cells (CPV cells) with ever increasing efficiency: 41.8% in 2009, 42.3% in 2010 and 43.5% in April 2011. They are based on multi-junction thin film processing of germanium semiconductor materials, whose structure has different and much superior qualities than c-Si or any of the conventional TFPV devices. The CPV solar cells are very small but very efficient and stable. A 1cm2 CPV cell can deliver over 20W of power under certain conditions. The HCPV tracking systems are well suited for desert operation due to their low temperature coefficient and annual power degradation. Because they track the sun constantly, however, they require more maintenance than fixed-mount PV systems. Nevertheless, we see HCPV technology dominating the large-scale power installations in the US deserts in the near future.

Patents

Patenting usually indicates the level of innovation, as well as the intensity and speed an industry or sector is growing. Patenting of photovoltaic technologies and processes grew rapidly during the past decade, with the number of PV patent applications published by the world's major patent offices more than tripling in the past 10 years. This is partly due to the generally growing trend of seeking intellectual property protection all over the world, as more and more individual inventors, companies and research institutions realize the importance and economic impact of patenting their inventions.

Growth in the number of PV patent applications has outpaced other sectors, reflecting the growing interest in renewable energy sources in general and recent advances in PV technology in particular.

As depicted in Figure 9-14, solar technology patents continued a dramatic rise in fits and starts through 2009 to reach a record high in the first quarter of 2010 after hitting a record low in the third quarter of 2008. Solar patents in the first quarter of 2010 reached a level second

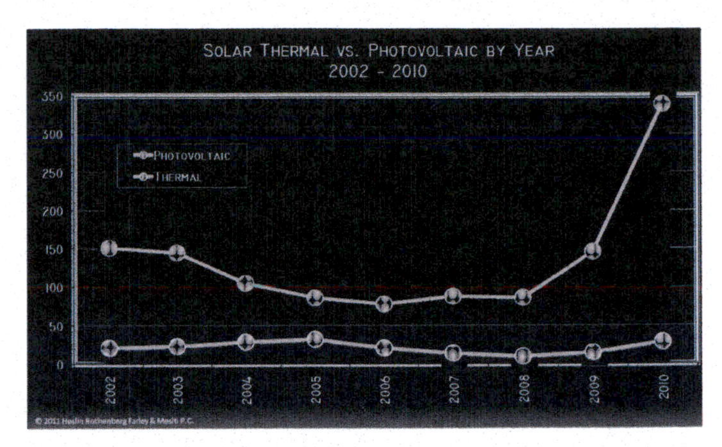

Figure 9-14. Patent applications 2002-2010 (18)

only to fuel cell patents which had led the other sectors since the beginning of 2002. Solar patents reached and exceeded the level of wind patents in 2009 which had been in second place to fuel cell patents since 2006.

This means that solar technologies are undergoing changes and innovations geared towards improving efficiency, practicality and cost. This also means that a lot of money is spent on R&D and testing of new materials, components, and systems.

We have no doubt that these improvements will continue in the future, bringing the solar industry to the level of the state of the art it deserves.

CONCLUSIONS

OK, we know all about solar energy now—its makeup, uses and practical applications—past, present and future. Based on these facts and all available information, our conclusions are demonstrated in Figure 9-15's pyramid.

At the top of the pyramid are the PV energy markets (and especially the commercial and utility-scale segments), which are dependent upon a number of factors. The figure reflects these in a condensed graphic form for ease of understanding. Here we see a number of factors affecting the PV products and installations which affect the world's energy markets.

The political winds and an unsustainable flood of money propelled solar energy onto the world stage, presenting it as a feasible solution to our energy and environmental problems. These winds are blowing in a different direction now, so that the future looks uncertain. The bitter lessons of the 1970s and 1980s taught us that we must not assume anything, and that we must fight to keep solar alive—for our sake and for that of future generations.

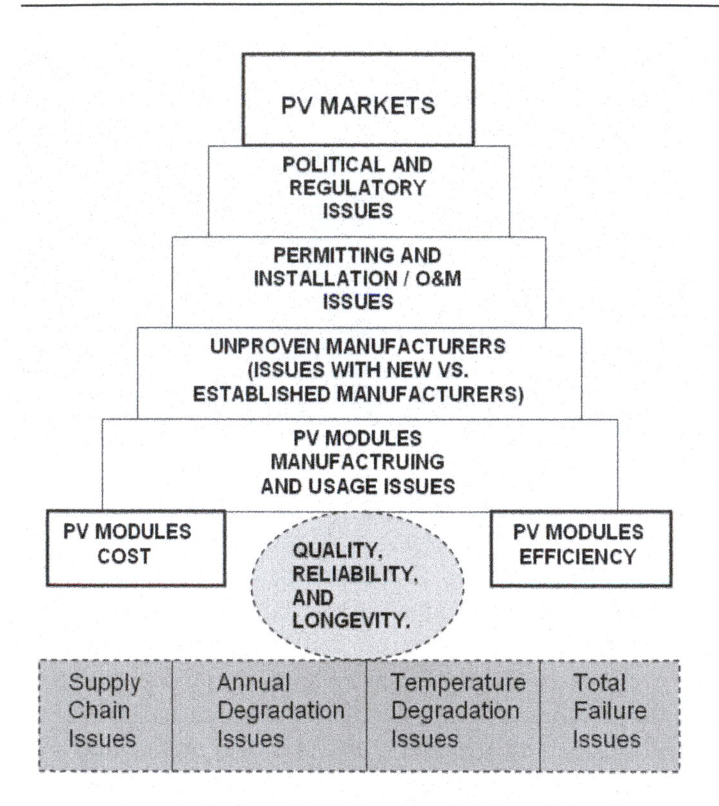

Figure 9-15. PV market issues

The political winds also drive the regulatory and logistical issues, which create confusion, delays and cancellations. This problem will be resolved only if and when we have functional national energy policy.

But even with a national energy policy the solar industry is only as good as its products. Currently we are relying mostly on cheap, unproven products made by foreign companies who have been in the solar business less than 3 years. Many of these companies operate in unhealthy and unstable political and economic environments. This is awkward and unsustainable reliance on foreign energy, which is just like jumping from the frying pan (Arab oil) into the fire (of Asian-made products). How does that solve our energy dependence? How does that solve our jobless situation? And how does that solve our drive for highest quality PV products in the desert power fields?

The critical point to note here is the fact that, in addition to the need to solve all other issues, the foundation of the PV market is laid on the three main pillars—cost, efficiency and quality. Cost and efficiency are being actively and ambitiously addressed as we speak, and we expect good results on both fronts. Quality and the related reliability and longevity factors, on the other hand, are still uncertain. These issues have been either ignored or insufficiently addressed by the manufacturers, and are absolutely and positively not even close to being solved.

The above assessment could be disputed, but the key issues and the wide gaps between the segments depicted in this graph are undeniable. Of all the vulnerabilities we have examined, the longevity of PV modules would be especially problematic and is the most ignored issue which needs to be openly discussed and solved.

So the public continues buying new, shiny PV modules without knowing what exactly is inside and what will happen during 30 years on-sun operation in the desert. The quality—and the related reliability and longevity of the PV modules and therefore the success of the large-scale projects—depend on a number of product-related factors, as seen in Figure 9-15, over most of which we (the customer) have no control.

Summarizing the factors at the base of the 9-15 pyramid:

1. The quality of supply chain materials and consumables is something that we have no access to, let alone control of, therefore we have no idea what kind or type of materials were used to manufacture our PV modules.

2. Annual degradation is a serious factor, and as can be seen from the field tests discussed earlier, it could signify 1.5% power loss per annum or higher, which is 50% higher than most manufacturers' specs and warranties allow.

3. Power loss due to exposure to extreme temperatures and the related defects which extreme IR and UV radiation could cause in long-term desert operations are also unpredictable and are expected to be higher than specified in many cases of desert use.

4. Total failure of PV modules in long-term desert operation depends on a number of factors related to quality of materials and processing, as well as proper installation and O&M. The reasons for this phenomenon are many and not well understood, and because of that they are simply not discussed as openly and thoroughly as needed to provide a comfortable level of confidence in these products and services.

All these factors influence the final project quality which, therefore, is also out of our control, since it hinges on the quality of the cheap product we received from a company which started operations a year or two back somewhere in Asia. How can we stand behind these products if we have no idea of their quality, which eventually affects their reliability and overall longevity in the field?

On the basis of what we have seen in this text, the total failure rates can be extrapolated to several percent in 5-10 years and much higher in 25 and 30 years. How high is anyone's guess…but high enough to be of concern, for sure!

So, the long-term, large-scale PV energy market hinges on solving all issues discussed above, but especially those related to the quality of the PV modules and their proper use. Quality, reliability and longevity concerns as outlined in this text and graphically represented in Figure 9-15, must be fully understood, addressed and solved. Is this doable? Perhaps, and only if the PV modules manufacturers choose to acknowledge the problems and work with customers, installers, operators and investors towards their gradual and permanent resolution.

A Final Word

The world's PV markets are alive and vibrant because of the ongoing concerns with energy and environment, and are driven by a flood of money poured into them by governments worldwide. Therefore:

1. Solar energy is here to stay…for now and for the foreseeable future, if governments continue to support its progress until it has stepped steadily on its own feet.
2. Solar energy's long-term future has some obvious uncertainties, but all indicators point towards sustainable growth of the energy markets, of which solar is and will be a part.
3. Most PV technologies are in their embryonic stages (as far as large-scale applications are concerned) and lots of work needs to be done to get them to maturity—high efficiency, low degradation rates, low cost and low environmental impact. Our estimate is that by year 2040-2050 the present-day solar embryo will have grown enough and walk on its own.
4. Silicon based PV technologies will dominate the PV energy market while government subsidies last, and will be around for a long while after that…until new, more efficient and reliable PV technologies replace silicon all together.
5. The major thin film PV technologies today (CdTe and CIGS) will play a major part in the energy markets for awhile, but their success in the long run is uncertain due to a number of factors related to their efficiency, materials, toxicity, etc.
6. a-Si PV technology might be best suited for large-scale PV applications, due to availability of materials and low cost, if and when its efficiency is increased sufficiently.
7. Lack of government subsidies will eventually force some PV technologies out of the market—starting with those with the lowest efficiency.
8. Technical and supply chain issues will force some of the existing PV technologies out of the market too, or will push them into specialized, narrower niche markets.
9. Environmental issues are paving the way to new possibilities in the world PV industry, and new ways to deal with the environment and energy use are on the horizon.
10. Energy independence means diversification and balanced use of all types of energy resources, which will provide new opportunities for developing countries.
11. Long-term overuse of any one of the available resources will eventually result in an imbalance of the energy markets, which would bring serious negative consequences.
12. Fossils and renewables alike must be wisely, efficiently and properly used, to meet current and projected energy demand, while minimizing the damage to the environment.
13. Solar energy has the potential of changing our energy future by making us less dependent on imports and improving the environment.
14. Solar (and PV in particular) technologies of tomorrow will be much more elegant, sophisticated, inexpensive and surely much more efficient and long lasting, without the number of issues presently found in some, and the related dangers and limitations these bring.
15. The solar future looks bright. We have the tools, and we have no other choice!

Yogi Berra wisely said, "Predictions are very hard to make, especially those for the future," so we will stop short of predicting the future of the PV energy markets. We will, however, dare to suggest a very simple, practical, and absolutely essential path to quickly resolving the issues at hand and opening wide the doors to fast implementation of existing PV technologies on the worldwide PV markets:

1. Focus now on "low-tech" PV technologies and systems, proven by the test of time. Find the ways to manufacture them efficiently and use them properly.

2. Don't waste time and money looking for miracles! We have enough know-how and expertise to ensure the quality, reliability, and longevity (and safety) of present-day technologies. Let's work on these while looking for something better.

3. More effort and investment are needed immediately in order to understand and specify the practical, safe, efficient, and reliable application of the available PV technologies in the proper locations.

4. Make sure each PV technology is tailored to each specific location and application. Don't put fragile, untested and unproven technologies in harsh environments, where they will fail and might even harm the environment.

5. Standardize manufacturing, installation and operation procedures of existing and proven PV technologies. This will go a long way toward obtaining good results in all areas of the PV industry and the energy markets.

6. The PV industry must form a consortium and start serious work towards ensuring proper manufacturing, installation, and operation procedures.

7. The PV industry should create a forum to openly discuss product quality issues, standardization, FiT markets, government and utilities incentives and subsidies, carbon credits and other such instruments. Efficient ways to optimize use of solar and avoid the related problems is urgently needed.

8. Most importantly, we should stop depending on foreign energy imports now! From Arab oil we went to importing Asian PV products! How sustainable is that? US manufacturing of PV products is a must, if we are to control our energy markets and achieve true energy independence. It is that simple!

Important update: In May 2011 DOE informed dozens of solar project managers that their loan guarantee will not be considered unless they close their respective loan guarantees by September 30. This is very short notice for accomplishing many months worth of work. This unexpected change represents a cut of ~$15 billion dollars (out of the $30 billion loan guarantee program), or nearly half of the 27 clean energy projects with $40 billion in secured private investment in danger of being cancelled. Thousands of jobs over 21 states will be cancelled too. The affected projects include utility-scale solar power plants in the southwestern US and solar manufacturing facilities in California, Colorado, and Indiana.

Forty applicants are on hold now, uncertain of the future of their projects, although they have spent millions of dollars already in the pre-construction stages. What will they do? Large companies will lick their wounds and keep on going, but many small owners and investors will lose their shirts. In all cases, government support and its credibility are questionable.

Remember the issues and dangers of pre-construction finance we discussed previously? We didn't even mention this one, because it is a new one—the US government going back on their word. What a boost for the solar industry...NOT!

On the other side of the solar pipeline, US utilities are also in a systematic retreat. At the 8th Annual Renewable Energy Finance Forum-Wall Street (REFF), co-hosted by the American Council on Renewable Energy (ACORE) and Euromoney Energy Events in June, 2011, the attendees were mostly financiers looking for PV projects and ready to invest. Money is available, the need is obvious, the synergy is there and they have only one requirement: credible developers with signed power purchase agreements (PPAs). They looked hard for these, but found few.

The absence of utilities (ANY UTILITIES) or other large power buyers at this meeting was obvious and provided a good picture of the situation—utilities are simply not interested. This event was like going to the opera with one of the main soloists not showing up. What a disappointment...

But it's a sign of the times. As mentioned on many occasions in this text, while some utilities are playing the game because they are forced into it by federal mandates, most of them are quietly staying on the sidelines, waiting for the solar hoopla to go away as it did in the 1970s, 1980s and 1990s. Some are even working towards that end by maneuvering around the mandates, buying out-of-state RECs to comply, instead of developing their own alternative power sources as mandated and expected by the people—and future generations.

Without the active participation of the US utilities—ALL OF THEM—the US solar energy revolution will not get very far.

The solar embryo is crying for help...again, but its Big Brother, the utility industry is looking the other way...again!

As a confirmation of the utilities' retreat from active participation in solar, one major utility will no longer allow commercial PV installations to be connected to the grid. The utility will only consider load offset, making PV installations impractical and financially unfeasible. A number of PV projects have been dropped, and few are planned for the near future.

By abandoning these projects we abandon hopes for substantial commercial power generation—at least until more reasonable conditions for PV development are implemented. Sadly, the utilities rule the solar game, and it appears that, as in the 70s and 80s, they are putting solar on the back burner...again!

On the bright side, the US and the world made a great effort during 2008-2010, achieving unprecedented progress in solar energy generation. Germany and Spain demonstrated that large amounts solar energy generation are possible, and as a result they now have many MW of PV power generators working for them. The US is following in their steps, and we foresee the energy revival continuing in the long run, provided the solar industry finds a way to operate without subsidies.

Do we still have unresolved issues? Yes, we do—many of them. But one way to resolve them is to work on them one by one, day after day. We cannot and should not stop now just because we don't have complete technological solutions, or full government support (which was the excuse of the earlier revivals). We have a chance to complete the effort we started in the 1970s and bring inexpensive, efficient and clean solar energy to the US and the world's energy markets. This is our ONLY hope for energy independence and a clean environment.

Table 9-17 shows the increasing participation of US PV installations in the world energy market. We do believe that the US has all prerequisites needed to develop and sustain a strong PV industry, capable of replacing the majority of conventional technologies.

Yes, the US of A will take the lead in world PV installations in the future; we have the sunshine and the infrastructure. We have enough efficient and reliable solar technologies. We have the know-how and the ability to use them efficiently and safely—today. Not tomorrow, but today!

Last, but not least, history shows that America has always faced challenges head-on and has succeeded in overcoming them every time. Sometimes it takes longer, or as Winston Churchill put it, "You can always count on Americans to do the right thing—after they've tried everything else." Embracing solar won't be an exception, it will take time to complete, but it will happen.

Let us be responsible citizens of Mother Earth and do all we can to protect her from further damage, while wisely using our available resources and looking for ways to obtain the energy needed for our survival and comfort.

PV technologies offer a real solution (one of many) to solving the energy and environmental issues we are faced with now. Let's not wait any longer!

Notes and References

1. Module Prices, Systems Returns & Identifying Hot Markets for PV in 2011, AEI, September, 2010
2. Fostering Renewable Electricity Markets in North America, Commission for Environmental Cooperation, Canada, 2007. http://www.cec.org/Storage/60/5230_Fostering-RE-MarketsinNA_en.pdf
3. 6 Key Drivers of Emerging Markets, by: Frank Holmes, CEO of U.S. Global Investors, April 25, 2010
4. PC Roadmap, http://photovoltaics.sandia.gov/docs/PVRMExecutive_Summary.htm
5. European Photovoltaic Technology Platform comment on "Towards a new Energy Strategy for Europe 2011-2020," http://www.eupvplatform.org/
6. Economic Impacts from the Promotion of Renewable Energy Technologies. The German Experience. Manuel Frondel, Nolan Ritter, Christoph M. Schmidt, Colin Vance http://repec.rwi-essen.de/files/REP_09_156.pdf

Table 9-17. World PV installations, 2010-2050

Year	Europe	Asia	USA	WORLD
	%	%	%	%
2010	81	11	5	3
2012	62	22	11	5
2015	53	26	14	7
2020	35	20	35	10
2050	20	15	50	15

7. Analysis of UK Wind Power Generation from November 2008 to 2010. Stuart Young Consulting, March 2011
8. Performance Degradation of Grid-Tied Photovoltaic Modules in a Desert Climatic Condition by Adam Alfred Suleske, ASU, Tempe, Arizona
9. CAISO Today's Outlook, http://www.caiso.com/outlook/SystemStatus.html
10. Fracking Problem: Shale Gas may be Worse for Climate than Coal, http://envirols.blogs.wm.edu/2011/04/25/fracking-problem-shale-natural-gas-may-be-worse-for-climate-than-coal/
11. NASA solar data. http://eosweb.larc.nasa.gov/cgi-bin/sse/grid.cgi?uid=3030
12. NREL SAM, https://www.nrel.gov/analysis/sam/download.html
13. http://www.greenpeace.org/international/Global/international/planet-2/report/2009/5/concentrating-solar-power-2009.pdf
14. Weather Durability of PV Modules; Developing a Common Language for Talking About PV Reliability, Kurt Scott, Atlas Material Testing Technology
15. An Eye on Quality. Intevac, July 2011 http://www.electroiq.com/articles/pvw/print/volume-2011/issue-4/features/an-eye-on-quality.html
16. http://en.wikipedia.org/wiki/Environmental_impact_of_the_Three_Gorges_Dam
17. http://www1.american.edu/ted/itaipu.htm
18. Clean Energy Patent Growth Index (CEPGI), http://cepgi.typepad.com/files/cepgi-1st-quarter-2011.pdf

Glossary of Terms

A

Absorber is a PV device, the material that readily absorbs photons to generate charge carriers (free electrons or holes).

AC is alternating current.

Acceptor is a dopant material, such as boron, which has fewer outer shell electrons than required in an otherwise balanced crystal structure, providing a hole, which can accept a free electron.

Air mass is the cosine of the zenith angle-that angle from directly overhead to a line intersecting the sun. The air mass is an indication of the length of the path solar radiation travels through the atmosphere..

Alternating Current (AC) is a type of electrical current, the direction of which is reversed at regular intervals or cycles. In the United States, the standard is 120 reversals or 60 cycles per second.

Ambient Temperature is the temperature of the surrounding area.

Amorphous Silicon is a thin-film, silicon PV cell having no crystalline structure.

Ampere (amp) is a unit of electrical current or rate of flow of electrons.

Ampere Hour (Ah) Meter is an instrument that monitors current with time.

Ampere-Hour (Ah/AH)is a measure of the flow of current (in amperes) over one hour

Angle of Incidence is the angle that a ray of sun makes with a line perpendicular to the surface.

Anode is the positive pole of a device where electrons leave and current enters the system;

Antireflection Coating (AR) is a thin film coating on a solar cell surface, intended to reduce light reflection and increases light transmission.

Array Current is the electrical current produced by a PV array

Array Operating Voltage is the voltage produced by a PV array connected to a load.

Autonomous System is a stand-alone system.

Availability is the availability of a PV system to provide power to a load.

Azimuth Angle is the angle between true South and a point on the horizon directly below the sun.

B

Balance of System (BOS) are all components other than PV modules

Band Gap is the energy difference between the conduction and valence bands.

Band Gap Energy (Eg) is the amount of energy (in eV) as required to transfer an electron from the valence to the conduction band.

Base Load is the average amount of electric power that a utility must supply in any period.

BIPV (Building-Integrated Photovoltaics) is design and integration of PV technology into buildings

Blocking Diode is a one-way valve that allows electrons to flow forwards, but not backwards.

Boron (B) is a chemical element used as the dopant in solar cells manufacturing.

Btu (British thermal unit) is the heat required to heat 1 lb. of water 1 degree F..

Bypass Diode is a device connected across a solar cell to protect it from overcurrent.

C

Cadmium (Cd) is a chemical element used in making certain types of solar cells and batteries.

Cadmium Telluride (CdTe) is a polycrystalline thin-film PV material.

Capacity Factor is the ratio of the power output to the capacity rating over a period of time.

Cathode is the negative pole of a device where electrons enter and current leaves the system;

Cell Barrier is a very thin region of static electric charge between the positive and negative layers in a solar cell.

Cell Junction is the area between the positive and negative layers of a solar cell

Charge Carrier is a free and mobile conduction electron or hole in a semiconductor.

Charge Controller controls the flow of current to and from a PV system.

Chemical Vapor Deposition (CVD) is thin film deposition method using heat and reactive gasses.

Cleavage of Lateral Epitaxial Films for Transfer (CLEFT) is a GaAs cells manufacturing process.

Cloud Enhancement is the increase in solar intensity caused

by nearby clouds.
Combined Collector is a PV module that provides heat energy in addition to electricity.
Concentrator is a PV module, using optics to concentrate sunlight onto the solar cells
Conduction Band is the energy band in which electrons can move freely
Conductor is the material through which electricity is easily transmitted.
Contact Resistance is the resistance between metallic contacts and a semiconductor.
Converter is a unit that converts certain DC voltage to higher or lower voltage level.
Copper Indium Diselenide ($CuInSe_2$, or CIS) is thin-film PV material.
Copper Indium Gallium Diselenide ($Cu(In,Ga)Se_2$ or CIGS) are the improved CIS devices
Crystalline Silicon is type of material silicon (c-Si) solar cells are made from.
Current at Maximum Power (Imp) is maximum power generated by a PV cell or module.
Cutoff Voltage is the voltage level when charge controller disconnects the PV array from the load.
Czochralski Process is a method of growing large size, high quality single crystal silicon ingots.

D

Dangling Bond is a disjointed bond hanging at the surface layer of a crystal.
Days of Storage is the time a solar system will generate power without solar energy input.
Dendrite is a slender threadlike spike of pure crystalline material, such as silicon. Dendritic Web Technique is a method for making sheets of polycrystalline silicon.
Depletion Zone is a thin region, depleted of charge carriers (free electrons and holes).
Derate factor is the DC power produced by a PV system vs. AC power delivered into the grid.
Design Month is the insolation and load requiring maximum energy from the PV array
Diffuse Insolation is sunlight received as a result of scattering due to clouds, fog, haze and dust.
Diffuse Radiation is sunlight received after ground reflection and atmospheric scattering.
Diffusion Furnace is used to create p-n junctions in semiconductor devices and solar cells.
Diffusion Length is the mean distance a free electron or hole moves before recombining.
Diode is an electronic device that allows current to flow in one direction only.
Direct Beam Radiation is sunlight received directly from the solar rays.
Direct Current (DC) is electricity that flows in one direction through the conductor.
Direct Insolation is sunlight falling directly upon a collector.
Disconnect is a switch gear used to connect or disconnect components in a PV system.
Distributed Energy Resources (DER) is a variety of small, modular power-generators, used to add to or improve the operation of the power delivery system.
Distributed Generation is localized or on-site power generation.
Distributed Power is any power supply located near the point where the power is used.
Distributed Systems are installed at or near the location where the electricity is used.
Donor is n-type dopant (Phosphorus) that puts an additional electron into an energy level very near the conduction band and is easily exited into the conduction band by incoming sunlight.
Donor Level is the level that donates conduction electrons to the system.
Dopant is a chemical element added a pure semiconductor the electrical properties.
Doping is the addition of dopant type chemicals to a semiconductor.
Downtime is the time when the PV system cannot provide power for the load.
Duty Cycle is the ratio of active time to total time, or operating regime of PV system loads.
Duty Rating is the amount of time an inverter can generate full rated power.

E

EOL is End-of-Life, or time to decommission the modules or the power field.
Edge-Defined Film-Fed Growth (EFG) is a method for making sheets of silicon for PV devices.
Electric Circuit is the path followed by electrons through an electrical system.
Electric Current is the flow of electricity in a conductor (wires), measured in amperes.
Electrical grid is an integrated system of electricity transmission and distribution.
Electricity is the energy resulting from the flow of electrons or ions.
Electrochemical Cell is a device containing two conducting electrodes of dissimilar materials immersed in a chemical solution that transmits positive ions from the negative to the positive electrode.
Electrode is a conductor used to make contact with the nonconductive part of a circuit.

Electrodeposition is a process where metal is deposited from a solution of its ions.

Electrolyte is a liquid used to carry ions to be deposited at the electrodes of electrochemical cells.

Electron is an atomic particle with a negative charge and a mass of 1/1837 of a proton.

Electron Volt (eV) is the amount of kinetic energy gained by an electron when accelerated through an electric potential difference of 1 Volt, equivalent to 1.603 x 10^-19.

Energy is the capability of bodies or systems to do work or convert energy into other forms. Energy Audit is a survey that shows how much energy is used in a place.

Energy Contribution Potential is the recombination occurring in the emitter region of a PV cell.

Energy Density is the ratio of available energy per pound in storage batteries.

Energy Levels is the energy represented by an electron in the band model of a substance.

Epitaxial Growth is the growth of one crystal on the surface of another crystal.

Equinox is the two times of the year when the sun crosses the equator and night and day are of equal length. These event occur on March 21st (spring equinox) and September 23 (fall equinox).

Extrinsic Semiconductor is the product of doping a pure semiconductor.

F

Fermi Level is the energy level at which the probability of finding an electron is one-half.

Fill Factor is the ratio of a PV cell's actual power to its maximum current and voltage levels.

Fixed Tilt Array is a PV array set in at a fixed angle with respect to horizontal.

Flat-Plate Array is a PV array that consists of non-concentrating PV modules.

Flat-Plate Module is an arrangement of PV cells or material mounted on a rigid flat surface with the cells exposed freely to incoming sunlight.

Float-Zone is a method of growing high-quality silicon ingots.

Frequency is the number of repetitions per unit time of a complete waveform, expressed in Hertz

Frequency Regulation is the variability in the output frequency.

Fresnel Lens is an optical device that focuses light like a magnifying glass.

Full Sun is the sunlight received on Earth's surface at noon on a clear day (about 1,000 W/m^2).

G

Gallium (Ga) is a metal used in making solar cells and semiconductor devices.

Gallium Arsenide (GaAs) is a compound used to make solar cells and semiconductor material.

Gassing (or outgassing) is the evolution of gas from different materials.

Gigawatt (GW) is a unit of power equal to 1 billion Watts; 1 million kilowatts, or 1,000 megawatts.

Grid Lines are metal contacts on the solar cell surface.

Grid-Connected Systems act like a central generating plants by supplying power to the grid.

Grid-Interactive System is similar as grid-connected system, but may be more flexible in use.

H

Harmonic Content is the number of frequencies in the output waveform in addition to the primary frequency (50 or 60 Hz.).

Heterojunction is a region of electrical contact between two different materials.

High Voltage Disconnect is the voltage at which a charge controller will disconnect the PV array. High Voltage Disconnect Hysteresis is the voltage difference between the high voltage disconnect setpoint and the voltage at which the full PV array current will be reapplied.

Hole is the vacancy where an electron would normally exist, behaving like a positive particle.

Homojunction is the region between an n-layer and a p-layer in a single material, PV cell.

Hybrid System is a PV system that includes other sources of electricity generation, such as wind. Hydrogenated Amorphous Silicon is a-Si with small amount of incorporated hydrogen.

I

Incident Light. Light that shines onto the face of a solar cell or module.

Indium Oxide is a semiconductor, used as a front contact or a component of a solar cell.

Infrared Radiation is electromagnetic radiation from 0.75 to 1000 micrometers wavelength.

Input Voltage is determined by the total power required by the alternating current loads and the voltage of any direct current loads.

Insolation is the solar power density incident on a surface of certain area and orientation.

Interconnect is a conductor (wire) that connects the solar cells electrically.

Intrinsic Layer is pure, undoped, semiconductor material

in the solar cell structure

Intrinsic Semiconductor is a pure, undoped semiconductor.

Inverter is a device that converts DC to AC electricity.

Ion is an electrically charged atom that has lost or gained electrons.

Irradiance is the direct, diffuse, and reflected solar radiation that strikes a surface.

Islanding is an unwanted condition where portion of the grid is energized by a local generator, while that portion of the grid is supposed to be disconnected.

ISPRA Guidelines are guidelines for the assessment of PV power plants, published by the Joint Research Centre of the Commission of the European Communities, Ispra, Italy.

I-Type Semiconductor is a material that is left intrinsic.

I-V Curve is a graphical presentation of the current versus the voltage from a PV device as the load is increased from the short circuit (no load) condition to the open circuit (maximum voltage).

J

Joule is a metric unit of energy or work; 1 joule per second equals 1 watt or 0.737 foot-pounds.

Junction is a region of transition between semiconductor layers, such as a p-n junction.

Junction Box is an enclosure on the module for connectors, wires and protection devices. Junction Diode is a semiconductor device passing current in one direction better than the other

K

Kilowatt (kW) is a standard unit of electrical power equal to 1000 watts.

Kilowatt-Hour (kWh) is energy measure of 1,000 watts acting over a period of 1 hour.

Langley (L) is unit of solar irradiance equal to 1 gram calorie/cm^2, or 85.93 kwh/m2.

L

Lattice is the regular periodic arrangement of atoms in a crystal of semiconductor material.

Levelized Cost of energy (LCOE) is the net present value of total life cycle costs of the project vs. the quantity of energy produced during the system's life. Life Cycle Cost (LCC) is the sum of all costs and expenses from the design to decommissioning stages of the project. Future costs must be calculated by considering the changing value of money.

Light Trapping is the trapping of light inside a semiconductor material by refracting and reflecting the light at critical angles; trapped light will travel further in the material, greatly increasing the probability of absorption and hence of producing charge carriers.

Light-induced Defects are defects, such as dangling bonds, induced in an amorphous silicon semiconductor upon initial exposure to light.

Line-commutated Inverter is an inverter that is tied to a power grid or line.

Load is the demand on an energy producing system; the energy consumption or requirement of a piece or group of equipment, expressed in terms of amperes or watts in reference to electricity.

Load Circuit is the wire, switches, fuses, etc. that connect the load to the power source.

Load Current (A) is the current required by the electrical device.

Load Resistance is the resistance presented by the external electrical load.

Low Voltage Cutoff (LVC) is the level at which a charge controller will disconnect the load.

Low Voltage Disconnect is the voltage at which a charge controller will disconnect the load.

Low Voltage Disconnect Hysteresis is the voltage difference between the low voltage disconnect set point and the voltage at which the load will be reconnected.

Low Voltage Warning is a warning signal indicating that a low voltage set point has been reached.

M

Majority Carrier is the current carriers (either free electrons or holes) that are in excess in a specific layer of a semiconductor material.

Maximum Power Point (MPP) is the point on the current-voltage (I-V) curve of a module under illumination, where the product of current and voltage is maximum.

Maximum Power Point Tracker (MPPT) is the means of a power conditioning unit that automatically operates the PV generator at its maximum power point under all conditions.

Maximum Power Tracking (Peak Power Tracking) is the operation a PV array at the peak power point of the array's I-V curve where maximum power is obtained.

Megawatt (MW) is 1,000 kilowatts, or 1 million watts. A measure of PV array generating capacity.

Megawatt-Hour is 1,000 kilowatt-hours or 1 million watt-hours. A measure of generated energy.

Metalorganic is a crystalline compound, consisting of metal ions or clusters coordinated to often rigid organic molecules to form one-, two-, or three-dimensional structures.

Microgroove is a small groove scribed in the solar cell' surface, filled with metal for contacts.

Minority Carrier is a current carrier, either an electron or a hole, that is in the minority in a specific layer of a semiconductor material.

Minority Carrier Lifetime is the average time a minority carrier exists before recombination.

Modified Sine Wave is a waveform that has at least three states (i.e., positive, off, and negative). Modularity is the use of multiple inverters connected in parallel to service different loads.

Module Derate Factor accounts for lower PV module output due for field operating conditions such as dirt accumulation on the module.

Monolithic means that it was fabricated as a single structure.

Movistor is a metal oxide varistor, used to protect electronic circuits from surge currents. Multicrystalline is a material composed of variously oriented, small, individual crystals. It is also referred to as poly-crystalline or semi-crystalline.

Multijunction (MJ) Device is a high-efficiency PV device containing two or more cell junctions. Multi-Stage Controller is a charging controller unit that allows different level charging currents.

N

National Electrical Code (NEC) contains guidelines for all types of electrical installations. The Article 690, "Solar Photovoltaic Systems" should be followed when installing a PV systems.

National Electrical Manufacturers Association (NEMA). This organization sets standards for some non-electronic products like junction boxes, frames and such.

Nickel Cadmium Battery contains nickel and cadmium plates and an alkaline electrolyte.

Nominal Voltage is a reference voltage, describing modules and systems (i.e., 12-volt system). Normal Operating Cell Temperature (NOCT) is the estimated temperature of a PV module when operating under 800 w/m^2 irradiance, 20°C ambient temperature and wind speed of 1 meter per second.

N-Type is a negative semiconductor material in which there are more electrons than holes.

N-Type Semiconductor is a semiconductor produced by doping an intrinsic semiconductor with an electron-donor impurity (e.g., Phosphorus in silicon).

N-Type Silicon is a silicon substrate that has been doped with a material that has more electrons in its atomic structure than silicon.

O

Ohm is a the electrical resistance of a material equal to the resistance of a circuit in which the potential difference of 1 volt produces a current of 1 ampere.

One-Axis Tracking is a system capable of rotating about one axis.

Open-Circuit Voltage (Voc) is the maximum possible voltage across a PV cell.

Operating Point is the current and voltage that a PV module produces when connected to a load. Orientation is placement of PV modules with respect to N, S, E, W direction

Outgas is the emission of gas from some materials under certain conditions.

P

p-n junction is a structure formed between a p- and n-type layer.

Packing Factor is the ratio of array area to actual land area of the PV system.

Parallel Connection is connecting positive leads together and negative leads together.

Passivation is a chemical reaction that eliminates the detrimental effect of electrically reactive atoms on a solar cell's surface.

Peak Demand/Load is the maximum energy demand or load in a specified time period.

Peak Power Current is the amperes produced by a PV module or array operating at the voltage of the I-V curve that will produce maximum power from the module.

Peak Power Point is the operating point of the I-V curve for a solar cell or PV module where the product of the current value times the voltage value is at a maximum.

Peak Sun Hours is the number of hours per day when the solar irradiance is at maximum.

Peak Watt is the maximum nominal output of a PV device, in watts (Wp) under STC.

Phosphorous (P) is a chemical element used as a dopant in making n-type semiconductor layers.

Photocurrent is an electric current induced by radiant energy.

Photoelectric Cell is a device for measuring light intensity.

Photoelectrochemical Cell is a type of PV device in which the electricity induced in the cell is used immediately within the cell to produce a chemical, such as hydrogen, which can then be used.

Photon is a particle of light that acts as an individual unit of energy.

Photovoltaic Array is an interconnected system of PV modules that function as a single electricity-producing unit.

Photovoltaic Cell (solar cell) is the smallest semiconductor element within a PV module to perform the immediate conversion of light into electrical energy.

Photovoltaic Conversion Efficiency is the ratio of the electric power produced by a PV device to the power of the sunlight incident on the device.

Photovoltaic Device is a solid-state electrical device that converts light directly into direct current.

Photovoltaic Effect is the phenomenon that occurs when photons in a beam of sunlight knock electrons loose from the atoms they strike, thus creating useful electric current.

Photovoltaic Generator is a PV array, or system, which are electrically interconnected.

Photovoltaic Module is the smallest environmentally protected, essentially planar assembly of solar cells and ancillary parts, such as interconnections, terminals, [and protective devices such as diodes] intended to generate direct current power under unconcentrated sunlight.

Photovoltaic Panel is often used interchangeably with PV module (especially in one-module systems), but more accurately used to refer to a physically connected collection of modules (i.e., a laminate string of modules used to achieve a required voltage and current).

Photovoltaic System is a complete set of components for converting sunlight into electricity by the PV process, including the array and BOS components.

Photovoltaic(s) (PV) pertains to the direct conversion of light into electricity via the photovoltaic effect.

Photovoltaic-Thermal (PV/T) System is a PV system that, in addition to converting sunlight into electricity, collects the residual heat energy and delivers both heat and electricity in usable form. Physical

P-I-N is a semiconductor PV device structure that layers an intrinsic semiconductor between a p-type semiconductor and an n-type semiconductor, used mostly with amorphous silicon PV devices.

Point-Contact Cell is a high efficiency silicon PV concentrator cell that employs light trapping techniques and point-diffused contacts on the rear surface for current collection.

Polycrystalline Silicon is a material used to make PV cells, which consist of many twisted and intersecting crystals unlike the uniform single-crystal silicon structure.

Power Conditioning is the process of modifying the characteristics of electrical power.

Power Conditioning Equipment is used to convert power from a PV array into a form suitable for subsequent use. A collective term for inverter, converter, battery charge regulator, and blocking diode.

Power Conversion Efficiency is the ratio of output power to input power of the inverter.

Power Density is the ratio of available power from a source to its mass (W/kg) or volume (W/l).

Power Factor (PF) is the ratio of actual power being used in an electric circuit, expressed in watts or kilowatts, to the power that is actually drawn from a power source in volt-amperes or kilovolt-amperes.

Projected Area is the net south-facing glazing area projected on a vertical plane.

P-Type Semiconductor is a semiconductor in which holes carry the current; produced by doping an intrinsic semiconductor with an electron acceptor impurity (e.g., boron in silicon).

Pulse-Width-Modulated (PWM) is a wave inverter that produce a high quality (nearly sinusoidal) voltage, at minimum current harmonics.

Pyranometer is an instrument used for measuring global solar irradiance.

Pyrheliometer measures direct beam solar irradiance with aperture of 5.7° to transcribe the solar disc.

Q

Quad is one quadrillion (1,000,000,000,000,000) Btu.

Qualification Test is a procedure applied to commercial PV modules involving the application of defined electrical, mechanical, or thermal stress tests in a prescribed manner and amount.

R

Rated Module Current (A) is the current output of a PV module measured at STC.

Rated Power of the inverter is the full, per spec, power an inverter can generate.

Reactive Power is the sine of the phase angle between the current and voltage waveforms in an alternating current system.

Recombination is the action of a free electron falling back into a hole. Radiative recombination processes are where the energy of recombination results in the emission of a photon. Nonradiative recombination processes are where the energy of recombination is given to a second electron which then relaxes back to its original energy by emitting phonons.

Rectifier is a device that converts AC into DC current.

Regulators prevent excess power by controlling charge cycle to conform to specific needs.

Remote Systems are same as stand-alone (non-grid connected) systems.

Reserve Capacity is the amount of generating capacity a central power system must maintain to meet peak loads as and when needed.

Resistance (R) is the electromotive force needed for a unit current flow, or the property of a conductor to oppose the flow of an electric current resulting in the generation of heat in the conductor.

Resistive Voltage Drop is the voltage developed across a cell by the current flow through the resistance of the cell.

Reverse Current Protection is any method of preventing unwanted current flow from the load to the PV array (usually at night).

Ribbon (PV) Cells is a type of PV device made in a continuous process of pulling material from a molten bath of PV material, such as silicon, to form a thin sheet of material.

Root Mean Square (RMS) is the square root of the average square of the instantaneous values of an AC output. For a sine wave the RMS value is 0.707 times the peak value. The equivalent value of alternating current, I, that will produce the same heating in a conductor with resistance, R, as a DC current of value I.

S

Sacrificial Anode is a piece of metal buried near a structure to protect it from corrosion.

Satellite Power System (SPS) is a concept for providing large amounts of electricity for use on the Earth from one or more satellites in geosynchronous Earth orbit.

Schottky Barrier is a cell barrier established as the interface between a semiconductor, such as silicon, and a sheet of metal.

Scribing is the cutting of a grid pattern of grooves in a semiconductor material, generally for the purpose of making interconnections.

Semiconductor is any material that has a limited capacity for conducting an electric current.

Semicrystalline is same as multicrystalline.

Series Connection is a way of joining PV cells by connecting positive leads to negative leads.

Series Resistance is parasitic resistance to current flow in a cell due to mechanisms such as resistance from the bulk of the semiconductor material, metallic contacts, and interconnections.

Short-Circuit Current (Isc) is the current flowing freely through an external circuit that has no load or resistance. It is the maximum current possible.

Shunt Controller is a charge controller that redirects or shunts the charging current away from the battery. The controller requires a large heat sink to dissipate the current from the short-circuited PV array.

Siemens Process is a commercial method of making purified silicon.

Silicon (Si) is a semi-metallic chemical element that makes an excellent semiconductor material for PV devices. It crystallizes in face-centered cubic lattice like a diamond.

Sine Wave is a waveform corresponding to a single-frequency periodic oscillation that can be mathematically represented as a function of amplitude versus angle in which the value of the curve at any point is equal to the sine of that angle.

Sine Wave Inverter is an inverter that produces utility-quality, sine wave power forms.

Single Junction (SJ) Device has one single p-n junction, vs. multi-junction (MJ) PV devices.

Single-Crystal Material is a material that is composed of a single crystal structure.

Single-Crystal Silicon is material with a single crystalline formation.

Single-Stage Controller is a charge controller that redirects all charging current as the load dictates.

Solar Cell are same as photovoltaic (PV) cells.

Solar Constant is the average amount of solar radiation that reaches the earth's upper atmosphere on a surface perpendicular to the sun's rays equal to 1353 W/m^2, or 492 Btu/ft^2.

Solar Cooling is the use of solar thermal energy or solar electricity to power a cooling appliance.

Solar Energy is the electromagnetic energy transmitted from the sun (solar radiation). The amount that reaches the earth is equal to one billionth of total solar energy generated, or the equivalent of about 420 trillion kilowatt-hours.

Solar Noon is the time of the day, at a specific location, when the sun reaches its highest, apparent point in the sky; equal to true or due, geographic south.

Solar Resource is the amount of solar insolation a site receives, usually measured in kWh/m^2/day, which is equivalent to the number of peak sun hours.

Solar Spectrum is the total distribution of electromagnetic radiation emanating from the sun.

Solar Thermal Electric Systems are solar energy conversion technologies that convert solar energy to electricity by heating a working fluid to power a turbine that drives a generator.

Solar-Grade Silicon or intermediate-grade silicon is used in the manufacture of solar cells.

Specific Gravity is the ratio of the weight of the solution to the weight of an equal volume of water at a specified temperature.

Spinning Reserve is the electric power plant or utility capacity on-line and running at low power in excess of actual load.

Split-Spectrum Cell is a compound PV device in which sunlight is first divided into spectral regions by optical means. Each region is then directed to a different PV cell optimized for converting that portion of the spectrum into electricity, thus achieving significantly greater overall conversion efficiency.

Sputtering is a process used to apply PV semiconductor material to a substrate by a physical vapor deposition process where high-energy ions are used to bombard the source material, ejecting vapors of atoms that are then deposited in thin layers on a substrate.

Square Wave is a waveform that has only two states, (i.e., positive or negative) and contains a large number of harmonics.

Square Wave Inverter is a type of inverter that produces square wave output.

Staebler-Wronski Effect is the tendency of amorphous Silicon PV devices to degrade (drop) efficiency upon initial exposure to light.

Stand-Alone System is an autonomous or remote PV system, not connected to a grid.

Standard Reporting Conditions (SRC) is a fixed set of conditions (including meteorological) to which the electrical performance data of a PV module are translated from the set of actual test conditions.

Standard Test Conditions (STC) is the conditions under which a module is tested in a laboratory, set at 1,000W/m^2 illumination and 25^0C temperature.

Standby Current is the amount of current used by the inverter when no input power is available. Stand-off Mounting is a technique for mounting a PV array on a sloped roof, which involves mounting the modules slightly above the pitched roof and tilting them to the optimum angle.

Storage Battery is a device capable of transforming energy from electric to chemical form and vice versa. These are uses to store energy for night use, or in case of cloudy conditions.

String is a number of PV modules or panels interconnected electrically in series to produce the operating voltage required by the load.

Substrate is the physical material upon which a PV cell is applied.

Subsystem is any one of several components in a PV system (i.e., array, controller, batteries, inverter, load, etc.).

Superconducting Magnetic Energy Storage (SMES) is a technology that uses superconducting characteristics of low-temperature materials to produce intense magnetic fields to store energy. It has been proposed as a storage option to support large-scale use of photovoltaics as a means to smooth out fluctuations in power generation.

Superconductivity is the abrupt and large increase in electrical conductivity exhibited by some metals as the temperature approaches absolute zero.

Superstrate is the cover (glass usually) on the top side of a PV module, providing protection for the solar cells from impact and environmental degradation while allowing maximum transmission of the incoming sunlight.

Surge Capacity is the maximum power, usually 3-5 times the rated power, that can be provided over a short time.

System Availability is the percentage of time (usually expressed in hours per year) when a PV system will be able to fully meet the load demand.

System Operating Voltage is the PV array output voltage under load.

T

Temperature Coefficient is used to determine decrease in PV module output at high temperature. Temperature Factor is used to determine decrease in the current carrying capability of wire at high temperature.

Thermophotovoltaic Cell (TPV) is a device where sunlight concentrated onto an absorber heats it to a high temperature, and the emitted energy is used by a PV cell designed for that purpose.

Thick-Crystalline Materials are semiconductor material, typically measuring from 200-400 microns thick, cut from ingots or ribbons.

Thin Film is a layer of semiconductor material, such as copper indium diselenide or gallium arsenide, a few microns or less in thickness, used to make PV cells.

Thin Film PV Module is a PV module constructed with sequential layers of thin film semiconductor materials.

Tilt Angle is the angle at which a PV array is set to face the sun relative to a horizontal position.

Tin Oxide is a wide band-gap semiconductor similar to indium oxide; used in heterojunction solar cells or to make a transparent conductive film, called NESA glass when deposited on glass.

Total AC Load Demand is the sum of the alternating current loads used in selecting an inverter.

Total Harmonic Distortion is the measure of closeness in shape between a waveform and it's fundamental component.

Total Internal Reflection is the trapping of light by refraction and reflection at critical angles inside a semiconductor device so that it cannot escape the device and must be eventually absorbed by the semiconductor.

Tracking Array is a PV array that follows the path of the sun to maximize the solar radiation incident on the PV surface. One axis tracker is where the array tracks the sun east to west, while two-axis tracking is where the array points directly at the sun at all times. Tracking arrays use both the direct and diffuse sunlight. Two-axis tracking arrays capture the maximum possible daily energy.

Transformer is an electromagnetic device that changes the voltage of alternating current electricity.

Tray Cable (TC) is used for interconnecting BOS components.

Tunneling is a Quantum mechanics concept whereby an electron is found on the opposite side of an insulating barrier without having passed through or around the barrier.

Two-Axis Tracking is a PV array tracking system capable of rotating independently about its two x and y axes (e.g., vertical and horizontal).

U

Ultraviolet. Electromagnetic radiation in the wavelength range of 4 to 400 nanometers.

Underground Feeder (UF) is used for PV array wiring if sunlight resistant coating is specified.

Underground Service Entrance (USE) is used for interconnecting BOS components. Uninterruptible Power Supply (UPS) is the designation of a power supply providing continuous uninterruptible service.

Utility-Interactive Inverter is an inverter that can function only when tied to the utility grid, and uses the prevailing line-voltage frequency on the utility line as a control parameter to ensure that the PV system's output is fully synchronized with the utility power.

V

Vacuum Evaporation is the deposition of thin films of semiconductor material by the evaporation of elemental sources in a vacuum.

Vacuum Zero is the energy of an electron at rest in empty space; used as a reference level in energy band diagrams.

Valence Band is the highest energy band in a semiconductor that can be filled with electrons.

Valence Level Energy / Valence State is the energy content of an electron in orbit about an atomic nucleus. It is also called bound state.

Vapor Deposition is a method of depositing thin films by evaporating the film materials.

Varistor is a voltage-dependent variable resistor, used to protect sensitive equipment from power spikes or lightning strikes by shunting the energy to ground.

Vertical Multijunction (VMJ) Cell is a compound cell made of different semiconductor materials in layers, one above the other. Sunlight entering the top passes through successive cell barriers, each of which converts a separate portion of the spectrum into electricity, thus achieving greater total conversion efficiency of the incident light. Also called a multiple junction cell.

Volt (V) is a unit of electrical force equal to that amount of electromotive force that will cause a steady current of one ampere to flow through a resistance of one ohm.

Voltage is the amount of electromotive force, measured in volts, that exists between two points. Voltage at Maximum Power (Vmp) is the voltage at which maximum power is available from a PV module.

Voltage Protection is used where many inverters have sensing circuits that will disconnect the unit from the load if input voltage limits are exceeded.

Voltage Regulation indicates the variability in the output voltage. Some loads will not tolerate voltage variations greater than a few percent.

W

Wafer is a thin sheet of semiconductor (PV material) made by cutting it from an ingot.

Watt is the rate of energy transfer equivalent to one ampere under an electrical pressure of one volt. One watt equals 1/746 horsepower, or one joule per second.

Waveform is the shape of the phase power at a certain frequency and amplitude.

Window is a wide band gap material chosen for its transparency to light. Generally used as the top layer of a PV device, the window allows almost all of the light to reach the semiconductor layers beneath.

Wire Types are reviewed in detail in Article 300 of National Electric Code.

Work Function is the energy difference between the Fermi level and vacuum zero, or the minimum amount of energy it takes to remove an electron from a substance into the vacuum.

X

x-y Drive is the gears and controls of a two-axis tracker, which determine the path the tracker follows during the day in tracking the sun.

Z

Zenith Angle is the angle between the direction of interest (of the sun, for example) and the zenith (directly overhead).

Index